Lecture Notes in Physics

Bisher erschienen/Already published

Vol. 1: J. C. Erdmann. Wärmeleitung in Kristallen, theoretische Grundlagen und fortgeschrittene experimentelle Methoden. II, 283 Seiten. 1969.

Vol. 2: K. Hepp, Théorie de la renormalisation. III, 215 pages. 1969.

Vol. 3: A. Martin, Scattering Theory: Unitarity, Analyticity and Crossing. IV, 125 pages. 1969.

Vol. 4: G. Ludwig, Deutung des Begriffs „physikalische Theorie" und axiomatische Grundlegung der Hilbertraumstruktur der Quantenmechanik durch Hauptsätze des Messens. 1970. Vergriffen.

Vol. 5: Schaaf, The Reduction of the Product of Two Irreducible Unitary Representations of the Proper Orthochronous Quantummechanical Poincare Group. IV, 120 pages. 1970.

Vol. 6: Group Representations in Mathematics and Physics. Edited by V. Bargmann. V, 340 pages. 1970.

Vol. 7: R. Balescu, J. L. Lebowitz, I. Prigogine, P. Résibois, Z. W. Salsburg, Lectures in Statistical Physics. V, 181 pages. 1971.

Vol. 8: Proceedings of the Second International Conference on Numerical Methods in Fluid Dynamics. Edited by M. Holt. 1971. Out of print.

Vol. 9: D. W. Robinson, The Thermodynamic Pressure in Quantum Statistical Mechanics. V, 115 pages. 1971.

Vol. 10: J. M. Stewart, Non-Equilibrium-Relativistic Kinetic Theory. III, 113 pages. 1971.

Vol. 11: O. Steinmann, Pertubation Expansions in Axiomatic Field Theory. III, 126 pages. 1976.

Vol. 12: Statistical Models and Turbulence. Edited by C. Van Atta and M. Rosenblatt. Reprint of the First Edition. VIII, 492 pages. 1975.

Vol. 13: M. Ryan, Hamiltonian Cosmology. VII, 169 pages. 1972.

Vol. 14: Methods of Local and Global Differential Geometry in General Relativity. Edited by D. Farnsworth, J. Fink, J. Porter, and A. Thompson. V, 188 pages.

Vol. 15: M. Fierz, Vorlesungen zur Entwicklungsgeschichte der Mechanik. V, 97 Seiten. 1972.

Vol. 16: H.-O. Georgii, Phasenübergang 1. Art bei Gittergasmodellen. IX, 167 Seiten. 1972.

Vol. 17: Strong Interaction Physics. Edited by W. Rühl and A. Vancura. V, 405 pages. 1973.

Vol. 18: Proceedings of the Third International Conference on Numerical Methods in Fluid Mechanics, Vol. I. Edited by H. Cabannes and R. Temam. VII, 186 pages. 1973.

Vol. 19: Proceedings of the Third International Conference on Nemerical Methods in Fluid Mechanics, Vol. II. Edited by H. Cabannes and R. Temam. VII, 275 pages. 1973.

Vol. 20: Statistical Mechanics and Mathematical Problems. Edited by A. Lenard. VIII, 247 pages. 1973.

Vol. 21: Optimization and Stability Problems in Continuum Mechanics. Edited by P. K. C. Wang. V, 94 pages. 1973.

Vol. 22: Proceedings of the Europhysics Study Conference on Intermediate Processes in Nuclear Reactions. Edited by N. Cindro, P. Kulišic and Th. Mayer-Kuckuk. XIV, 329 pages. 1973.

Vol. 23: Nuclear Structure Physics. Proceedings 1973. Edited by U. Smilansky, I. Talmi, and H. A. Weidenmüller. XII, 296 pages. 1973.

Vol. 24: R. F. Snipes, Statistical Mechanical Theory of the Electrolytic Transport of Non-electrolytes. V, 210 pages. 1973.

Vol. 25: Constructive Quantum Field Theory. The 1973 "Ettore Majorana" International School of Mathematical Physics. Edited by G. Velo and A. Wightman. III. 331 pages. 1973.

Vol. 26: A. Hubert, Theorie der Domänenwände in geordneten Medien. XII, 377 Seiten. 1974.

Vol. 27: R. K. Zeytounian, Notes sur les Ecoulements Rotationnels de Fluides Parfaits. XIII, 407 pages. 1974.

Vol. 28: Lectures in Statistical Physics. Edited by W. C. Schieve and J. S. Turner. V. 342 pages. 1974.

Vol. 29: Foundations of Quantum Mechanics and Ordered Linear Spaces. Advanced Study Institute, Marburg 1973. Edited by A. Hartkämper and H. Neumann. VI, 355 pages. 1974.

Vol. 30: Polarization Nuclear Physics. Proceedings 1973. Edited by D. Fick. IX, 292 pages. 1974.

Vol. 31: Transport Phenomena. Sitges International Schools of Statistical Mechanics, June 1974. Edited by G. Kirczenow and J. Marro. XIV, 517 pages. 1974.

Lecture Notes in Physics

Edited by J. Ehlers, München, K. Hepp, Zürich, and
H. A. Weidenmüller, Heidelberg, and J. Zittartz, Köln
Managing Editor: W. Beiglböck, Heidelberg

56

Current Induced Reactions

International Summer Institute on Theoretical Particle Physics in Hamburg 1975

Edited by J. G. Körner, G. Kramer, and D. Schildknecht

Springer-Verlag Berlin Heidelberg GmbH

Editors
J. G. Körner
D. Schildknecht
DESY
Notkestieg 1
D–2000 Hamburg 52

G. Kramer
II. Institut für Theoretische Physik
Universität Hamburg
Luruper Chaussee 149
D–2000 Hamburg-Bahrenfeld

Library of Congress Cataloging in Publication Data

International Summer Institute on Theoretical Physics, 7th, Hamburg, 1975.
Current induced reactions.

(Lecture notes in physics ; 56)
1. Particles (Nuclear physics)--Addresses, essays, lectures. 2. Nuclear reactions--Addresses, essays, lectures. I. Körner, J. G., 1939- II. Kramer, Gustav, 1932- III. Schildknecht, D., 1934- IV. Title. V. Series.
QC793.I556 1975 539.7'21 76-29733

ISBN 978-3-540-07866-1 ISBN 978-3-540-38097-9 (eBook)
DOI 10.1007/978-3-540-38097-9

Originally published by Springer-Verlag Berlin Heidelberg New York in 1976.

CONTENTS

PREFACE

In this Volume of the Lecture Notes in Physics we present the Proceedings of the 1975 International Summer Institute on Theoretical Physics held at the Deutsches Elektronen-Synchrotron DESY in Hamburg. This Institute was the seventh in a series of summer schools devoted to particle physics and organized by universities and research institutes in the Federal Republic of Germany.

The topic of the summer school has been "Current Induced Reactions" with particular emphasis on the physics of the new particles which had just been discovered during the preceding year. The summer institute included lectures on the experimental situation as well as the theoretical interpretation of neutrino induced and electron induced reactions in the space- and/or time-like regions. Quite naturally, theories based on quarks had a prominent place.

The present volume contains the notes of fourteen lectures as prepared by the authors. The lectures by Dr. G. Wolf (DESY) "Recent Experimental Results on e^+e^- Annihilation with Emphasis on the New Particles" and Dr. J. von Krogh (University of Wisconsin) "Neutrino Interactions-Experimental Results" were not available at the time of printing. The editors were responsible for preparing the manuscripts for print. In this connection we gratefully acknowledge the assistance of Frau Fischer and W. Knaut. Special thanks go to Frau Stuckenberg, Fräulein Laudien, Frau Schmöger, Frau Siemer and Frau Platz for their excellent typing of the manuscripts.

The Institute took place at the Deutsches Elektronen-Synchrotron DESY. On behalf of all the participants we thank the officials and the administration of this institution for their cooperation and help before and during the Institute.

The Institute was sponsored by the NATO Advanced Study Institute Programme and generously supported by the Bundesministerium für Forschung und Technologie in Bonn and the Senat der Freien und Hansestadt Hamburg.

J. G. Körner

G. Kramer

D. Schildknecht

June 1976

INELASTIC ELECTRON SCATTERING

by

Hinrich Meyer
Deutsches Elektronen-Synchrotron DESY, Hamburg

In this short review some of the more recent data from electroproduction experiments will be discussed. Part I is concerned with the shadow effect in photo- and electroproduction. Some remarks about scaling follow. We than turn to some specific channels and discuss mostly the dependence on Q^2. In the final part (IV) data on charged particle multiplicities will be presented.

I. The Shadow Effect

a) Real Photons

The hadronic nature of photon hadron interactions is a well established experimental fact. As a more general consequence of it a shadow effect for the total photoproduction cross section on nuclei (of mass number A), $\sigma(\gamma A)$, is expected to exist.

For hadronic beams, the experimental material can be very well parametrised by

$$\sigma(A) = \sigma_0 \cdot A^x \qquad (1)$$

where $\sigma(A)$ is the nuclear absorption cross section and σ_0, x are parameters depending on the nature of the incident particle. For pions we have $x = 0.72$ and $\sigma_0 = 32$ mb for momenta larger than 5 GeV.

In order to derive a mass number dependence for $\sigma(\gamma A)$ similar to (1) we need a model. A suitable frame work is provided by the vector dominance model. It in fact has been used to provide numerical predictions for a shadow effect in $\sigma(\gamma A)$.[1] At higher photon energies, $\nu > 10$ GeV, a relation similar to (1), with $\sigma_0 \sim 120\ \mu b$ and $x \sim 0.90$ results.

Most of the data on the shadow effect is at rather low energies.[2] Here the predictions are more model dependent and it is more appropriate to use the ratio

$$S = \sigma(\gamma A)/N \cdot \sigma(\gamma D) - (N - Z)\sigma(\gamma P) \qquad (2)$$

for a nucleus with N neutrons and Z protons.

A complete model calculation involves data on the nuclear shape, the energy dependence of the vector meson-nucleon cross section $\sigma(VN)$, the ratio of the real to the imaginary part of the vector meson-nucleon forward scattering amplitude Re/Im(f(0)) and assumptions about the correlations of nucleons inside the nucleus. Clearly, at low energies $\sigma(VN)$ and Re/Im(f(0)) are poorly known and not to much should be expected for the numerical predictions.

Finally, it is well known, that the Compton scattering sum rule for the total cross section $\sigma(\gamma N)$ is not yet saturated by the well established vector mesons and therefore in calculations not taking care of this discrepancy, a shadow effect which is to large is predicted. Previous data[2] did show a shadow effect, but somewhat smaller than the more naive calculations have predicted.[3]

New data, a measurement of Compton scattering

$$\gamma + A \rightarrow \gamma + A$$

on several nuclei and photon energies of 3 and 5 GeV is now available from DESY.[4] The data is shown in Fig.1, 2. The angular distribution at 5 GeV exhibits an indication of a diffraction pattern, as expected for a hadron like scattering process on nuclei. A detailed model of the nuclear shape has been used to fit the data as function of t and to provide the necessary extrapolation to $t = 0$. An example of such a fit is shown in Fig.3 for Ag at 5 GeV photon energy.

From the extrapolated values at $t = 0$ and total cross section data on proton and neutrons[5] the ratio S (see (2) above) has been calculated using the optical theorem

$$d\sigma/dt\ (\gamma A \rightarrow \gamma A)_{t=0} = \frac{1}{16\pi}\sigma^2(\gamma A)\ [1 + (\mathrm{Re/Im}(f(0)))^2] \qquad (3)$$

This is shown in Fig.4. The curves result from a calculation within the frame work of the vector dominance model. Again a shadow effect exists, but somewhat less then predicted. (curve labeled 1)

b) Virtual Photons

In inelastic scattering of electrons and muons on nuclei the shadow effect is expected to disappear with increasing Q^2, the virtual photons mass squared. Reliable numerical predictions are difficult to get, the most recent calculations being done by Ditsas, Read and Shaw.[6] Qualitatively, the kinematic condition for a shadow effect in electroproduction is given by

$$0.1\ \text{GeV} > \frac{m_v^2}{2\nu} + m_p \cdot \frac{1}{x} \tag{4}$$

where $r \equiv$ virtual photon energy, $m_p \equiv$ proton mass, m_v = vector meson mass, and $1/x = \omega = 2m_p\nu/Q^2$ the scaling variable. This follows from a consideration of the coherence length of vector mesons with mass m_v inside the nucleus. It is a necessary condition, and implies the requirement $x < 0.1$.

The region $x > 0.1$ is very unlikely to show a shadow effect for a different reason as well. The ratio of the virtual photon neutron and proton cross sections is found to decrease with x and is significantly different from 1 already for $x = 0.1$. This indicates a strong incoherence of the virtual photon over the diameter of one nucleon and no coherence is therefore expected to exist inside a nucleus.

The data available so far[7] show little shadowing, even at very small Q^2. The results of the various experiments are shown in Figs.5, 6. In order to compare different experiments, two specific energies $\nu = 4$ GeV and $\nu = 8$ GeV and two nuclei, Cu and Au, have been chosen. It is seen in Fig.5, 6 that the data are not inconsistent with each other, especially in view of the fairly large systematic errors to be expected in this kind of experiment. The curves are theoretical estimates of the Q^2-dependence of the shadow effect from Ref.8. First, I consider it important, that at least some shadow effect has now been seen. But also, there are some puzzling subtleties in the data, for example, there is a complete lack of ν dependence in the Cornell data (Fig.7), in fact for the measurement at $\nu = 5$ GeV, no shadow effect is observed at all. In the SLAC data (Fig.8) at intermediate ν a trend for an "antishadow" is seen. This indicates unknown sources of systematic effects and more detailed experiments and more refined analysis is needed to establish the correct Q^2, W dependence of a shadow effect in electroproduction.

II. Scaling Violations

At London Conference (1974)[9] for the first time some indications of scaling violations have been reported. The experiments have better data now and more insight into the nature of the effect can be gained. From new data, using the 8 GeV spectrometer at SLAC,[9a] we see for fixed $x = 1/\omega > 0.3$ a definite decrease with Q^2 for both $2\,MW_1$ and νW_2. (Fig.9) The authors state that by using $\omega' = \omega + m_p^2/Q^2$ instead of ω the deviations from scaling are weaker. Scaling can even be retained by using a new variable $\tilde{\omega} = \omega + M^2/Q^2$, with $M^2 = 1.5\ \text{GeV}^2$.

Under the restriction of $Q^2 > 2$ GeV we are still left with a rather interesting and significant scaling violation. For $x = 1/\omega < 0.2$ the NAL data points are consistently higher by about 15-30%. This trend also appears in the comparison of the structure function as measured in neutrino scattering if compared to the SLAC scaling function, and in fact by about the same amount and in the same x intervall. One may speculate this to be due to a new threshold in W, but it can not yet be inferred from the data. The new effect (in terms of the quark-parton model) would appear in collisions with quarks from the sea, (small x) and possibly is due to a new variety of quarks.

III. Two Body Channels

In this part, specific two body channels in electroproduction will be discussed. They are

(1) $e\, p \to e\, p$

(2) $e\, p \to e\, n\, \pi^+$

(3) $e\, p \to e\, \Delta^{++}\, \pi^-$

which have been used to determine various transition form factors. Furthermore we will mention

(4) $e\, p \to e\, p\, \pi^0$

because this reaction shows a very peculiar Q^2 dependence. Finally in this part, data and results of a detailed analysis of the elastic production of vector-mesons will be presented

(5) $e\, p \to e\, p\, \rho^0$

a) The Reactions $e\, p \to e\, p$ and $e\, p \to e\, n\, \pi^+$

A compilation of previous and recent SLAC data on reaction (1), elastic ep scattering, is shown in Fig.13.[10] Although not entirely conclusive, the data are very suggestive of a dependence proportional to $(Q^2)^2$ for $Q^2 > 6$ GeV^2. Therefore we have at large Q^2

$$G_M/\mu_p \times Q^4 \simeq 0.4 \qquad (5)$$

for the proton.

Reaction (2) at small t gets a sizeable contribution from the one pion exchange term. Analysis of reaction (2) as function of Q^2 then determines the pion form factor $F_\pi(Q^2)$ from the Q^2-behavior of the $\gamma\pi\pi$ vertex. A Harvard group working at Cornell has extended the measurements out to $Q^2 \simeq 4\ GeV^2$.[11] A compilation of the data is shown in Fig.14. From this one would infer a Q^2 dependence at $Q^2 > 1.5\ GeV^2$ of about

$$F_\pi(Q^2) \cdot \times Q^2 \simeq 0.4 \ . \qquad (6)$$

The difference for the proton and pion in powers of Q^2 is expected in models, that describe large Q^2 scattering by a point like photon-quark scattering process, sometimes called the dimensional counting rule.[12] In that model, at large Q^2, $F(Q^2) \sim (Q^2)^{1-n}$ where n is the minimum number of fundamental quark fields involved in the scattering process. For pions $n = 2$ and for the proton $n = 3$, which is consistent with the new data quoted.

b) $e\ p \to e\ \Delta^{++} \pi^-$

Reaction (3) near threshold can be used to extract the nucleon axial vector form factor. The procedure to be applied has been described in full detail by Adler and Weisberger.[13] PCAC is assumed to hold. The new data comes from an streamer chamber experiment carried out by a DESY-Glasgow collaboration at DESY.[14] The Q^2 dependence of the axial vector form factor extracted from the data is shown in Fig.15. Also included is data from reaction (2) which very near threshold can be used to determine the axial vector form factor. The full curve, assuming a dependence proportional to

$(1/1 + m_A^2/Q^2)$ uses a value of $m_A = 1.16 \pm 0.03$ GeV. This is larger then the value determined from the reaction[15]

$\nu d \to \mu^- \ p \ p_S$ which gives $m_A = 0.89 \pm 0.08$.

It is not yet clear, if this discrepancy is due to a limited applicability of PCAC. Relaxing in the neutrino data the CVC requirement results in $m_V = 0.92 \pm {}^{.05}_{.11}$ and $m_A = 0.75 \pm {}^{.21}_{.10}$,[15] thereby increasing the discrepancy but also reducing the statistical significance of it.

c.) The Reaction $e\ p \to e\ p\ \pi^0$

For reaction 4 measurements have been proposed[16] as a test of various mechanisms that generate a dip in the angular distribution around $t=0.5\ GeV^2$. An

An experiment at DESY made available data at $W \sim 2.55$ GeV and three Q^2 intervals for $0.2 < t < 1.4$ GeV^2.[17] The observed behaviour is rather dramatic and totally unexpected as can be seen in Fig.16. The cross section for $t > 0.8$ GeV^2 drops by a factor of 10 between $Q^2 = 0$ and $Q^2 = 0.22$ GeV^2 and shows little variation with Q^2 up to $Q^2 = 0.85$ GeV^2. At smaller t the slope of the t distribution changes somewhat with Q^2 to smaller values then in photon production and the Q^2 dependence approximately follows a ρ^0-propagator. No explanation has been put forward so far for the extremely rapid drop at higher t, between $Q^2 = 0$ and $Q^2 = 0.22$ GeV^2.

d) Vector Meson Production (ρ^0)

Vector meson production in electron scattering $e\,p \to e\,p + V^0$ has been studied in a number of expriments. The interest in this reaction partly arises from the very clear predictions made in the frame work of the vector dominance model.[18] In detail we expect for ρ^0 meson electroproduction (at fixed W)

$$\sigma(e\,p \to e\,p\,\rho^0) = f(t_{min}) \times f(Q^2, \omega^2) \times \left(\frac{1}{1+Q^2/m_\rho^2}\right) \times \left(1 + \varepsilon\,\xi^2\,\frac{Q^2}{m_\rho^2}\right) \times \sigma(\gamma p \to \rho^0 p) \qquad (7)$$

the function $f(t_{min})$ describes a simple kinematic correction, due to the minimum momentum transfer required, enters to take into account the different flux factors for the virtuell photons and vector mesons respectively. The Q^2 dependence is given by the ρ^0 propagator and the contribution due to longitudinal photons is predicted to increase proportional to Q^2. In Fig.17 the Q^2 dependence of the data from the streamer chamber experiment at DESY is shown for 3 intervals in W.[19] The full lines indicate that equation (7) describes the data very well for $W > 2$ GeV, if we allow for a sizeable longitudinal contribution to the cross section. This will be discussed further below.

For the t dependence of the ρ^0 cross section at $t < 0.5$ GeV^2 we expect an exponential shape. In Fig.18 the data from Ref.19 are shown and for comparison the data a $Q^2 = 0$.[20] No significant change in the slope of the t distribution with Q^2 can be observed. This does not yet contradict expectations for a shrinkage of the t-distribution, in fact, much higher Q^2 at fixed W are required. There is data for the slope also from UCSC-SLAC collaboration[21] (Fig.19), again no significant change with Q^2 is seen.

For a determination of the parameter ξ^2 in equation (7) either measurements at fixed Q^2 and W^2 but different ε are required, or one has to use the decay angular correlations for the ρ^0 with the assumption of s-channel helicity conservation (SCHC).

The data are quite consistent with SCHC, as can be inferred from a study of of the spin density matrix elements. From the data then the matrix element r^{04}_{00} is determined which is related to ξ^2 by

$$\xi^2 = \frac{m_\rho^2}{Q^2} \times \frac{1}{\varepsilon} \times \frac{r^{04}_{00}}{1-r^{04}_{00}}$$

The results from Ref.19 are shown in Fig.20. It can be observed that with increasing Q^2 more longitudinal ρ^0 are produced as predicted by VDM.[18] Also at low W the longitudinal ρ^0 cross section seems to be even dominant.

From the data also the phase δ between the longitudinal and transverse ρ^0 production amplitude can be determined. The results are shown in Fig.21. In the low W region both amplitudes are out of phase, but for $\omega > 2.5$ GeV they seem to come in phase as expected for diffractive ρ^0 electroproduction.[18]

IV. Charged Particle Multiplicities

As one of the very general features of high energy interactions, the average charged particle multiplicity $\langle n \rangle$ has been determined in track chambers[21-23] and a pure counter experiment.[24] The variables of interest are Q^2 and W^2, the four momentum squared at the lepton vertex and the total energy squared at the hadron vertex, pictorially shown below

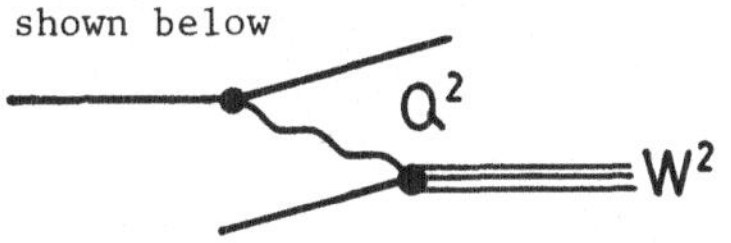

In Fig.22 we show data for $\langle n \rangle$ from the DESY Glasgow streamer chamber collaboration at an average Q^2 of 0.78 GeV^2 as function of $\ln s$. For comparison, photoproduction data[25] are shown (full line). The electroproduction points are lower than photoproduction for 3.5 GeV$^2 < s <$ 10 GeV^2, the dependence on s being about linear in $\ln s$, as in photoproduction.

In Fig.23 this data is shown in comparison with other experiments and now for three Q^2 intervals. Different experiments agree fairly well.

In Fig.24 the data are shown for fixed W intervals as function of Q^2.[23] We observe a drop from the photoproduction value to a level, which does not change with Q^2; the lowest measured points is $Q^2 = 0.3$ GeV^2. The transition therefore takes place at surprisingly low Q^2 with now further change with Q^2.

In the W region discussed only 1 prong, 3 prong and 5 prong events occur, and in what follows we discuss the fractional prong cross sections σ_n/σ_{tot} as function of Q^2 and s. The prong cross sections are related to the mean multiplicity by

$$\langle n\rangle \quad \sum n \cdot \sigma_n/\sigma_{tot}$$

From Fig. 25 we see that the lower average multiplicity is due to an increasing one prong cross section, the three prong cross section being reduced accordingly and the five prong cross section showing little change. This behaviour does not depend on Q^2 dependence at all, even out to a Q^2 of 6 GeV^2, see Fig. 26

The dependence of σ_n/σ_{tot} on s , now averaged over Q^2 , is shown in Fig. 27. Up to $s = 10\ GeV^2$ the various fractional cross section show strong variations with s , much the same way as observed in photoproduction (full lines). In the following figure the electroproduction data are compared to proton-proton data,[26] when only one proton has been excited, usually called "Pomeron" proton scattering. The "Pomeron" now takes the part of the virtual photon.

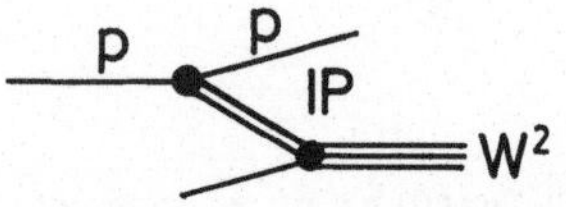

The proton data are shown as full lines Fig. 28. Again here a very similar pattern (see Fig. 27) arises.

There is a new data available on charged particle multiplicities from neutrino proton scattering. Only charged current events have been taken into account. At high neutrino energies E_ν high average Q^2 is achieved because

$$\langle Q^2\rangle \sim 0.17 \cdot E_\nu$$

As can be seen from Fig. the mean charged particle multiplicity in neutrino (antineutrino) interactions follows essentially the same empirical law

$$\langle n\rangle \sim a + b \ln W^2$$

with values for a, b close to the values found in electroproduction. (see Fig. 22)

We may conclude this section with the observation that the mean charged multiplicity is very insensitive to the incoming particle. In fact it is about the same for the electromagnetic current, the weak current and apparently also the "Pomeron" current. Also it does <u>not</u> depend on the Q^2 values of the currents.

Clearly, the amount of scaling violation depends on the choice of the scaling variable.

Also at SLAC, new data have been taken with the 1.6 GeV spectrometer at large angles, (50° and 60°) on both hydrogen and deuterium.[10] At this large angles essentially $2MW_1$ is being measured. In addition to proton data also the structure function for neutrons has been extracted. For the interval $0.5 < x' < 0.7$ for both protons and neutrons $2MW_1$ is shown in Fig.10. Surprisingly enough, for protons scaling seems to be violated, but very little violation is seen for the neutron. The amount of scaling violation seen in this experiment is consistent with the data from the 8 GeV spectrometer.[9a] The neutron and proton cross sections seem to approach each other, a very clear trend for a deviation from the so far universal σ_n/σ_p ratio (Fig.11).

The high energy experiment at NAL using 56 GeV and 150 GeV muons on an iron target also has yielded new data on scaling violations.[9b] The kinematic region is extended by that experiment beyond the SLAC region to much larger ν (from $\nu = 15$ to $\nu = 100$ GeV) and slightly larger Q^2 (from 16-24 GeV^2 to $\sim 40\ GeV^2$). Therefore tests of scaling is mostly extended into the larger ω region. The result of the NAL experiment is shown in Fig.12 as the ratio of measured cross sections and cross section values predicted by a Monte Carlo programm for the NAL set up, using as input a parametrisation of the SLAC scaling functions. Taken at face value at large ω the structure function increases with Q^2 and at small ω it decreases by approximately the same amount as observed in the SLAC data (but with much less statistical significance).

For a most critical test of scaling one would tend not to use the data points for $Q^2 < 2$ GeV for two reasons. As we have seen in the discussion of the shadow effect, there is now evidence for some shadow at $\omega > 10$. We take $\omega > 10$ as a necessary requirement for a shadow effect to exist. Otherwise, we assume the shadow effect to be a function Q^2 only, this implies that the shadow effect is mainly caused by the intermediate ρ^0 mesons. I am aware of the fact that this assumption is at variance with the generalized vector dominance model.[6,8] If shadow were also present in the NAL data the low Q^2 points should have been increased for a comparison with data on nucleons by about 15%, which is roughly the amount needed to cancel the Q^2 dependence in the data at large ω.

At low ω and low Q^2, W is small and scaling very critically depends on the choice of the scaling variable.

REFERENCES

1. For a summary see K. Gottfried in Proceedings of the 1971 International Symposium on Electron and Photon Interactions at High Energies. Cornell University, Ithaca, N.Y. 1971, p.221.

2. E. Gabathuler in Proceedings of the 6th International Symposium on Electron and Photon Interactions at High Energies, Bonn. North-Holland Publishing Co. 1974, p.299.

3. S. J. Brodsky and J. Pumplin; Phys. Rev. 182, 1794 (1969).

4. L. Criegee et al., Contribution to the 1975 SLAC Conference, Abstract No.169 and Th. Kahl, Thesis 1976 unpublished.

5. T. A. Armstrong et al., Nucl. Phys. B41, 445 (1972) (for γd) and T. A. Armstrong et al., Phys. Rev. D5, 1640 (1972) (for γp).

6. P. Ditsas, B. J. Read and G. Shaw; Nucl. Phys. B99, 85 (1975).

7. J. Eickmeyer et al., Phys. Rev. Letters 36, 289 (1976)
S. Stein et al., Phys. Rev. D12, 1884 (1975)
M. May et al., Phys. Rev. Letters 35, 407 (1975)
G. Brookes et al., Phys. Rev. D8, 2826 (1973)
D. O. Caldwell et al., Phys. Rev. D7, 1362 (1973).

8. D. Schildknecht; Nucl. Phys. B66, 398 (1973).

9. Proceedings of the XVII International Conference on High Energy Physics, London, July 1974, p.IV-57 and IV-149.

9a. E. M. Riordan et al., SLAC-PUB-1634, August 1975.

9b. C. Chang et al., Phys. Rev. Letters 35, 901 (1975)
Y. Watanabe et al., Phys. Rev. Letters 35, 898 (1975).

10. R. E. Taylor in Proceedings of the 1975 International Symposium on Lepton and Photon Interactions at High Energies, p.679.

11. G. Wolf in Proceedings of the 1975 International Symposium on Lepton and Photon Interactions at High Energies, p. 795.

12. S. J. Brodsky and G. Farrar; Phys. Rev. Letters 31, 1153 (1973).

13. S. L. Adler and W. I. Weisberger; Phys. Rev. 169, 1392 (1968).

14. P. Joos et al., DESY Report 76/09 (1976).

15. D. H. Perkins in Proceedings of the 1975 International Symposium on Lepton and Photon Interactions at High Energies, 571.

16. H. Harari in Proceedings of the 1971 International Symposium on Electron and Photon Interactions at High Energies. Cornell University, Ithaca, N.Y. (1971) p.299.

17. F. W. Brasse et al., DESY Report 75/23 (1975).

18. H. Fraas and D. Schildknecht; Nucl. Phys. B14, 543 (1969).

19. P. Joos et al., DESY Report 76/17 (1976).

20. J. Ballam et al., Phys. Rev. D5, 545 (1972).

21. K. Bunnell et al., Contribution to the 1975 SLAC Conference, Abstract No.272.

22. J. Ballam et al., Phys. Letters 56B, 193 (1975).

23. DESY-Glasgow Collaboration, contribution to the 1975 SLAC Conference Abstract No.242.

24. P. H. Gabrincius et al., Phys. Rev. Letters 32, 328 (1974)
A. J. Sadoff et al., Phys. Rev. Letters 32, 995 (1974).

25. K. C. Moffeit et al., Phys. Rev. D5, 1603 (1972).

26. Bonn-Hamburg-München Colaboration see B. Schwarz, DESY Internal Report F1-74/4 (1974).

27. J. W. Chapman et al., FERMILAB-Pub-75/85-EXP, 7300.045.

FIGURE CAPTIONS

Fig.1 Differential cross section for Compton scattering of real photons an various nuclei at 3 GeV. The data are from Ref.4

Fig.2 As in Fig.1 but at 5 GeV photon energy. (Ref.4)

Fig.3 At 5 GeV photon energy a fit to the differential cross section for Compton scattering of silver. (Ref.4)

Fig.4 The ratio of the forward (t = 0) cross section for Compton scattering on nuclei to the forward cross section on nucleons as function of the mass number A. The t = 0 point on nucleons has been calculated from total cross section data (Ref.4) using the optical theorem. The curves labeled 1-6 represent the results of model calculations indicating the dependence on various input parameters. For detail see Ref.4.

Fig.5 A_{eff}/A vs. Q^2 for Cu and Pb, Au at r = 4 GeV. The data are from Ref.7, S. Stein et al. Ref.7, M. May et al. Ref.7, I. Eickmeyer et al. Ref.4,7, G. Brooks et al. and D. O. Calwell et al.

Fig.6 Same as Fig.5 but at r = 8 GeV. The curve has been taken from Ref.8.

Fig.7 The total cross section for electron scattering on nuclei normalized to the cross section on nucleons as function of the mass number A. Q^2 is small for this data $\sim$ 0.1 GeV^2, r is the energy of the virtual photon. (I. Eickmeyer et al., Ref.7).

Fig.8 Electron scattering data on nuclei at fixed angle (4°) and incident energies of 13 and 20 GeV. From the left hand part of the figure, the strong influence of radiative corrections at high r can be seen. The right hand part shows the cross sections on nuclei, normalised to the nucleon cross section. (S. Stein et al., Ref.7).

Fig.9 The structure functions νW_2 and $2MW_1$ on protons and deuterium, for various values of $x = 1/\omega = Q^2/2M\nu$. (Ref.9a).

Fig.10 The structure function $2MW_1$ vs. Q^2 for protons and neutrons. (Ref.10).

Fig.11 The ratio of neutron to proton cross sections vs. $x' = 1/\omega' = Q^2/2M\nu+M^2$. (Ref.10).

Fig.12 Data for μ Fe scattering at 150 GeV incident energy normalized to predictions from SLAC scaling (in ω') data. For perfect scaling at NAL energies this ratio should be 1 at all ω and Q^2. (Ref.9b).

Fig.13 The magnetic form factor for $e\,p \to e\,p$, multiplied with Q^4 vs. Q^2. At larger Q^2 a dependence of the form $1/Q^4$ is indicated. (Ref.10).

Fig.14 The pion form factor, deduced from $e\,p \to e\,\pi^+\,n$. For $Q^2 > 1\ \mathrm{GeV}^2$ the data are consistent with a $1/Q^2$ dependence. (Ref.11).

Fig.15 Data for the cross section $e\,p \to e\,\pi^-\Delta^{++}$ close to threshold (upper part) and the Q^2 dependence of the nucleon axial vector form factor deduced from this data (lower part of the figure). See Reference 14.

Fig.16 Data for the reaction $e\,p \to e\,p\,\pi^0$. The full line indicates the behavior for photoproduction. At finite Q^2 no dip is observed in the data in contrast to photoproduction. (Ref.17).

Fig.17 Cross section for electroproduction of ρ^0 mesons as function of Q^2. The full line is the dependence on Q^2 as expected from the vector dominance model. (Ref.19).

Fig.18 The differential cross section for ρ^0 electroproduction (full points) and ρ^0 photoproduction (open points). See Reference 19.

Fig.19 The slope A of the differential cross section for ρ^0 electroproduction. (Ref.21).

Fig.20 Data on the ratio of longitudinal to transverse ρ^0 electroproduction. (Ref.19).

Fig.21 The phase between the production amplitude for longitudinal and transverse ρ^0 electroproduction. (Ref.19).

Fig.22 The mean multiplicity for the electroproduction of charged particles vs. ℓn s. The total center of mass energy squared (full points). The full line interpolates through photoproduction data. (Ref.23).

Fig.23 Comparison of various experiments for the mean multiplicity in electroproduction. (Ref.11).

Fig.24 The Q^2 dependence of $\langle n_{ch} \rangle$ for three W intervalls. (Ref.23).

Fig.25 The Q^2 dependence of fractional prong cross sections. (Ref.23).

Fig.26 Prong cross sections as function of Q^2, for $Q^2 < 2.5$ GeV^2 Ref.21 and $Q^2 > 2.5$ GeV^2 Ref.24.

Fig.27 The s dependence of fractional prong cross sections in electroproduction (data points) as compared to photoproduction (full lines). (Ref.23).

Fig.28 The same data points as in Fig.26, but the full lines are now from pp data. (Ref.23).

Fig.29 Data on $\langle n_{ch} \rangle$ in νp collisions, (♦), and pomeron proton scattering (✱) (○) from Reference 27.

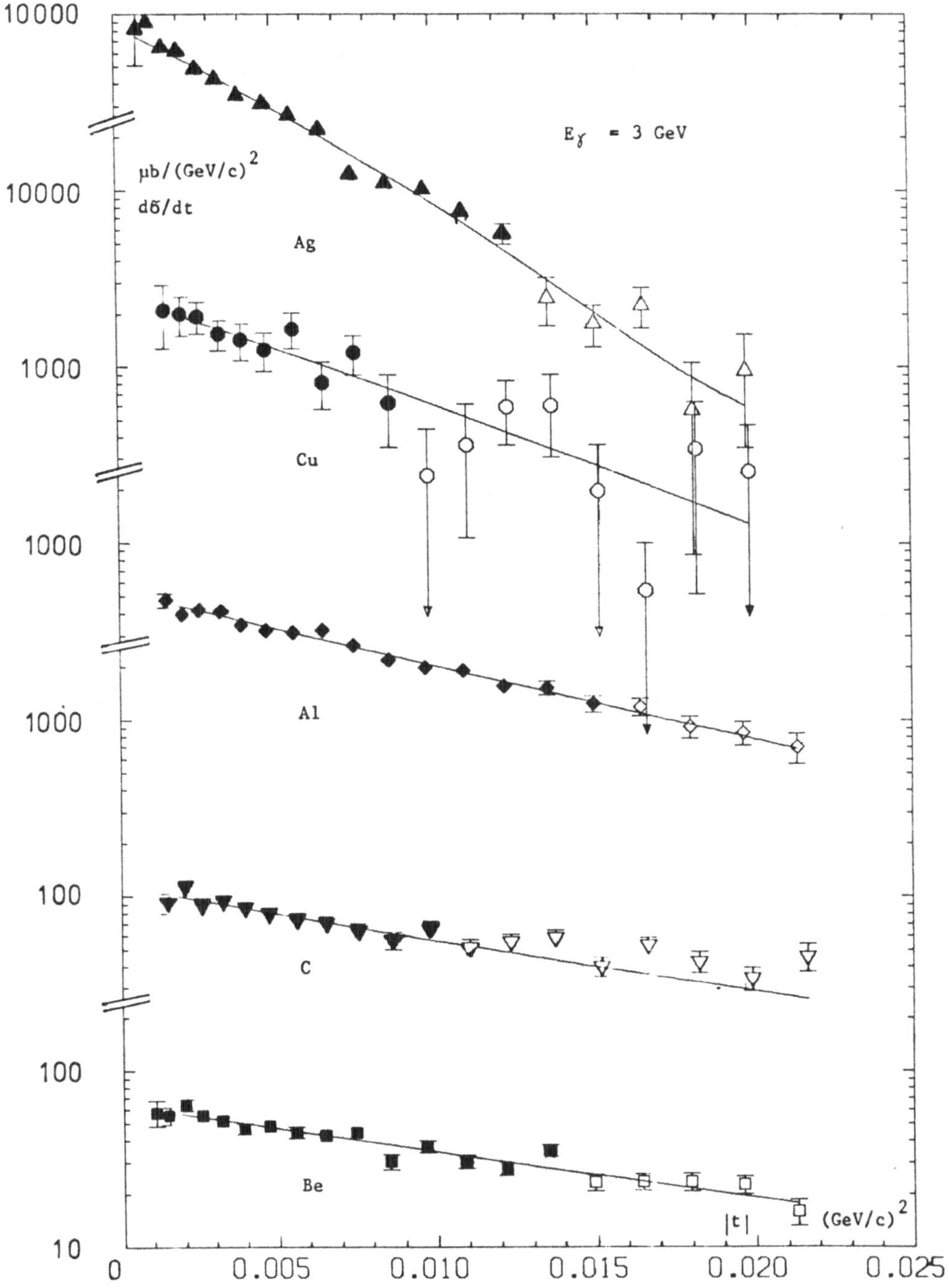
10000
10000
1000
1000
1000
100
100
10
μb/(GeV/c)²
dσ/dt
Eγ = 3 GeV
Ag
Cu
Al
C
Be
|t| (GeV/c)²
0
0.005
0.010
0.015
0.020
0.025

FIG.1

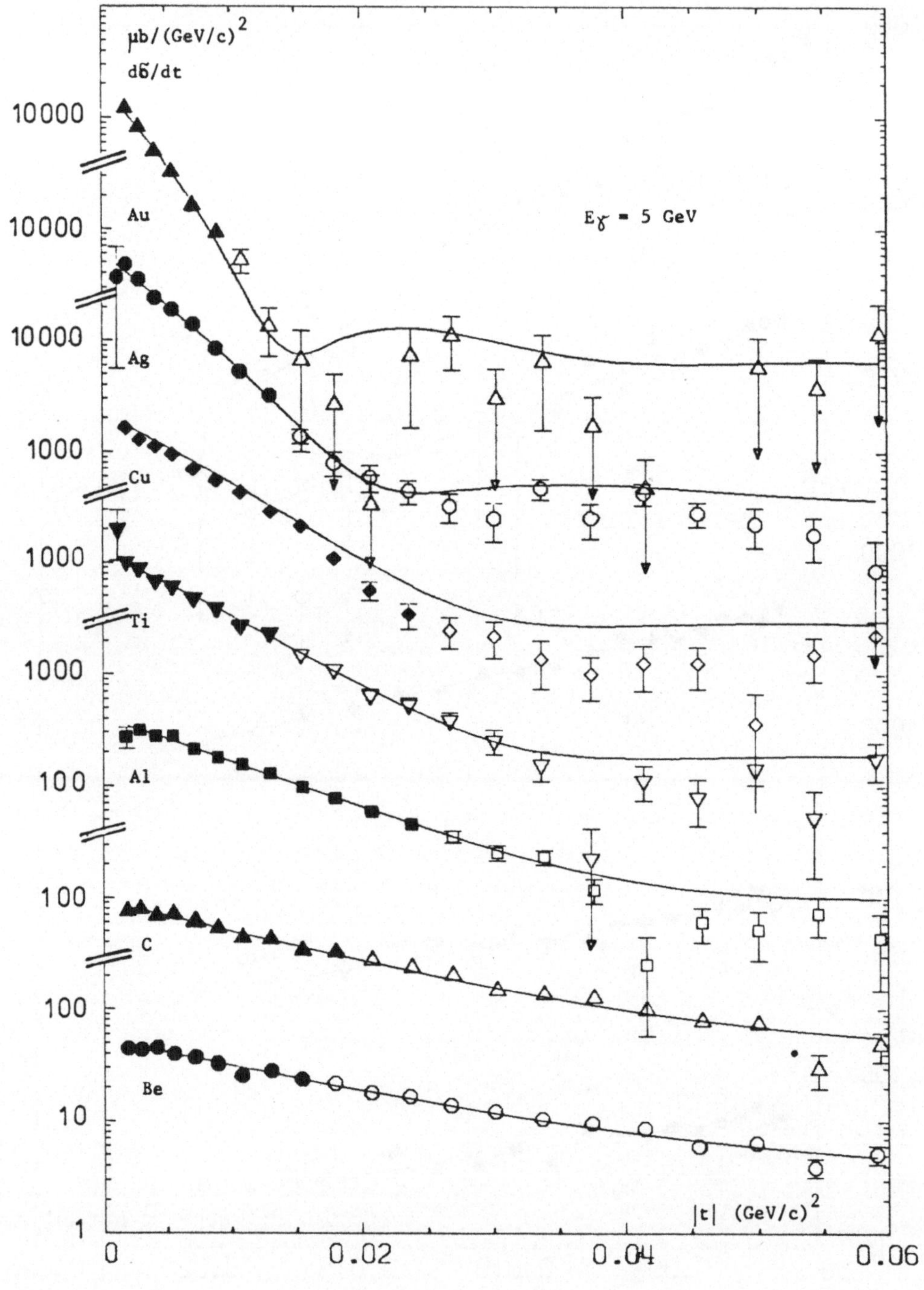

μb/(GeV/c)^2
dσ/dt
10000
10000
10000
1000
1000
1000
100
100
100
10
1
Au
Ag
Cu
Ti
Al
C
Be
Eγ = 5 GeV
|t| (GeV/c)^2
0
0.02
0.04
0.06

FIG.2

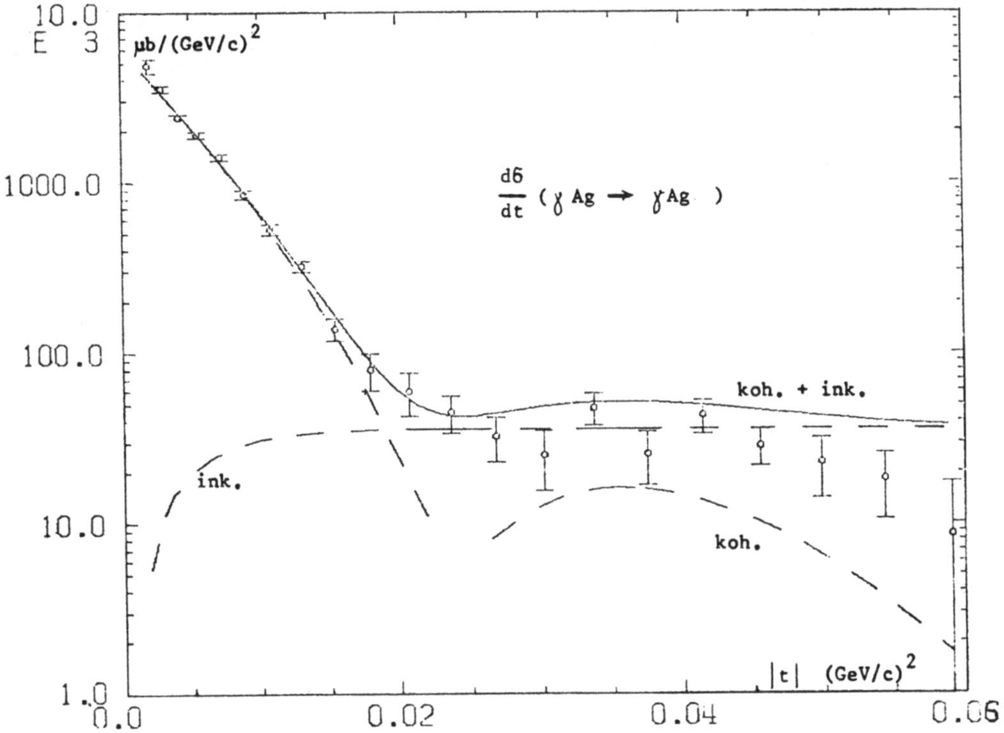

10.0
E 3
μb/(GeV/c)²
1000.0
100.0
10.0
1.0
dσ/dt (γ Ag → γAg)
koh. + ink.
ink.
koh.
|t| (GeV/c)²
0.0
0.02
0.04
0.06

FIG.3

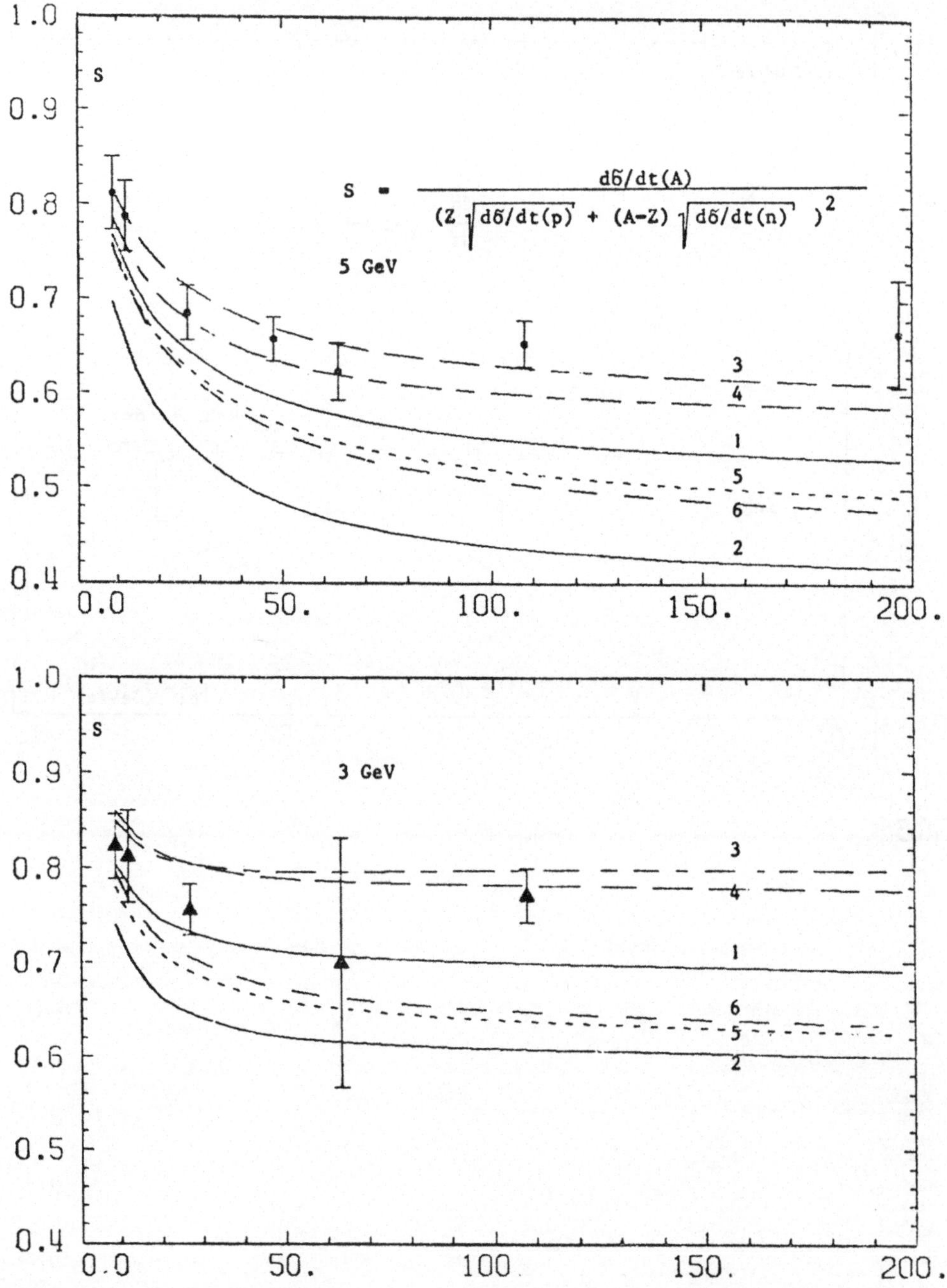
S
S = dσ/dt(A) / (Z √dσ/dt(p) + (A−Z) √dσ/dt(n))²
5 GeV
3
4
1
5
6
2
1.0
0.9
0.8
0.7
0.6
0.5
0.4
0.0
50.
100.
150.
200.
S
3 GeV
3
4
1
6
5
2
1.0
0.9
0.8
0.7
0.6
0.5
0.4
0.0
50.
100.
150.
200.

FIG.4

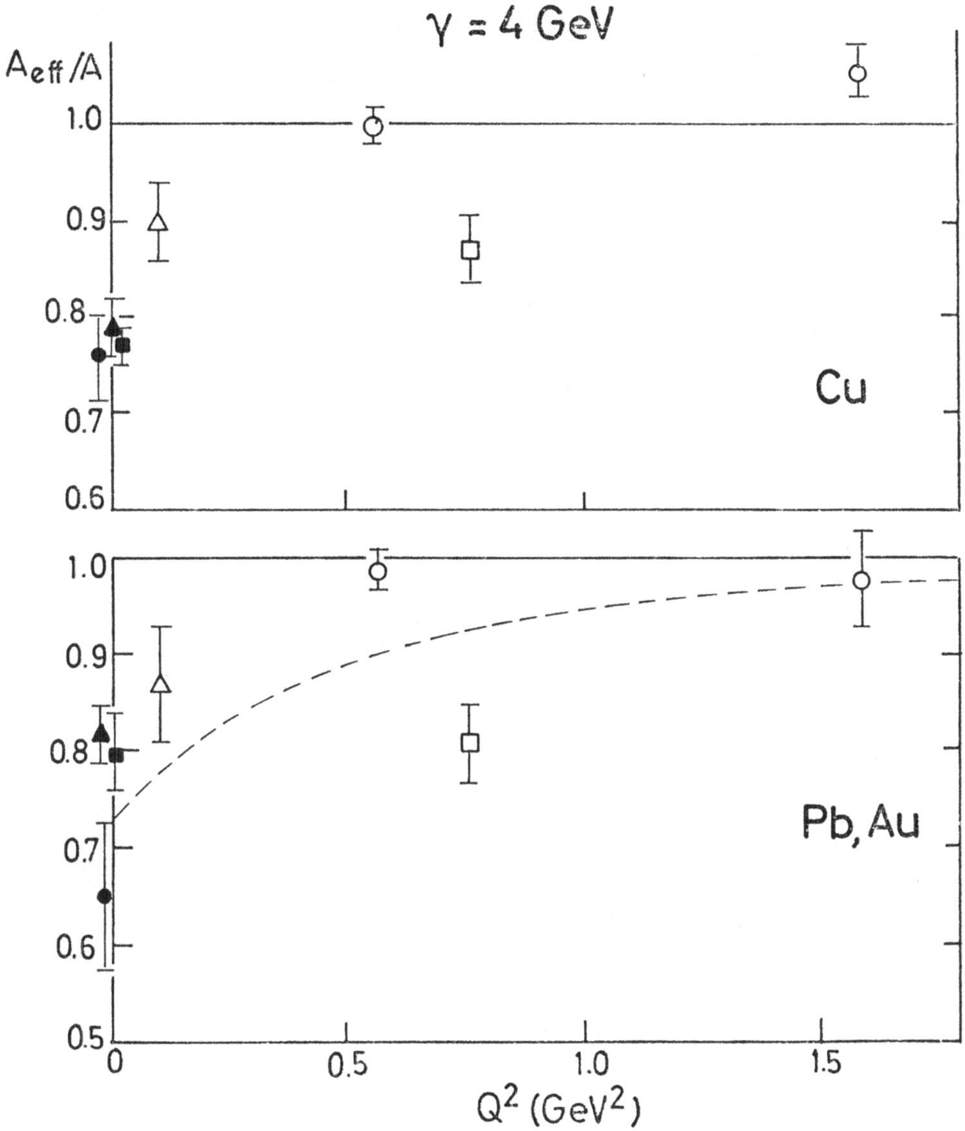
γ = 4 GeV
Aeff /A
1.0
0.9
0.8
0.7
0.6
Cu
1.0
0.9
0.8
0.7
0.6
0.5
Pb, Au
0
0.5
1.0
1.5
Q² (GeV²)

Fig.5

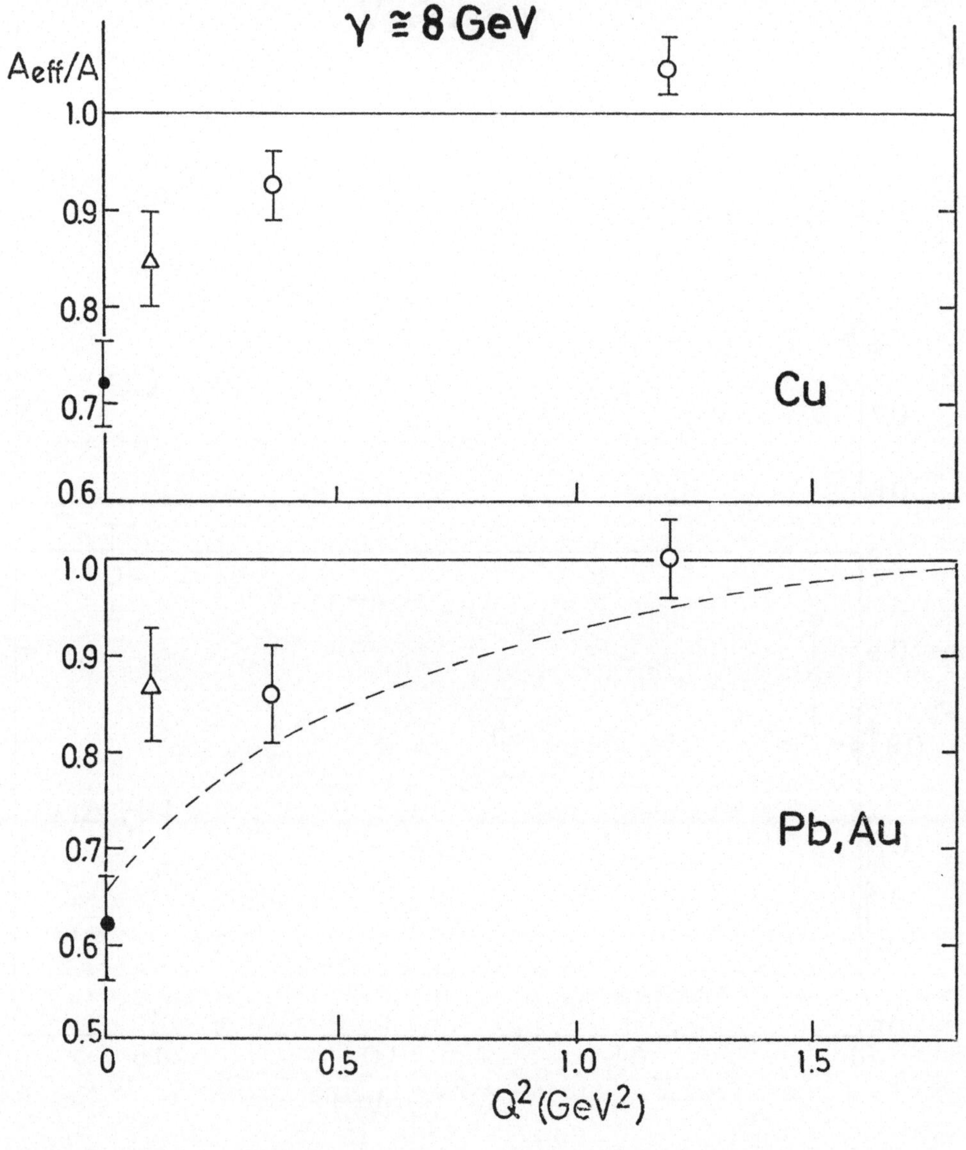

γ ≅ 8 GeV
Aeff/A
1.0
0.9
0.8
0.7
0.6
Cu
1.0
0.9
0.8
0.7
0.6
0.5
Pb, Au
0
0.5
1.0
1.5
Q2 (GeV2)

Fig.6

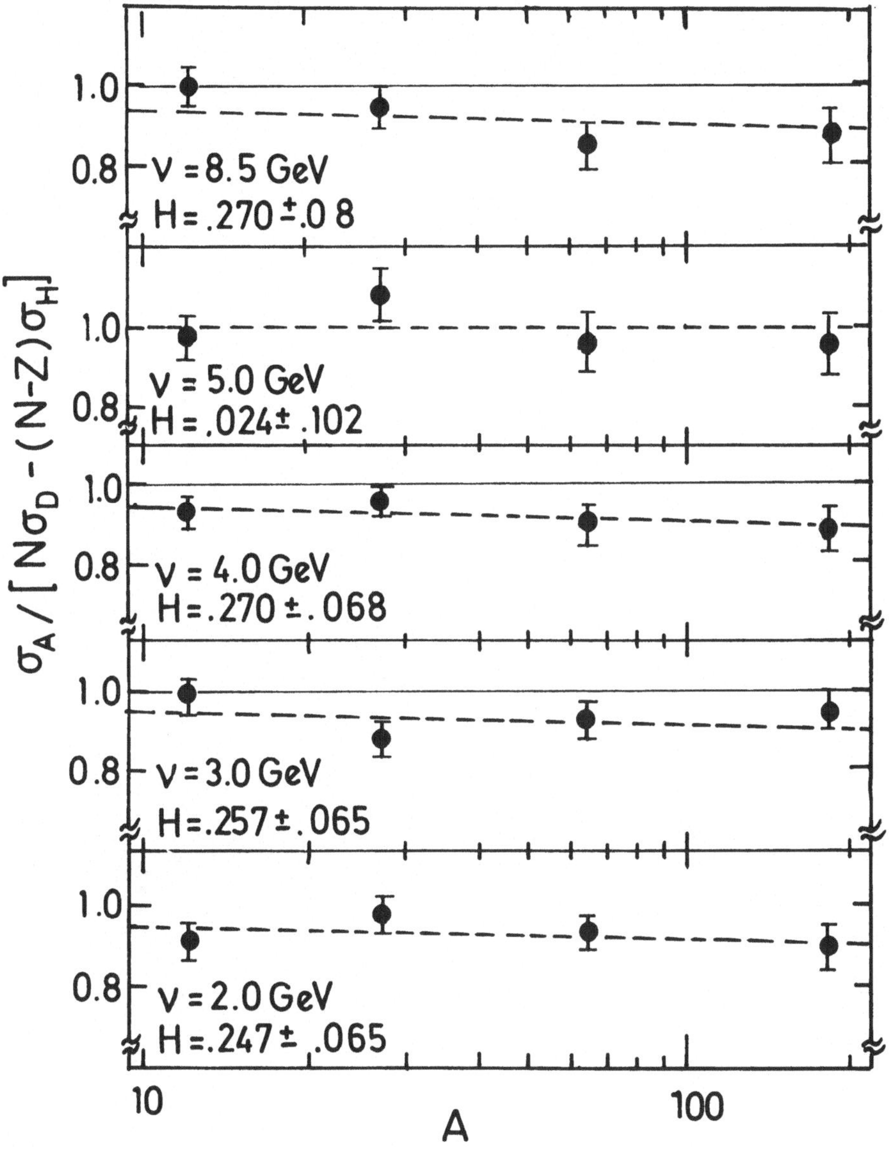
1.0
0.8
ν = 8.5 GeV
H = .270 ± .0 8
1.0
0.8
ν = 5.0 GeV
H = .024 ± .102
1.0
0.8
ν = 4.0 GeV
H = .270 ± .068
1.0
0.8
ν = 3.0 GeV
H = .257 ± .065
1.0
0.8
ν = 2.0 GeV
H = .247 ± .065
σ_A / [Nσ_D - (N-Z)σ_H]
10
100
A

Fig.7

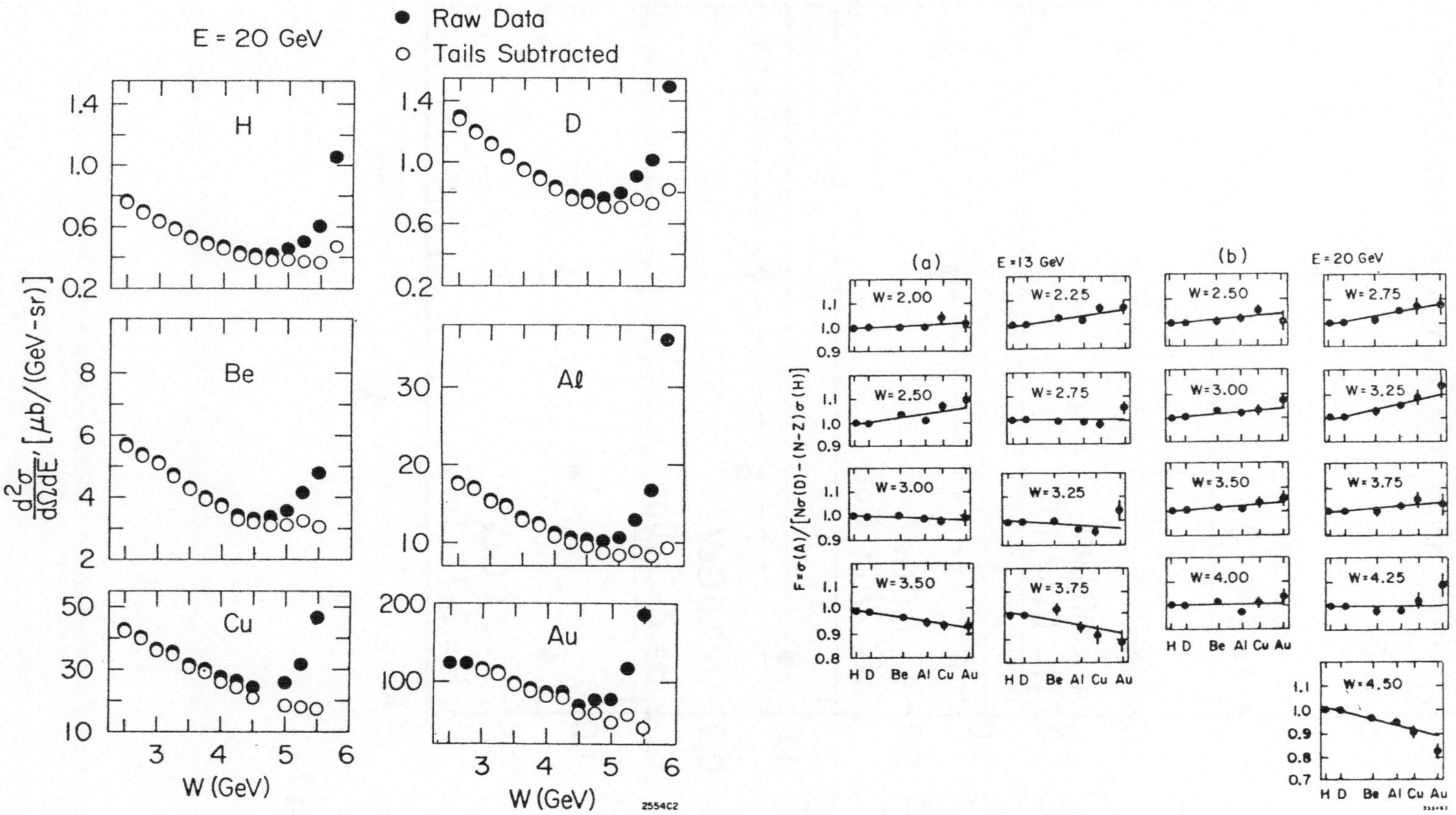

E = 20 GeV
Raw Data
Tails Subtracted
H
D
Be
Aℓ
Cu
Au
W (GeV)
2554C2
(a)
E = 13 GeV
(b)
E = 20 GeV
W = 2.00
W = 2.25
W = 2.50
W = 2.75
W = 3.00
W = 3.25
W = 3.50
W = 3.75
W = 4.00
W = 4.25
W = 4.50
H D Be Al Cu Au

FIG.8

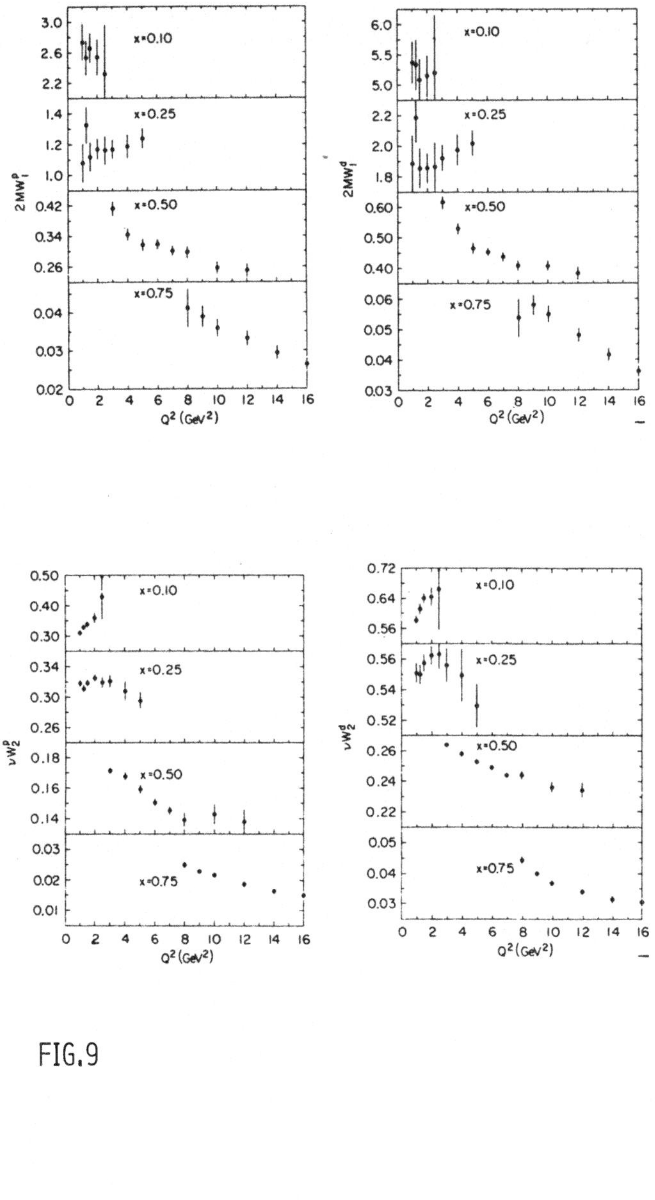

FIG.9

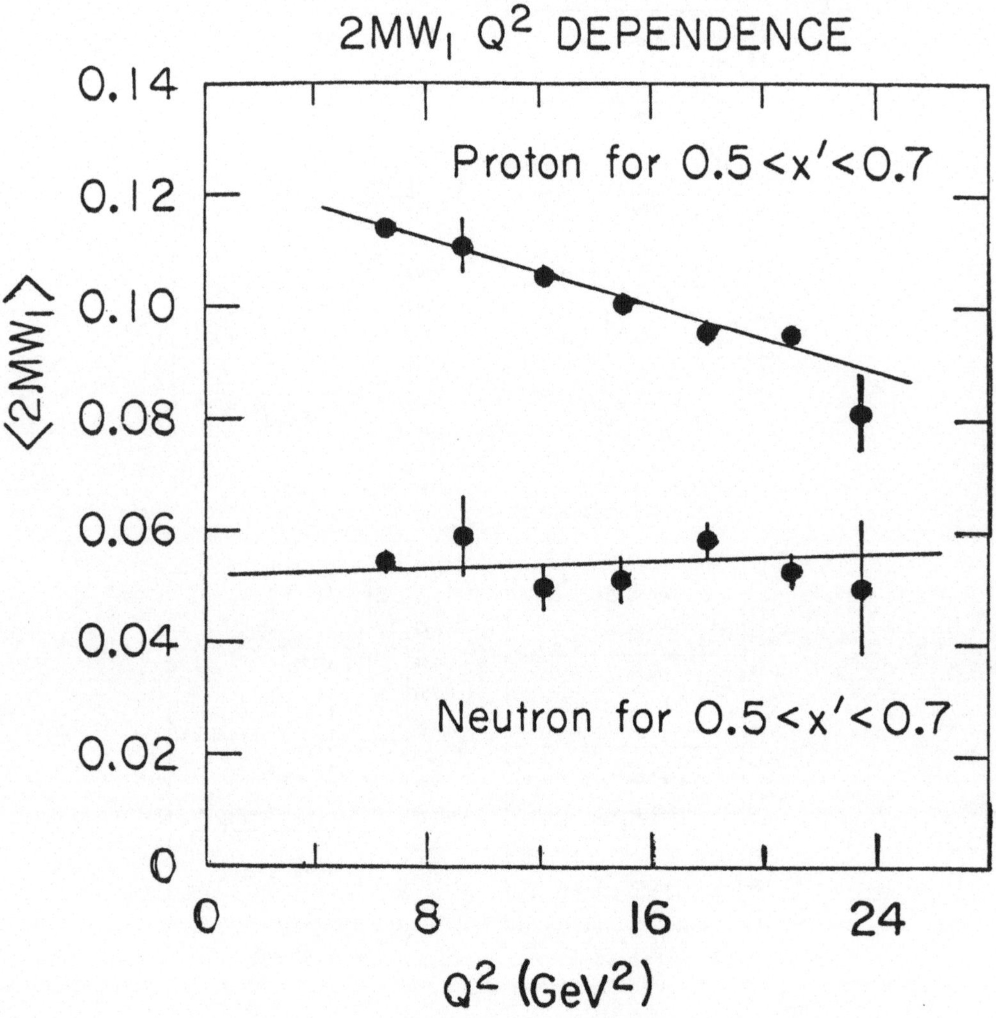
2MW₁ Q² DEPENDENCE
0.14
0.12
0.10
0.08
0.06
0.04
0.02
0
⟨2MW₁⟩
Proton for 0.5<x'<0.7
Neutron for 0.5<x'<0.7
0
8
16
24
Q² (GeV²)

FIG.10

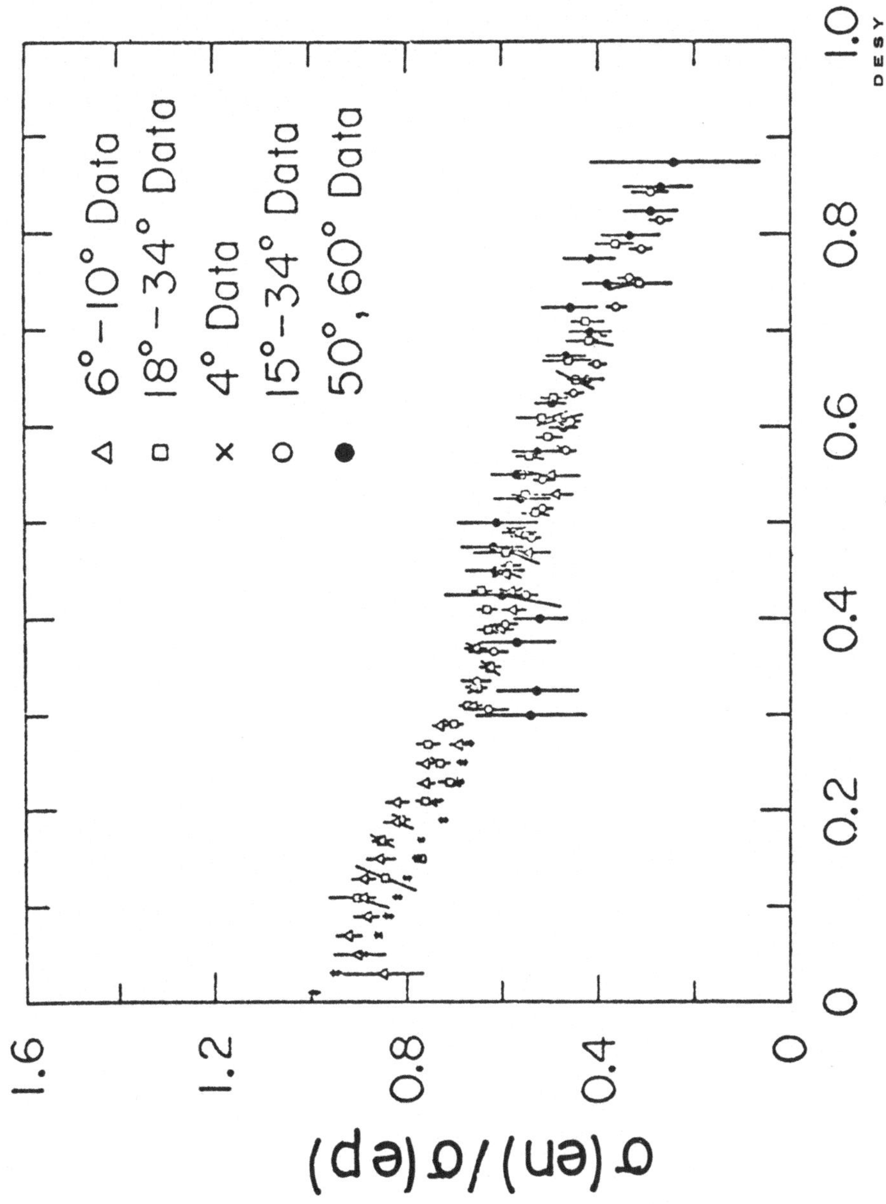

DESY 24197

FIG. 11

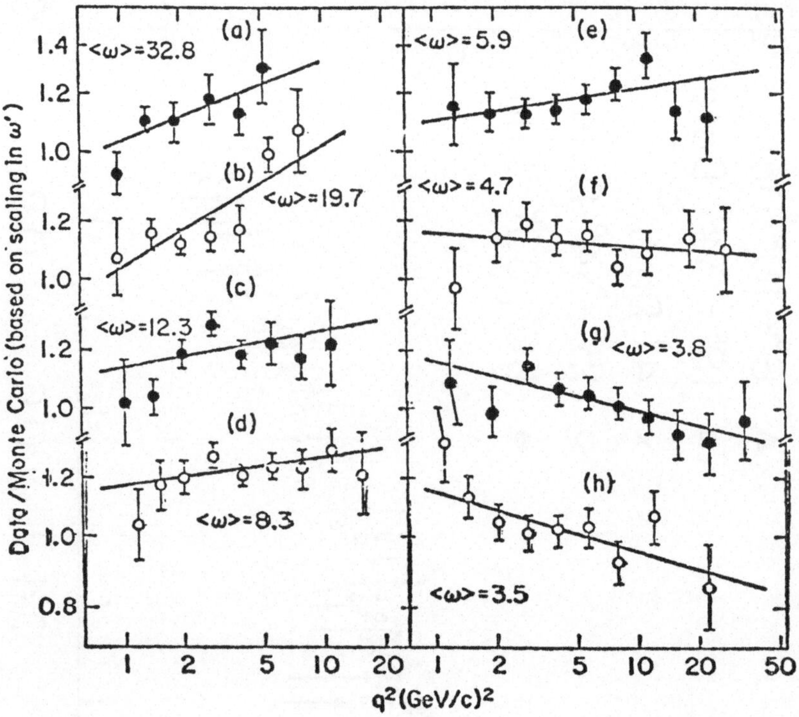

Data / Monte Carlo (based on scaling in ω')
q²(GeV/c)²
(a)
<ω>=32.8
(b)
<ω>=19.7
(c)
<ω>=12.3
(d)
<ω>=8.3
(e)
<ω>=5.9
(f)
<ω>=4.7
(g)
<ω>=3.8
(h)
<ω>=3.5
1.4
1.2
1.0
0.8
1
2
5
10
20
50

FIG.12

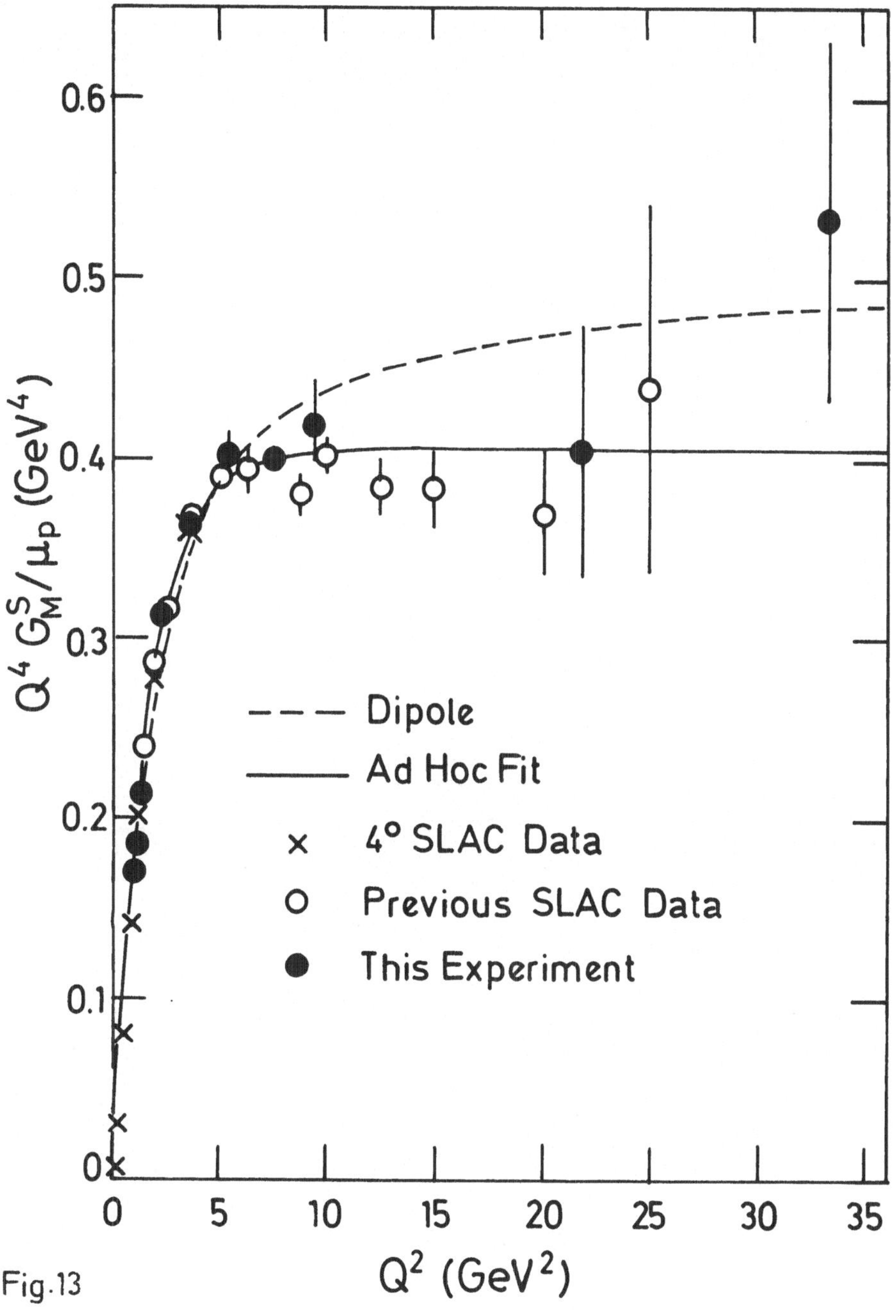

0.6
0.5
0.4
0.3
0.2
0.1
0
$Q^4 G_M^S/\mu_p$ (GeV^4)
Dipole
Ad Hoc Fit
4° SLAC Data
Previous SLAC Data
This Experiment
0
5
10
15
20
25
30
35
Q^2 (GeV^2)

Fig.13

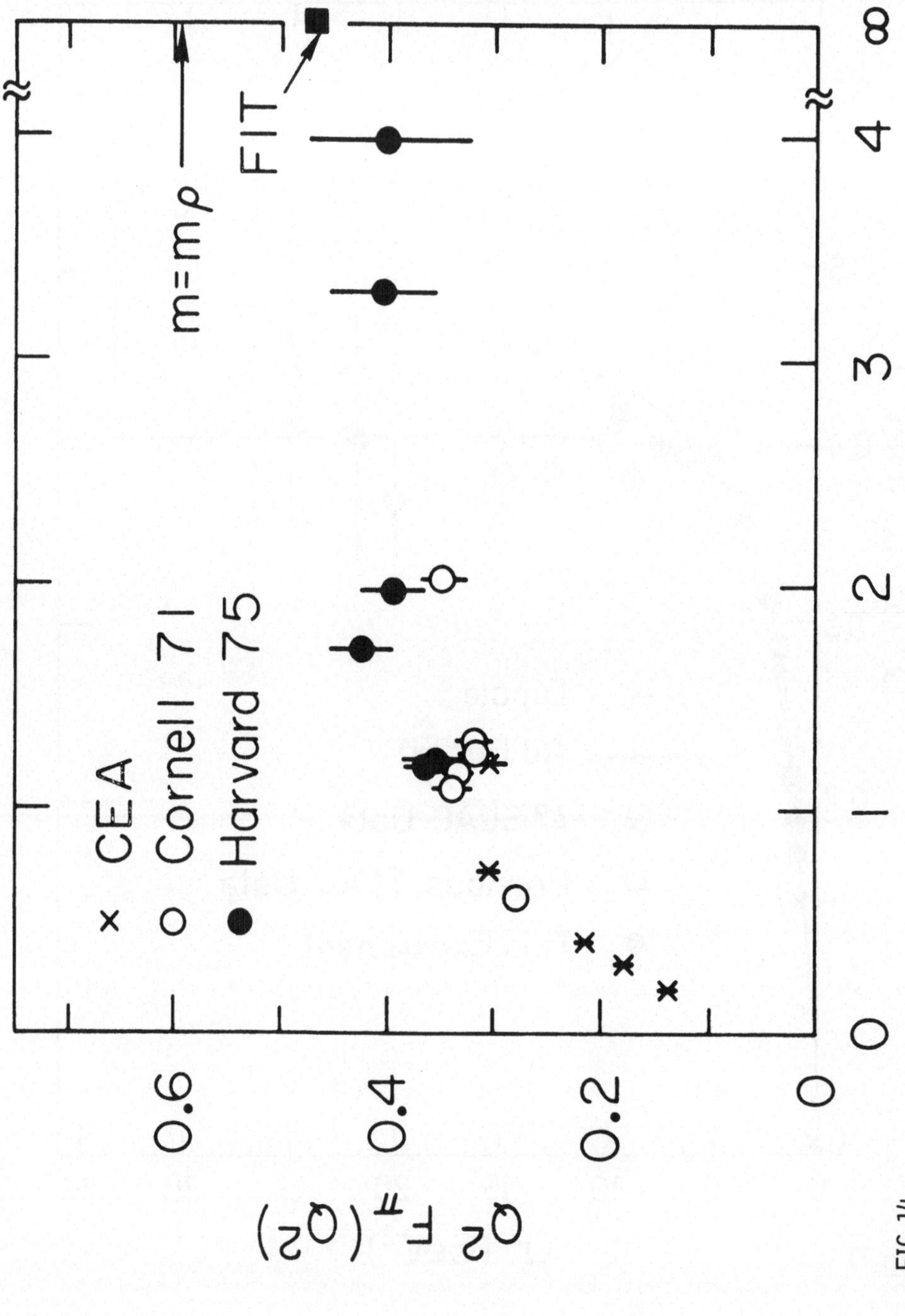
× CEA
○ Cornell 71
● Harvard 75
m = m$_\rho$
FIT
$Q^2 F_\pi(Q^2)$
0
0.2
0.4
0.6
0
1
2
3
4
∞
Q^2 (GeV^2)
2788A1

FIG.14

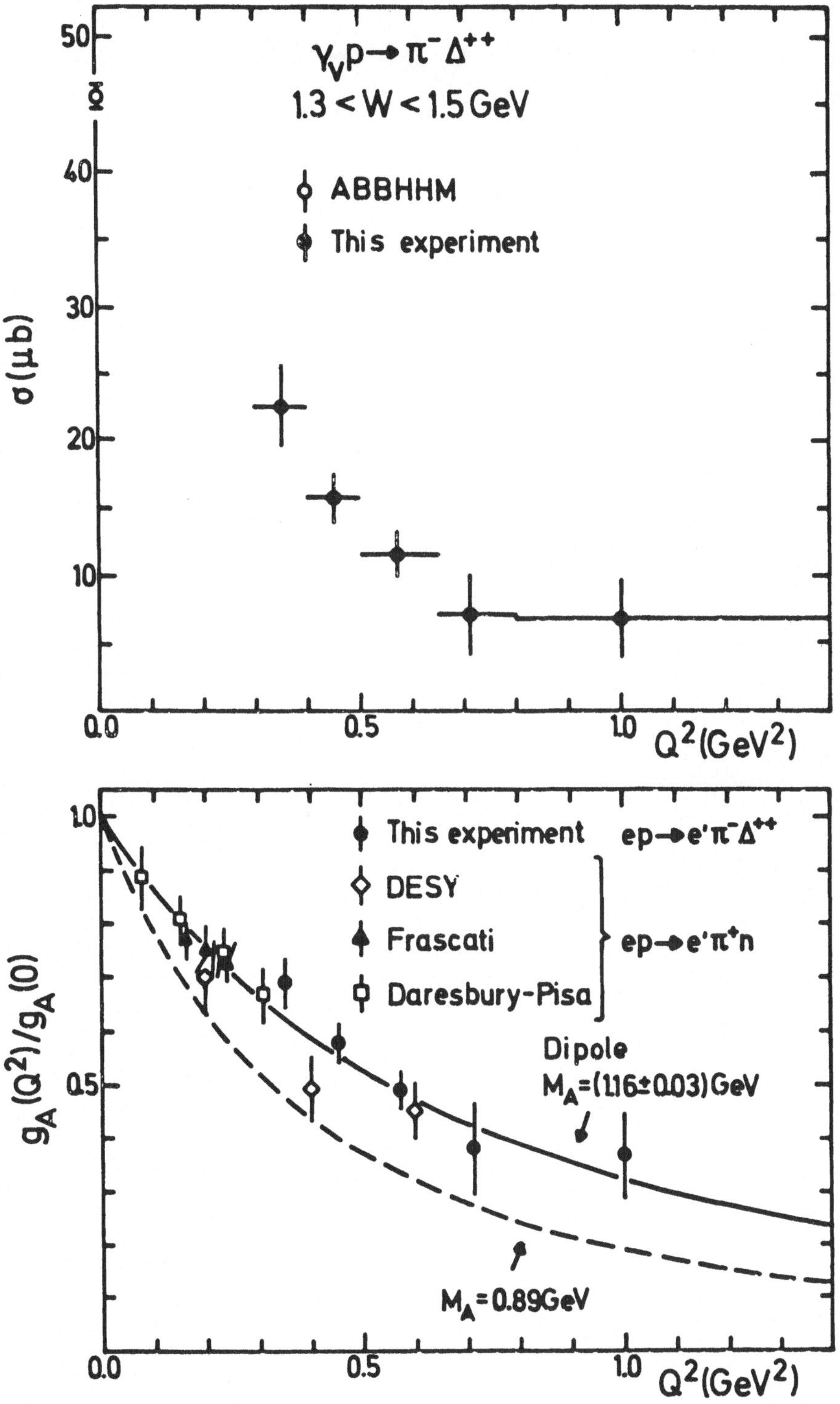

$\gamma_V p \to \pi^- \Delta^{++}$
1.3 < W < 1.5 GeV
ABBHHM
This experiment
σ(μb)
Q²(GeV²)
This experiment ep→e'π⁻Δ⁺⁺
DESY
Frascati
Daresbury-Pisa
ep→e'π⁺n
Dipole
M_A=(1.16±0.03) GeV
M_A=0.89GeV
g_A(Q²)/g_A(0)
Q²(GeV²)

FIG.15

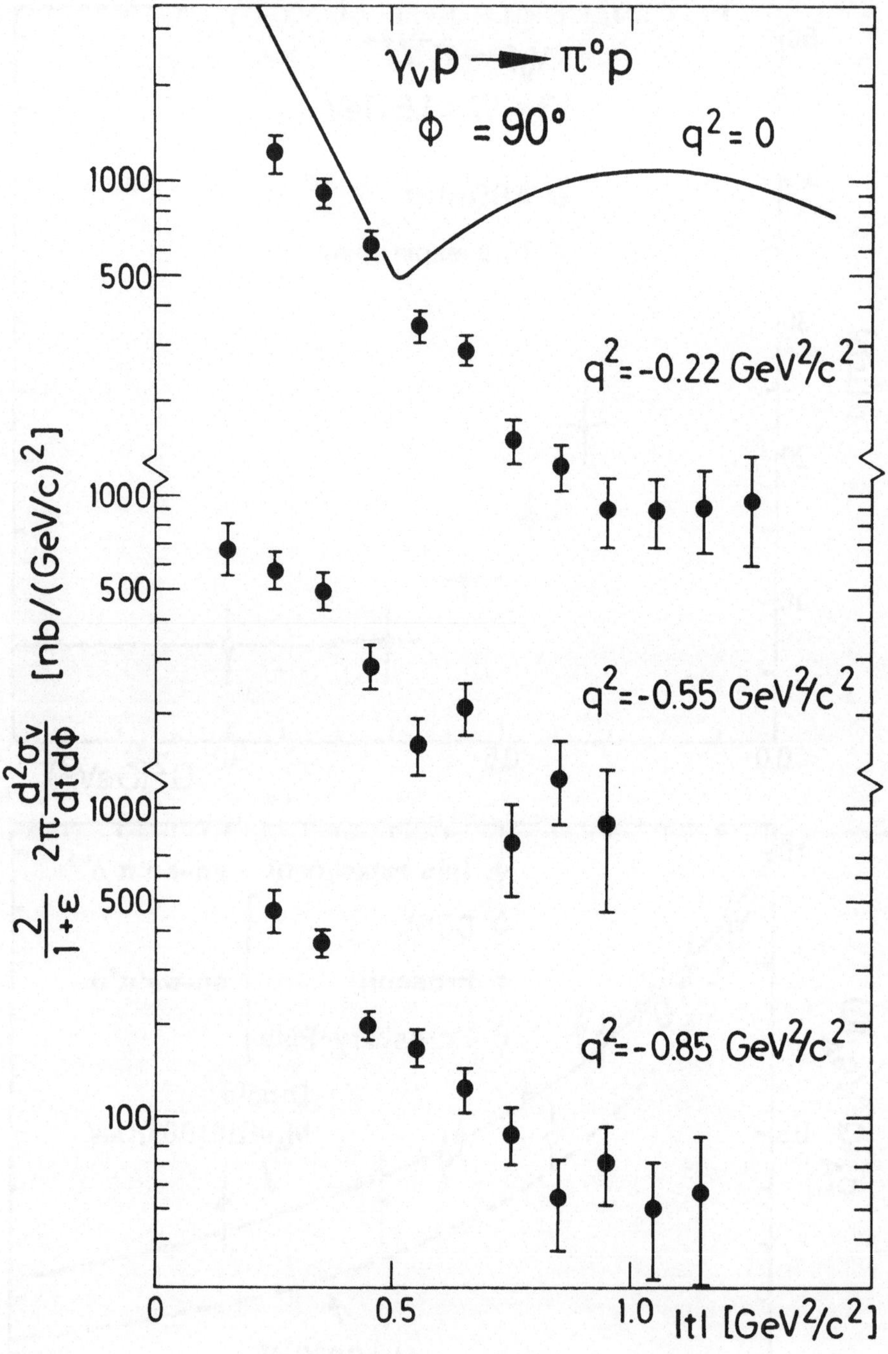

$\gamma_V p \rightarrow \pi^o p$
$\Phi = 90°$
$q^2 = 0$
$q^2 = -0.22\ GeV^2/c^2$
$q^2 = -0.55\ GeV^2/c^2$
$q^2 = -0.85\ GeV^2/c^2$
1000
500
1000
500
1000
500
100
$\frac{2}{1+\varepsilon}\ 2\pi \frac{d^2\sigma_V}{dt\,d\Phi}$ [nb/(GeV/c)^2]
0
0.5
1.0
$|t|$ [GeV^2/c^2]

FIG.16

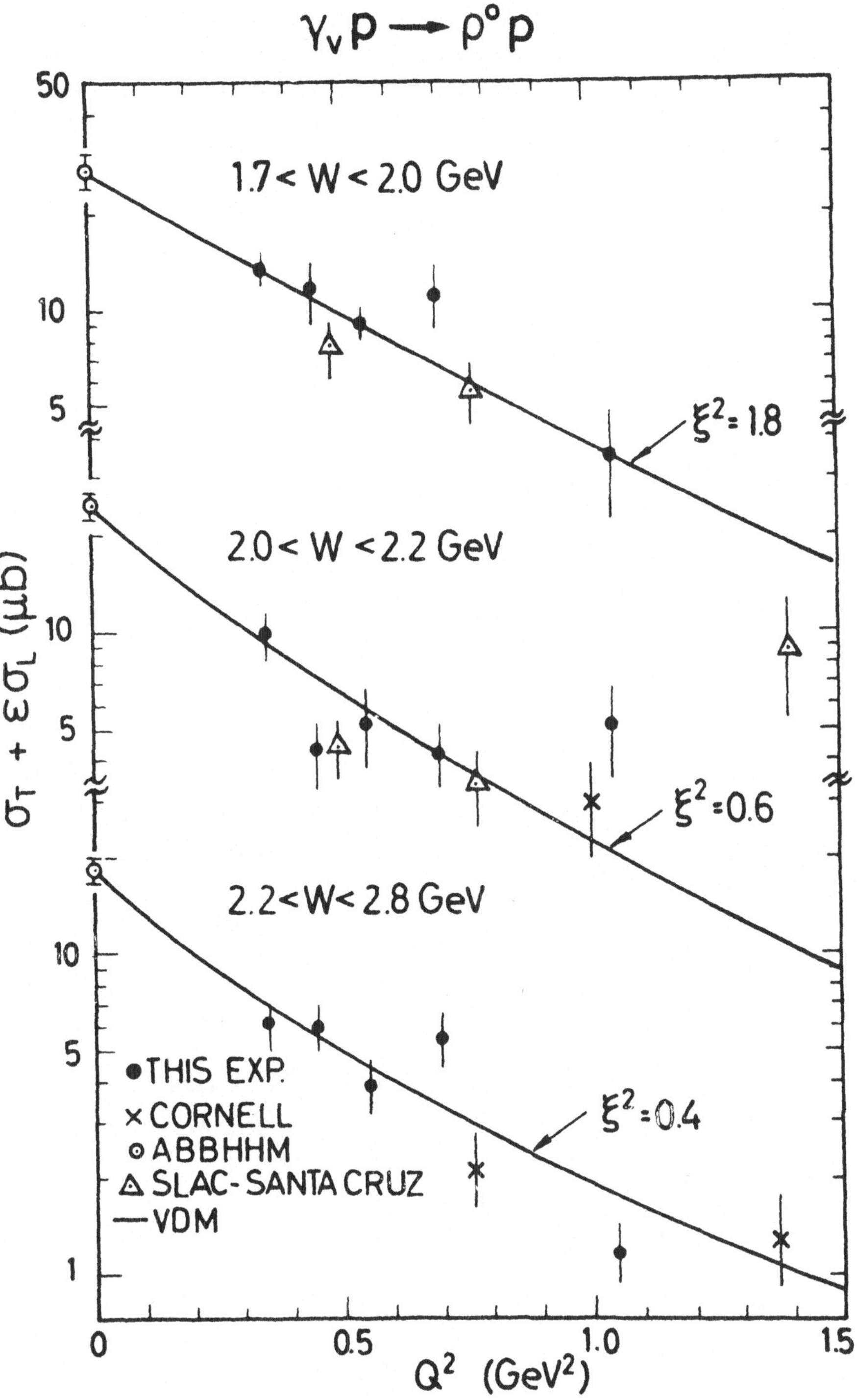

$\gamma_v p \longrightarrow \rho^\circ p$
1.7< W < 2.0 GeV
2.0< W <2.2 GeV
2.2<W< 2.8 GeV
ξ^2= 1.8
ξ^2=0.6
ξ^2=0.4
●THIS EXP.
× CORNELL
⊙ ABBHHM
△ SLAC-SANTA CRUZ
— VDM
$\sigma_T + \varepsilon\sigma_L$ (μb)
Q^2 (GeV^2)
50
10
5
1
0
0.5
1.0
1.5

FIG.17

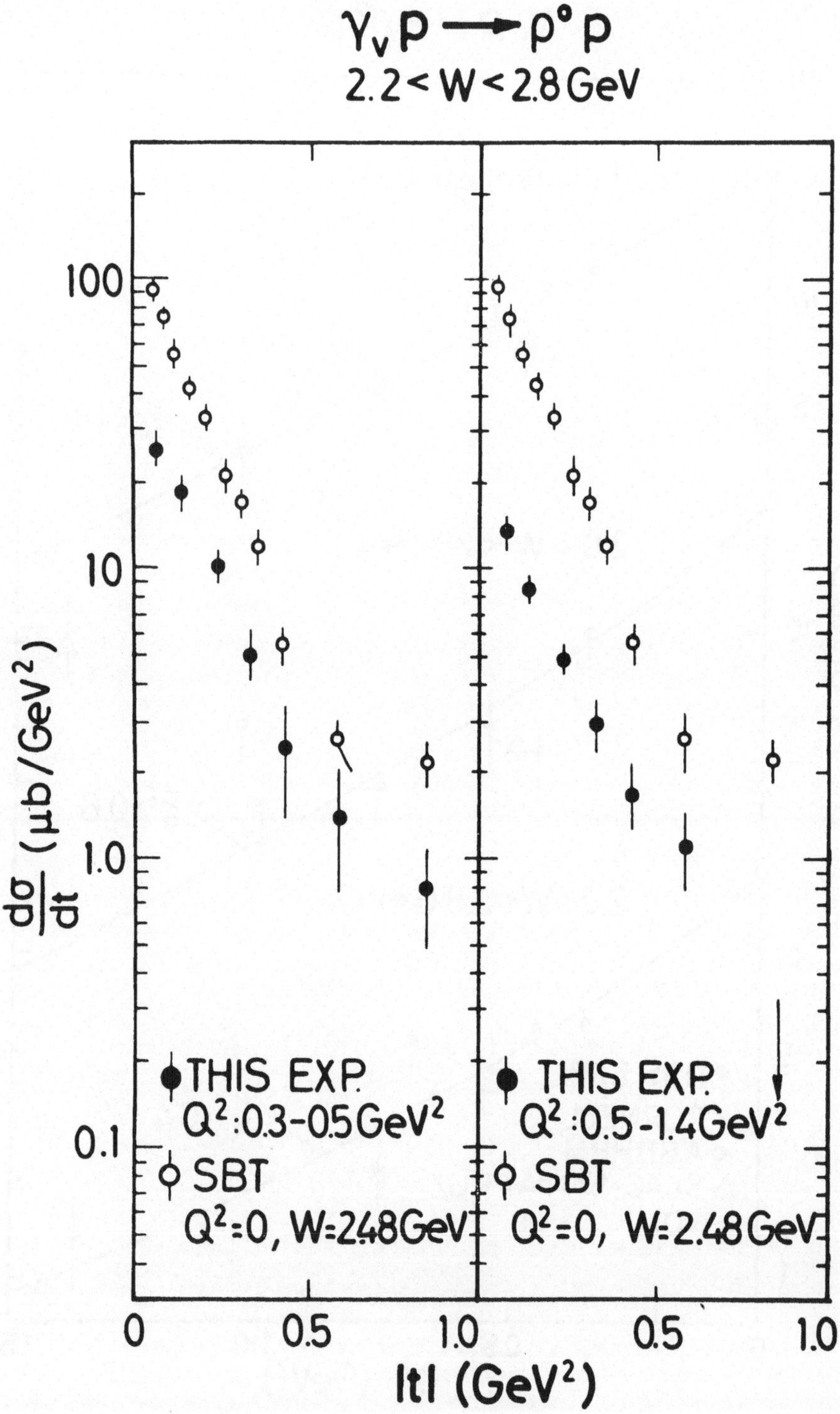

$\gamma_v p \longrightarrow \rho^o p$
2.2 < W < 2.8 GeV
$\frac{d\sigma}{dt}$ (μb/GeV²)
100
10
1.0
0.1
THIS EXP.
Q²:0.3-0.5GeV²
SBT
Q²=0, W=2.48GeV
THIS EXP.
Q²:0.5-1.4GeV²
SBT
Q²=0, W=2.48GeV
0
0.5
1.0
0.5
1.0
|t| (GeV²)

FIG.18

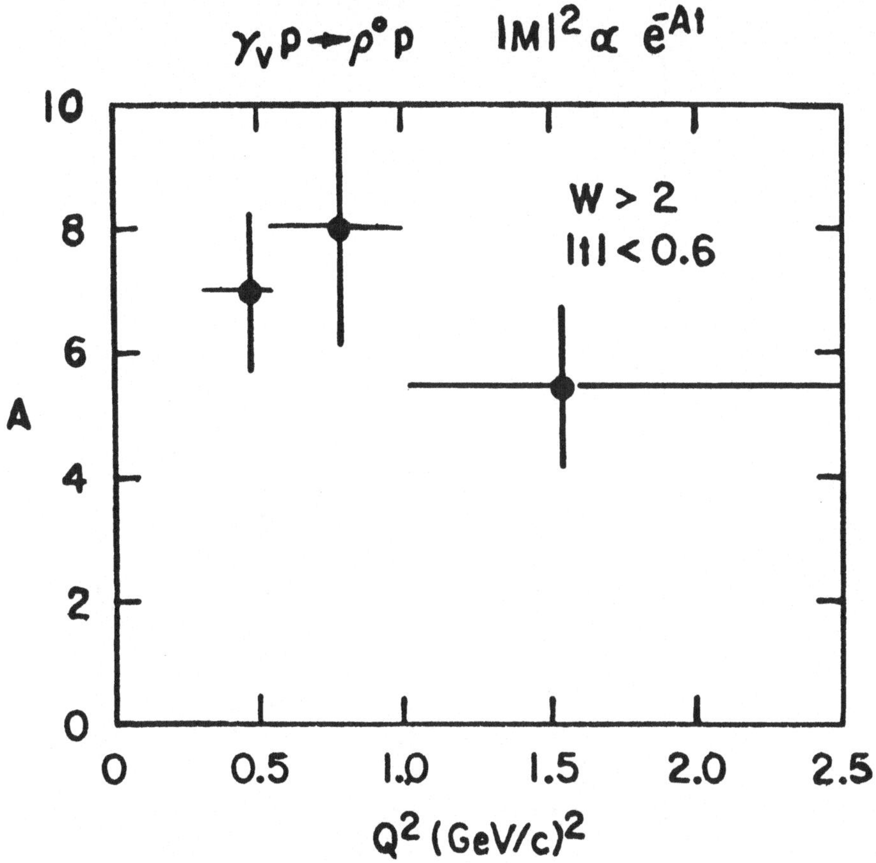

$\gamma_V p \to \rho^0 p$ $|M|^2 \propto e^{-At}$
W > 2
|t| < 0.6
A
10
8
6
4
2
0
0
0.5
1.0
1.5
2.0
2.5
Q^2 (GeV/c)^2

FIG.19

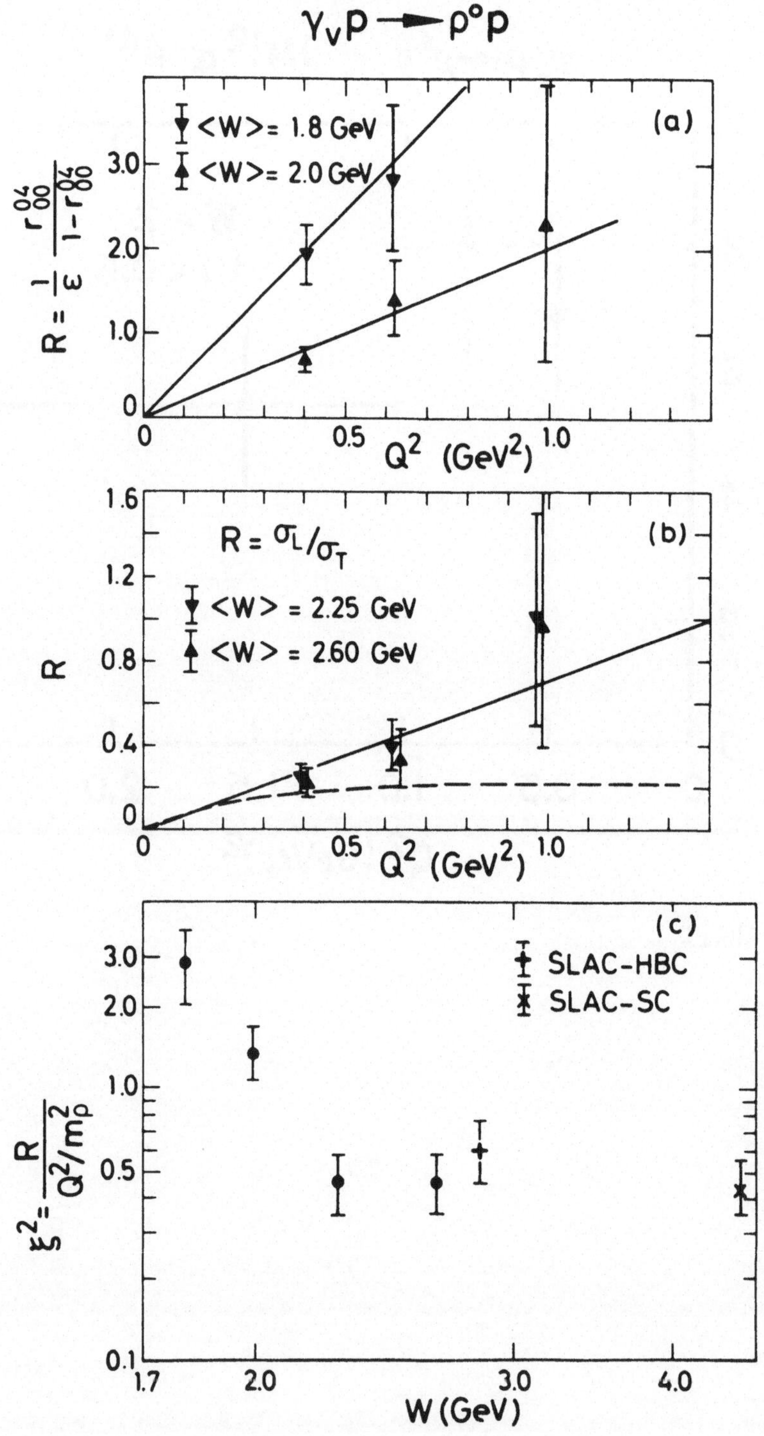
γ_V p → ρ°p
<W> = 1.8 GeV
<W> = 2.0 GeV
(a)
R = 1/ε r₀₀⁰⁴/(1−r₀₀⁰⁴)
Q² (GeV²)
R = σ_L/σ_T
<W> = 2.25 GeV
<W> = 2.60 GeV
(b)
R
Q² (GeV²)
(c)
SLAC-HBC
SLAC-SC
ξ² = R/(Q²/m_ρ²)
W (GeV)

FIG.20

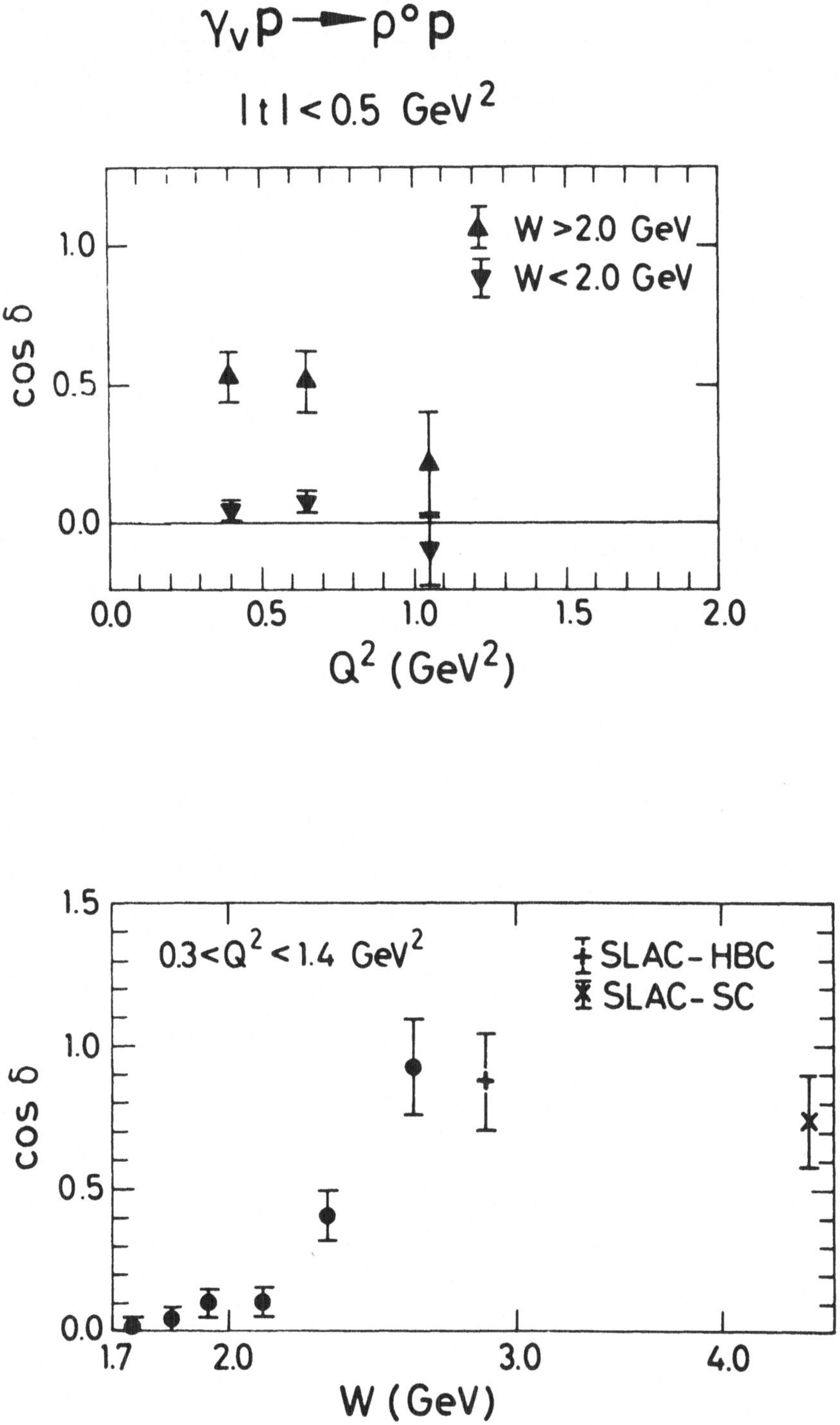

$\gamma_V p \to \rho^\circ p$
$|t| < 0.5$ GeV^2
W >2.0 GeV
W<2.0 GeV
cos δ
1.0
0.5
0.0
0.0
0.5
1.0
1.5
2.0
Q^2 (GeV^2)
$0.3<Q^2<1.4$ GeV^2
SLAC-HBC
SLAC-SC
1.5
1.0
0.5
0.0
1.7
2.0
3.0
4.0
W (GeV)

FIG.21

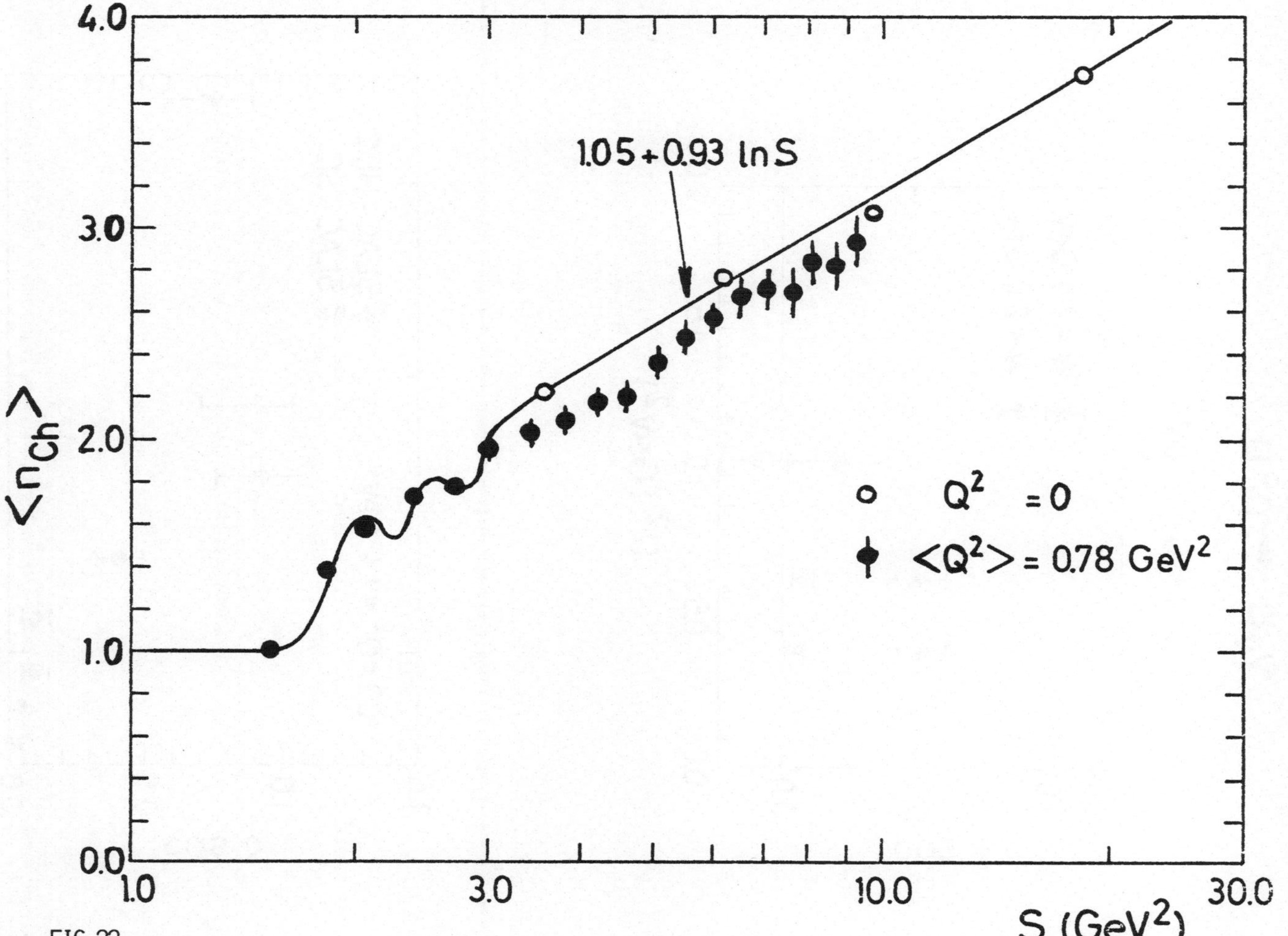

4.0
3.0
2.0
1.0
0.0
1.0
3.0
10.0
30.0
$\langle n_{ch} \rangle$
S (GeV^2)
1.05+0.93 lnS
Q^2 = 0
$\langle Q^2 \rangle$ = 0.78 GeV^2

FIG.22

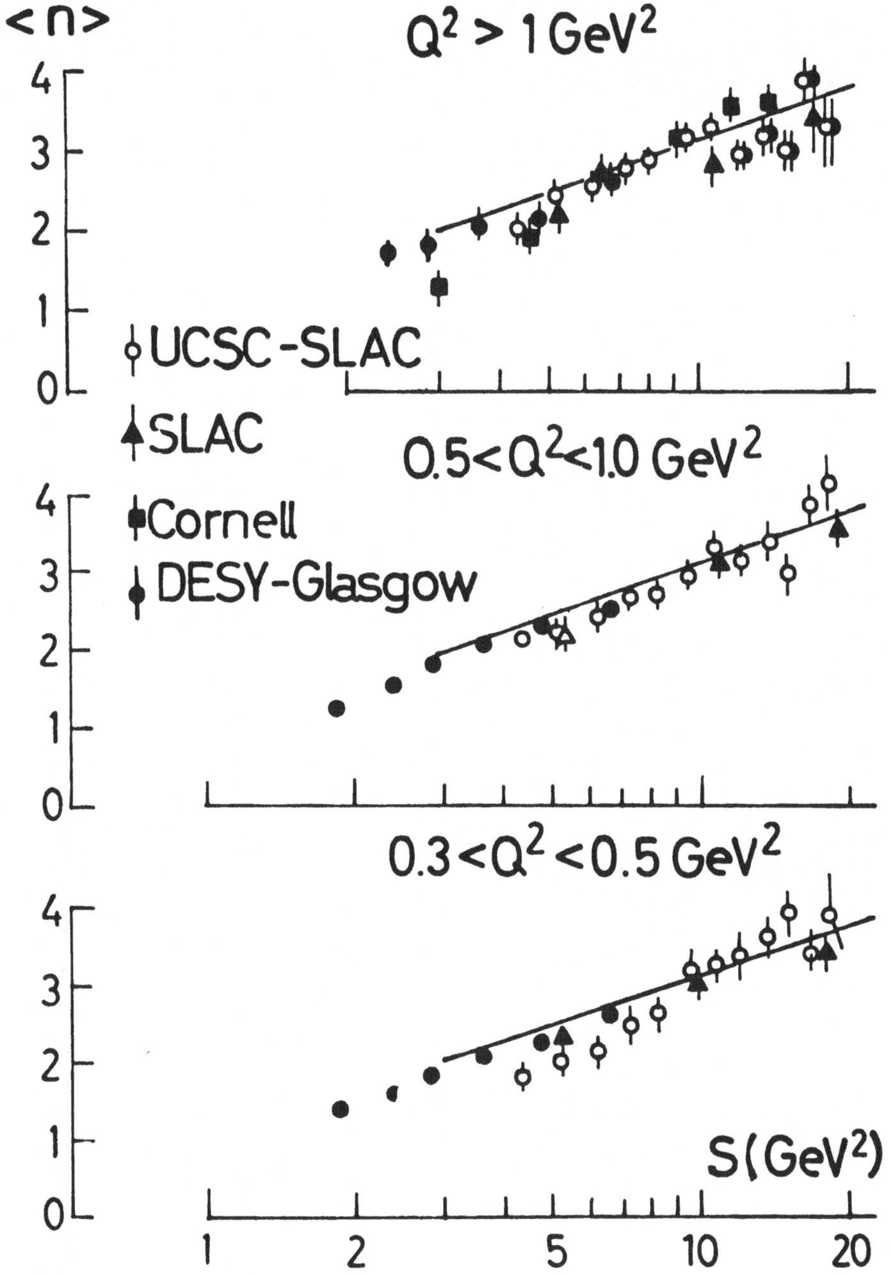

<n>
Q² > 1 GeV²
UCSC-SLAC
SLAC
Cornell
DESY-Glasgow
0.5<Q²<1.0 GeV²
0.3<Q²<0.5 GeV²
S(GeV²)
0
1
2
3
4
1
2
5
10
20

FIG.23

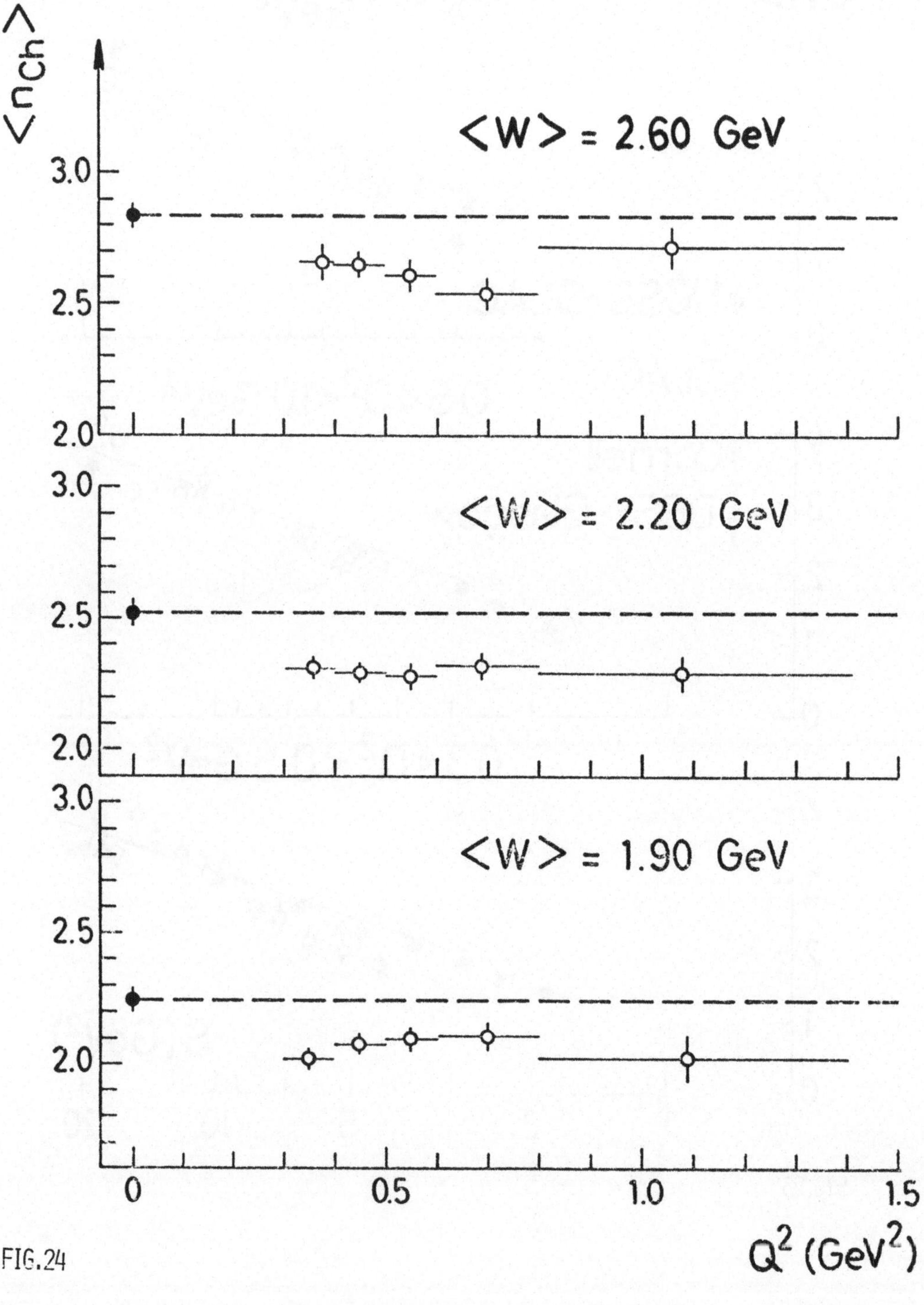

FIG.24

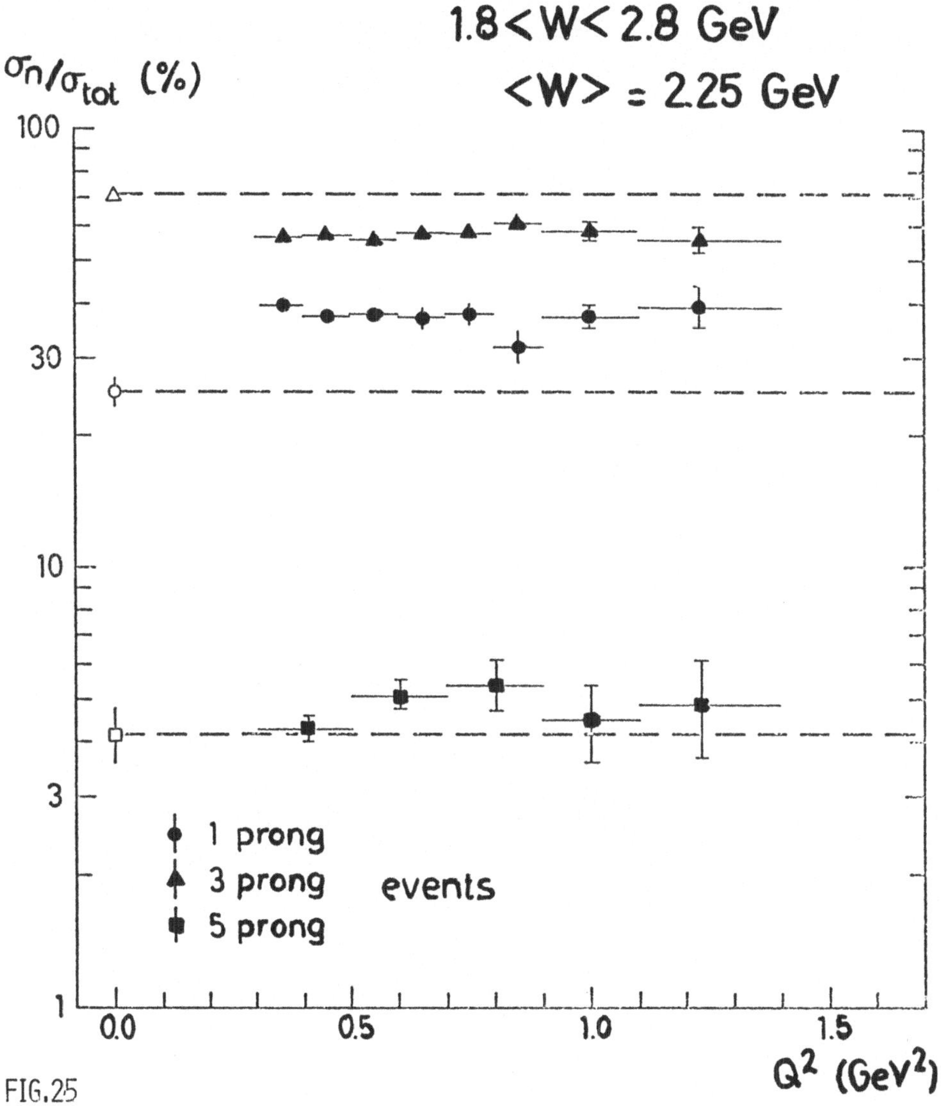
1.8<W<2.8 GeV
<W> = 2.25 GeV
σn/σtot (%)
100
30
10
3
1
1 prong
3 prong
5 prong
events
0.0
0.5
1.0
1.5
Q2 (GeV2)

FIG.25

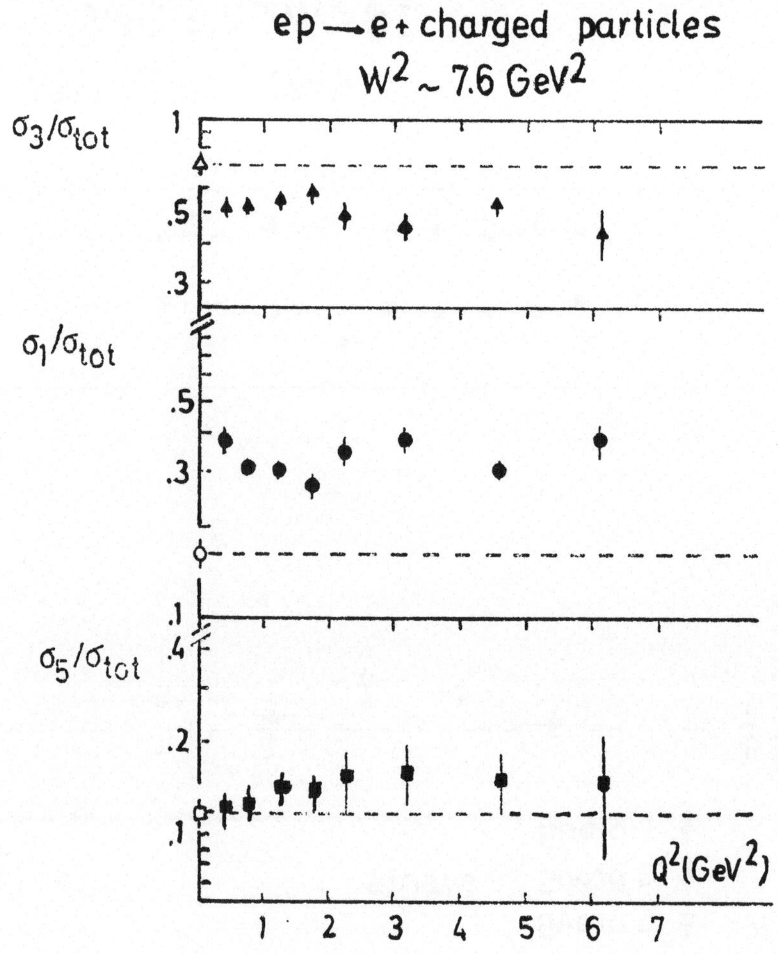
ep → e + charged particles
$W^2 \sim 7.6$ GeV^2
σ_3/σ_{tot}
σ_1/σ_{tot}
σ_5/σ_{tot}
1
.5
.3
.5
.3
.1
.4
.2
.1
Q^2(GeV^2)
1
2
3
4
5
6
7

FIG.26

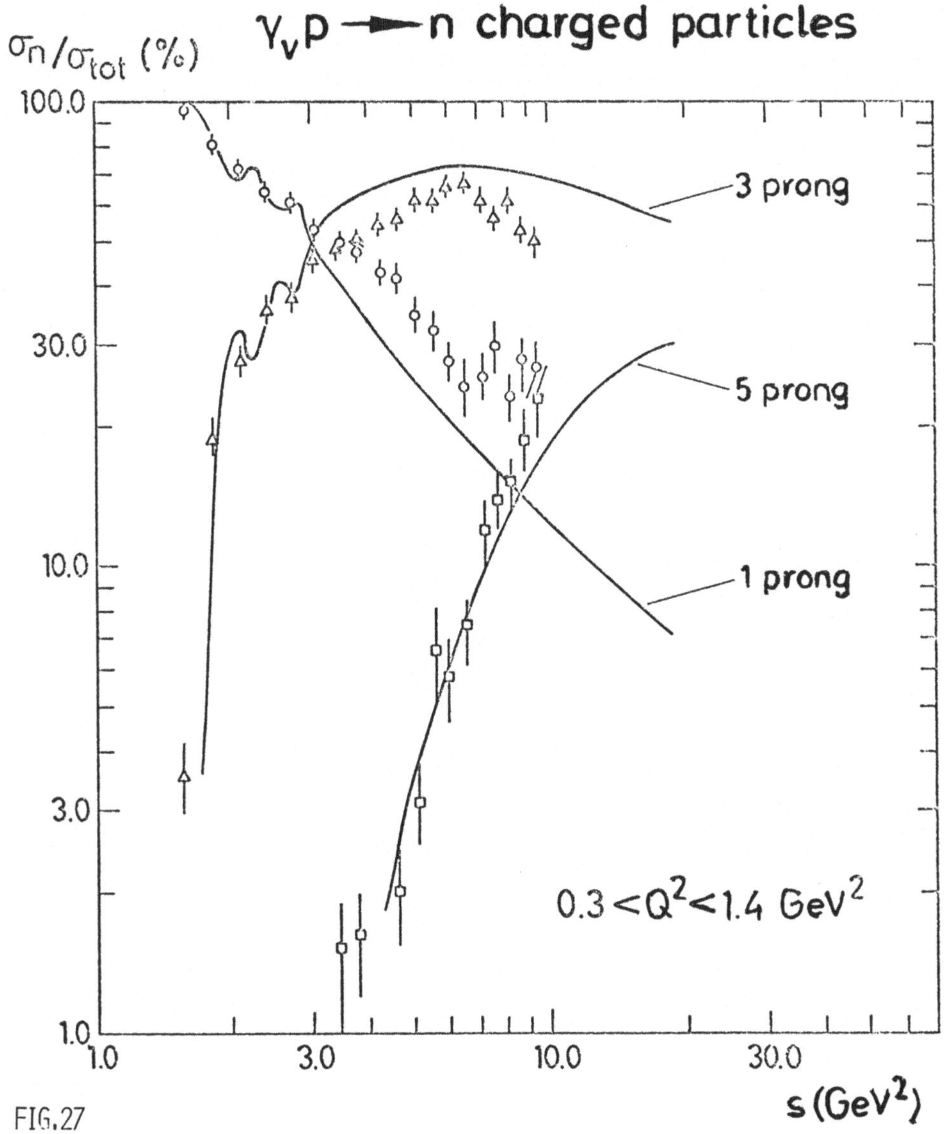

$\gamma_v p \rightarrow n$ charged particles
σ_n/σ_{tot} (%)
100.0
30.0
10.0
3.0
1.0
1.0
3.0
10.0
30.0
s (GeV^2)
3 prong
5 prong
1 prong
$0.3 < Q^2 < 1.4$ GeV^2

FIG.27

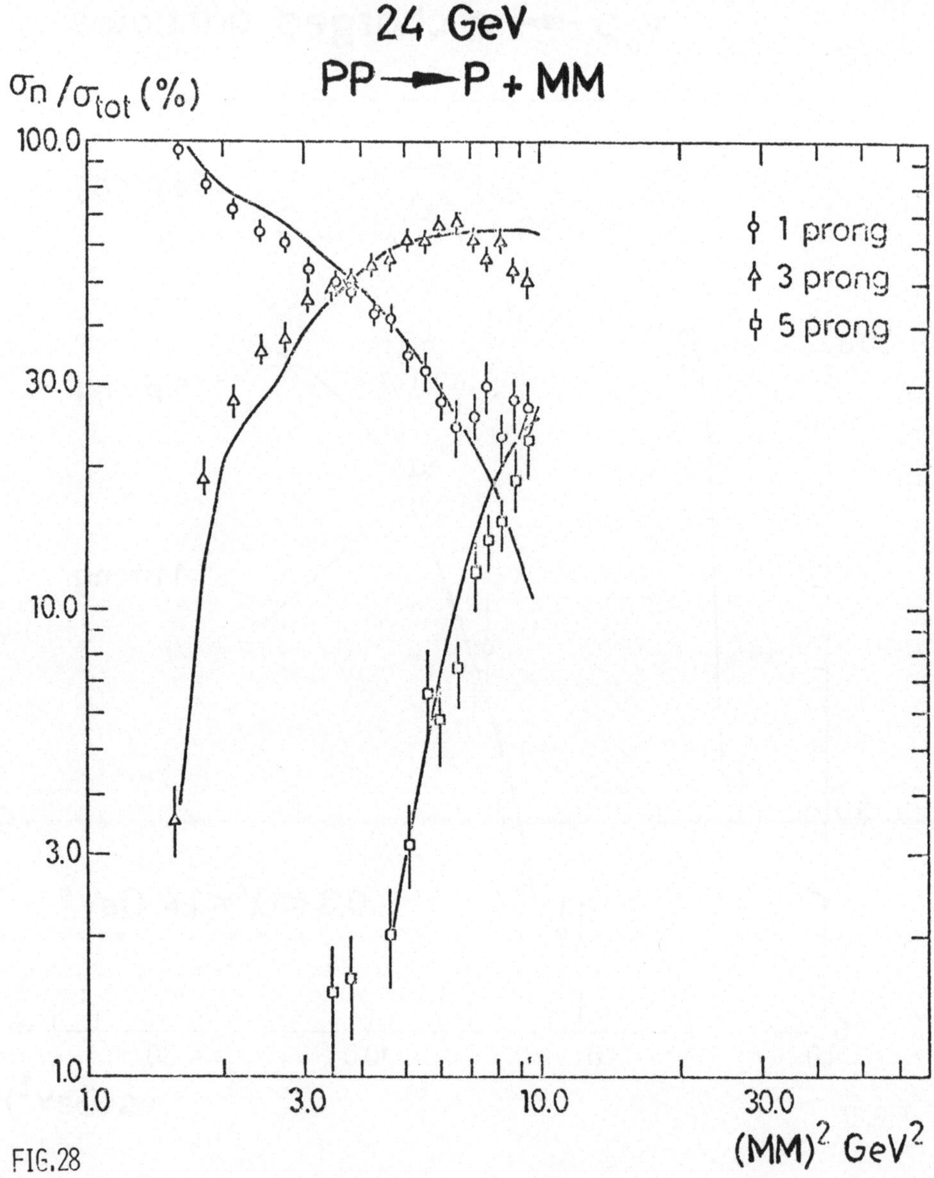
24 GeV
PP → P + MM
σn /σtot (%)
100.0
30.0
10.0
3.0
1.0
1.0
3.0
10.0
30.0
(MM)² GeV²
1 prong
3 prong
5 prong

FIG.28

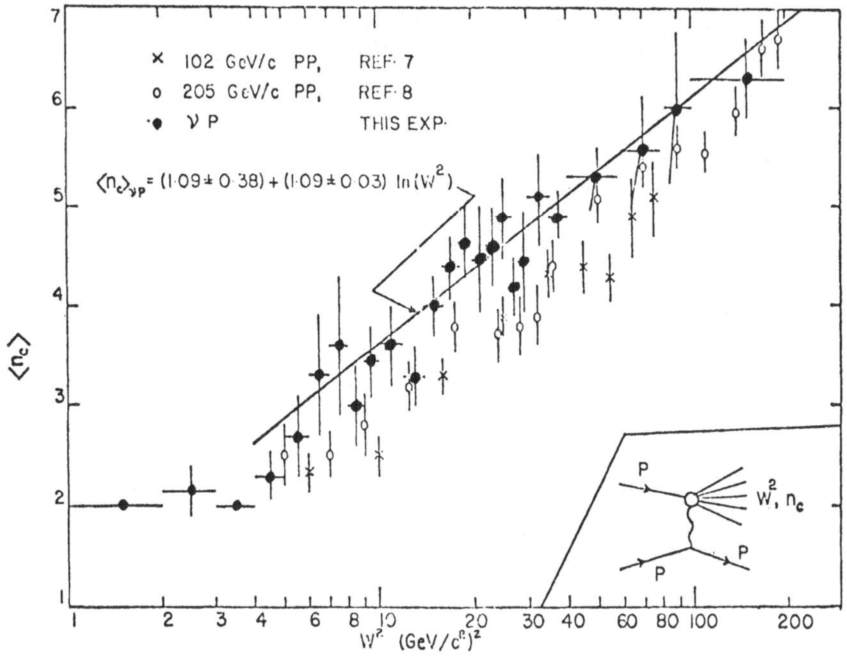
× 102 GeV/c PP, REF 7
o 205 GeV/c PP, REF 8
• νP THIS EXP
⟨n_c⟩_νP = (1·09 ± 0·38) + (1·09 ± 0·03) ln(W²)
⟨n_c⟩
W² (GeV/c²)²
P
W², n_c
P
P

FIG.29

DESCRIPTIONS OF HADRONIC STRUCTURE

by

Y. Nambu

The University of Chicago
The Enrico Fermi Institute
5640 Ellis Avenue
Chicago, Illinois 60637, USA

I would like to devote this lecture to discussing two topics: the evolution of the quark model and the problem of quark confinement. By way of introduction, however, I must stress that after ten years of relative stability, high energy physics is again in a state of flux. At this point, therefore, I had better deal only with generalities. A few significant points characterize the general pattern of the present situation. First there is an evolution of the quark model leading to a proliferation of fundamental constituents. This is of course related to the discovery of new particles. But quantum numbers alone do not seem enough this time, unlike the days of strange particles. New dynamics may also be necessary. An example is the Zweig-Iizuka rule, but asymptotic freedom, quark confinement, etc., must also be included as part of the general problem. One can even speculate that a new post-quantum mechanics is hidden somewhere, although this may be going too far.

The two quantum numbers, color and charm, have been theoretically anticipated for some time. The motivation for color comes from the problem of statistics in the baryon wave function, and that for charm comes, logically speaking, from the weak interactions. There is now more supporting evidence for either concept, but so far the evidence is not direct. This might change any minute. But if color should turn out to be a completely hidden quantum number as some theorists desire, the reality of colored quarks will remain a highly theoretical, almost fictitious one. Even if color becomes visible, the quarks could still be confined, retaining their unreal nature. But of course we should not object to such possibilities from our philosophical prejudices. It is an experimental question to be settled with the aid of a theory.

Clearly there is a mismatch between the strong and weak interactions as we know them now. The former exhibits a pattern of SU(3), the latter a pattern of SU(2). It makes our life easy to assign them separate spaces, color and flavor. Each space, especially the flavor, may require further escalation. At the moment a strong and immediate theoretical motivation for it, like those for color and charm, is lacking. But further experimental evidence, such as heavy leptons and more ψ's, may force us into that direction. Eventually we will have to worry about a synthesis of all interactions, and how many fundamental particles there are. In this context Gürsey's ideas of using exceptional groups and nonassociative algebras are interesting, albeit highly speculative. They even offer room for post-quantum mechanics. There are also other schemes, such as Pati and Salam's, which admit the possibility of integrally charged, non-confined quarks.

How can one distinguish between the various schemes? Though the ψ particles and neutrino-induced reactions will obviously be crucial in this respect, at least at the moment no single theoretical scheme can be definitely ruled out or ruled in. This is because, for one thing, our spectroscopic data are still very meager, and besides, we don't quite understand the dynamics such as decay widths and branching ratios. Even if the narrow ψ's are to be identified with charm, color excitations could still show up at higher energies ($\gtrsim$ 4 GeV). Since it will not be profitable to keep on speculating without making specific predictions, I will end this part of the discussion by proposing one extreme color model which M.-Y. Han and I have considered. Assume the quarks to have integral charges and baryon numbers. More specifically, the assignments are:

	Red	Green	Blue
Q(charge)	(-1,0)	(1,0)	(1,0)
Baryon no.	0	0	0
Lepton no.	± 1	∓ 1	0

There must be an appreciable mixing between q and $\bar{q}$ having the same Q, B and L, which makes the quarks unstable. The quarks with $L = \pm 1$ are hybrid lepton-quarks which will undergo semileptonic and leptonic decays. The diagonal isospin I' + I" (or perhaps I' and I" separately) is assumed to be a good symmetry. The quark masses will be in the range ~ 1.9 - 2.0 GeV.

There will be two-body color excitations of type qq (color triplet), $q\bar{q}$ (octet) and qq (sextet) arranged in this order according to the Casimir operator of color SU(3). The triplet qq is a bound state, but the octet and sextet are unbound resonances.

The ψ are ψ' identified with the color triplet-antitriplet configuration $qq\text{-}\overline{qq}$ ($^3P_1$, I=0, C=-1) made of B=0, L = ± 1 quark pairs. By assumption, the $q\text{-}\bar{q}$ mixing is momentum dependent, being large for large momenta, which could be induced, for example, by a short range interaction between quarks. This allows the wave function to have an appreciable $q\bar{q}$ (octet) admixture at the origin, which enables it to couple to the photon. But the outer part of the wave function is purer, and the decays to ordinary hadrons (with or without a γ) are strongly suppressed. The broad resonances above 4 GeV could be interpreted as genuine color octet. Furthermore, the μ-e events reported from SPEAR could be due to pair produced free lepton-quarks.

I will now come to the topic of quark confinement. Its primary motivation derives from the absence of fractionally charged particles, but of course there are also various supporting indications, both experimental and theoretical, which make it an attractive and challenging problem.

As theoretical possibilities for confinement, one may list three categories.

a. Formal arguments like the existence of Dirac monopoles with $g^2/4\pi$ = = 137/4 or 137, and the consequence of non-associative algebras (post-quantum mechanics).

b. Quantum mechanical confinement due to infrared slavery.

c. Confinement picture which works already at the classical level, such as strings and bags.

Common to b) and c) is the use of color gluon gauge fields. Skipping the case a), let me comment on b) and c). The case b) is characterized by unbroken color gauge symmetry which has the elegant feature of combining ultraviolet (asymptotic) freedom and infrared slavery. Unfortunately, the latter aspect is not well understood yet. The behaviour of confining forces and the description of hadronic wave functions are not explicit. In this context, the often assumed linear potential ($\propto$ r), which has a dimensional constant, and the infrared catastrophe (e.g., Cornwall and Tiktopoulos) are two different mechanisms of confinement. It appears to me that since hadrons have a finite size, the main part of confining mechanism must depend on a dimensional parameter in an essential way, even if there are additional effects which defy classical analogy.

This brings me to case c), which offers a priori various degrees of confinement:

1. Color invisible
2. Color visible in zero triality sector (broken or unbroken color symmetry).
3. Color visible in all sectors, which means free quarks and no confinement.

In an Abelian model system, the analogy with superconductivity, with the addition of monopoles, offers an attractive mechanism because it is based on spontaneous symmetry breakdown following the long chain of theories, Ginzburg-Landau-BCS-Higgs-Abrikosov-Nielsen-Olesen, and leads to the concept of strings with finite thickness, which combines both strings and bags in a natural way. The former is appropriate to excited hadrons, the latter to ground states. The geometrical configuration of baryons in the string model has been an unsolved question, but recently K. Johnson and T. Eguchi (both private communications) argue that excited baryons will take a resonating pattern of q-qq string configurations.

The bold approach of Wilson and Susskind seems rather different except for their use of stringlike paths. But it may actually be a combination of b) and c). Unless they can carry out their program successfully, there remains the basic problem of formulating a consistent non-Abelian analog of the monopole theory. Even the Abelian theory of Dirac has difficulties with quantum mechanical treatment. The MIT bag theory seems to be essentially equivalent to the picture in which quarks move in a cavity made of an electrically superconducting material, but a precise description of the role of the superconductor is lacking. On the other hand, a semiclassical non-Abelian Dirac string leads to severe and unreasonable constraints on the solution (Eguchi).

I would like now to propose the following formulation based on a conventional and at least superficially renormalizable field theory. Let us define an antisymmetric tensor $G^a_{\mu\nu}$ (a ist the color index which will be suppressed below). It is the analog of Dirac's string field. Then write the defining equation

$$m^2 G_{\mu\nu} = mF_{\mu\nu} + [D_\mu(A)B_\nu - D_\nu(A)B_\mu]^* \qquad (1)$$

m is a dimensional parameter which makes dim $G \sim 1/L$, like that of a potential. $F_{\mu\nu}$ is the usual gauge field derived from a vector potential A_μ. B_ν is an independent magnetic potential, and $D_\mu(A)$ is a covariant derivative with respect to A_μ. The symbol * means taking the dual tensor.

Next we relate B_μ to $G_{\mu\nu}$ by

$$B_\mu = - D_\nu G^*_{\nu\mu} \qquad . \qquad (2)$$

Since the right-hand side defines a covariant magnetic current ($\equiv 0$ for $F^{*}_{\mu\nu}$), this is an analog of the London ansatz, or the field-current identity. Substitution of (2) in Eq.(1) makes the latter into a second order wave equation for $G_{\mu\nu}$, with $F_{\mu\nu}$ as its source.

The third equation to be postulated is

$$D_{\mu}(a)[F_{\mu\nu} - m\, G_{\mu\nu}] = j_{\nu}(q) + j_{\nu}(G) \tag{3}$$

where $j_{\nu}(q)$ is the "electric" current due to the quarks, and $j_{\nu}(G)$ is a similar contribution from the G field.

Eqs.(1)-(3) follow from a Lagrangian (with $G_{\mu\nu}$ and A_{μ} as independent variables)

$$L = \frac{1}{2}\,(D_{\mu}G^{*}_{\mu\nu})^{2} - \frac{1}{4}\,(mG_{\mu\nu} - F_{\mu\nu})^{2} - j_{\mu} \cdot A_{\mu} \tag{4}$$

which resembles those considered by Kalb and Ramond, by Cremmer and Scherk, and by Kyriakopoulos with various motivations. Suppose $m = \text{const} \neq 0$. Inspection of Eqs.(1) and (2) shows that the combination mG-F satisfies a source-free equation, hence may be taken = 0, which also means $B_{\mu} = 0$. But then it leads to $j_{\nu}(G) = j_{\nu}(q) = 0$. As a consequence, quarks cannot exist in such a medium. This seems to be related to the Dirac "veto" in his monopole theory, namely the condition that an electric charge cannot touch a string. In our case the magnetic charge is carried by the field $G_{\mu\nu}$ or B_{μ} . Since a field in the classical sense represents continuous charge distribution, the electric quarks have no place to live due to Dirac's veto.

On the other hand, $G_{\mu\nu}$ and $F_{\mu\nu}$ are decoupled if $m = 0$, and we go back to the usual Yang-Mills theory as far as $F_{\mu\nu}$ is concerned. Now the two cases can be bridged if we replace m by a neutral scalar Goldstone field ϕ which develops a nonzero expectation value under normal circumstances. The quarks, however, will create a living space around themselves by making $< \phi > = 0$, which will then be the MIT bag, or the Nielsen-Olesen string, as the case may be. The roles of electric and magnetic charges somehow get switched, and the quarks acquire physical electric strings attached to them.

ON THE PHENOMENOLOGY OF MULTIHADRON PRODUCTION IN e^+e^- ANNIHILATION

H. Satz
Department of Theoretical Physics
University of Bielefeld
Germany

ABSTRACT:

The phenomenological foundations of the jet and cluster approaches to e^+e^- annihilation are investigated, and the critical implications of these pictures are presented both asymptotically and at finite energies, where they are compared with recent data. Corresponding predictions of other approaches (hydrodynamical model, dual models) are considered, and the interrelation between characteristic multihadron features (Feynman scaling, ultimate temperature) and other dynamical concepts (Bjorken scaling, constituent nature of hadrons) is discussed.

I. INTRODUCTION

The basic question to be considered here is schematically illustrated in Fig.1: what can we say about the mechanism of multihadron production in e^+e^- annihilation? As indicated, we shall examine the hadronic decay of a "heavy" virtual photon, assuming that one-photon exchange either dominates the whole process, or that it may be separated out experimentally, e.g., by large angle measurements.

The sources of information we have available to predict the properties of e^+e^- annihilation are on one hand the experimental results from hadron-hadron interactions and from photo- and electroproduction reactions, on the other an instrumentarium of various theoretical concepts, such as partons, quarks, fireballs, scaling, and so on.

Two aspects in particular, however, make the usefulness of these sources an open question. The hadronic system in Fig.1 is of large mass, but of spin one, while all other hadronic systems so far studies were highly anisotropic (spin/mass ~

constant). - The e^+e^- interaction is pointlike, suggesting that the produced hadronic system is, at least in its formation stage, not of the canonical hadronic size, determined by the range of strong interactions ($R \sim m_\pi^{-1}$). Nevertheless, since we do not have a theory for e^+e^- annihilation, we can only attempt to extrapolate our pictures of hadron-hadron and lepton-hadron interactions: thus we shall either find a suitable description of annihilation, or learn about the limits of our concepts.

The experimental results on hadron-hadron interactions presently considered as most relevant for e^+e^- annihilation are: (i) the universal transverse momentum bound at small p_t, (ii) scale-invariant distributions ("Feynman scaling") in longitudinal momenta, and (iii) a slower (power-like?) fall-off at large p_T with a scale-invariant dependence on the incident energy (i.e., s-dependence only through $p_T/\sqrt{s}$). The phenomenological interpretation of these results has led to two opposite points of view.

In the cluster or fireball picture, the characteristic features of the strong interaction are assumed to be those seen in transverse distributions; longitudinal scaling is taken as a consequence of collision kinematics, of the specific initial conditions provided by a hadron-hadron collision. Models in this vein go back as far as Heisenberg's wave equation approach[1]; they include Landau's hydrodynamical model[2] and, as extension of the Fermi model[3], the thermodynamical picture of Hagedorn[4]. Their common feature is a strongly localized interaction region, Lorentz-contracted along the beam axis. For a simple prototype[5], consider

$$e^{-\vec{r}^2/R^2} \tag{1.1}$$

as general coordinate space distribution of hadronic matter, with R as the range of strong interactions. The Lorentz-contraction of a hadron-hadron collision modifies this to

$$e^{-[\vec{r}_T^{\,2} + \vec{r}_L^{\,2}/\gamma]/R^2} \tag{1.2}$$

with $\gamma = 2m/\sqrt{s}$, m being the mass of the incident hadron. Fourier transforming, we obtain with

$$e^{-[\vec{p}_T^{\,2} + 4m^2 x_L^2]R^2/4} \tag{1.3}$$

a momentum distribution giving both a p_T bound and longitudinal scaling.

The jet picture, on the other hand, takes the peripheral structure with scaling distributions along the jet axis as the essential aspect of strong interactions; the p_T bound, corresponding to the jet diameter, is assumed to simply reflect the hadronic size. Along this line we find Heisenberg's infrared analogy[6)], later Bremsstrahlung models[7)] and the scaling considerations of Wilson[8)] and Feynman[9)] as well as the more detailed dynamical approach of Amati, Fubini and Stanghellini[10)], together with its reggeized[11)] and dual[12)] extensions. The joint element in these various schemes is the two-jet structure of hadronic systems, with scaled emission energies and, in hadron-hadron collisions, sufficient formation memory to line up the jet axis. Through duality[13)] or urbaryon[14)] diagrams, the multiperipheral approach is moreover linked to a constituent (quark, parton, urbaryon) view of hadrons.

Large p_T results seem to provide further support for a constituent picture of hadrons, bringing us to a regime where not the hadron-hadron interaction, but instead that among constituents[15)] ("gluon exchange") becomes relevant (cf. Fig.2). Whether this interaction still has common features with conventional hadronic forces, e.g., whether jets produced by reggeon, gluon or photon exchanges have the same "dimensions", is a very important though presently unanswerable question.

Consider now the extension of cluster and jet pictures to e^+e^- annihilation. If we suppress the conceptual difficulties in connection with the hadronic volume, we have for the cluster approach an uncontracted interaction region. Among the models of this type applied to e^+e^-, we find mainly the statistical bootstrap[16-18)], the hydrodynamical[19-21)] and the free gas ("Fermi") model[17,22)]. - The jet picture, on the other hand, remains essentially unaltered when taken over to e^+e^- annihilation: only the orientation of the jet axis now becomes random[23)] (except for possible small effects due to the spin of the constituents). If we could, event by event, arrange e^+e^- annihilation reactions such that the jet axes in each case become aligned, then we should obtain a situation like that in hadron-hadron interactions[16)]. Most parton[24)], quark[25)] and similar constituent models[26)] for e^+e^- annihilation provide such a picture.

Deep inelastic e-p scattering led to Bjorken scaling, in which the mass M of the excited hadronic system is scaled by the mass $\sqrt{-q^2}$ of the spacelike virtual photon (cf. Fig.3a), yielding distributions depending on $-q^2/M^2$. Extrapolated to timelike photons (cf. Fig.3b), this implies annihilation spectra which depend on $x = 2p_o/\sqrt{q^2} \simeq 1 - M^2/q^2$. Though such an extrapolation is in general not theoretically justifiable, it resulted in the anticipation[27)] of scaling distributions in annihilation. Moreover, the early onset of (Bjorken) scaling for spacelike photons ($-q^2 \gtrsim 1 - 2\ \text{GeV}^2$, $M \gtrsim 2$ GeV) had led many to also look for early (Feynman) scaling in spectra from timelike photons. Experimentally, this does not seem to be the case and we shall see later on that for systems of hadronic dimensions such early scaling is in fact kinematically forbidden, even if the systems asymptotically provide scaling distributions.

The similarity between the hadronic jet picture and photo- as well as fixed q^2 electroproduction appears plausible both in terms of a general constituent picture (cf. Fig.2a and Fig.4) and through vector meson dominance. The connection between deep inelastic electron-hadron interactions and e^+e^- annihilation is, however, much less clear, also on a phenomenological level: in one case, the (spacelike) photon penetrates the hadronic bag to interact with the constituents, in the other the (timelike) photon creates a constituent-anticonstituent pair, which now has to "inflate" the bag to hadronic size. Thus the "hadronic volume" problem[27,28] already mentioned arises in a partonic jet approach as well as in the cluster picture.

Summarizing, we find that jet and cluster views of hadron-hadron and lepton-hadron interactions can, though with some reservations, be extended to e^+e^- annihilation. While difficult to distinguish in collision processes, the predictions of the two approaches are expected to become very different for annihilation reactions.

The subsequent material is organized as follows. In section II, we study simple "extremal" models of each type and derive their asymptotic properties; in section III, the finite energy behaviour (approach to asymptopia) of the models is investigated, while in section IV some alternative approaches and interrelations with other dynamical schemes are discussed.

II. FIREBALL vs. JET APPROACH TO e^+e^- ANNIHILATION

Experiments on e^+e^- annihilation provide information about two main and a priori independent aspects of multihadron production: the q^2 dependence of the total cross section $\sigma_{tot}(e^+e^- \to \text{hadrons})$, and the distribution of this cross section over the allowed inelastic channels, expressed e.g. in the form of normalized single particle spectra

$$F(q^2,\vec{p}) \equiv \frac{2p_0}{\sigma_{tot}} \frac{d^3\sigma}{d^3\vec{p}} \tag{2.1}$$

with $p^2 = \sqrt{p^2+m^2}$ denoting the CMS energy of the observed secondary, m its mass. While some models (in particular those assuming hadrons to be made up to pointlike constituents) discuss both features together, we shall in this section restrict ourselves to a study of the relative structure of hadron production channels independently of the q^2 behaviour of σ_{tot}. This gives us the advantage of being able to deal with the many-particle aspects in as general a fashion as possible; we must pay for it by leaving open the question of connecting our results for timelike photons with those for spacelike photons of deep inelastic e-p interactions. In section III, we shall briefly return to this question.

As indicated above, we may phenomenologically divide the various theoretical descriptions proposed for e^+e^- multihadron annihilation into two types. For this, we write the normalized CMS energy distribution

$$F(q^2, p_0) \equiv \int d^2\Omega \, F(q^2, \vec{p}) \tag{2.2}$$

(with Ω describing the angular orientation of the observed secondary) in the form[29)]

$$F_\nu(q^2, p_0) = g_\nu(q^2)\, \phi_\nu(2p_0 R^\nu/(q^2)^{(1-\nu)/2}) \tag{2.3}$$

Here $0 \leq \nu \leq 1$, and R is a scale parameter (dimension length). Most models based on the creation of a constituent-anticonstituent pair lead for $q^2 \to \infty$ to a scale-invariant distribution

$$F_0(q^2, p_0) = \phi_0(2p_0/\sqrt{q^2}) \tag{2.4}$$

Although this result is easily visualized in terms of a two-jet picture with randomly oriented axis[23)], such a structure is not the only one to obtain equation (2.4)[17)]. Also, it is certainly possible to construct constituent-anticonstituent models which provide neither jet behaviour nor scaling[30)]. The distribution form (2.3) for $\nu > 0$ is generally obtained in cluster models of various types. Here one should remember, however, that one could also construct jet models leading to $\nu > 0$, since p_T bounds and scaling are a priori independent features of multihadron production. - The special case $\nu = 1$ is realized in the statistical bootstrap model

$$F_1(q^2, p_0) = g_1(q^2)\, \phi_1(2p_0 R) \tag{2.5}$$

and will therefore be called a fireball distribution.

Scaling and fireball spectra are thus in a sense opposite possible extremes for single particle distributions in e^+e^- annihilation. To gain an understanding of the range of possible behaviours it seems to us best to study a simple model of each of these extremes. The results of such a program[17,31,32)] will form the main content of this and a large part of the next section. In each case we shall present the predictions for the most commonly studied observables, in particular the secondary energy spectrum $F(q^2, p_o)$, the multiplicity $\bar{\bar{N}}(q^2)$ and the correlation integral $f_2 = \overline{N(N-1)} - \bar{N}^2$. The spectrum satisfies, because of normalization and energy conservation, the sum rules

$$\frac{1}{2}\int dp_o\, p\, F(q^2, p_o) = \bar{N}(q^2) \tag{2.6}$$

$$\frac{1}{2}\int dp_o\, p_o\, p\, F(q^2, p_o) = \sqrt{q^2} \tag{2.7}$$

where we restrict ourselves for the moment to one type of secondary only.

A. The Fireball Picture[16,17,31)]

The statistical bootstrap model[4,33)] provides us quite generally with a decay scheme for isotropic excited hadronic systems ("fireballs"), and as such is an ideal candidate for the application to e^+e^- annihilation. The model is statistical in the sense that transition probabilities $P(\alpha \to \beta)$ between "inclusive" states α and β are fully determined by the corresponding level densities $\tau(\alpha)$ and $\tau(\beta)$:

$$P(\alpha \to \beta) \sim \tau(\beta)/\tau(\alpha) \tag{2.8}$$

It is a bootstrap model in requiring that the relevant level density of the system is identical to the mass spectrum of its constituents. The level density of a hadronic system of mass $\sqrt{q^2}$ can be written

$$\tau(q^2) = \delta(q^2-m^2)\Theta(q_o) + \sum_{N=2}^{\infty} \frac{\lambda^{N-1}}{N!} \int \prod_{i=1}^{N} d^4Q_i\, \rho(Q_i^2)\, \delta^{(4)}\Big(\sum_{i=1}^{N} Q_i - q\Big) \tag{2.9}$$

if the system consists of "free" hadrons with a mass spectrum $\rho(Q^2)$, possessing $Q^2 = m^2$ as lowest and discrete state. The coupling parameter λ is in a free gas picture proportional to the volume of the system, considered here to be of hadronic size[35)].

The bootstrap condition $\tau(q^2) = \rho(q^2)$ requires that "fireballs are made of fireballs" and yields

$$\tau(q^2) = \delta(q^2-m^2)\Theta(q_o) + \sum_{N=2}^{\infty} \frac{\lambda^{N-1}}{N!} \int \prod_{i=1}^{N} d^4Q_i\, \tau(Q_i^2)\, \delta^{(4)}\Big(\sum_{i=1}^{N} Q_i - q\Big) \tag{2.10}$$

as the bootstrap equation. Its solution can be written as an expansion in phase space integrals of the stable observed secondaries[34)]

$$\tau(q^2) = \sum_N \lambda^{N-1} g_N \Omega_N(q^2) \tag{2.11}$$

$$\Omega_N(q^2) = \int \prod_{i=1}^{N} \frac{d^3p_i}{2p_{i0}} \delta^{(4)}\left(\sum_{i=1}^{N} p_i - q\right) \tag{2.12}$$

Equations (2.11/2.12) can be obtained simply through elimination of the intermediate fireballs by resubstituting (1.10) in itself; graphically, the corresponding cascade decay is illustrated in Fig.5.

From equations (2.11/2.12), together with (2.8), we have immediately

$$F(q^2, p_0) = \frac{4\pi}{\tau(q^2)} \frac{\partial}{\partial\lambda}\left[\lambda\tau((q-p)^2)\right] \tag{2.13}$$

for the secondary energy spectrum,

$$\bar{N}(q^2) = \lambda \frac{\partial}{\partial\lambda} \ln\left[\lambda\tau(q^2)\right] \tag{2.14}$$

for the multiplicity, and

$$f_2(q^2) = \lambda^2 \frac{\partial^2}{\partial\lambda^2} \ln\left[\lambda\tau(q^2)\right] \tag{2.15}$$

for the correlation integral. The behaviour of these observables is thus fully fixed once the level density $\tau(q^2)$ is explicitly given.

The asymptotic solution of equation (2.10) can be obtained by Laplace transform techniques[36,37]; as result we have the familiar linearly exponential rise in fireball mass

$$\tau(q^2) \simeq \text{const.}\ (q^2)^{-3/2}\, e^{\sqrt{q^2}/T_0} \tag{2.16}$$

where the ultimate temperature $T_o = 1/\beta_o$ is given through the relation

$$\frac{2m\pi\lambda}{\beta_0} K_1(m\beta_0) = 2\ln 2 - 1 \tag{2.17}$$

in terms of the secondary mass m and the coupling parameter λ. Equation (2.17) takes on a particularly simple form if we neglect the mass of the secondaries[16]: for $m \to 0$, we have

$$\frac{2\pi\lambda}{\beta_0^2} = 2\ln 2 - 1 \tag{2.18}$$

as the determining equation for the ultimate temperature. Inserting this in equation (2.16) yields as the asymptotic $(q^2 \to \infty)$ form

$$F(q^2, p_0) \propto \text{const.}\ \sqrt{q^2}\ e^{-p_0/T_0} \tag{2.19}$$

for the spectrum,

$$\bar{N}(q^2) \simeq \sqrt{q^2}/2T_0 \tag{2.20}$$

for the multiplicity, and

$$f_2(q^2) \simeq -\sqrt{q^2}/4T_0 = -\tfrac{1}{2}\bar{N}(q^2) \tag{2.21}$$

for the correlation integral. Corresponding to (2.20) we have

$$\bar{\omega} = 2T_0 \tag{2.22}$$

as the asymptotic average energy per secondary. - These results for $m = 0$ remain in form valid also for $m \neq 0$; the most significant change is that now

$$\bar{\omega} = \frac{2}{\beta_0} + m\frac{K_0(m\beta_0)}{K_1(m\beta_0)} \tag{2.23}$$

gives the asymptotic secondary energy, with β_0 fixed by equation (2.17); the multiplicity is changed correspondingly.

Equations (2.19) through (2.23) summarize the essential asymptotic predictions of the statistical bootstrap model for e^+e^- annihilation: an exponential, q^2-independent form for the secondary energy distribution, an asymptotically constant energy per secondary, and a multiplicity rising linearly in photon mass. The entire description contains (for one type of secondary) only one "open" parameter, the hadronic volume λ . Assuming this to be same in e^+e^- and hadron-hadron interactions, or, equivalently, assuming a universal ultimate temperature T_0 , we may fix λ from the transverse momentum distribution of secondaries in hadron-hadron interactions. If we take the fireballs produced in proton-proton collisions at ISR energies as already asymptotic, then a fit to $\exp(-p_T/T_0)$ yields $T_0 \simeq 160$ MeV. For pions produced in e^+e^- annihilation, this means

$$\bar{\omega} = 412\ \text{MeV} \tag{2.24}$$

as asymptotic average energy. - In section III we shall return to the fireball picture to consider its predictions at finite energies.

B. The Uncorrelated Jet Picture[16,23,32)]

The uncorrelated jet model for particle production in hadronic collisions[38,39)] is again of statistical nature in the sense of expression (2.8), but it brings in dynamics "ad hoc" by calculating level densities from phase space containing the empirically observed transverse momentum restrictions. These restrictions represent not only one of the main features of the production process, but through duality they may even, in some average sense, reflect the bulk of collision dynamics.

The model for hadron-hadron interactions is extended[32)] to e^+e^- annihilation by describing the fully exclusive decay of a virtual photon into N secondaries of momenta p_i, for simplicity chargeless pions, in the form

$$d^N\sigma \sim \frac{x^N}{N!}\prod_{i=1}^{N}\frac{d^3p_i}{2p_{io}}\,\delta^{(4)}\Big(\sum_{i=1}^{N}p_i-q\Big)\int d^2\hat{e}\prod_{i=1}^{N}e^{-B|\vec{p}_i\times\hat{e}|} \tag{2.25}$$

Here $\hat{e}$ is a unit vector describing the jet axis in each event. In the form (2.25) we have assumed that the photon decay is independent of its polarization and hence $\hat{e}$ is randomly oriented. In specific models, this may be different: thus a jet picture based on the inital pair creation of spin 1/2 constituents would imply asymptotically a factor $(1+\cos^2\Theta_i)$ in the integration over $\hat{e}$, where Θ_i denotes the angle between $\hat{e}$ and p_i.

In terms of the longitudinal phase space

$$\Omega(\hat{e},q) = \sum_{N=2}^{\infty}\frac{x^N}{N!}\int\prod_{i=1}^{N}\frac{d^3p_i}{2p_{io}}\,e^{-B|\vec{p}_i\times\hat{e}|}\,\delta^{(4)}\Big(\sum_{i=1}^{N}p_i-q\Big) \tag{2.26}$$

the normalized single particle spectrum (2.1) is then given by

$$F(q^2,\vec{p}) = \Big\{\int d^2\hat{e}\times e^{-B|\vec{p}\times\hat{e}|}\,\Omega(\hat{e},q-p)\Big\}\Big/\int d^2\hat{e}\,\Omega(\hat{e},q) \tag{2.27}$$

The corresponding secondary energy distribution (2.2) can be written

$$F(q^2,p_o) = 4\pi F(q^2,\vec{p}) = \frac{x}{\sqrt{x^2-4m^2/q^2}}\;\frac{16\pi}{\sqrt{q^2}}\;\frac{d\sigma}{\sigma_{tot}\,dx} \tag{2.28}$$

introducing the scaled variable $x = 2p_o/\sqrt{q^2}$.

The physical significance of the two parameters B and κ is quite evident: B^{-1} determines the jet width and κ characterizes the multiplicity distribution, an increase in κ giving higher weight to larger particle numbers.

The evaluation of the level density (2.26) can be carried out using saddle point methods[32,40,41]; once $\Omega(\hat{e},q)$ is given, we obtain the energy spectrum through (2.27), the multiplicity from

$$\bar{N}(q^2) = \kappa \frac{\partial}{\partial \kappa} \ln \left\{ \int d^2\hat{e}\, \Omega(\hat{e},q) \right\} \tag{2.29}$$

or through sum rule (2.6) from the spectrum, and the correlation integral f_2 from

$$f_2(q^2) = \kappa^2 \frac{\partial^2}{\partial \kappa^2} \ln \left\{ \int d^2\hat{e}\, \Omega(\hat{e},q) \right\} \tag{2.30}$$

This provides us asymptotically $(q^2 \to \infty)$ with the predictions

$$\bar{N}(q^2) \simeq \tilde{\kappa} \ln q^2 \tag{2.31}$$

for the multiplicity,

$$f_2(q^2) \simeq 1 - 2\tilde{\kappa}\psi^{(1)}(\tilde{\kappa}) \tag{2.32}$$

for the correlation integral, and

$$\frac{1}{\sigma_{tot}} \frac{d\sigma}{dx} \simeq 2\tilde{\kappa}(1-x)^{\tilde{\kappa}-1}/x \tag{2.33}$$

for the spectrum, with $\tilde{\kappa} = \pi\kappa/B^2$. Expressions (2.31) - (2.33) contain only the leading term in q^2; they summarize the asymptotic behaviour of the jet picture: logarithmic multiplicity, scaling spectra, and constant correlation integral.

To fix the numerical values of B and κ we take recourse to data from hadron-hadron collisions, i.e., we assume the jets produced in e^+e^- annihilation to be of "hadronic dimensions". With the jet axis now fixed in the beam direction, we have asymptotically

$$\left[2p_0 \frac{d^3\sigma}{d^3p} \right]_{p_L \approx 0,\ p_T \text{ small}} \sim e^{-B|\vec{p}_T|} \tag{2.35}$$

$$\bar{N}(s) = \tilde{\varkappa} \ln s \tag{2.34}$$

for the transverse momentum distribution and the multiplicity of secondaries from a hadron-hadron collision at squared CMS energy s. The parameter B^{-1} thus is the same as the ultimate temperature of the fireball model and hence $B^{-1} \simeq 160$ MeV from ISR data[42]; for the second parameter, a value $\kappa \simeq 3$ appears to fit the observed charged multiplicity behaviour[43].

C. Phase Space and Dynamics

We conclude section II with the predictions of the simplest possible statistical picture, the ideal gas model[3]. Both fireball and jet pictures include, of course, interaction dynamics: the former by taking into account resonance structure[44,45], the latter through peripherality. The results of an ideal gas picture for e^+e^- annihilation[17,22] are thus of interest both in order to show that dynamical features exist and remain essential at high energies, and to demonstrate again that fireball and jet approaches depart in opposite directions.

The ideal gas model is obtained from equation (2.26) by removing the p_T bound: $B = 0$. The evaluation of the phase space volume

$$\Omega(q^2) = \sum_{N=2}^{\infty} \frac{\varkappa^N}{N!} \int \prod_{i=1}^{N} \frac{d^3p_i}{2p_{i0}} \delta^{(4)}\left(\sum_{i=1}^{N} p_i - q\right) \tag{2.36}$$

is again possible by standard methods of statistical physics[46]. We obtain asymptotically with

$$\Omega(q^2) \sim (q^2)^{-4/3} e^{3[\pi\varkappa q^2/2]^{1/3}} \tag{2.37}$$

a level density increasing weaker than the bootstrap form (2.16), and consequently a multiplicity

$$\bar{N}(q^2) \simeq [\pi\varkappa q^2/2]^{1/3} \tag{2.38}$$

growing as a power less than linear in photon mass. Equivalently the average secondary energy grows indefinitely

$$\bar{\omega}(q^2) \simeq (2/\pi K)^{1/3} (q^2)^{1/6} \tag{2.39}$$

as $q^2 \to \infty$. The spectrum

$$F(q^2, p_0) \sim \varkappa e^2 \left[(\pi K/2)^{1/3} p_0/(q^2)^{1/6} \right] \tag{2.40}$$

is of the form (2.3) with $\nu = 2/3$.

In Table I we summarize the asymptotic predictions of the three approaches considered in section II. It is clear that the fireball model provides essentially the soft, the jet model the hard secondary limit of the possible behaviours classified by equation (2.3).

III. FINITE ENERGIES AND THE APPROACH TO ASYMPTOPIA

A. Finite Energy Predictions

As the presently available e^+e^- energies correspond to hadronic collision energies of $p_{lab} \lesssim 20$ GeV/c, two roads seem possible. Either we argue that jet structure remains fundamental, and then we can no more expect asymptotic behaviour at $\sqrt{q^2} = 6$ GeV than we can in central region spectra for secondaries in hadronic collisions[47] - which increase by more than a factor two in going from 6 to 60 GeV incident CMS energy. Or we believe that all of the 6 GeV virtual photon mass go into an intrinsically spherical system - then asymptotic behaviour may be possible. These arguments can easily be made more quantitative.

We show in Figs.6-12 the results of numerical calculations[17,32] performed for the fireball, jet and ideal gas models, with "hadronic" parameters as indicated in section II; in particular, the ideal gas model is taken as the $B = 0$ limit of the jet picture.

From the multiplicity prediction in Fig.6, it is clear that the fireball model very quickly becomes asymptotic, essentially for $\sqrt{q^2} \simeq 2 - 3$ GeV. The same conclusion is obtained from the q^2 dependence of the average secondary energy (Fig.7) and of the spectrum (Fig.8).

The jet model, on the other hand, leads at present energies to a multiplicity closer to that predicted by the ideal gas picture than to it own asymptotic form (Fig.9). Similarly, the resulting spectra deviate strongly from the scaling limit <u>at all x</u> (Fig.10), although the smallness of q^2 interval makes the q^2

dependence of the spectra not very pronounced at large x ("simulated" scaling?). Hence: present energies do not allow a really definite manifestation of jet dynamics in inclusive distributions - if such jets are of hadronic dimensions.

The correlation integral, from Fig.11, appears not very suitable to distinguish dynamics for $q^2 \lesssim 100\ \text{GeV}^2$.

B. Predictions vs. Data[48,49,50,51]

A comparison between models and data is at present still beset by two difficulties: the "neutral energy crisis" and the lack of identified high energy spectra over a fairly large momentum interval. Nevertheless, the most recent data do seem to provide rather strong hints in favour of a jet picture, and we shall come back to this evidence after some comments on the two problems just mentioned.

The excess of neutral energy presently reported[48,49] (cf. Fig.13) makes a study of the overall multiplicity impossible without further assumptions which rather leave the more or less statistical framework in which we have worked here; clearly, such a framework would prefer asymptotically $E_{ch}/E = 2/3$, although the inclusion of photons or even leptons in a statistical scheme is in principle possible and could give different values for this ratio. - The present value of E_{ch}/E given, we can either assume that the (unobserved) neutral particles carry the same momentum as the (observed) charged ones, or that the neutral multiplicity grows in the same way as the observed $\bar{N}_{ch}$. The first assumption gives more neutrals and a faster growing $\bar{N}$, the second more energetic neutrals and a faster growing secondary energy ω - both in comparison with the observed behaviour of charged secondaries. Hence we have two possible assumptions with rather opposite consequences. - Apart from this difficulty, accurate multiplicity measurements could in principle allow a distinction between the jet and fireball extremes, once $\sqrt{s} \gtrsim 6$ GeV.

In Fig.14 we show a comparison between measured spectra for charged secondaries[48] and predictions for pion production in the three models considered above. When taken over the whole available range of secondary energy, the jet model seems to give the best agreement. However, one expects even in a fireball picture that kaons and baryons carry a higher average momentum than pions, so that the elimination of such events from the data may well shift the large momentum part of the spectrum down. Such a change would be enhanced, if the $\pi : K : N$ ratio increases with increasing secondary energy beyond the value 100 : 7 : 2.5 used in the fireball calculation of Ref.17). In a small range of secondary energy, where identified π^+ spectra exist, a fit to the form $\exp - (p_o/T_o)$, with $T_o = 160$ MeV, seems to give good agreement [48].

Considering now the most recent experimental results[49,50,51], we find two rather strong supports in favour of a jet picture. The average energy per track[49] (cf. Fig.15) reaches at $\sqrt{s} = 7.4$ GeV a value of about 650 MeV; even if one does include rather strong K and N production[50], this value appears considerably too high to be accommodated in a fireball picture with the universal hadronic temperature. - To test for jets on an event-by-event basis, one may introduce[49] the "sphericity"

$$S = \frac{3}{2} \frac{\sum_i \vec{p}_{iT}^{\,2}}{\sum_i \vec{p}_i^{\,2}} \tag{3.1}$$

relative to a jet axis reconstructed by minimalizing $\sum_i \vec{p}_{iT}^{\,2}$, with the sum running over all particles. As only charged particles are observed, this implies a Monte Carlo reconstruction of the neutrals. In the case of a secondary distribution isotropic event-by-event, one expects asymptotically $S = 1$, while a jet picture predicts $S = 0$. At higher energies the experimental results[49] (cf. Fig.16) are clearly seen to deviate from phase space towards an agreement with a jet picture of hadronic dimensions. It should be noted here, though, that this conclusion could be made even more decisive by considering pion production only: kaons and baryons produced at finite energies would, if they receive more average momentum than pions, by momentum conservation shift the sphericity towards lower values.

Before leaving this section, we comment briefly on the question of scaling. We show in Fig.17 the measured <u>normalized</u> distributions $(1/\sigma_{tot})d\sigma/dx$ in $x = 2\omega/\sqrt{s}$ for various incident energies[51]. It is clear that at present energies one does not observe scaling - in accord with a jet picture of hadronic dimensions. - As σ_{tot} in most of the presently available energy range does not yet follow the $1/q^2$ behaviour asymptotically expected from a pointlike parton picture, the energy dependence of the spectra in Fig.17 can be partially compensated through multiplication by $q^2\sigma_{tot}(q^2)$[49]. Such a compensation would, however, appear to have little connection with scaling both in a general jet picture or in specific parton considerations.

C. Bjorken vs. Feynman Scaling

We have seen above that for a jet picture of hadronic dimensions, scaling at present energies is kinematically forbidden at all x. How can one then be led to expect scaling from extrapolation arguments of deep inelastic behaviour, and, in turn, what implications could our results have on deep inelastic features? Let us attempt to shed some light on these problems by considering Table II, together with

Fig.3 and 18.

Pointlike constituents in a "naive" hadron picture provide us both with deep inelastic (Bjorken) scaling and with the $1/q^2$ decrease of σ_{tot} $(e^+e^- \to$ hadrons), as then the photon propagator is the only way to bring in an explicit q^2 dependence.

On the other hand, a jet picture is the common basis of (Feynman) scaling for secondary energy distributions in e^+e^- annihilation as well as for inclusive electroproduction (Regge limit). Hence here, and only here, should we expect parameters interpretable as jet dimensions to determine the approach to scaling.

Extrapolation arguments from $q^2 < 0$ to $q^2 > 0$ can, from this point of view, at best connect annihilation spectra with deep inelastic structure functions when $|q^2| \to \infty$, $M^2 \to \infty$. At finite q^2, the approach to scaling is governed phenomenologically by quite different mechanisms. The deep inelastic structure functions essentially measure the "size" of a hadron as seen by a virtual photon. Such a quantity is like a total cross section, and this can, as also seen in pp interactions, reach an asymptotic form long before the opening of new channels ceases to be a significant feature and jet structure becomes decisive.

Besides this, even a cross section extrapolation from one physical region to another may lead to entirely different nonasymptotic aspects: thus the total cross sections for pp and $\bar{p}p$ scattering, though asymptotically equal through the Pomeranchuk theorem, can attain this equality only when the effect of $\bar{p}p$ annihilation into mesons has vanished - a process which has no counterpart in pp interactions.

In summary: There appears to be little physical reason to expect any connection in the approach to scaling of deep inelastic structure functions and e^+e^- spectra, even if these quantities should asymptotically be related. And we would like to emphasize again that if scaling should really be observed in any x region for $q^2 \lesssim 6 - 8$ GeV, then this must imply some nonhadronic features in the underlying jet dynamics.

IV. EXTENSIONS AND ALTERNATIVES

One immediate extension of our considerations up to here is obvious: the inclusion of the internal quantum numbers of the secondaries. The resulting modifications have been studied in particular detail for the fireball picture: different types of secondaries, isospin structure and the production of neutrals, the exclusion of exotic states in the bootstrap equation and its solution, the bootstrap approach for quarks and the effect of Bose-Einstein statistics have all been investigated in this context. Although certainly of importance for a detailed description of precision data, the inclusion of these features does not significantly alter the picture we have sketched above, so we refer the reader to the original literature (Ref.31, 52, 53) for further discussion. Instead, we shall consider here mainly three topics: the hydrodynamical model, implications of constituent interactions on hadronic size, and the relation between jet and fireball descriptions on one hand and dual models for e^+e^- annihilations on the other.

A. The Hydrodynamical Model[19,20,21]

Originally proposed[2] to account, in the framework of a cluster picture, for the observed jet structure of hadron-hadron collisions, the hydrodynamical model provides a three-step description of multihadron production. In step one, the total collision energy is deposited in a hadronic interaction volume, which is Lorentz-contracted along the beam direction. The hadronic matter in this volume is governed by an equation of state - in the original version, that of an ultra-relativistic gas. In step two, the system expands according to the hydrodynamic equations of an ideal relativistic fluid (isentropically, no viscosity, no turbulence). Because the pressure gradient is largest along the collision axis, the motion of expansion is greatest in that direction. Step three finds the overall volume so large that all constituents of the system are separated by a distance R, the range of nuclear forces. As these now vanish, the system breaks up into free hadrons. A rather natural variation[54] proposes instead a break-up into clusters of hadronic matter, which then decay "thermally", i.e. with a local spectrum given by the fireball picture. In either case, the low energy limit of the hydrodynamical model (no hydrodynamical expansion) yields the picture proposed by Pomeranchuk[55] - and which is equivalent to the statistical bootstrap approach[35].

The extension from hadron-hadron collisions to e^+e^- annihilation is, as already indicated in section I, quite straightforward: no more Lorentz-contraction, the same ("hadronic") initial volume. Given the equation of state of hadronic matter,

$$p = c_0^2 \epsilon \tag{4.1}$$

with p, ε and c_o denoting pressure, CMS energy density and speed of sound, respectively, one must now solve the hydrodynamic equations

$$\partial_\mu T^{\mu\nu} = 0 \tag{4.2}$$

where $T^{\mu\nu}$ is the energy-momentum tensor of the ideal fluid. Because of the spherical symmetry of the problem, expression (4.2) contains only two independent equations. Their solution gives us the expansion velocity of hadronic matter, which, either directly or when folded with the local thermal decay spectrum, yields the secondary momentum or energy distribution.

The predictions of the model are of two kinds: the multiplicity, determined by the entropy of the system, can be obtained from equation (4.1) alone, without solving (4.2), as the hydrodynamic expansion is by assumption isentropic. To obtain spectra, on the other hand, one must solve equations (4.2) - a rather complex task even in the spherical symmetric case[20,21]. The asymptotic multiplicity corresponding to (4.1) is given by[56,57]

$$\bar{N}(q^2) \sim (q^2)^{\frac{1}{2}(1+c_o^2)} V_o^{c_o^2/(1+c_o^2)} \tag{4.3}$$

with V_o for the initial interaction volume. The equation of state of an ultra-relativistic gas gives $c_o^2 = 1/3$ and hence with

$$\bar{N}(q^2) \sim (q^2)^{3/8} \tag{4.4}$$

a behaviour quite similar to that of the phase space model (cf. equation (2.38)). In hadron-hadron collisions, however, the choice $c_o^2 = 1/3$ appears to lead to some difficulties[57], since it there predicts an average transverse momentum $\bar{p}_T \sim s^{1/12}$, a behaviour not compatible with the constancy observed in present accelerator data[58]. If this constancy persists for $s \to \infty$, it would imply $c_o^2 \to 1$, which in hadronic collisions allows the multiplicity to increase at most logarithmically, as in a jet picture. For e^+e^-, $c_o^2 \to 1$ yields

$$\bar{N}(q^2) \sim (q^2)^{1/4} \tag{4.5}$$

for the multiplicity, in contrast to Eq.(2.31) of the jet picture. The situation is summarized in Table III.

The prediction of secondary spectra requires, as mentioned, the solution of Eq.(4.2) and generally implies at least some numerical work; for the results, see Ref.20) and 21). Here we note only some general features.

For $c_o^2 \to 1$, the spectra do not become scale invariant. Nevertheless, locally (i.e., for a restricted region around zero in the CMS rapidity $y = \frac{1}{2} \ln [(p_o + p_L)/(p_o - p_L)]$) we do have an approximately flat distribution in y[20] ("plateau" behaviour). This "plateau", however, rises with increasing incident energy - in contrast to the genuine plateau of the jet picture, which is energy independent.

For $c_0^2 \to 1$, we obtain two distinct situations, as already indicated by the multiplicity behaviour. Because of the Lorentz-contraction, the spectra for secondaries from hadron-hadron collisons become independent of the incident energy, so that here the hydrodynamical model becomes equivalent to a jet picture. For e^+e^-, on the other hand, the spectra retain an energy dependence[20,21], in accord with Eq.(4.5). Hence the hydrodynamical model can, for $c_0^2 \to 1$, provide a rather simple unified picture containing Feynman scaling for single particle distributions in ep and hadron-hadron interactions, yet non-scaling behaviour (even asymptotically) for e^+e^- spectra[59]. In section IV.C we shall encounter another such case.

B. Constituents and Hadronic Size

The basic common feature of the e^+e^- models we have considered so far is their use of hadronic dimensions as an input. In this section we want to consider some possible tests of this assumption.

Deep inelastic electron-proton interactions provide one of the main bases for a constituent view of hadrons, with a pointlike photon-parton interaction. Electroproduction experiments should then allow us to investigate if jets initiated by parton excitation differ from those obtained through hadron excitation (cf. Fig.18). Consider inclusive pion production through virtual photons, $\gamma_V p \to \pi$ + anything. Does the transverse momentum distribution

$$d\sigma / dx_F \, dp_T^2 \sim e^{-B\sqrt{p_T^2 + m^2}} \; ; \; x_F = 2p_L / \sqrt{s} \qquad (4.6)$$

in the photon fragmentation region $(0 < x_F \leq 1)$ depend on q^2? There are some[60], though perhaps not conclusive indications that $B(q^2)$ decreases with increasing $|q^2|$. Large q^2 data for this would be of great interest - present measurements have $-q^2 \lesssim 2 - 3$ GeV^2. Complementary information might come from the slope parameter dependence on q^2 for exclusive momentum transfer measurements.

Another possible point of departure from hadronic size is the large p_T behaviour in hadron-hadron collisions. A breakdown of the "universal" small p_T form (4.6) was predicted[15] on the basis of a charged constituent picture: the electromagnetic interaction of the constituents, mediated by photon exchange with pointlike coupling and hence falling as p_T^{-4}, must for sufficiently large p_T dominate the rapidly falling hadronic form (4.7). The observed deviations from Eq.(4.7) were in fact far too large to be explanable as electromagnetic effects;

hence "gluon" exchange with a pointlike but strong coupling was introduced[61]. Another way of studying the properties of jets from constituent excitation is thus provided by multiparticle production associated with a large p_T secondary in hadronic collisions, and it appears[62] that e.g. the observed multiplicity in fact differs from that in the small p_T region.

Both examples considered here involved constituent excitation by spacelike (photon or gluon) exchange, and it is of course completely open whether the constituent-anticonstituent pair originating from a timelike photon leads to jets of similar form. But if the "hadronic dimensions" are altered for large spacelike q^2, it appears dangerous to assume them as unchanged for large timelike values.

C. Duality and e^+e^- Annihilation

A dual amplitude by construction takes into account crossing symmetry, and hence the dual description of a process involving a real or virtual photon should be applicable to electron-hadron interactions both in the Regge $(s \to \infty$, q^2 fixed) and the Bjorken $(s \to \infty$, s/q^2 fixed) limit as well as to e^+e^- annihilation. Such a picture can thus be subjected to more empirical tests than any other hadron production process, and it is perhaps not surprising that a satisfactory dual γ-hadron amplitude is still lacking. A detailed discussion of the problems in this context is beyond the scope of the present survey; we shall here only indicate some of the features which arise and refer the reader to Ref.63) to 66) for further discussion.

There are at present two possible roads leading to a dual description of inclusive spectra from e^+e^- annihilation: the generalized optical theorem and an explicit construction of states in the operator formalism.

In a unitary scheme, the inclusive spectrum arising from the decay of a virtual photon, $\gamma \to c$ + anything, can be expressed[63] as discontinuity over the missing mass $s = M_x^2$ of the forward photon-hadron amplitude (cf. Fig.19):

$$F(q^2, p_0) \sim \operatorname{disc}_{(s)} T_{\gamma\bar{c} \to \gamma\bar{c}}(s, t=0; q^2) \qquad (4.7)$$

Besides the specific difficulties arising in the construction of a dual photon-hadron amplitude[64,65], there also remain problems which exist already in a purely hadronic world, in particular those connected with the non-unitarity of B_N. Ignoring all these, we write

$$T_{\gamma\bar{c}\to\gamma\bar{c}}(s,t) = c_1[V(s,t)+V(u,t)] + c_2 V(s,u) \tag{4.8}$$

corresponding to the three possible $\gamma - \bar{c}$ channels, with

$$\begin{aligned} s &= (q-p_c)^2 = q^2+m^2-2p_0\sqrt{q^2} \\ u &= (q+p_c)^2 = q^2+m^2+2p_0\sqrt{q^2} \\ t &= (q-q^1)^2 \end{aligned} \tag{4.9}$$

The first two terms in (4.8) give rise to a scaling-type spectrum in the most naive $\gamma\bar{c}$ B_4 amplitude[63], and such a behaviour is maintained also in more realistic formulations[64]. By of scaling type we mean that - apart from a possible q^2-dependence of the γ-hadron coupling - these terms produce Feynman scaling

$$F(q^2,p_0) \sim \phi(2p_0/\sqrt{q^2}) \tag{4.10}$$

for e^+e^- spectra, Bjorken scaling

$$\nu W_2(q^2,s) \sim \bar{\phi}(s/q^2) \tag{4.11}$$

in the deep inelastic region. - On the other hand, the simplest B_4 model for the (s,u) term gives

$$F(q^2,p_0) \sim e^{-4\alpha' p_0^2} \tag{4.12}$$

for the annihilation spectrum, with α' as the (universal) Regge slope, while generally leading to a vanishingly small contribution in the deep inelastic region.

Hence all three terms lead to Bjorken-type scaling for the deep inelastic structure functions; the behaviour of annihilation spectra, however, depends on which term dominates in (4.8) - a question which cannot be decided through duality considerations alone, as c_1 and c_2 may depend on q^2.

This conclusion is reinforced by looking at the calculation of inclusive e^+e^- spectra in the operator formalism of duality. Here the intermediate ("anything") states are explicitly constructed in terms of oscillator creation and annihilation operators, as is the initial photonic system of spin one and mass $\sqrt{q^2}$. In dual models such systems are, as is well known, highly degenerate - with a level degeneracy

$$d(q^2) \sim e^{c\sqrt{q^2}} \tag{4.13}$$

growing with increasing mass in the same way as found in the statistical bootstrap model (cf. Eq.(2.16)). Hence the essential question is to which of the degenerate states the incident photon couples, and how strongly.

A statistical approach[67], assuming equal coupling and random phases for all $d(q^2)$ degenerate states, reproduces the behaviour of the fireball picture, i.e., we obtain

$$F(q^2, p_0) \sim e^{-cp_0} \tag{4.14}$$

as well as all other fireball featurs, such as $\bar{N} \sim \sqrt{q^2}$, etc.- It should be noted here that the "dual" result (4.12) is not of this type, as there the cut-off, though of fireball nature (i.e., independent of q^2), is much stronger than can be accounted for by the level degeneracy (4.14) alone.

If, contrary to the statistical assumptions, some dynamical mechanism decouples the incident photon from all but a small subset of the degenerate states, then the result (4.14) can be completely altered - in fact, there exist also states leading to a spectrum of the scaling type (4.10)[29]. By judicious choice of the coupling between the photon and the initial degenerate dual states it thus appears possible to generate any type of secondary spectrum for e^+e^- annihilation. This is perhaps the most transparent way of showing that dual dynamics alone do not determine e^+e^- annihilation.

REFERENCES:

1) W. Heisenberg; Z. Physik 113, 61 (1939)
W. Heisenberg; Z. Physik 133, 65 (1952)

2) L. D. Landau; Izv. Akad. Nauk SSSR 17, 51 (1953)

3) E. Fermi; Progr. Theoret. Physics (Japan) 1, 570 (1950)

4) R. Hagedorn; Nuovo Cim. Suppl. 3, 147 (1965)

5) H. Satz, in Proceedings of the IX. Balaton Symposium on Particle Physics, Vol. II, 149 (1974)

6) W. Heisenberg, in Die kosmische Strahlung, Springer Verlag, Berlin 1953 (1943), p.148

7) L. Stodolski; VIIth Rencontre de Moriond, Vol.II, p.3 (1972)

8) K. Wilson; Acta Phys. Austr. 17, 33 (1963)

9) R. P. Feynman; Phys. Rev. Lett. 23, 1415 (1969)

10) D. Amati, S. Fubini and A. Stanghellini; Nuovo Cim. 26, 896 (1962)

11) N. F. Bali, G. Chew and A. Pignotti; Phys. Rev. Lett. 19, 614 (1967)

12) For references, see e.g. M. Jacob (ed.), Dual Theory, North Holland Publishing Co., 1974

13) H. Harari; Phys. Rev. Lett. 22, 562 (1969)
J. L. Rosner; Phys. Rev. Lett. 22, 689 (1969)

14) M. Imachi, T. Matsuoka, K. Ninomiya and S. Sawada; Progr. Theoret. Physics (Japan) 4o, 353 (1968)
K. Kinoshita and N. Noda; Progr. Theoret. Physics (Japan) 48, 877 (1972) and further references there

15) S. M. Berman, J. D. Bjorken and J. B. Kogut; Phys. Rev. D4, 3388 (1971). For further references, see e.g. J. D. Bjorken, Journ. de Physique 34, C-1 385 (1973)

16) J. D. Bjorken and S. J. Brodsky; Phys. Rev. D1, 1416 (1970)

17) J. Engels, H. Satz and K. Schilling; Nuovo Cim. 17A, 535 (1973)
J. Engels, H. Satz and K. Schilling; Phys. Lett. 49B, 171 (1974)

18) B. Margolis, W. J. Meggs and S. Rudaz; Phys. Rev. D8, 3945 (1973)
S. Rudaz; Lett. Nuovo Cim. 6, 292 (1973)
H. J. Möhring, J. Kripfganz, E. M. Ilgnefritz and J. Ranft; Nucl. Phys. B85, 221 (1975)
H. J. Möhring; Nucl. Phys. B87, 509 (1975)

19) E. V. Shuryak; Phys. Lett. 34B, 509 (1971)

20) M. Chaichian and E. Suhonen; Bielefeld Preprint Bi 74/04 (1974)
J. Baacke; Dortmund Preprint Feb. 1974

21) F. Cooper, G. Fry and E. Schonberg; Phys. Rev. D1
P. D. Morley; Nucl. Phys. B85, 471 (1975)

22) W. S. Lam and E. Suhonen; Phys. Lett. 50B, 453 (1974)
Meng Ta-chung and E. Moeller; Berlin Preprint FUB-HEP May 74/6

23) N. Cabibbo, G. Parisi and M. Testa; Lett. Nuovo Cim. 4, 35 (1970)

24) S. D. Drell, D. J. Levy and T. M. Yan; Phys. Rev. D1, 1617 (1970)

25) R. Gatto and G. Preparata; Nucl. Phys. B67, 362 (1973)

26) See e.g.:
J. D. Bjorken, in Proceedings of the VIth International Colloquium on Electron and Photon Interactions at High Energies, Bonn 1973
C. H. Llewellyn-Smith, Erice Lectures 1974 (to be published)
S. Ellis, in Proceedings of the 17th International Conference on High Energy Physics, London 1974

27) J. D. Bjorken, in Proceedings of the Vth International Colloquium of Electron and Photon Interactions at High Energies, Cornell 1971

28) G. Grammer, H. T. Nieh and Y. P. Yao; Phys. Lett. 57B, 66 (1975)
cf. also R. W. Griffith, Bonn Preprint PI 2-135, May 1973

29) I. Andrić, I. Dadić and H. Satz; Nucl. Phys. B89, 326 (1975)

30) I. Montvay, N. Cabibbo, private communication

31) J. Engels, K. Fabricius and K. Schilling; Nuovo Cim. 23A, 581 (1974)
J. Engels, K. Fabricius and K. Schilling; Phys. Lett. 53B, 65 (1974)

32) R. Baier, J. Engels, H. Satz and K. Schilling; Nuovo Cim. 28A, 455 (1975)

33) S. Frautschi; Phys. Rev. D3, 2821 (1971)

34) J. Yellin; Nucl. Phys. B52, 583 (1973)

35) M. I. Gorenstein, V. A. Miransky, H. Satz, V. P. Shelest and G. M. Zinovjev; Nucl. Phys. B76, 453 (1974)

36) W. Nahm; Nucl. Phys. B45, 525 (1972)

37) R. Hagedorn and I. Montvay; Nucl. Phys. B59, 45 (1973)

38) L. Van Hove; Rev. Mod. Phys. 36, 655 (1964)

39) A. Krzywicki; Nuovo Cim. 32, 1067 (1964)

40) A. Bassetto, M. Toller and L. Sertorio; Nucl. Phys. B34, 1 (1971)

41) E. H. de Groot; Nucl. Phys. B48, 295 (1972)

42) B. Alper et al., Phys. Lett. 47B, 75 (1973)

43) E. L. Berger; Erice Lectures 1973 (to be published)

44) E. Beth and G. E. Uhlenbeck; Physica 4, 915 (1937)
S. Z. Belenkij; Nucl. Phys. 2, 259 (1956)

45) H. Satz; Phys. Lett. 44B, 353 (1973)

46) H. Satz; Nuovo Cim. 37, 1407 (1965)

47) H. Bøggild et al., Contribution to the 1975 Palermo Conference
E. Lillethun; Acta Phys. Polon. B4, 769 (1973)

48) B. Richter, in Proceedings of the 17th International Conference on High Energy Physics, London 1974

49) R. Schwitters, Report at the 1975 Lepton-Proton Symposium, Stanford, California, August 1975

50) B. Wiik, Report at the 1975 Lepton-Photon Symposium, Stanford, California, August 1975

51) R. Hollebeek, University of California/Berkeley Preprint LBL 3874, May 1975

52) E. M. Ilgenfritz and J. Kripfganz; Nucl. Phys. B62, 141 (1973)
E. M. Ilgenfritz and J. Kripfganz; Phys. Lett. 48B, 329 (1974)

53) F. Csikor, I. Farkas, Z. Katona and I. Montvaj; Nucl. Phys. B74, 343 (1974)
F. Csikor, F. Niedermayer and I. Montvay; Phys. Lett. 49B, 47 (1974)
F. Csikor, I. Farkas and I. Montvay; Nucl. Phys. B (in press)

54) G. A. Milekhin and I. L. Rozental; ZETP 33, 197 (1957)

55) I. Ya. Pomeranchuk; Doklad. Akad. Nauk 78, 889 (1951)

56) J. Enkenberg, K. E. Lassila, S. Sohlo and E. Suhonen; Phys. Rev. Lett. 31, 15667 (1973)

57) M. Chaichian, H. Satz and E. Suhonen; Phys. Lett. 50B, 362 (1974)

58) A. M. Rossi et al., Nucl. Phys. B84, 269 (1975)

59) M. Chaichian and E. Suhonen, loc. cit. Ref. 20), revised version to appear soon

60) F. Brasse; in Proceeding of the VIth International Colloquium on Electron and Photon Interactions at High Energies, Bonn 1973

61) S. M. Berman and M. Jacob; Phys. Rev. Lett. 25, 1683 (1970)

62) See e.g. A. M. Smith, in Proceedings of the VIth International Symposium on Many Particle Hadrodynamics, Leipzig 1974, and references there

63) H. Satz; Nuovo Cim. 12A, 205 (1972)

64) P. V. Landshoff; Phys. Lett. 32B, 57 (1970)
P. V. Landshoff, and J. C. Polkinghorne; Phys. Lett. 35B, 50 (1971)

65) E. Cremmer and J. Scherk; Nucl. Phys. B58, 557 (1973)

66) D. Amati and S. Fubini; Phys. Lett. 49B, 293 (1974)
D. Amati, S. D. Ellis and J. H. Weis; CERN-TH 1885 (1974)

67) M. I. Gorenstein, V. A. Miransky, V. P. Shelest, B. V. Struminsky and G. M. Zinovjev; Phys. Lett. 45B, 475 (1973).

TABLE I:

Inclusive Predictions of Fireball, Phase Space and Jet Model for e^+e^- Annihilation

	FIREBALL	PHASE SPACE	JET
$\bar{N}(q^2)$	q^2	$(q^2)^{1/3}$	$\ln(q^2)$
$w(q^2)$	const.	$(q^2)^{1/6}$	$\sqrt{q^2}/\ln(q^2)$
$-f_2(q^2)$	$\bar{N}$	$\bar{N}$	const.
$F(p_o,q^2)$	e^{-cp_o}	$e^{-cp_o/(q^2)^{1/6}}$	$(1-x)^c/x$ $x=2p_o/\sqrt{q^2}$

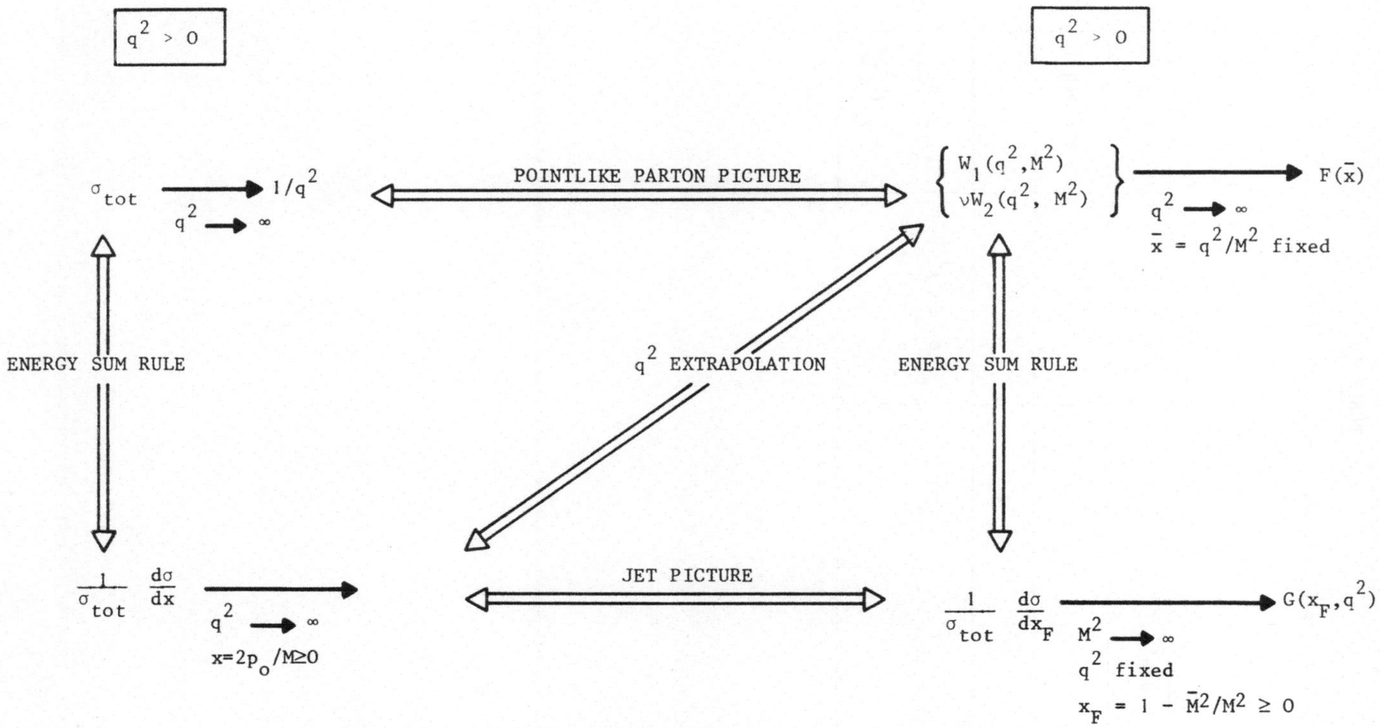

TABLE II: Schematic View of Current Region Relationships

TABLE III

Hydrodynamical Predictions for Hadron-Hadron Interactions and e^+e^- Annihilation

		$c_o^2 = 1/3$	$c_o^2 \to 1$
MULTIPLICITY	e^+e^-	$(q^2)^{3/8}$	$(q^2)^{1/4}$
	hadron-hadron	$s^{1/4}$	$\ln s$
SPECTRUM	e^+e^-	non-scaling	non-scaling
	hadron-hadron	non-scaling	scaling

FIGURE CAPTIONS:

Fig.1: The one-photon contribution to e^+e^- annihilation into hadrons

Fig.2: Interaction regimes: (a) hadronic regime with Regge exchange; (b) constituent regime with gluon exchange

Fig.3: Current induced reactions: (a) e-p scattering; (b) e^+e^- annihilation

Fig.4: Multihadron photoproduction

Fig.5: Cascade decay

Fig.6: Charged pion multiplicity in the fireball model (T_o=160 MeV); asymptotic (solid line) and finite energy (dahsed line) prediction

Fig.7: Secondary pion energy: the approach to asymptotic constancy

Fig.8: Secondary pion spectrum (T_o=193 MeV); predictions for 3 GeV (solid line), 3.8 GeV (dashed) and 4.8 GeV (dash-dot) incident CMS energy

Fig.9: Pion Multiplicity in the jet model at finite energies (dashed line) and asymptotic form (solid line) and in the phase space model

Fig.10: Secondary pion spectra in the jet model at 3 GeV (dash-dot), at 3.8 GeV (dashed) and at 4.8 GeV (solid) incident CMS energy, compared with the scaling limit

Fig.11: Correlation integral for the phase space model (solid line), for the fireball model (dash-dot) and for the jet model at finite energy (dashed) and asymptotically (heavy solid)

Fig.12: Pion multiplicity for the fireball model (dashed line) and the jet model (solid line)

Fig.13: Total charged secondary energy ($E_{charged}$) compared to the incident CMS energy (W), from Ref.49.

Fig.14: Charged secondary data (from Ref.48) compared with predicted pion spectra from the jet model (solid line), the fireball model (dashed line), and from phase space (dash-dot).

Fig.15: Energy per charged secondary as function of the incident CMS energy, from Ref.49.

Fig.16: Sphericity distributions, from Ref.49.

Fig.17: Normalized secondary spectra at incident CMS energies of 3 ( ), 3.8 ( ) and 4.8 ( ) GeV, from Ref.51.

Fig.18: Inclusive electroproduction.

Fig.19: Generalized optical theorem.

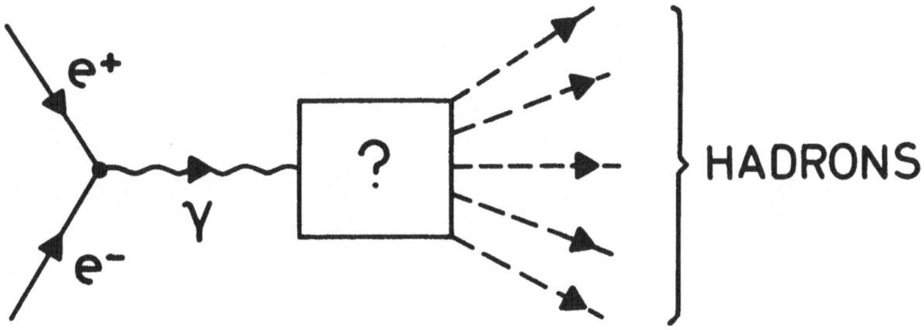

Fig. 1

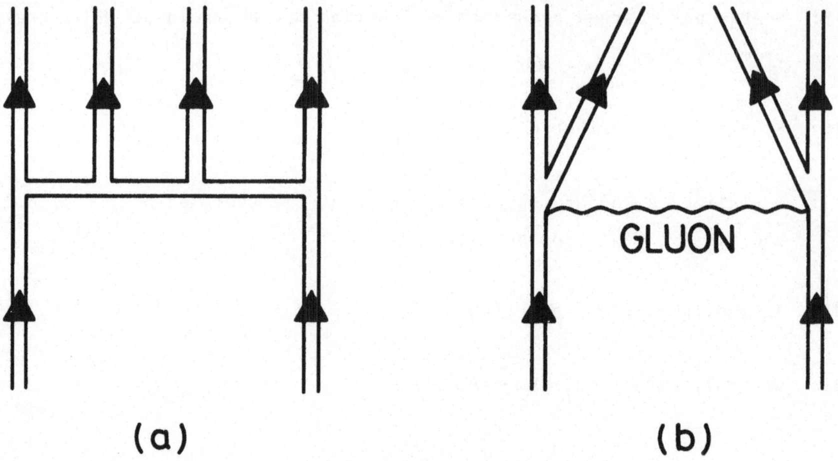

Fig. 2

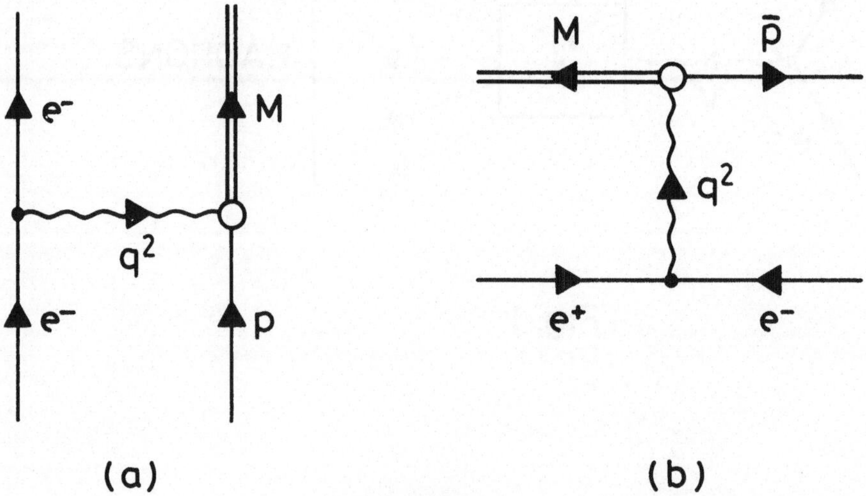

Fig. 3

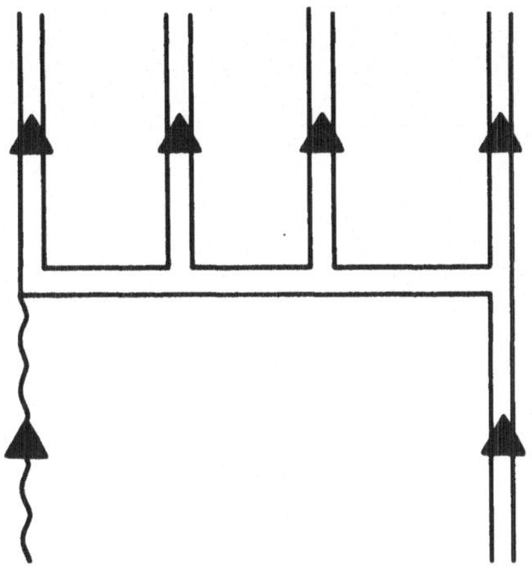

Fig. 4

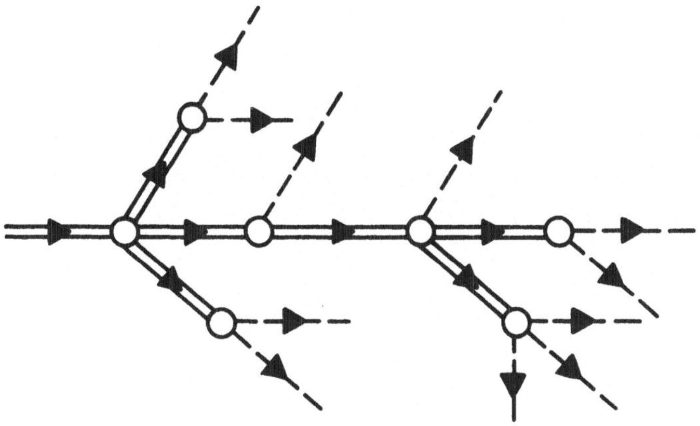

Fig. 5

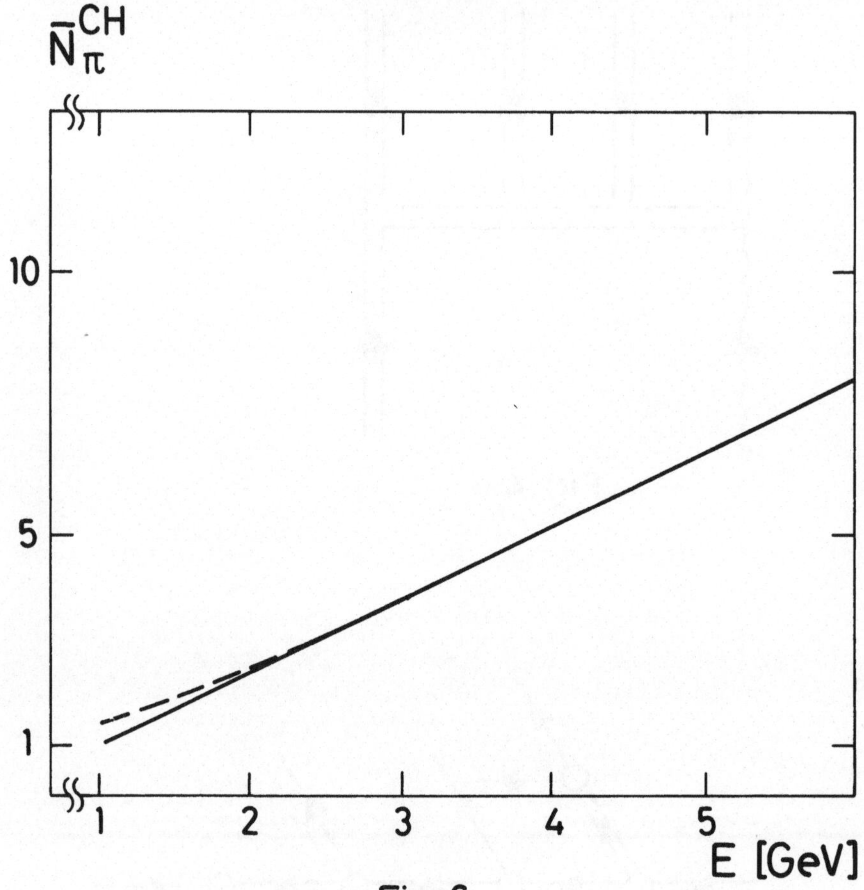

Fig. 6

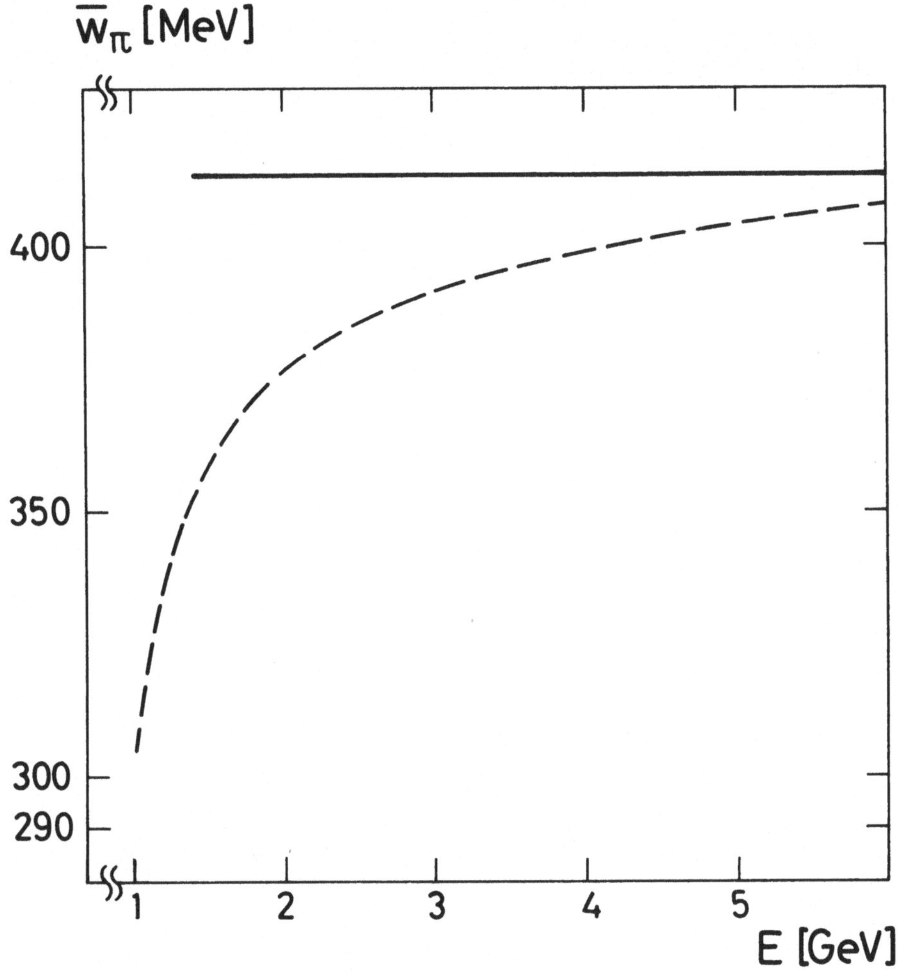

Fig. 7

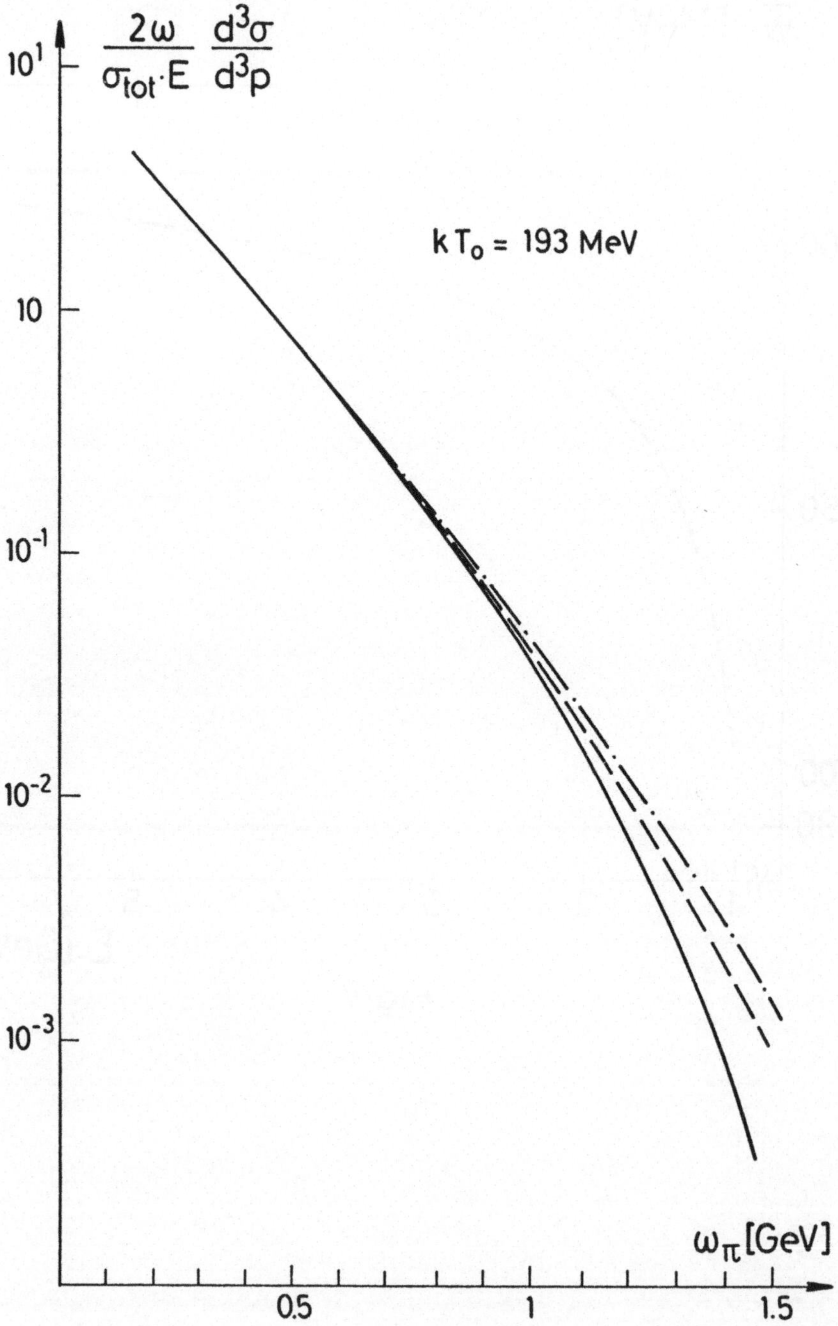

$\frac{2\omega}{\sigma_{tot} \cdot E} \frac{d^3\sigma}{d^3p}$
10^1
10
10^{-1}
10^{-2}
10^{-3}
kT_o = 193 MeV
ω_π[GeV]
0.5
1
1.5

Fig. 8

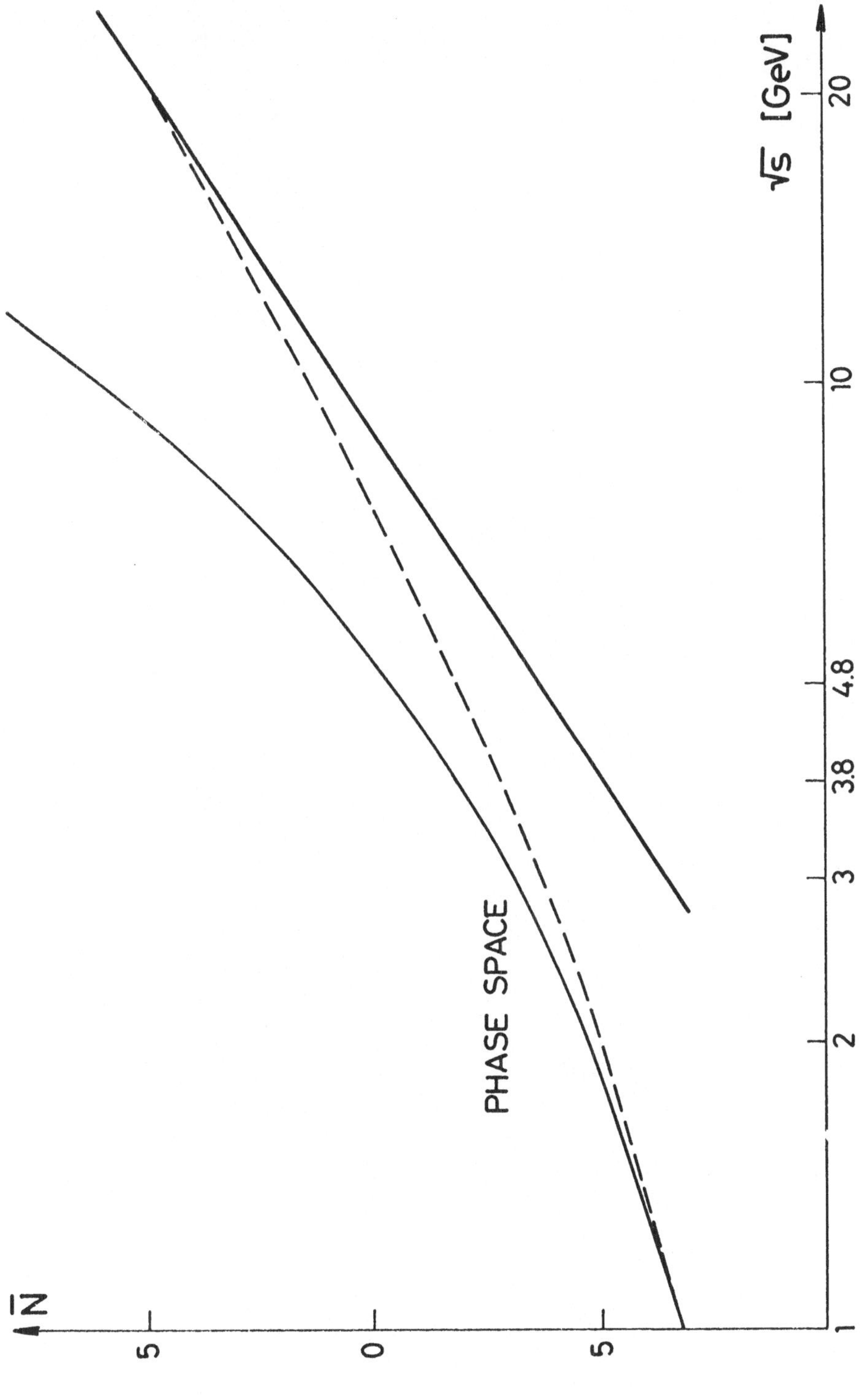

Fig. 9

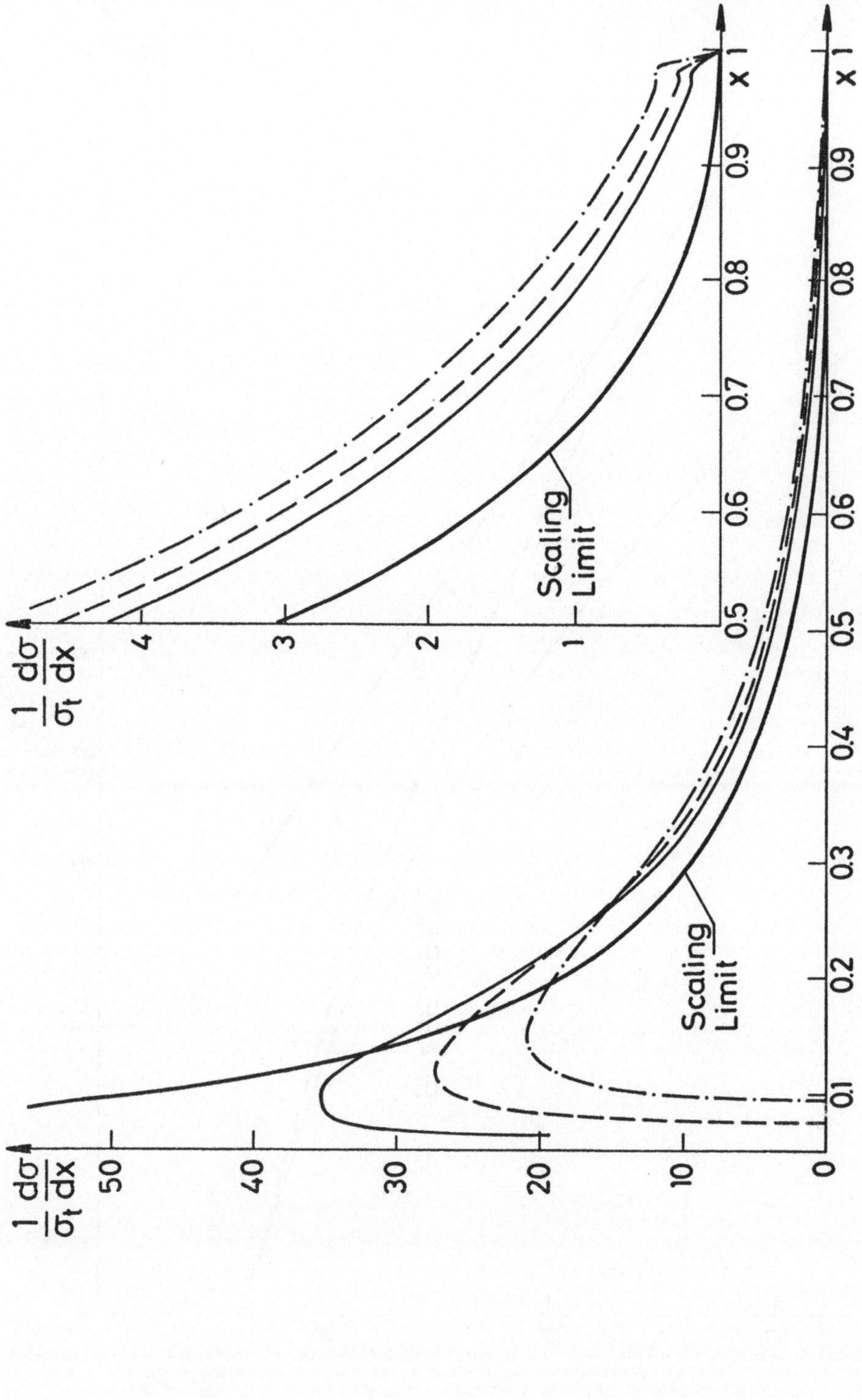
$\frac{1}{\sigma_t}\frac{d\sigma}{dx}$
50
40
30
20
10
0
0.1
0.2
0.3
0.4
0.5
0.6
0.7
0.8
0.9
x 1
Scaling Limit
$\frac{1}{\sigma_t}\frac{d\sigma}{dx}$
4
3
2
1
0.5
0.6
0.7
0.8
0.9
x 1
Scaling Limit

Fig. 10

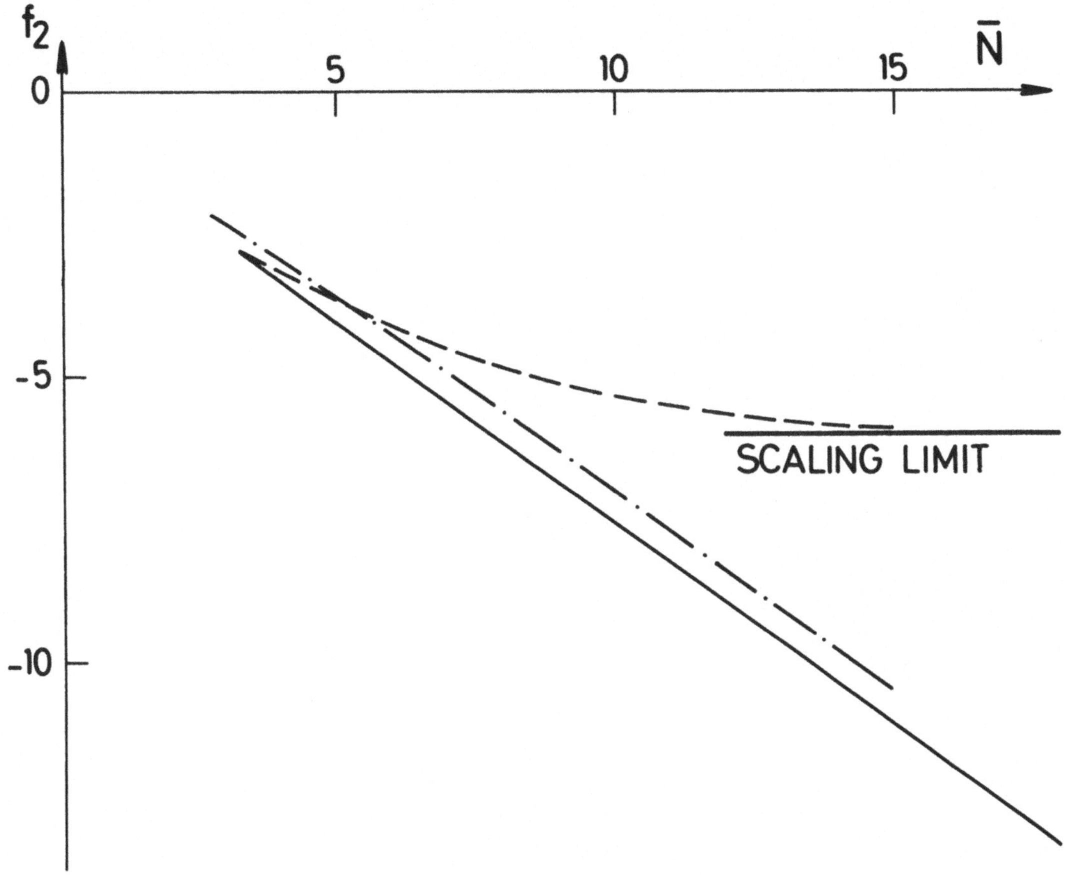

Fig. 11

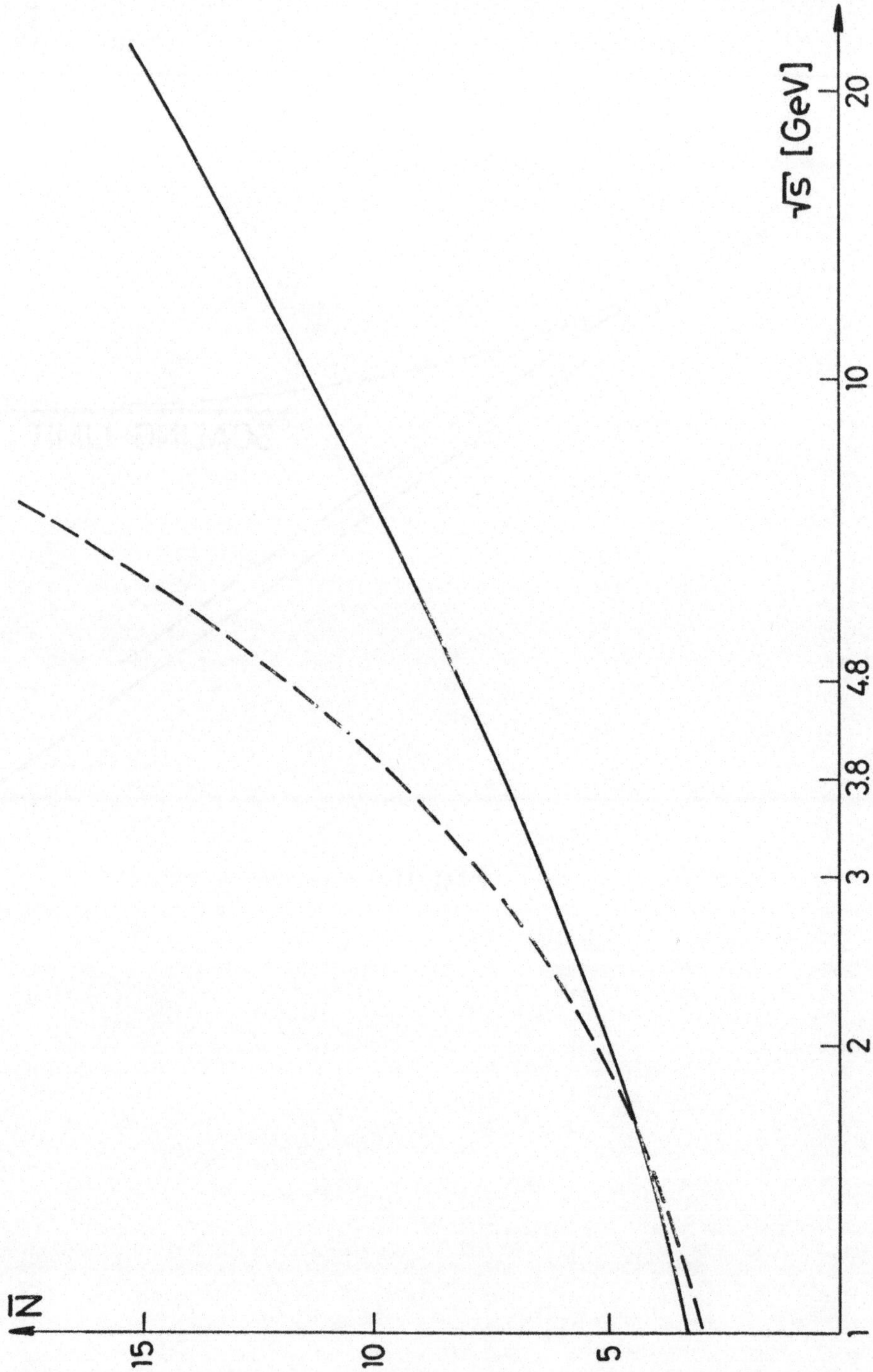

Fig. 12

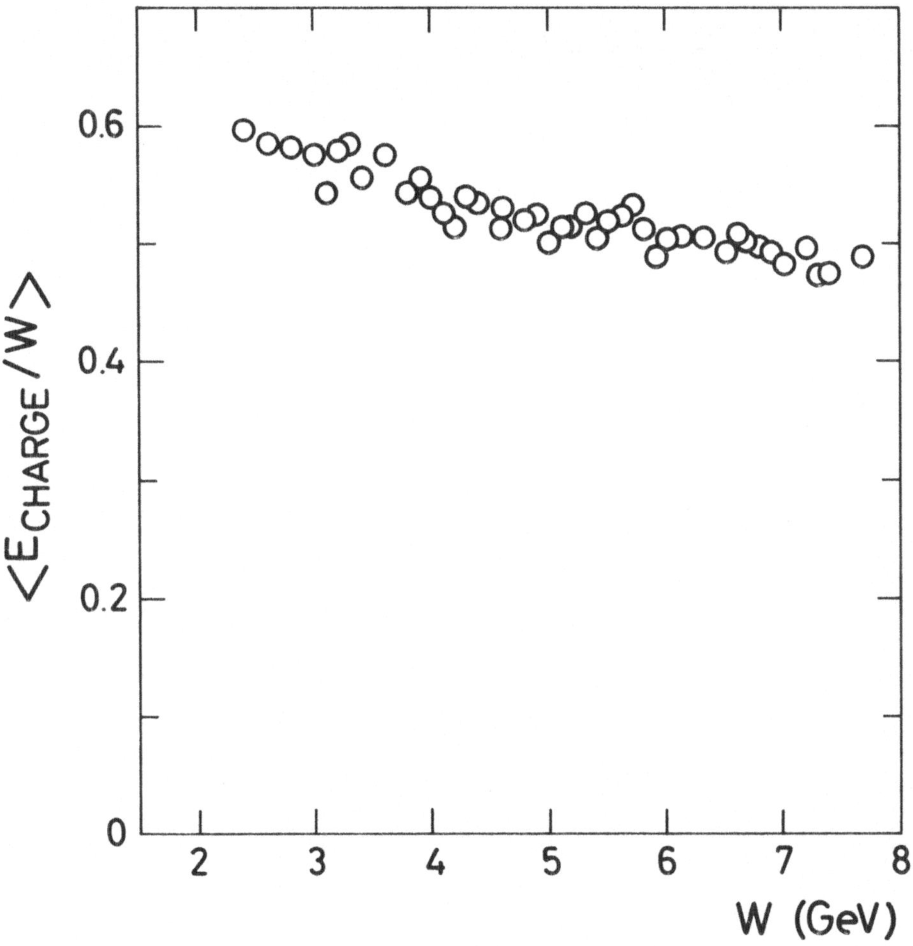

Fig. 13

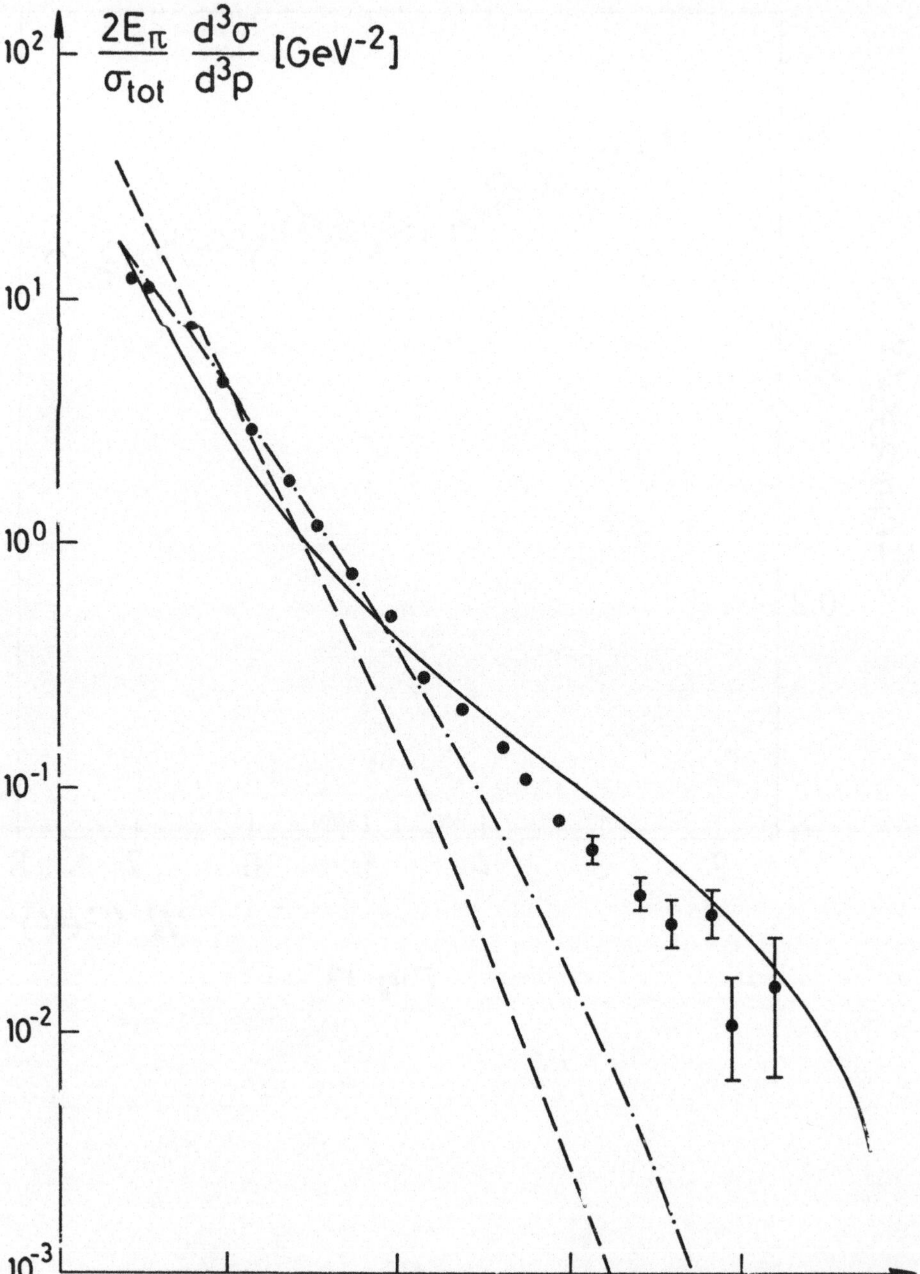

$\frac{2E_\pi}{\sigma_{tot}} \frac{d^3\sigma}{d^3p}$ [GeV^{-2}]
10^2
10^1
10^0
10^{-1}
10^{-2}
10^{-3}
0.5
1.0
1.5
2.0
ω_π [GeV]

Fig. 14

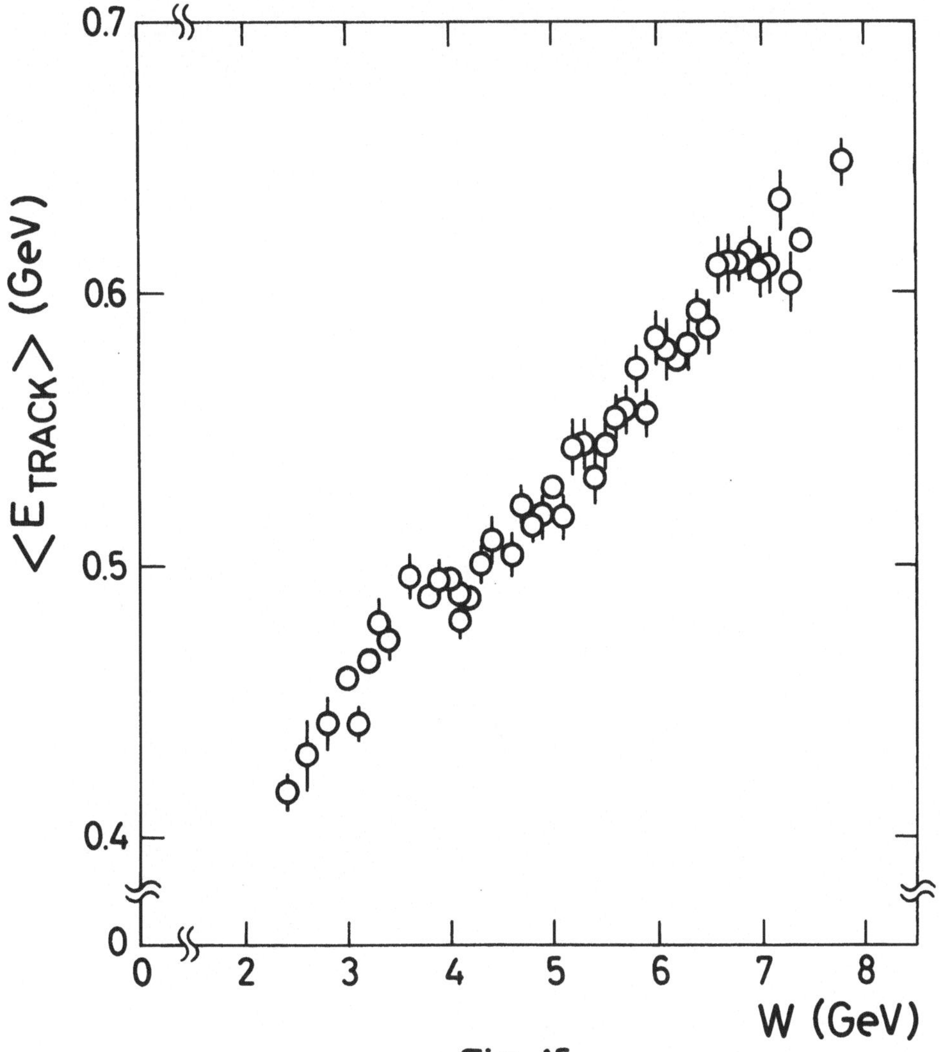

Fig. 15

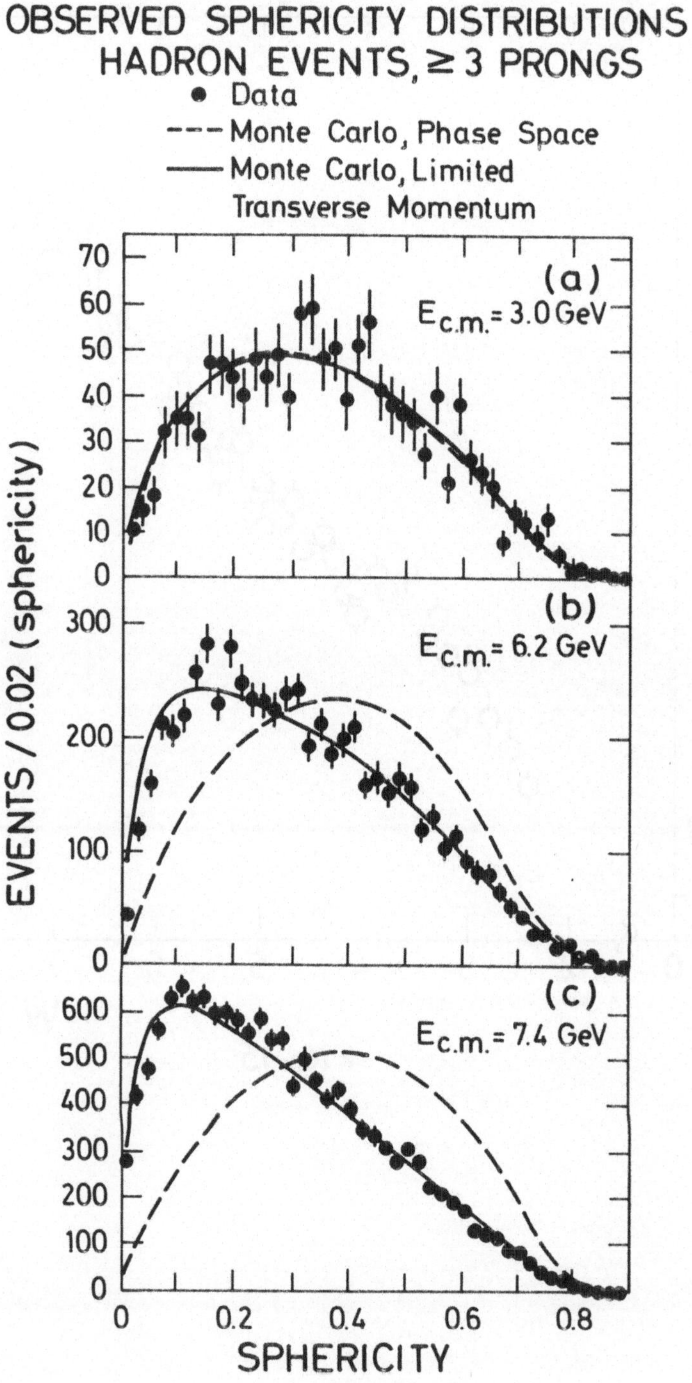
OBSERVED SPHERICITY DISTRIBUTIONS
HADRON EVENTS, ≥ 3 PRONGS
Data
Monte Carlo, Phase Space
Monte Carlo, Limited Transverse Momentum
(a)
$E_{c.m.}$ = 3.0 GeV
(b)
$E_{c.m.}$ = 6.2 GeV
(c)
$E_{c.m.}$ = 7.4 GeV
EVENTS / 0.02 (sphericity)
SPHERICITY
0
10
20
30
40
50
60
70
100
200
300
400
500
600
0.2
0.4
0.6
0.8

Fig. 16

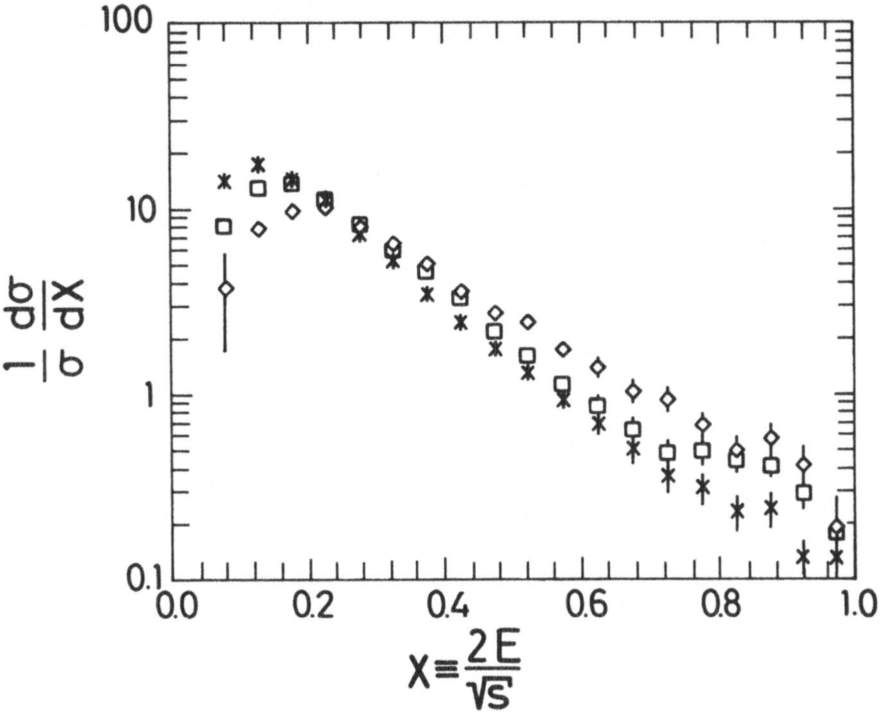
100
10
1
0.1
0.0
0.2
0.4
0.6
0.8
1.0
$\frac{1}{\sigma}\frac{d\sigma}{dX}$
$X \equiv \frac{2E}{\sqrt{s}}$

Fig. 17

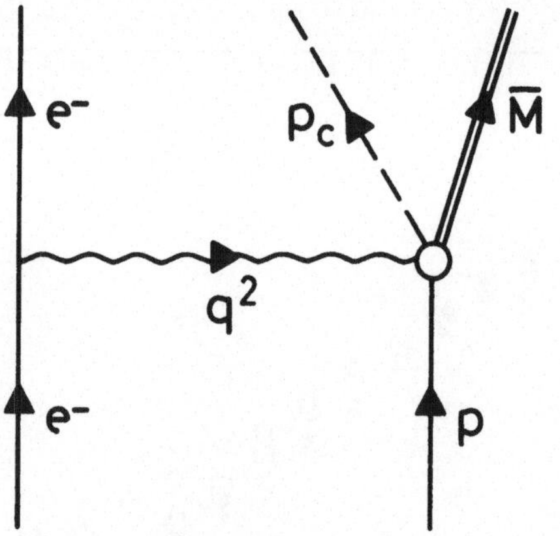

Fig. 18

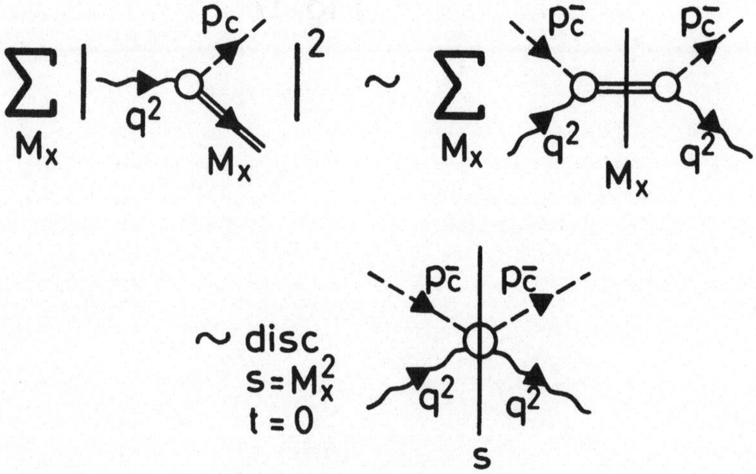

Fig. 19

HADRON FINAL STATES IN DEEP INELASTIC PROCESSES*

J. D. Bjorken
Stanford Linear Accelerator Center
Stanford University, Stanford, California 94305

I. Introduction

These lectures[1] deal mainly with old physics; in particular the description and discussion of hadron final states in electroproduction, colliding beams, and neutrino reactions from the point of view of the simple parton model, such as one finds described in detail in Feynman's book.[2] We first describe the standard parton model formalism and predictions for distributions of final state hadrons in processes $e^+e^- \to$ hadrons, $e^-p \to e^-$ + hadrons, $\mu^-p \to \mu^-$ + hadrons, $\nu p \to \mu^-$ + hadrons, $\bar{\nu}p \to \mu^+$ + hadrons, etc. Once having described in broad terms the predictions and (briefly) the status of the experimental situation we then explore in more detail the various regions of phase space: the so-called fragmentation regions and plateau regions and see how they differ for deep inelastic processes as compared to ordinary processes. Then we consider all this in a more general context which Kogut and I call correspondence.[3] Correspondence is closely related to duality ideas. One argues that for processes in new regions of high Q^2 any hadron distribution is smoothly connected to the regions of low Q^2, and that inclusive processes are always smoothly connected to exclusive processes, as in the Drell-Yan-West connection in electroproduction.[4,5,6] Kogut and I concluded that the smoothest possible behavior consistent with our intuition about exclusive processes, photoproduction processes, or

*Work supported by the Energy Research and Development Administration.

other low-Q^2 processes is that at high Q^2 nothing changes drastically as Q^2 is increased as long as the hadron center of mass energy is held fixed, and as long as the distributions one is talking about are normalized to the total cross section. We will see this in more detail later on. Thus far, the correspondence idea seems to be borne out by the data as well.[7] It also meshes nicely with parton model ideas but does not necessarily require them.

As the second major topic of these lectures we will discuss the space-time evolution of final states in the parton model. This is important, in that it bears upon very basic questions of the dynamics of deep inelastic processes. The central question is why, when quarks are struck by leptons or other currents and one would expect to see them in the final states, one does not and only ordinary hadrons come out. Looking in detail at the space-time evolution of the process, say in electroproduction, from the time when the quark is hit to the time when the final hadrons emerge as asymptotic states one should learn something about the dynamics of confinement. In order to do this we first look at the space-time evolution of ordinary collisions in the parton model framework or the framework of theories possessing only short range correlations in rapidity. Then we look, as an aside, at collisions which involve nuclei as targets or as projectiles. These are of interest these days in high energy hadron nucleus collisions. Nuclei may eventually be important in electroproduction at large ω, where the problems of shadowing or antishadowing are significant, fundamental, and at present in a confused state. Finally we look at the processes $e^+e^- \rightarrow$ hadrons, and deep inelastic electroproduction or neutrino reactions.

II. The Parton Model Picture

In the parton picture for deep inelastic collisions (or, for that matter for any collision), one views an energetic hadron as equivalent to a beam of partons. That is, for the purpose of calculation the incident hadron beam is replaced by a parton beam. The partons are assumed to be point-like constituents within the hadrons. If, as in electroproduction or neutrino reactions, the other incident projectile is a lepton one assumes that the lepton parton interaction is point-like. In other words the parton has no more structure than a lepton would have. Finally, as an independent hypothesis, it is assumed that the struck parton or final state parton of high transverse momentum (the high transverse momentum essentially defines what one means by deep inelastic processes) evolves into hadrons in a manner which is independent of the rest of the environment, i.e., those partons and/or produced hadrons which are distant from the struck parton in phase space. We now examine this picture in more detail.

A. Colliding Beams: $e^+e^- \to$ hadrons

The easiest example of this is in the colliding beam processes such as $e^+e^- \to$ hadrons [or, equivalently, the decay of a heavy intermediate boson, or the process $e^- + \bar{\nu}_e \to$ hadrons, which of course is a little hard to realize experimentally]. In the colliding beam reaction one first of all presumes that the e^+e^- system annihilates through a virtual photon into a parton-antiparton pair, assumed in fact to be quark-antiquark. Then just after the collision, one has a free quark and antiquark which begin to recede from each other. What happens next is less clear. But at much later times the quark and antiquark have been replaced by a system of hadrons, which are assumed to have the following properties.[2,8,9] First of all, relative to the direction of momentum of the originally produced quarks, there is limited p_T for the produced hadrons.

Secondly, there is assumed to be scaling behavior in the longitudinal momentum variable, as given by the following formula

$$\frac{dN_h}{dz} = \frac{\sum_i e_i^2 D_i^h(z)}{\sum_i e_i^2} \tag{2.1}$$

where, in the laboratory frame

$$z = p/p_{max} = p/p_{quark} \quad \text{and} \quad zD_i^h(z) \equiv g_i^h(z) \tag{2.2}$$

Finally, if the energies are high enough (and this means extremely high—at least 30 GeV in the center of mass) there should be a central rapidity plateau. That is, if for a given event one defines the z axis along the direction of the originally produced parton antiparton pair and defines the appropriate rapidity variable, then the rapidity distribution of produced nonleading hadrons should be uniform, just as it is in ordinary reactions. We shall call this flat rapidity distribution of hadrons the "current plateau" to distinguish it from the corresponding plateau region found in ordinary hadron-hadron collisions (which we will call a hadron plateau). The regions near the boundaries of phase space contain the highest momentum hadrons in the event. It is those regions which obey the scaling behavior in Eq. (2.1), and they define the parton fragmentation regions as shown in Fig. 1.

The experimental evidence for the kind of behavior embodied in the above hypotheses is reasonably good. First of all, the longitudinal scaling behavior as in Eq. (2.1) has been checked in the SPEAR experiments.[10] Jet structure has been discovered thanks to the existence of transverse beam polarization. This picture also demands a relatively low multiplicity, logarithmically growing with energy as in ordinary hadron collisions. That is also observed in the colliding beam reactions.

B. Electroproduction and Neutrino Reactions

Next in complexity, with regard to the description of hadron final states, is deep inelastic electroproduction or neutrino reactions. We assume familiarity with the formalism and scaling picture for these processes[11] when hadron final states are summed over. We will first look at the reactions from a relatively naive level, valid for small ω (large x), say $\omega < 10$. We choose to look at the process in the lab frame; in that frame we see first of all that inclusive cross section for deep inelastic electron scattering is given by the formula (for electroproduction)

$$\nu \frac{d\sigma}{dQ^2 d\nu} \cong \frac{4\pi\alpha^2}{Q^4}\left(1 - y + \frac{y^2}{2}\right) \sum_i e_i^2 f_i(x) \tag{2.3}$$

with

$$\begin{aligned} Q^2 &= 4EE' \sin^2 \frac{\theta}{2} = -q^2 \\ \nu &= E - E' \\ x &= Q^2/2m\nu = \omega^{-1} \\ y &= \nu/E \end{aligned} \tag{2.4}$$

In the parton picture, this describes the probability of a fast parton of given type being produced in the collision. Given this ordinary deep inelastic cross section for production of fast partons, we then assume that the produced hadrons evolve from the fast partons in a way that depends only on the nature of the struck parton. Therefore the distribution of hadrons must be very similar to the distribution of the fast hadrons that we just discussed for the colliding beam reaction. Specifically, defining z as equal to the momentum of the produced hadron divided by ν, the virtual photon energy in the laboratory frame, we should have Feynman scaling for the inclusive distribution of fast hadrons measured along the direction of the virtual photon:

$$\frac{dN_h}{dz} = \sum_i \epsilon_i(x)\, D_i^h(z) \tag{2.5}$$

where

$$z = p_{hadron}/\nu = (p/p_{max})_{lab} \tag{2.6}$$

and

$$\epsilon_i = \frac{e_i^2 f_i(x)}{\sum_i e_i^2 f_i(x)} = \text{probability the struck parton is of type i.} \tag{2.7}$$

Secondly, because the u quark has charge 2/3 it should be the quark that is predominantly produced in electroproduction reactions.

$$\epsilon_{u} + \epsilon_{\bar{u}} \geq 2/3 \tag{2.8}$$

Also for large x (small ω) there should be many more u-quarks than $\bar{u}$-quarks, because we are in the valence region. Likewise for neutrino reactions the u-quarks again dominate, because the predominant subprocess for charged-current reactions is a neutrino striking a d-quark, producing a μ^- and a u-quark. On the other hand, for antineutrino induced reactions the d-quarks should dominate:

$$\bar{\nu}_\mu + u \rightarrow \mu^+ + d \quad . \tag{2.9}$$

The probability of finding a hadron of certain type emerging from a quark of a certain type are given by the D functions of Eq. (2.1). For large z (that is, for leading hadrons) we would expect

$$D_u^{\pi^+}(z) >> D_u^{\pi^-}(z) \tag{2.10}$$

Furthermore, from isospin conservation there follows

$$D_u^{\pi^+}(z) = D_d^{\pi^-}(z)$$
$$D_i^{\pi^o}(z) = \frac{1}{2}\left(D_i^{\pi^+}(z) + D_i^{\pi^-}(z)\right) \qquad (2.11)$$

Therefore, for low ω (large x) we expect that at large z the charge ratio (i.e., the ratio of positively charged produced hadrons to negatively charged hadrons) should strongly favor the positives. In neutrino reactions this is seen very clearly.[12, 13]

For electroproduction at very large ω, the distribution of leading hadrons should be the same function as in e^+e^- annihilation into hadrons, because all partons participate with the same weight as in e^+e^- annihilation. There is one exception: at high energies in the colliding beam reactions about half the total cross section is in new physics. The new physics may not be present in electroproduction to the same degree at presently attainable Q^2 (although for $Q^2 \gg 10\ \text{GeV}^2$ it may in fact contribute in the same proportion as for e^+e^- physics). Therefore it may be best to only consider the hadrons produced by old-physics mechanisms in colliding beams, and compare that inclusive distribution with that in electroproduction. Furthermore, to zeroth order of accuracy there should be a plateau structure for small values of z near zero, say $z \ll 0.1$. This however is a more complicated situation than in colliding beams because of the presence of hole fragmentation, a concept we return to later on. However for small ω this is not a complication, and the plateau structure must be the same as the plateau structure which one finds in colliding beams. Finally there should be limited transverse momentum of produced hadrons relative to the virtual photon direction (in the lab frame), just as was in the case for the jets from colliding beams (or for that matter for ordinary processes).

By now the evidence for these properties is in general quite good. In electroproduction there are data[14] from DESY, Cornell, SLAC, and Fermilab which shows approximate scaling of the inclusive distribution of fast hadrons, at least at the factor-of-two level and probably somewhat better. The recent Fermilab data from the 15-foot bubble chamber,[12, 13] with incident neutrinos and antineutrinos show a very large positive charge ratio for leading (large z) hadrons in νp interactions at high Q^2 and ν. In electroproduction, at SLAC energies (10 - 20 GeV) with moderate ω, one sees a positive charge excess for leading hadrons, and in some circumstances even when the target is a neutron. This is in fact in line with parton model expectations. At higher energies (~ 100 GeV) the FNAL data[15] indicate much less, if any such effect. That data is at considerably larger ω, and a smaller excess is in fact expected, but it is not clear that the situation is good from a quantitative point of view.

Additional evidence that at least semiquantitatively the parton model ideas are working is that the mean multiplicity is roughly independent of Q^2 at fixed W^2, W^2 being the total hadron energy in the final state. This was first seen in a nice experiment from Cornell[16] for $Q^2 \lesssim 8\ \mathrm{GeV}^2$. The recent data from Fermilab neutrino reactions in the 15-foot bubble chamber[12] have extended this observation to much higher values of ν and also of Q^2. At the highest energies, the mean Q^2 is ~ 20 or 30 GeV^2, and the multiplicity seems to be typical of that found in ordinary hadron processes. Finally there seems to be some evidence that the mean transverse momentum of produced hadrons is not too different from that in ordinary processes. If the p_T distribution is parametrized as

$$\frac{dN}{dp_\perp^2} \sim e^{-bp_\perp^2} \tag{2.12}$$

then one finds $b \sim 4$ to 6 GeV^{-2} in experiments from DESY, from Cornell, from SLAC, and from Fermilab.[14, 15] There is only one exception experimentally: Preliminary data[17] from the Santa Barbara group, who measure π^o electroproduction at SLAC energies (10 - 20 GeV lab energy). They find for Q^2 the order of 5 - 7 GeV^2 a much lower value of b, $\lesssim 1$ GeV^{-2}. This result should soon be checked for the charged particle distributions, which do not yet reach that far in Q^2.

C. Fragmentation Regions

Now we turn in yet more detail to the properties of the plateau and fragmentation regions in the deep inelastic processes.[2, 18, 19]

As we mentioned, for the process $e^+e^- \rightarrow$ hadrons one should be able to define a jet axis along which hadron momenta tend to be large and scale with the incident beam energies and transverse to which the transverse momenta are limited. The rapidity distribution of produced hadrons relative to the jet axis should have the behavior shown in Fig. 1. At the boundaries of the phase space there are the parton fragmentation regions and then in between (if the energy is high enough) the current plateau.

Next, in preparation for examining electroproduction processes, we look at an ordinary hadron-hadron collision, which has a similar kind of structure. Let us take as an example

$$\rho + N \rightarrow \text{hadron} + \text{anything} \quad .$$

The distribution in rapidity of produced hadrons consists again of three regions. The highest rapidity particles are found in the rho fragmentation region. Then there is a hadron plateau of intermediate rapidities. Finally the slow particles in the laboratory frame are fragments of the proton and belong to the proton fragmentation region.

Now for photoproduction of hadrons by real photons, vector dominance would imply a similar distribution as in ρ-nucleon interactions. We may now replace the real photon by a virtual photon, and look at large ω first. Then we find, first of all, that the nucleon fragmentation region and the hadron plateau regions should still be present because all we are changing are the properties of the incident projectile. Therefore by the hypothesis of short range rapidity correlation in ordinary processes, the other regions should not be affected by a change in projectile properties.

However the virtual-photon fragmentation region is different from the real-photon fragmentation region. Even its length in rapidity space changes. Why is this so? It is because at small ω there is no ordinary Pomeranchuk trajectory exchange possible, i.e., no hadron plateau can exist. The Pomeranchuk trajectory has properties similar to a ladder graph in a superrenormalizable theory. We can ask, "Under what circumstances can we exchange ladders in deep inelastic processes?" If we look, for example, at elastic processes, such as virtual $\gamma + N \rightarrow \rho + N$ we can ask for fixed energy at what value of q^2 can we no longer coherently produce the rho. This occurs when the minimum momentum transfer

$$\Delta_{min} = \frac{m_\rho^2 - q^2}{2\nu} = \frac{m_\rho^2 + Q^2}{2\nu} \approx Mx \tag{2.13}$$

exceeds a few hundred MeV. And this occurs when ω becomes smaller than some fixed amount, say 3 or 10 or so. Therefore, when ω is small the hadron plateau must disappear. It follows that its length is of order $\log \omega$, as shown in Fig. 2. This leaves for the photon fragmentation region a length in the rapidity space of order $\log Q^2$. This is satisfying inasmuch as the length of the photon fragmentation region should be only a function of Q^2 and therefore we find consistency. These considerations do not require the parton model.

However, assuming the parton model, we find more detailed structure within the virtual photon fragmentation region. In particular, the photon fragmentation region itself divides up into three pieces.[18] For the largest y there is the parton fragmentation region, which is familiar from the colliding beam discussion. Adjacent to it is the current plateau. However, there must be a transition region between the current plateau and the hadron plateau, which we call the hole fragmentation region. The reason for this name is that it is the location in phase space of the original parton before it was struck. This is shown in Fig. 3.

We can see this picture is sensible. In the limit of small ω, the hole fragmentation region merges with the target fragmentation region, leaving the current plateau to separate target and parton fragmentation regions. In the limit of small Q^2, the hole fragmentation region merges into the parton fragmentation region, leaving the hadron plateau to separate the target fragmentation from the photon fragmentation region. An interesting question is whether the rapidity distribution of hadrons in the hadron plateau is equal to the rapidity distribution of hadrons in the current plateau. It is not clear that this should be so. In fact the mean p_T of the produced hadrons may not be the same either. However, from experiment it appears that the height of the hadron plateau and the current plateau are approximately the same, although perhaps the mean p_T may be somewhat larger for the current plateau than it is for the hadron plateau.[13]

III. Vector Dominance

Further insight into these fragmentation regions can be obtained by looking at large-ω electroproduction from the point of view of vector dominance. After all, this concept works well for high energy real-photon physics, and we might expect a generalization to exist for virtual photons as well. This is especially the case in the light of the arguments of Ioffe[20] that large longitudinal distances (proportional to ω) are important in electroproduction at high Q^2 and ω. In considering the vector dominance approach we use the "diagonal approximation" shown in Fig. 4, assuming the intermediate vector states m and n in the forward virtual Compton amplitude to be the same.[21] A simple way of calculating this vector dominance contribution for real photons was given by Gribov[22] several years ago: the probability for the photon to be absorbed by the nucleon is given by the probability that the incident photon fluctuates into a hadron intermediate state, multiplied by the probability that the hadron interacts with the nucleon. These factors are shown below:

$$\begin{aligned} \sigma_{\gamma N} &= (\text{probability } \gamma \text{ is hadron}) \times (\text{probability hadron interacts}) \\ &= \left(\int_0^{\sim s} \frac{dm^2}{m^2} \frac{\alpha}{3\pi} R(m^2) \right) \times \sigma(m^2) \qquad (3.1) \\ &\cong (1 - Z_3)\, \sigma(m^2) \end{aligned}$$

where Z_3 is the hadronic charge renormalization constant (the probability a physical γ = bare γ), and where R is again the familiar ratio that appears in the colliding beam cross sections:

$$R = \frac{\sigma(e^+e^- \to \text{hadrons})}{\sigma(e^+e^- \to \mu^+\mu^-)} \qquad (3.2)$$

For virtual photons we go through the same calculation as schematically shown below

$$\sigma_{\gamma N} \sim \sum_n <0|j^\mu \epsilon_\mu |n> \frac{1}{\Delta E_n} \sigma_n \frac{1}{\Delta E_n} <n|j^\mu \epsilon_\mu |0> \qquad (3.3)$$

We see the only change is in the energy denominators. For real photons they are $1/m^2$, while for virtual photons they become $1/(Q^2+m^2)$. Therefore the absorption cross section for a transverse virtual photon is given by the previous formula with this modification made:

$$\sigma^T_{\gamma^* N} = \frac{\alpha}{3\pi} \int_0^{\sim s} \frac{dm^2 \, m^2}{(Q^2+m^2)^2} R(m^2)\, \sigma(m^2) \qquad (3.4)$$

However, if we follow our intuition and assume that the absorption cross section for the virtual hadron state n is a constant σ_h independent of n, we reach a disaster:

$$\sigma_{\gamma^* N}(Q^2, \nu) \sim \frac{\alpha}{3\pi} \bar{R} \sigma_h \log \omega \qquad (3.5)$$

Because the absorption cross section σ_h contains an intrinsic dimension (essentially πR^2), there is no scaling of the deep inelastic cross section. Instead of σ_T falling as $1/Q^2$, it behaves roughly like a constant. What went wrong and what must be done to remedy the situation?

There are two main options. The first has been considered in particular by Greco,[23] Schildknecht and Sakurai,[24] and is simply that the absorption cross section for a massive virtual intermediate state is smaller than that for a not-so-massive intermediate state, and falls off like $1/m^2$. Suppose this is the case. Given that the e^+e^- annihilation process yields two jets in the final state, then vector dominance would imply that these jets would also be electroproduced at large ω. Then we should eventually see double jets in electroproduction at large ω. Also, because the inclusive distribution of the hadrons within the jets obeys scaling, the leading hadrons should contain a finite fraction of the $p_\perp$ of the jet. The mean $p_\perp$ of such a jet is of order its mass, which will typically be $\sim\sqrt{Q^2}$. Therefore it is predicted that the mean $p_\perp$ of leading hadrons ($z \sim 0.5$ seems optimum) should be $\sim\sqrt{Q^2}$, if the double-jet option is correct.[25] This is actually the option that is chosen by quantum electrodynamics, as shown by Cheng and Wu.[26] There is a second option. It is that these jets which appear in the virtual intermediate state must be aligned along the virtual photon direction.[27] The argument proceeds as before using vector dominance. But instead of the preceding assumptions, it is assumed that if the transverse momentum of the jets is large (in other words that the partition of longitudinal momentum is balanced between the two jets), then the cross section for absorption of such a system is very small. However, if the transverse momentum of the jets is small relative to the virtual photon direction (with consequent imbalance in the partition of longitudinal momentum), then they are absorbed by the target in the usual way, with a geometrical absorption cross section.

In their own rest frame the virtual-photon jets must have a roughly isotropic angular distribution, because they were produced by a spin 1 photon. Therefore the probability that the transverse momentum of the jets is small is $\Delta\Omega \sim \theta^2_{max} \sim (\langle p_\perp\rangle/\sqrt{Q^2})^2 \sim \mathrm{const}/Q^2$. It follows that in Eq. (3.4) instead of $\sigma(m^2)$ appearing we have to replace it by $\sigma(m^2)$ times the probability that the jet be aligned, and that product is of order $1/m^2$.

What does this option accomplish? First of all, the deep inelastic scaling survives. Secondly, one obtains low-transverse-momentum hadrons in electroproduction at large ω as the parton model would suggest. Finally the hole fragmentation region and the parton fragmentation region can be identified (using vector dominance) from the parton fragmentation regions in collding beams.

I find it very satisfying that the vector-dominance and parton-model ideas can be made to interlock in such a self consistent way. However, one must be on the lookout experimentally for high-p_T leading hadrons in electroproduction, especially at large ω, in the light of these two competing production mechanisms.

IV. Correspondence

The use of the word correspondence I discuss goes back to the idea used by Bohr, in inferring properties of quantum physics from the demand that quantum mechanics must have a smooth limit to classical physics as $h \to 0$. Knowing what the classical physics regime looks like puts constraints on how the quantum mechanical regions behave. In the present case we want to explore unknown regions in phase space—in particular the deep inelastic high Q^2 regions. The physics in these unknown regions most likely has smooth connections with the physics in the known regions. An example of this is duality in strong interactions, where the resonance region is connected smoothly to the continuum region of the Regge trajectories at higher energies. In electroproduction there is the example of the inclusive-exclusive connection of Drell, Yan,[4] West,[5] Bloom, Gilman,[6] and others which connects the shape of the deep inelastic scaling curve for the continuum contribution (extrapolated into the resonance region) to the behavior of the resonance contributions. A simple way of obtaining the above connections is to demand that neither resonance nor continuum should dominate in the resonance region: The ratio of resonance signal to continuum "noise" should be of order one under all circumstances. This is reasonable inasmuch as, if one is given some kind of continuum production mechanism which can be extrapolated into the resonance region, there are a limited number of open channels and partial waves that can comprise it. The amplitude in a resonant channel will be enhanced by some finite factor; hence one must have the ratio of the resonance signal to the total extrapolated inclusive background in the resonance region in general of order 1.

What Kogut and I tried to do for electroproduction and colliding beams processes[3] was to look at all the general connections of this nature that we could think up: inclusive vs. exclusive, a smooth connection between low Q^2 and high Q^2, smoothness in looking at the processes from both the point of view of vector dominance or a parton picture, smoothness in connecting low-s resonances with the high-s Regge behavior. We found that if one demanded all of these connections to be smooth we were almost forced into a unique picture of how the hadron distributions in the deep inelastic processes should look. Although we arrived at this in a rather roundabout way, the answer is in fact direct and very simple. It can be stated as follows:

The smoothest solution for hadron distributions for electroproduction and neutrino reactions at high Q^2 consistent with the known or expected behavior of the processes in the boundary regions (e.g., exclusive processes at high Q^2 and inclusive at low Q^2) is that the normalized distributions of almost anything are independent of Q^2 if W^2 is kept fixed. This assertion is meant to be taken only at the factor of 2 level of accuracy; i.e., there is no systematic behavior with Q^2 (like Q^2 to some power) in any normalized quantity. A second inference Kogut and I found is that the hadron plateau height should be approximately the same as the current plateau height. Finally, at large ω the photon fragmentation region should contain hadron distributions which are the same as those found in e^+e^- annihilation.

There is not the time (and it is probably not so interesting) to go through this logical route that Kogut and I took, but only to point out a couple of surprises that came along the way. First of all, there was a surprise in the behavior of two-body exclusive channels. As an example consider exclusive electroproduction of a π^+ from hydrogen: $\gamma^* p \to \pi^+ n$. If we go to the real-photon limit at $Q^2=0$, we expect a Regge-behaved cross section

$$\sigma(s) \sim s^{2\alpha-2} \qquad (4.1)$$

which is (according to ordinary duality) on the average true for all values of $s=W^2$. The prediction of correspondence is that at large Q^2 we should have essentially the same behavior provided we normalize the exclusive cross section to the total cross section with the same value of Q^2 and s. Is this reasonable? For very large ω the Regge behavior should be still good, and only for large ω. This implies that the cross section should go as follows:

$$\frac{\sigma_{tot}(\gamma^* p \to \pi^+ n)}{\sigma_{tot}(\gamma^* p \to \text{all})} \sim s^{2\alpha-2} \equiv (W^2)^{2\alpha-2} \tag{4.2}$$

Hence

$$\sigma_{tot}(\gamma^* p \to \pi^+ n) \sim \frac{1}{Q^2}\left(\frac{W^2}{Q^2}\right)^{2\alpha-2} (Q^2)^{-2+2\alpha} = \frac{1}{Q^2}(\omega')^{2\alpha-2}(Q^2)^{-2+2\alpha} \tag{4.3}$$

The first two factors can be considered a "pointlike" cross section for producing the π by a Reggeon exchange. The γ-π-Reggeon vertex is given a pointlike value and the Regge factor is $\omega' \sim s/Q^2$ to a power rather than s to a power. This leaves the last factor in Eq. (4.3) to be interpreted as the square of the form factor of the Reggeon-pion vertex. This implies that the form factor has the behavior

$$F_\pi(Q^2) \sim (Q^2)^{-1+\alpha} \tag{4.4}$$

Therefore we find a connection between the Regge intercepts that couple to the pion and the transition form factor of the pion to the various mesons on the Regge trajectory. Evidently this result isn't completely trustworthy quantitatively. But at least qualitatively we see that if the Regge intercept were at zero the form factor would have a monopole behavior $\sim 1/Q^2$. On the other hand, if we look at backward photoproduction with the exchange of a baryon, the Regge intercept would be of the order of $-1/2$ and the form factor would fall more rapidly with Q^2:

$$F \sim (Q^2)^{-3/2} \tag{4.5}$$

If it were possible to be more careful with spin and had better control of the dynamics beyond just this correspondence argument we could expect these to be precise results. But even so the exponents in the above formulae are probably uncertain to no more than 1/2 a unit or so. And just qualitatively, we find the conclusion that baryon form factors should fall off with Q^2 faster than meson form factors because their trajectories lie lower than the meson trajectories. This connection of the behavior of Reggeon intercepts with form factor behavior is found in the massive quark model of Preparata.[28] But beyond that there is no connection with ordinary Reggeon dynamics. We now turn to small ω. For fixed small s we have from correspondence

$$\frac{\sigma_{tot}(\gamma^*p \rightarrow \pi^+n)}{\sigma_{tot}(\gamma^*p \rightarrow \text{all})} \sim s^{2\alpha-2} \sim \text{const } 0(1) \text{ in the resonance region} \tag{4.6}$$

and this leads back to the Bloom-Gilman type of relationship because at small s the exclusive channels contain a finite fraction of the total cross section. Therefore the situation is consistent and the correspondence arguments work. We can summarize the behavior of this particular exclusive channel at large Q^2 as follows: for small ω (say $\lesssim 3$) we get a total exclusive cross section behavior as

$$\sigma_{tot}(\gamma^*p \rightarrow \pi^+n) \sim \frac{s^{2\alpha-2}}{Q^2}(\omega'-1)^3 \sim \frac{s^{2\alpha-2}}{Q^2}\left(\frac{s}{Q^2}\right)^3 \sim \frac{s^{2\alpha+1}}{(Q^2)^4} \tag{4.7}$$

For very large ω we find at fixed s a slow falloff with Q^2, $\sim Q^{-2}$. The net behavior is that roughly Q^6 times the exclusive cross section should have scaling behavior, as shown in Fig. 5.

We found another surprise. There is an argument why the hadron plateau should be approximately the same as the current plateau. This uses the assumption of negligible correlation between the hadron secondaries. Then the distribution in multiplicity may be used to connect the exclusive process with the total cross section. The exclusive process provides the low $\bar{n}$ tail of the multiplicity distribution, and is presemed to join smoothly with the bulk of the distribution. We choose large ω (as in Fig. 3a) and ask for the cross section for producing n_h hadrons in the hadron plateau and n_c in the current plateau. This will be given by the usual formulae as follows:

$$\frac{\sigma(n_h, n_c)}{\sigma_{tot}} \sim \left(\frac{\bar{n}_h^{\,n_h} e^{-\bar{n}_h}}{n_h!}\right)\left(\frac{\bar{n}_c^{\,n_c} e^{-\bar{n}_c}}{n_c!}\right) \tag{4.8}$$

The exclusive process can be read off from this formula and is given by

$$\begin{aligned} \frac{\sigma_{excl}}{\sigma_{tot}} \sim \frac{\sigma(0,0)}{\sigma_{tot}} &\sim e^{-(\bar{n}_h+\bar{n}_c)} \\ &\sim e^{-\left(c_h \log\omega + c_c \log Q^2\right)} \\ &\sim s^{-c_h}\,(Q^2)^{-(c_c-c_h)} \end{aligned} \tag{4.9}$$

However correspondence demands that the ratio of the exclusive cross section to the total cross section should be independent of Q^2:

$$c_c = c_h \;(\pm 1??) \tag{4.10}$$

This argument clearly isn't very precise but may at least be indicative that if one were to try to make the hadron plateau height and the current plateau height wildly different that there could be trouble in joining the multiplicity distributions smoothly onto the exclusive limits.

In any case we can draw the following conclusions from this line of argument. Most importantly, the correspondence idea plus the hypothesis of short range correlation in rapidity for low Q^2 processes, and power-law dependencies on Q^2 (and a little bit more now and then) implies the same qualitative picture of hadron final states as the parton model itself. In other words, it may well be that the predictions may have considerably more generality than that of the parton-model. Under such circumstances, can one therefore claim that the experimental support of the general picture of hadron final states which we have discussed really implies experimental support for the parton picture? Probably the strongest specific support for partons as opposed to the more general ideas is the existence of the jets in colliding beams and, most importantly, that their angular distribution is the same as the angular distribution for the μ pairs, indicating spin 1/2 parton parents. In addition the strongly positive charge ratios in electroproduction (and also in the neutrino reactions) at low ω and high hadron momentum are also strong support. However, whatever the ultimate dynamical picture turns out to be, I believe that the correspondence technique is general. Even if our views of dynamics of hadron states change, there still should be the same correspondence connections between various regions and types of processes as we have discussed here. The specific predictions of course would be different.

V. Space-Time Description of High Energy Processes

The first question to ask about this subject is "Why bother with it?" For someone who was raised in the shadow of the Berkeley bootstrap school, where the S matrix in momentum space contains all of physics, this is not an idle question. However I think there are significant reasons for bothering. First of all the importance of large distances and long time intervals at high energies is well established. This goes back as far as the Landau hydrodynamic model[29] and the work of Landau, Pomeranchuk, and others[30] starting in the 1950's. Therefore one should be able to map the geography of these high energy reactions on a distance scale greater than or the order of a fermi, and this geography should be largely independent of the dynamical details. Indeed models as different as the short-range-correlation models or parton models and the Landau model give very nearly the same space-time evolution.

Secondly because of these large distance and time scales at high energies, nuclear effects become very interesting. By putting additional nuclei downstream of the collision site one probes the structure of fragments of the collision at times short compared to the times required for them to reach their asymptotic configuration of free hadrons.

Third, the problem of quark confinement is relevant. The problem here is not why quarks remain confined in deep inelastic interactions (especially the colliding beam reaction $e^+e^- \to$ hadrons) but how they remain confined. Again it should be possible to trace out the geography of confinement on the large distance and time scales (if indeed such large distance and time scales can be shown to be relevant) for the deep inelastic processes as well as for ordinary collisions. In order to be reasonably specific we shall whenever possible again assume short-range-rapidity correlations only; that is, the parton, multiperipheral, etc. type of picture. However the main results probably have a generality beyond these considerations. We shall first look at ordinary collisions, and then at the effects of the collision occurring in nuclear matter (and therefore the currently interesting problems of multiplicity and inclusive distributions in nucleon-nucleus collisions and nucleus-nucleus collisions). Equipped with this space-time picture of ordinary processes, we will attack colliding beams and then electroproduction, first at low ω and then at high ω. As might be expected, with five different regions in rapidity space for high-ω electroproduction, this turns out to be the most subtle and interesting of the processes with which we deal.

Another way of seeing the need for studying the geography of the final states in space time is by looking at the nonrelativistic prototype of deep inelastic scattering with confinement, that of the scattering of an electron from a single charged particle in a potential well for which $V(x) \to \infty$ as $|x| \to \infty$. The structure function W for this case is a sum of contributions of the square of transition form factors from the ground state to the various discrete excited states. It obeys a scaling law of the following form[31]

$$qW(q^2, \nu) = F\left(\frac{\nu - q^2/2m}{q}\right) \tag{5.1}$$

where

$$W(q^2, \nu) = \sum_n \left| \int d^3x \, \psi_n^*(x) \, e^{i\vec{q}\cdot\vec{x}} \, \psi_0(x) \right|^2 \delta(E_n - \nu) \tag{5.2}$$

provided we make a coarse-grained average in energy over a group of excited states n. Thus for computing the scaling function W only the region of space where the ground state wave function is nonvanishing is important. The exact levels n can be replaced with free-field levels because the kinetic energy of excitation ν is large compared with the variation in potential energy ΔV over the region of space where the ground state wave packet is nonvanishing. However if we want to know about the final states we must know the behavior of the wave functions ψ_n all the way out to the turning point. Equivalently we must follow the time dependence of the ground state wave function multiplied by $e^{i\vec{q}\cdot\vec{x}}$. It is a wave packet of approximate momentum $\vec{q}$ which is distributed into the final states n. The problem then becomes "What is $\psi(x,t)$?" It will be a packet bouncing back and forth between the classical turning points. Therefore the description of the state becomes semiclassical and also dependent on rather large distances compared to what is required for the structure function W.

A. Structure of a Hadron

First of all, in order to describe ordinary collisions from a short-range-correlation or parton picture, we must have a description in space-time of an individual hadron. For a proton at rest, this is familiar: it is just the nonrelativistic quark model, with 3 quarks confined to a region of the order of a cubic fermi, which have certain levels of excitation with a level spacing typically a few hundred MeV. If one were so naive as to talk about the rapidity distribution of the quarks in the proton when the proton is at rest, it would of course just be concentrated in a low rapidity region, say $|y| \lesssim 1$-2. Naively, we expect that once we know what the wave function of the proton is in the rest frame this would suffice to describe the proton when it is accelerated or when it has high momentum. We just boost the proton; the spherical volume in which the 3 quarks resided would turn into a flattened, pancake shaped region with a thickness varying as $1/p$, where p is the momentum of the proton ($p >> 1$ GeV). Furthermore the excited levels of this system will all be at large energy but nevertheless for given large fixed p the level density would be higher. This follows simply from kinematics; if $E_n = \sqrt{p^2+m_n^2}$, then the spacing of the levels for a system of given total momentum is

$$\Delta E_n = \frac{\Delta m_n^2}{\sqrt{p^2+m_n^2}} \sim \frac{\Delta m^2}{p} = 0\left(\frac{1}{p}\right) \tag{5.3}$$

If for example $\gamma \sim 3$ the level spacings instead of being 300 MeV for a proton at rest would be reduced to the order of 100 MeV. This increase of level density with increasing momentum will turn out to be important in following through the dynamics of the high energy collision. In the boosted frame the rapidity distribution of the quarks within the proton is shifted, just displaced to higher values, $\sim \log \gamma$ compared to what they were for a proton at rest.

However, according to the parton picture, this isn't the whole story. In addition to the original three partons we must also consider vacuum fluctuations, let us say quark-antiquark pairs. The important vacuum fluctuations which couple to the boosted proton and which will be relevant for a description of high energy collisions are of relatively low energy, with the excitation energy of the order a few hundred MeV. We will assume that there is a hierarchy of such fluctuations with level spacings of the order of again a few hundred MeV. These low momentum vacuum fluctuations can couple to the boosted proton, provided the proton does not have too much momentum, let us say 3 GeV or less. The moving proton then consists of this conglomerate of an uncontracted vacuum fluctuation coupled to the Lorentz-contracted pancake of original three quarks. Now let us boost to a momentum of the order of 9 GeV. The vacuum fluctuation attached to the proton will now get Lorentz-boosted and Lorentz-contracted as well as the original 3-quark system. Furthermore its level density increases by a factor ~ 3. However this composite system of vacuum fluctuation and original

3 quarks again can couple to uncontracted vacuum fluctuations which have low momentum in this new laboratory frame, with again the level spacing of a few hundred MeV. In this way we can generalize: for a very energetic proton with momentum p we will have the original 3 quarks plus a number (of order log p) of vacuum fluctuations of geometrically decreasing momentum which are sequentially coupled to the original 3 quark system. The lowest-momentum set of vacuum fluctuations (the wee fluctuations) will be uncontracted. A vacuum fluctuation with momentum $\sim p'$ will have a level density inversely proportional to p' and a thickness in the longitudinal direction inversely proportional to p'. The rapidity distribution of the partons comprising the system of coupled vacuum fluctuations is assumed uniform. Thus the highly energetic proton becomes a quite complex object whose description depends upon its momentum. These extra additional vacuum fluctuations comprise the dx/x spectrum of partons of Feynman. Suppose we were to go back to the laboratory frame. Where do all these extra vacuum fluctuations go? We didn't have them there when we started, but suddenly we have attached them. They will occupy very large longitudinal regions as we undo the Lorentz boost in returning to the lab frame. However, the important feature of the vacuum fluctuations so obtained is that they will have a very low level density. The boost now has the effect of decreasing the level density. The first vacuum fluctuation which attached to the proton may have a first excited level ~1 GeV above its lowest state, whereas other vacuum fluctuations which occupy even larger regions of longitudinal configuration space have even larger excitation energies. Thus these excitation energies are too high to be relevant for ordinary spectroscopic processes involving low excitations and low energies. If there is a high energy projectile available, then they may be relevant and in fact part of the description of the

collision. However as we shall see below it is then more natural to associate these fluctuations with the incoming projectile.

Let us try to summarize all this. First of all we observe that rapidity is a concept which can be used <u>both</u> in momentum space and configuration space. In configuration space it measures the typical longitudinal extent of the system involved; the higher the rapidity the smaller the longitudinal extent as a consequence of Lorentz contraction. For a proton of very high momentum p, we assume the phase space of the partons comprising that proton can be broken up into approximately n cells ($n \sim \log p$). Each cell is labeled by its rapidity y. In the cell with rapidity y we have

Momentum $p \sim e^{y}$

Energy $E \sim e^{y}$

Level spacing $\Delta E \sim e^{-y}$

Natural time scale $\tau \sim e^{-y}$ (time dilation)

Thickness (or lag λ) $\sim e^{-y}$ (Lorentz contraction)

There is of course some dynamics involved in this picture. The guiding principle for such a dynamics is the hypothesis of short range correlation in rapidity: (i) couplings should exist only between cells which are neighboring in rapidity, (ii) the couplings between such cells are weak enough so that the concept of levels in the individual cells makes sense, and (iii) in collisions excitation of levels are possible only in those rapidity-cells which are in the same region of phase space (that means <u>both</u> momentum and configuration space). Finally no large amount of momentum or energy is exchanged in any frame at any stage. The subprocesses are as soft as possible. Therefore the flow and exchange of momentum is always $\lesssim 1$ GeV per fermi of elapsed time. [We are as usual letting c=1.]

B. Hadrons in Collision

Consider a high energy hadron-hadron collision in the overall center-of-mass frame. Before the collision there is a right-mover and a left-mover consisting of the previously described conglomeration of partons and their levels. Only the lowest levels in the various cells are occupied. As the hadrons collide with each other we see that only the uncontracted (wee) rapidity cells of the two projectiles will significantly overlap in phase space (momentum space in particular). We may, according to the Feynman picture, expect wee partons to be exchanged between the two hadrons and therefore excitation of the levels existing in those wee rapidity-cells. After the hadrons pass through each other we may expect, say, after a time of the order of 2 or 3 fermis, that these excited levels will have been de-excited, since 1-2 fermis is a natural time scale associated with those excitations. This de-excitation can proceed by the emission of wee hadrons emitted more or less isotropically. However in addition to that mechanism of de-excitation there is also possible de-excitation by excitation of levels in the neighboring cell of higher rapidity. This follows because there is coupling between neighboring rapidity cells and because the level density in the neighboring cell is higher than in the wee cell. Therefore, after this passage of time of 2 or 3 fermis when the hadrons have escaped each other, the event is not over. The next-to-wee rapidity cells of the incoming projectiles are excited and wee hadrons have been emitted. This process can now repeat itself. These not-so-wee cells in rapidity de-excite by emission of not-so-wee hadrons which are emitted primarily along the beam directions because of the large momenta involved and the motion of these cells along the beam directions. Again this is not the only mechanism of de-excitation. There can be de-excitation by excitation of the neighboring cell of still higher rapidity. Notice that this proceeds only in the outward direction in y-space because only in that direction does the level density increases with rapidity. There is negligible re-excitation of the wee cells, because their level density is too coarse. Also, this process of de-excitation and re-excitation takes place on a longer time scale than the original first process of the wee excitation and de-excitation because of time dilatation. The clock is running slower because of the large γ of this subsystem of the proton. The time scale for the de-excitation of a cell is proportional to the momentum of the constituents therein. This process now repeats itself on an exponentially increasing time scale. At a given time t the region of excitation is among those partons of momentum $p \sim (\text{const}) \cdot t$, the constant being $\lesssim 1$ GeV per fermi. Finally after a time proportional to the center-of-mass energy in the collision the leading partons will get excited and then de-excited (on the same time scale as their excitation), after which time the process is over, and the asymptotic hadrons recede toward the detection apparatus.

What are the messages from all of this? The first is that excitation occurs by parton interchange. Secondly, the de-excitation of the excited levels in the various cells occurs by two mechanisms. One is the emission of hadrons and the second the excitation of neighboring cells. This is made possible by a level density which increases with the energy or rapidity of the cells. Third (if the picture makes sense at all) the initial excitation by the collision of the two projectiles is really equivalent to excitation by the "heating" of a neighboring rapidity cell. After all we can look at the process in various frames; some of the cells which belong to one hadron in one frame will belong to the other hadron in another frame. But in any case in any frame <u>the slowest hadrons</u> (that is, the wee hadrons) <u>emerge first from the collision</u>. <u>The fastest, most energetic hadrons emerge last</u>. Relativity is very important. There is no simultaneity in the process and the geography is in fact spread out over a long time scale. Hadrons are emitted and partons are excited at a time t which is proportional to their momentum. Everything happens locally in rapidity; namely if the partons at momentum p at any given time are excited then partons of momenta much different from p are at that time unexcited.

It is important that the description above be a covariant description. We should be able to look at this process in different frames of reference and, even though there will be rather different evolutionary pictures, must still find that the distribution of hadrons which emerge, including the time sequence, be self-consistent. This is an important test of any model which I believe is necessary before it becomes credible. This is in fact one of the weaker points of the Landau hydrodynamic model. It picks out a special frame for the initial conditions, even though the evolution of the dynamics after the initial conditions are set is treated covariantly.

It is in fact instructive to look at what we just went through in another frame which I like to call the Fool's ISR (FISR). The FISR is a double storage ring consisting of very high energy protons going in almost the same direction and colliding with each other with a very small crossing angle 2θ. If we look in at the collision from the upstream direction what is seen is two circular disks of hadronic matter slowly moving through each other with a velocity $\sim\theta$. What happens this time? The rapidity distributions of the partons of the two incoming projectiles look identical; however the transverse momenta in the leading rapidity cells differ. How do we decide which partons are excited as the disks move through each other? First of all, the leading partons will not be immediately excited; because although their distributions overlap in longitudinal phase space they do not overlap in transverse-momentum phase space.

Therefore we must find longitudinal momenta sufficiently small so that the transverse momenta are of the typical several-hundred-MeV and that there is overlap in transverse-momentum space as well as in longitudinal momentum space. It is easily checked that this occurs for those partons which, in the center-of-mass description, we found were first excited by the collision. But now we may ask about the partons which have less rapidity than those which were excited: will they be excited as well? The answer is no, because they are moving through each other very slowly and, although the transverse momentum distributions overlap, the level densities of these very-low-rapidity cells are so small that there can be no excitation. There simply isn't enough kinetic energy imparted by the partons as they go through each other to significantly excite those levels. The situation in fact has a very nice analogue in atom-atom collisions. Collisions between leading partons correspond to very energetic electrons in two atoms colliding with each other. Such collisions are rare because the interaction is weak ($0(\alpha)$) when the relative velocity is very large. The strongest interaction in atom-atom collisions occurs at velocities for which the two atoms can be considered as two Thomas-Fermi gases interpenetrating and for which their momentum-space distributions overlap significantly and for which there can be electron exchange and transfer of momentum as well (and thereby excitation). This case is analogous to the Feynman wee parton exchange. The collisions of the very wee partons with each other are analogous to atom-atom collisions at very low velocities (much less than the velocities of the electrons within the atom). Under such circumstances the adiabatic approximation for the collision is applicable. As the atoms go through each other the position of levels change. But they change so slowly and so slightly that if the systems were in their ground state before the collision they would remain in the ground state after the collision, provided there are no level crossings.

The space time aspects of this picture can be summarized in terms of space-time diagrams where the world lines of initial projectiles and all final produced hadrons lie approximately along the light cone. That is, they will be rays emerging in various directions along the light cone since most produced hadrons are travelling at the speed of light, and there are no special features associated with the finite velocities of the hadrons which we have used. Therefore if we look from the top of the light cone (this is shown in Fig. 6) we see all the rays emanating in the transverse directions. The real dynamical activity occurs near the t-z plane (defined as x=y=0). Define the region $|x_\perp| = \sqrt{x^2+y^2} \leq 1f$ as the interaction cylinder. Rays outside the interaction cylinder describe essentially asymptotic hadrons, unless two rays are so close to each other that one must wait a little longer for them to separate by a fermi from each other. Projecting the rays of emitted hadrons onto the zt plane, we see that the time at which the hadrons emerge from the interaction cylinder as asymptotic particles is when their proper time τ is ~1 fermi:

$$\tau^2 \sim (t-z)(t+z) - x_\perp^2 \sim \frac{1}{p} \cdot p - x_\perp^2 \sim 0(1) \tag{5.4}$$

Thus the boundary surface that distinguishes the asymptotic hadrons from the still-interacting hadrons is a hyperbola given by the approximate equation

$$t^2 - z^2 \sim 1 \text{ fermi}^2 \quad . \tag{5.5}$$

Thus at some intermediate time, we see the picture as in Fig. 6a. The hadrons going along the beam direction within 1 fermi of the beam direction or so will be still excited. The thickness of the excited region will be inversely proportional to the time at which we observe this intermediate state; this time is assumed large compared to 1 fermi. Momenta in the excited region will be proportional to the time t. The highese momentum of the emitted hadrons will also be proportional to t. The rapidity distribution of the produced hadrons will be a plateau starting from the inside out (wees first) and growing at the ends to larger and larger rapidities as time goes on, as shown in Fig. 6c. The rapidity distribution of the partons in the original projectiles is shown in Fig. 6b, and the excited parton will have momentum proportional to the time. This picture of the collision is geometrical and what one would expect given only classical considerations. But it also seems to be consistent with anything quantum mechanical that one imposes upon it.

We now turn to nuclear collisions and see what the implications of this picture are in that case.

C. Nucleon-Nucleus Collisions

The nucleus at rest is generally considered a collection of nucleons inside of which reside a collection of quarks of low rapidity. When this nucleus is boosted to sufficiently high rapidities, naively one obtains again a pancake-shaped object containing all the quarks. If the boost is sufficiently high, the thickness of the pancake will be $\lesssim 10^{-13}$ cm. Under such circumstances we must again consider the effect of vacuum fluctuations on the structure of the moving nucleus; in such a case we must attach one layer of wee vacuum fluctuations over the entire surface of the moving disk. As the nucleus is boosted still further, the same process as for a single nucleon repeats itself. The original vacuum fluctuations attach to vacuum fluctuations of lesser rapidity, until the last layer of vacuum fluctuations are again the wees. An important consequence of this is that the thickness of a moving nucleus (in the longitudinal direction) never is less than a fermi, just as the case for a single nucleon. Thus the number of wee and not-so-energetic partons will be proportional to the surface area of the nucleus $\sim A^{2/3}$ rather than strictly proportional to A. This occurs for boosts greater than a critical velocity or γ, with the critical gamma $\gamma_c \sim 2.4\ A^{1/3}$, as follows just from our knowledge of the nuclear size. Partons of rapidity $y \gtrsim y_{max} - \log \gamma_{crit}$ belong to the "naive" nucleus ($n \sim A$), while the other partons belong to the vacuum polarization "fur" ($n \sim A^{2/3}$) which coats the surface of the Lorentz-contracted nucleus.

Now let us look at a very high energy nucleon-nucleus collision in some center-of-mass frame for which the γ of the nucleus exceeds the critical value. Then the nucleon projectile is excited by the layer of wee partons around the "naive" nucleus in the same way as it would be excited were there only a single nucleon as target instead of the entire nucleus. Wee partons in the projectile are excited and wee hadrons in the lab frame are emitted. As far as the fragments of the nucleon projectile are concerned, the same sequence of events then occurs as for an ordinary nucleon-nucleon collision. Therefore the inclusive spectrum of leading hadrons from the nucleon projectile is at sufficiently high energies the same as for an ordinary hadron-hadron collision. This is shown in Fig. 7. This situation persists into the central-plateau region, until a critical rapidity is reached, corresponding to the critical gamma which we described before. Then for rapidities beyond this critical value we must change our considerations and study afresh the distribution of these particles, which comprise the nucleus-fragmentation region. This is easiest to do in the rest frame of the nucleus and is most interesting in the somewhat artificial case of dilute nuclear matter (interaction length >> 1 f) where we can watch in some detail the time evolution of the process. What happens? In the rest frame of the nucleus the nucleon projectile enters and interacts with stationary nucleons within the nucleus. Wee partons in the projectile are excited, wee hadrons are emitted into the nuclear matter at large angles, and the excited projectile continues on its way. After a few fermis the wee partons are de-excited, re-exciting not-so-wee partons in the projectile, which in turn emit not-so-wee hadrons, which excite even more energetic partons in the projectile and so on. This continues on until a subsequent collision occurs downstream with another nucleon in the nuclear matter. At that time there will be re-excitation of the wee partons in the projectile (recall that those wee partons which were excited in the first encounter have already cooled off). Wee hadrons will be emitted at large angles from this second nuclear site, and the process will repeat itself: the wee partons in the projectile excite the not-so-wee partons, etc., until the parton excitation from the second collision overtakes (in rapidity-space) the parton excitation still remaining from the first collision. The two excitations merge into one and proceed onward toward higher rapidity until the next nuclear collision occurs and the process is repeated. This is all shown in Fig. 8, which illustrates the configuration as the excited nucleon leaves the nucleus. There will be various stars, separated on the average by an interaction length and containing the large-angle low-rapidity hadrons which were previously emitted. In the forward cone, as roughly defined by an angle $\lesssim$ 1 fermi/(nuclear diameter), there will exist the excited projectile which will have not yet reached its asymptotic cooled-off condition. It will contain excited partons which have momenta or rapidity of the order of the critical rapidity ($\sim \log \theta^{-1} \sim \log A^{1/3}$) which we discussed previously. Therefore by looking in these two reference frames, we have accounted for the entire hadron distribution. We see that the multiplicity of slow nucleons emitted in the nucleus will be proportional to $A^{1/3}$ (the number of stars) times an additional factor coming from the secondary cascade of the slow hadrons as they proceed in a transverse direction through the nuclear matter. This must be calculated by a straightforward but complex cascade model. But in terms of the multiplicity of particles which within the nuclear matter leave the interaction cylinder (radius ~1 fermi and axis along the trajectory of the incident projectile), we have a multiplicity, relative to an ordinary nucleon collision, of the order $A^{1/3}$.

D. Nucleus-Nucleus Collisions at Fantastic Energies

Before leaving this subject it is fun to consider the collision of two nuclei at energies sufficiently high so that in addition to the fragmentation regions, a central plateau region can develop. Let us consider a central collision of a relatively small nucleus, say carbon, with a big one, say lead. Let us look at this collision in a center-of-mass frame for which the rapidities of both of the nucleus projectiles exceeds the critical rapidity. In such a frame they both possess the fur coat of wee-parton vacuum fluctuations. In such a central collision we see that the collision initially occurs between the fur of wee partons in each of the projectiles. Therefore the number of independent collisions will be of order of the area of overlap of the two projectiles; namely the cross-sectional area of the smaller nucleus.

It follows that the number dN/dy of emitted pions will also be of the order of the area of the smaller nucleus; namely $A_<^{2/3}$, where $A_<$ is the atomic number of the smaller nucleus. This determines the height of the hadron plateau in the central region, relative to what it is in an ordinary nucleon-nucleon collision. There are in addition the two target fragmentation regions. The easiest way to study their properties is to go into the appropriate rest frames. For the fragmentation region of the smaller nucleus we go to its rest frame. In this frame we see the big pancake of the larger nucleus (with its fur coat of wee partons) sweeping through the entire volume of the smaller nucleus. Therefore all nucleons in the smaller nucleus can be excited and the distribution of produced hadrons will therefore be proportional to $A_<^{2/3}$, multiplied by the distribution of produced hadrons for a nucleon-nucleus collision. For the fragmentation region of the larger nucleus, we go into its rest frame. What we see there is a smaller pancake (with wee partons attached) sweeping through the central volume of the large nucleus. The volume swept will be proportional to the cross-sectional area of the smaller nucleus times the diameter of the larger nucleus. Therefore the rapidity distribution of slow hadrons in the fragmentation region of the big nucleus will again be given by $\sim A_<^{2/3}$ times the distribution of slow hadrons for a nucleon incident on the larger nucleus. This can be made a little more quantitative by just estimating the number of independent vacuum fluctuations that can be attached to the incoming nuclei so that there is no overlap in the transverse directions. Doing that I found the semiquantitative guess for the rapidity distribution which is shown in Fig. 9. Much more professional studies along the same line of initial assumptions can be found in the work of O. Kancheli,[32] E. Lehman and G. Winbow,[33] J. Koplik and A. Mueller,[34] and A. Goldhaber.[35]

E. <u>Colliding Beams</u>

We now turn to the real subject at hand, which is the space-time evolution of produced hadrons in the deep inelastic processes, first for the colliding beams. We assume the existence of the jet-like structures discussed earlier and suggested by correspondence or the parton model. Specifically the hadron final states are assumed to be similar to, say $\pi\pi \rightarrow$ hadrons, with the axis of the $\pi\pi$ collision distributed almost isotropically. This is supported experimentally both by the jet observations and by the scaling behavior of the inclusive cross sections. [In making this comparison we must, however, leave out $\pi\pi$ final states involving diffractive excitation.] If the final state in colliding beams really is so similar to that in a $\pi\pi$ collision, it is very natural to suppose that the time evolution of the final state is similar in detail as well, at least at the geographical level that we have discussed. However, we run into a problem. In the colliding beam process, just after the e^+e^- collision, there exist only two produced partons, and they have widely different rapidities. There is no time before they escape from each other for vacuum fluctuations to attach to them or for anything else to happen. The natural time scale for these energetic partons to do something is, as we discussed, proportional to their momentum which in turn is proportional to the center-of-mass energy and can be made in principle as large as we like by going to higher and higher energy. Furthermore we cannot let the intrinsic time scale for these partons to do something be arbitrarily short because we want to protect the free field behavior of the parton propagation at short distances in order that the total cross section behavior comes out right; namely R (the ratio of hadron production to μ-pair production) to be a constant with energy. The simplest picture that comes to mind is that the emission of the mesons be sequential. First an energetic meson is emitted and then a not so energetic one, etc., i.e., the cascade starts with the leading parton dividing, then redividing and so on. I advocated this for a while,[25] but now I think it is wrong. It was also advocated by Drell and Yan,[8] who found it to be a consequence of their cutoff field-theory parton model. However, the trouble with this is that first of all there is no confinement. If there is just division of the original quark and the original antiquark without any communication between them some part of the final state jets will contain fractional charge. Furthermore a finite hadron multiplicity even at infinite energies is predicted. This is unacceptable because if the multiplicity were really finite at infinite energies and the total cross section scaled, then some partial cross section would have to comprise a finite fraction of the total cross section. Therefore some exclusive channel, composed of a finite number of hadrons would have to exhibit scaling behavior with respect to its total cross section. Then the problem is which hadron final state is it? If it did exist we would essentially find the partons directly.

What about adding a ladder between the two groups of partons associated with the quark and the antiquark? There is still trouble. This was discovered by Kogut, Sinclair, and Susskind,[36] and by Craigie, Kraemmer, and Rothe.[37] Again the trouble is the large time scale for the original parent partons to do something: they are too far apart by the time they start emitting their ladders. When one studies the diagrams in momentum space, what happens is that in ordinary collisions the ladder graph with n rungs corresponds to an amplitude behaving as $(\log s)^n$. Upon summation one generates the Regge trajectory. However in the case of colliding beams the $i\epsilon$'s in the propagators are in the wrong places to generate these logarithms, which come from pinches in certain integration contours. There are no $(\log s)^n$ factors generated in nth order, and in fact the ladders are rather inconsequential corrections to the lowest order diagram. The fact that the $i\epsilon$'s are in the wrong place is related to causality considerations, which again have to do with the space-time development. In the case of hadron-hadron collisions, there is a long time before the collision for virtual processes to prepare the long chain of virtual partons (including the wees) which initiate the kind of final state generation which we already discussed. However when everything starts at $t=0$ instead of $t=-\infty$, there is no time for all of that preparation.

What about other very exotic alternatives? One can imagine elastic rubber bands connecting the partons, which first separate a long way, emit some hadrons, come to a stop, then get pulled together by the rubber band, then oscillate back and forth with some damping, eventually annihilating. In terms of the space-time development, that looks bizarre relative to what we have considered. One might think about violent discharge in the vacuum, something analogous to lightning bolts. That again looks a little ridiculous. The best bet seems just to imitate as best as possible the ordinary processes we discussed, where the wee hadrons emerge first, the not-so-wee later on, etc., with all the activity concentrated on a proper time surface as shown in Fig. 6. But we must look at this option in more detail to see what the implications are. At a time, after the birth of the parton anti-parton pair, small compared to a fermi there is nothing but original quark-antiquark: no produced hadrons and no produced partons. After a time of the order of a fermi we somehow have wee hadrons produced, if we are to imitate what happens in ordinary collisions. If wee hadrons are produced, there can also be produced an extra wee quark-antiquark pair which follow respectively the originally produced antiquark and quark directions and begin to screen the fractional charge of the antiquark-quark. As time goes on, by hypothesis the time evolution is the same as discussed for ordinary collisions. Therefore later on, not-so-wee hadrons are produced and the original polarization clouds of produced parton and antiparton which were initially following the parent antiquark-quark get accelerated; as time goes on they gain momentum in such a way as to be in the same region of phase space as excited partons in an ordinary collision. This is all illustrated in Fig. 10. The time scale for this kind of evolution is assumed to be the same as for the ordinary collision. Therefore the polarization cloud of partons which follow the original parents attains a rapidity (or momentum) comparable to their parents at a time proportional to the original momentum of the partons; namely proportional to the center-of-mass energy. At that time they can annihilate into a final leading hadron system.

What are the messages this time? First of all, we see that in this process there must necessarily be long range correlations in rapidity in order to screen the fractional charge. This is in fact true even were we to invoke rubber bands or lightning bolts. There seems to be no way of avoiding it. Secondly, confinement of a quark occurs at a time proportional to the momentum of the quark, when the polarization cloud overtakes (in momentum space) its parent sufficiently to annihilate it. In the intermediate stage there is always a polarization cloud accompanying the quark. It will have a lag or a typical longitudinal thickness inversely proportional to its momentum. This cloud is gently accelerated by the leading quark as time goes on. In the c.m.s. frame of the quark and the polarization cloud, fractional charge is never separated by more than a fermi. As time goes on the leading quark gets slightly decelerated in order to feed energy into the produced hadrons as well as to feed energy into the accelerating polarization cloud so that at any given time energy can be approximately conserved. Since hadrons (i.e., quark pairs) are being emitted from the polarization cloud, it need not be a unique quark in the cloud which is accelerated. The current in the cloud is polarization current.

The rate of change of the momentum of the leading quark with time dp/dt is a gentle constant, say ~1 GeV per fermi of travel, involving no large mass scale. Furthermore in order of magnitude this is also the time rate of change of the momentum of the polarization cloud as well as the time rate of change of the total energy of the produced hadrons. The free-field behavior at short distances is clearly protected because the number of soft interactions per unit of time is of the order of a constant. We need at most a few soft interactions per fermi in order to change the momentum of the quark or create the produced hadrons, and this number is independent of the quark momentum. As the total energy W increases one only needs the free field behavior to be true at distances of the order of W^{-1} in order to protect the total cross section behavior. Finally, if the only effect on the original parent quark is to be gently decelerated by the polarization cloud which it is dragging along and by the hadrons which are being emitted, then the wave function of the parent quark should just be a free field wave function times an eikonal phase:

$$\psi_{\text{quark}}(x,t) \sim \psi_{\text{free}}(x,t)\, e^{iS(x,t)} \tag{5.6}$$

This is suggestive of an underlying gauge theory (however, in real life we may well want that gauge theory to be non-abelian; the presence of simple eikonal phases is a property more of Abelian rather than non-Abelian theories). But in any case a gauge theory for the confinement mechanism, either Abelian or non-Abelian, certainly is strongly suggested by this line of argument, both by the above feature of the gentle deceleration being associated with eikonal phases and also by the existence in the dynamics of long range correlations in rapidity, again suggesting the existence of spin 1 quanta in the field theory.

There is some formal support for this handwaving from the behavior of two-dimensional quantum electrodynamics, as found by Casher, Kogut and Susskind.[38] There is some question as to whether that theory is too simple to really be a good prototype, but in any case it does give some encouragement. Furthermore, one-dimensional concepts may be relevant even in the three-dimensional case, because we saw that for the system of the quark and polarization cloud the transverse extent is of the order of 1 fermi whereas the longitudinal extent of the important components of the system is much smaller than 1 fermi. The pancake-shaped polarization cloud may in fact produce something like a uniform electric field which then acts upon the leading quark. The relevance of such one-dimensional motions and approximations has already been exhibited in the Landau hydrodynamic model, which has in fact a realistic space-time development similar (but not identical) to what we have described.

Before leaving the subject of colliding beams, it will be useful to check the consistency of this picture by looking at the same process in the Fool's ISR. In this case electron and positron have comparable longitudinal momentum and collide at a small crossing angle θ at fantastically high energies. To describe this situation we simply have to change the old solution by boosting it in the transverse direction. Before we boost it, the characteristic time at which the first hadrons emerged from the collision was of the order of 1 fermi. Now in the boosted frame this time will be increased by the γ of the boost; $\gamma \sim \theta^{-1}$. Then the characteristic time will be 1 fermi/θ; much longer. No appreciable number of hadrons are emitted which are wee. The wee hadrons in the original frame now have momentum which are $\sim$(hundreds of MeV)/θ. An analog in very high energy electrodynamics is the Chudakov effect,[39] which inhibits the ionization produced in electromagnetic pair production at sufficiently high energy because the electron and positron do not have sufficient transverse separation to create a dipole moment strong enough to ionize the atoms until considerably downstream from the production point. For our case the quark and the antiquark must have a transverse separation (or an invariant separation in space-time) of the order of a fermi before there is any appreciable hadron production. In the FISR configuration that leads to a delay in the appearance of the hadrons until a time considerably longer than 1 fermi.

F. Electroproduction (or Neutrino Reactions)

The simplest case to study first is for small ω, where there is a reasonably straightforward generalization of what we have just discussed. We look at the process in the laboratory frame. The incident virtual photon strikes a wee quark and imparts to it all of its momentum ν. Just after this happens, the parton rapidity distribution will consist of wee partons and the struck parton of high energy, with a large rapidity gap ($\sim \log \nu$) between the two systems. Therefore the parton leaves the proton by a distance of ~1 fermi, wee hadrons and the wee polarization cloud accompanying the struck parton will again be produced. At some later time (~10 f. ?) we will have what is shown in Fig. 11, with the rapidity distribution of produced hadrons moving from the wee region outward. The rapidity distribution of partons in the polarization cloud terminates at a momentum comparable to the momentum of the most energetic produced hadron which in turn is proportional to the elapsed time. Finally the original parent parton (the one that was struck by the virtual photon) is still isolated in phase space (i.e., momentum space) and has been essentially unaffected except to be gently decelerated by the polarization cloud.

What are the messages? First of all, the time evolution resembles the colliding beam evolution. The polarization cloud exists and therefore for the rapidity distribution of produced hadrons there should be the same current plateau as there was in e^+e^- annihilation. Secondly the reaction is complete only when the elapsed time in the laboratory frame is of order ν, the energy of the virtual photon. Therefore if we were dealing with electron-nucleus scattering at large ν, the reaction terminates only after the quark has left the nucleus. I don't have a complete picture of what will happen under these circumstances. But at least it is clear that an experimental study of the

A-dependence of the spectrum of energetic hadrons produced in electroproduction could turn out to be quite interesting.

Finally, let us again look at electroproduction (in the laboratory frame) but when ω is very large (>> 100?). This is a difficult case; the problem is how the hole fragmentation phenomenon appears and whether we can see how vector dominance is involved. For large ω, the general considerations of Ioffe[20] imply that the process of virtual-photon dissociation into quark-antiquark will start before the virtual photon arrives at the target proton. There is some time for the quark-antiquark system to at least become partially dressed before arrival at the target. Let us look at this in detail. The kinematics is shown in Fig. 12a. At the naive classical level (which should not be too bad for very light quarks) we might expect the following picture: First of all let us suppose that the virtual photon dissociates into a quark-antiquark pair with comparable longitudinal momenta. If the virtual photon has virtual (spacelike) mass $\sim Q$, the typical mass (now timelike) of the quark-antiquark system will be again $\sim Q$. Therefore, for a symmetrical dissociation the transverse momenta of the dissociated quark-antiquark will also be $\sim Q/2$. The time Δt this fluctuation lives cannot be indefinitely long because we have not conserved energy. In the usual way we can estimate using the uncertainty relations

$$z \sim \Delta t \sim \frac{1}{\Delta E} \sim \frac{2\nu}{Q^2 + m_n^2} \sim \frac{\omega}{M} \qquad (5.7)$$

The distance z (or the time) the fluctuation survives is proportional to ω: when ω is large we have this kind of upstream phenomenon, when ω is small we do not.

We may also estimate the separation of the quark-antiquark in impact parameter when it arrives at the nucleon. The calculation gives

$$\Delta x_{\perp} \sim z\theta \sim z\left(\frac{p_{\perp}}{\nu}\right) \sim \left(\frac{\nu}{Q^2}\right)\left(\frac{\sqrt{Q^2}}{\nu}\right) \sim \frac{1}{\sqrt{Q^2}} \tag{5.8}$$

This is the "shrinking photon". When Q^2 is large the transverse separation is much less than a fermi and decreases with increasing $Q^2 \sim Q^{-1}$. This is too small to produce any wee partons because of the "Chudakov effect" which we discussed in the previous section. With no wee partons in the virtual state there is no way that this system can be absorbed by the target nucleon and nothing happens, in line with the discussion in Section I. However, if the fluctuation is asymmetric in the parton momenta then we can obtain a transverse separation ~1 fermi within the duration of the fluctuation. The kinematics is shown in Fig. 12b, and we see that the pair mass can be found by computing $E\text{-}p_{\parallel}$ for the pair and its components

$$E - p_{\parallel} \cong \frac{m^2}{2\nu} \cong \frac{p_{\perp}^2}{2p} + \frac{p_{\perp}^2}{2(\nu - p)} \tag{5.9}$$

With a pair mass $m \sim Q$ and $p \ll \nu$ we can calculate the transverse momentum $p_{\perp}$ of the partons; it follows from Eq. (5.9)

$$p_{\perp}^2 \sim \frac{p}{\nu} Q^2 = \frac{2Mp}{\omega} \tag{5.10}$$

Now we repeat the calculation leading to Eq. (5.8). The important longitudinal distance depends only on the virtual mass of the quark-antiquark system and on Q^2 and therefore is again proportional to ω. The transverse separation $\Delta x_{\perp}$ is given by

$$\Delta x_{\perp} \sim z\theta = z\left(\frac{p_{\perp}}{p}\right) = \frac{\nu}{pQ^2}\sqrt{\frac{2Mp}{\omega}} \sim \sqrt{\frac{\omega}{2Mp}} \tag{5.11}$$

and the condition $\Delta x_{\perp} \gtrsim 1f$ leads to

$$p \lesssim \omega \qquad (5.12)$$

From Eq. (5.10) we see, not unexpectedly, that this is equivalent to the condition $p_{\perp} \lesssim$ const. Probably (5.10) is the most reliable numerical estimate of p, using $\langle p_{\perp}^2 \rangle \sim 0.1 - 0.2\ \text{GeV}^2$. Therefore the typical rapidity for the slow quark is going to be of order log $\omega/10$. If the rapidity is larger than this, the transverse separation of the quark-antiquark at arrival is too small and we expect very little interaction. If, on the other hand, the rapidity of the slow quark is less than the characteristic value ~log $\omega/10$ there is no phase space. The slow quark has not a dx/x spectrum, but a dx spectrum corresponding to an isotropic decay of a parent (namely the virtual photon) into two secondaries. This situation leads to an exponentially decreasing contribution for smaller rapidities. We conclude that it is necessary for the slow parton to have a characteristic momentum ~(ω/10) GeV in order to induce some partons out of the vacuum by arrival time. If parton pairs are induced they will also have momentum $\lesssim(\omega/10)$ GeV. And there is enough time for all of these virtual quarks to have coupled to a sequence of ordinary vacuum fluctuations just like an ordinary hadron would. Therefore when this whole system arrives at the nucleon the virtual photon has the structure shown in Figs. 12c and 12d. We see that the ordinary vacuum fluctuations will interact just as in an ordinary hadron collision, provided the momentum scale is $\lesssim(\omega/10)$ GeV and the distance scale downstream is $\lesssim(\omega/10)$ fermi. At collision the cloud of vacuum fluctuations excites the nucleon. Then wee hadrons are emitted and again excitations in the virtual-photon cloud move upward in momentum (or rapidity) space in the way appropriate to a normal hadron-hadron collision. This continues until a characteristic time t ~ω/10 f.

Remember that (at sufficiently large ω) this is a long time and therefore in an electron-nucleus collision this mechanism should govern what is going on in the nuclear matter.

Finally for $t > \omega/10$ f, the leading parton in the polarization cloud (of momentum $\sim\omega/10$ GeV) begins to get accelerated by the leading quark up to momentum ν. In this period of time the hadron emission occurs in the same way as for small ω, because the polarization cloud of fractional charge is being accelerated by the energetic quark. This process generates a "current plateau" and completes the picture of hadron production in a way which is totally consistent with what we discussed in more general terms without use of the space-time picture. In particular the jet picture again emerges. There is limited transverse momentum of the produced hadrons, there is a current plateau for produced-hadron rapidities large compared with log $(\omega/10)$, and there is a hadron plateau for rapidities $\ll$ log $(\omega/10)$. The hole fragmentation region divides the two plateaux, and we have even identified some of the dynamics associated with the hole fragmentation region itself.

G. Conclusions

In summary, we have found that the picture of space-time evolution of hadron final states in deep inelastic processes isn't totally trivial, and also have found that it can be made consistent with the hypotheses of the parton model (and therefore also with the trends of the data at present). This is only the case provided that there exist long-range rapidity correlations in the deep inelastic dynamics. This is in fact what is indicated in two dimensional quantum electrodynamics, as discovered by Casher, Kogut and Susskind. It strongly suggests the relevance of some kind of gauge theory as a necessary element in the interpretation of the confinement problem. Finally it is clear that the discussion above is so qualitative that while hopefully providing some insight into the space-time geography of what is going on, it is a far cry from what we should like to have in order to make quantitative comparisons with experiment. In fact, I see the above description as a solution to a problem which hasn't been well defined. The problem remaining is to try to sharpen the above considerations by specifying precisely what the problem is that is to be solved. Perhaps it is a variant of the Landau hydrodynamic model, but one which incorporates the concepts of short range correlation in rapidity instead of Landau's concept of total arrest of hadronic matter in hadron-hadron collisions. If such a hydrodynamic picture could be worked out it might lead to some further insights into the dynamics of confinement and of deep inelastic processes.

REFERENCES

1. Much of these lectures are also presented in the 1973 Proceedings of the Summer Institute on Particle Physics, edited by M. Zipf, SLAC-167 (1973).
2. R. P. Feynman, Photon-Hadron Interactions (W. A. Benjamin, New York, 1972).
3. J. Bjorken and J. Kogut, Phys. Rev. D **8**, 1341 (1973).
4. S. Drell and T. Yan, Phys. Rev. Letters **24**, 181 (1970).
5. G. West, Phys. Rev. Letters **24**, 1206 (1970).
6. E. Bloom and F. Gilman, Phys. Rev. Letters **25**, 1140 (1970).
7. In addition to the contributions to this volume, which we do not explicitly cite, a valuable source of recent data is found in the Proceedings of the 1975 International Symposium on Lepton and Photon Interactions at High Energies, ed. W. T. Kirk, Stanford Linear Accelerator Center (1975).
8. S. Drell, D. Levy and T. Yan, Phys. Rev. D **1**, 1617 (1970).
9. S. Berman, J. Bjorken and J. Kogut, Phys. Rev. D **4**, 3388 (1971).
10. R. Schwitters, Ref. 2, p. 5.
11. A review of the kinematics for these processes can be found in F. Gilman, Phys. Reports C **4**, 98 (1972).
12. B. Roe, Ref. 2, p. 551.
13. A recent and informative discussion is found in J. Van der Velde; talk presented at IVth International Winter Meeting on Fundamental Physics, Salardu, Spain (1976); University of Michigan preprint.
14. G. Wolf, Ref. 2, p. 795. Also R. Mozley, Ref. 2, p. 783.
15. L. Mo, Ref. 2, p. 651.
16. P. Garbinicus *et al.*, Phys. Rev. Letters **32**, 328 (1974); A. J. Sadoff *et al.*, Phys. Rev. Letters **32**, 955 (1974).

17. See the report of G. Wolf, Ref. 2, especially p. 846.

18. J. Bjorken, Phys. Rev. D 7, 2747 (1973).

19. R. Cahn and E. Colglazier, Phys. Rev. D 9, 2658 (1974).

20. B. L. Ioffe, Phys. Letters B 30, 123 (1969).

21. The relaxation of this assumption is studied by H. Fraas, B. Read and D. Schildknecht, Nucl. Phys. B 86, 346 (1975).

22. V. Gribov, Fourth Winter Seminar on the Theory of the Nucleus and the Physics of High Energies, 1969, Ioffe Institute of Engineering Physics, Acad. Sci., USSR.

23. M. Greco, Nucl. Phys. B 63, 398 (1973).

24. J. J. Sakurai and D. Schildknecht, Phys. Letters B 40, 121 (1972); 41, 489 (1972); 42, 216 (1972).

25. J. Bjorken, Proceedings of the International Conference on Duality and Symmetry in Hadron Physics, ed. E. Gotsman (Weizmann Science Press, Jerusalem, 1971).

26. H. Cheng and T. T. Wu, Phys. Rev. 183, 1324 (1969).

27. J. Bjorken, Particle Physics (Irvine Conference, 1971), AIP Conference Proceedings No. 6, Particles and Fields Subseries, No. 2, ed. M. Bander, G. Shaw and D. Wong (American Institute of Physics, New York, 1972).

28. G. Preparata, Phys. Rev. D 7, 2973 (1973) and these proceedings.

29. L. D. Landau, Izv. Acad. Nauk. SSSR 17, 31 (1953).

30. E. Feinberg and I. Ya. Pomeranchuk, Nuovo Cimento Suppl. 3, Series 10, 652 (1956).

31. A recent discussion of the relationship of parton concepts to nonrelativistic quantum physics is given by G. West, Physics Reports 18, 263 (1975).

32. O. Kancheli, JETP Letters 18, 274 (1973).

33. E. Lehman and G. Winbow, Phys. Rev. D 10, 2962 (1974).

34. J. Koplik and A. Mueller, Phys. Rev., Preprint CO-2271-59 (to be published).

35. A. Goldhaber, Phys. Rev. Letters 33, 47 (1974).

36. J. Kogut, D. Sinclair, and L. Susskind, Phys. Rev. D 7, 3637 (1973).

37. N. Craigie, A. Kraemmer, and K. Rothe, Z. Phys. 259, 1 (1973).

38. A. Casher, J. Kogut, and L. Susskind, Phys. Rev. Letters 31, 792 (1973).

39. A. E. Chudahov, Isvestia Akad. Nauk. SSSR, Ser. Fiz. 19, 650 (1955).

FIGURE CAPTIONS

1. Expected rapidity distribution of hadrons produced in very high energy e^+e^- annihilation or ω-decay, as measured along the jet axis.

2. Expected rapidity distribution of electroproduced or neutrino-produced hadrons at very high ν: (a) Q^2 small or zero; (b) ω, Q^2 large; (c) ω small, Q^2 large.

3. Rapidity distribution of electroproduced or neutrino-produced hadrons according to the parton model: (a) ω large; (b) ω small.

4. Virtual forward Compton amplitude according to the vector dominance model.

5. Estimated scaling behavior of single-pion electroproduction according to correspondence arguments.

6. Space-time diagram of a hadron-hadron collision: (a) view down the light cone from time t; (b) view in the z-t plane ($x_\perp$ small); (c) parton rapidity distribution at time t; (d) rapidity distribution of produced hadrons at time t.

7. Rapidity distribution of produced hadrons in a nucleon-nucleus collision at very high energy.

8. Intermediate state of a nucleon projectile interacting in nuclear matter: (a) picture of the collision; (b) rapidity distribution of projectile partons; (c) rapidity distribution of produced hadrons.

9. Estimated rapidity distribution of produced hadrons for a central collision of a light nucleus of atomic weight $A_<$ and a heavy nucleus of atomic weight $A_>$.

10. Space-time picture of hadron production in e^+e^- annihilation: (a) view down the light cone at time t; (b) view in the z-t plane; (c) parton rapidity distribution at time t; (d) rapidity distribution of emitted hadrons at time t.

11. Electroproduction at small ω (laboratory frame): (a) configuration of produced hadrons at time t; (b) rapidity distributions of partons at time t; (c) rapidity distribution of produced hadrons.

12. Electroproduction at large ω: (a) dissociation of virtual γ into $q\bar{q}$ pair of comparable longitudinal momenta; (b) dissociation of virtual γ into $q\bar{q}$ pair with asymmetric partition of longitudinal momenta; (c) structure of virtual photon in rapidity space upon arrival at target (for asymmetric momentum partition); (d) structure of virtual photon in impact parameter (x-y) space upon arrival at target; (e) rapidity distribution of partons and produced hadrons at a time t after collision, $t \ll \omega$; (f) rapidity distribution of partons and produced hadrons at a time t after collision with $t \gg \omega$.

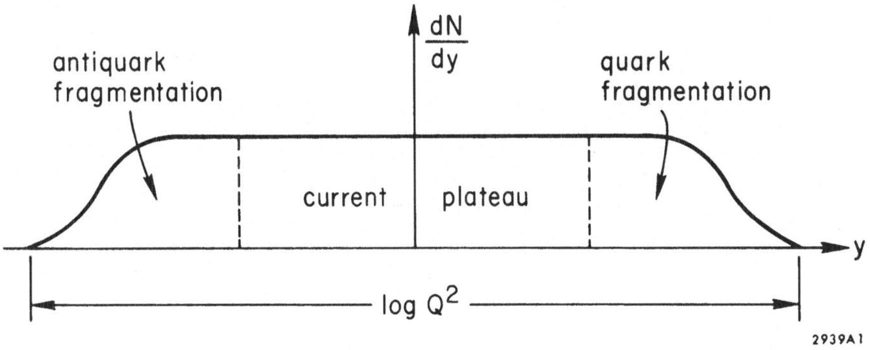
dN
dy
antiquark
fragmentation
quark
fragmentation
current
plateau
y
log Q2
2939A1

Fig. 1

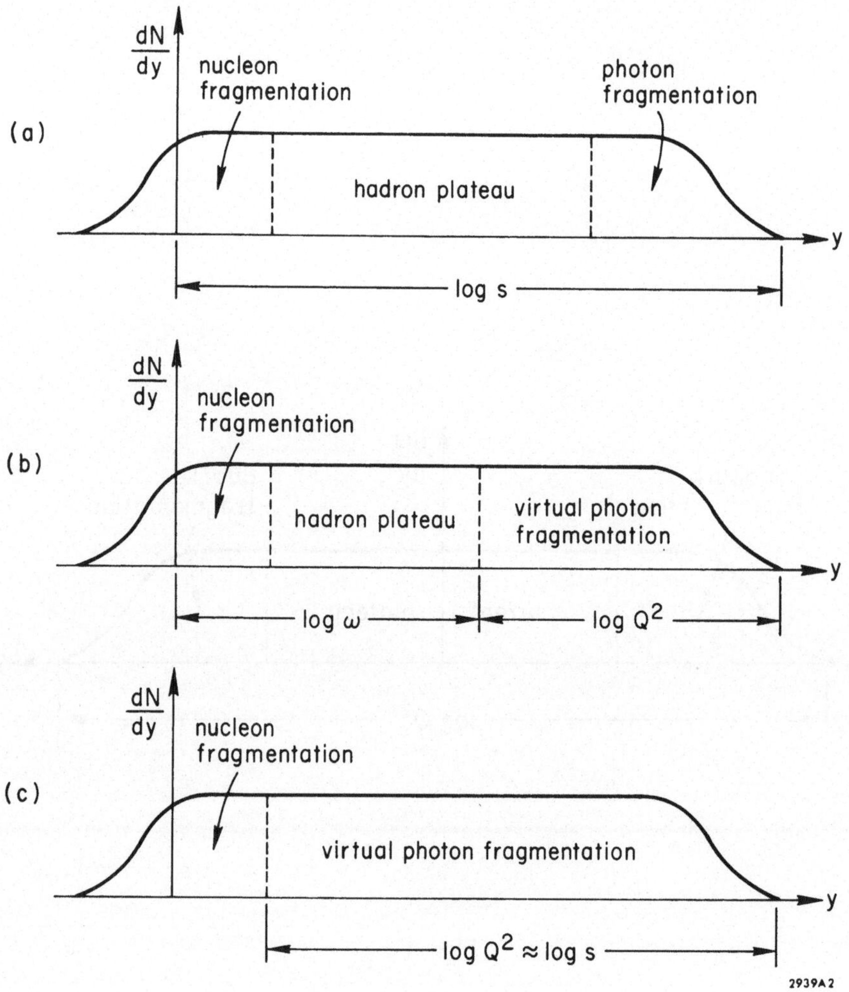
dN/dy
(a)
nucleon fragmentation
photon fragmentation
hadron plateau
y
log s
(b)
nucleon fragmentation
hadron plateau
virtual photon fragmentation
log ω
log Q2
(c)
nucleon fragmentation
virtual photon fragmentation
log Q2 ≈ log s
2939A2

Fig. 2

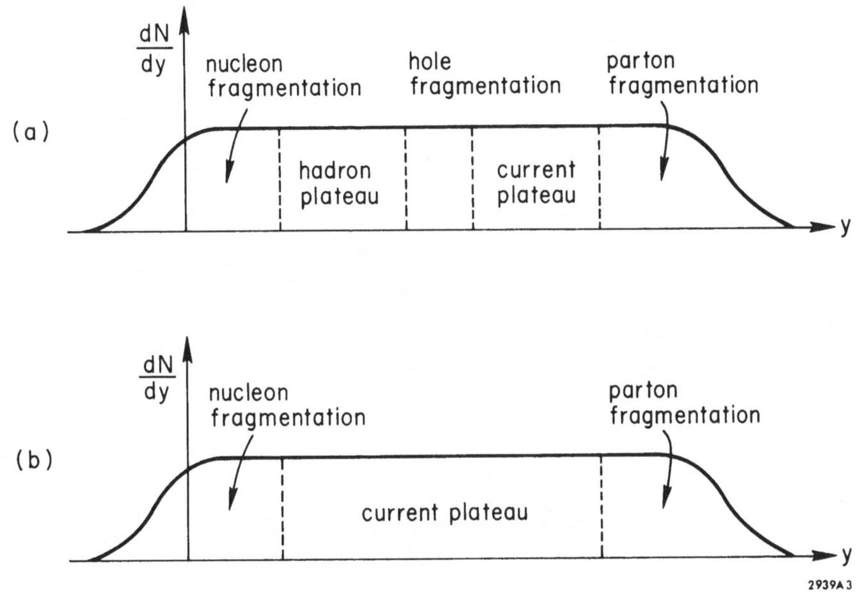
dN/dy
nucleon fragmentation
hole fragmentation
parton fragmentation
(a)
hadron plateau
current plateau
y
dN/dy
nucleon fragmentation
parton fragmentation
(b)
current plateau
y
2939A3

Fig. 3

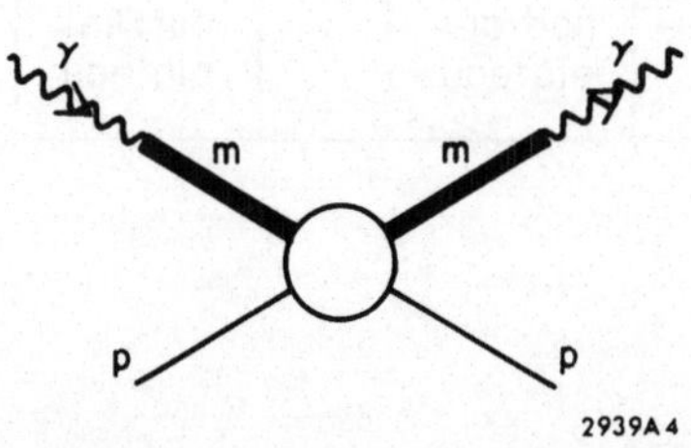

Fig. 4

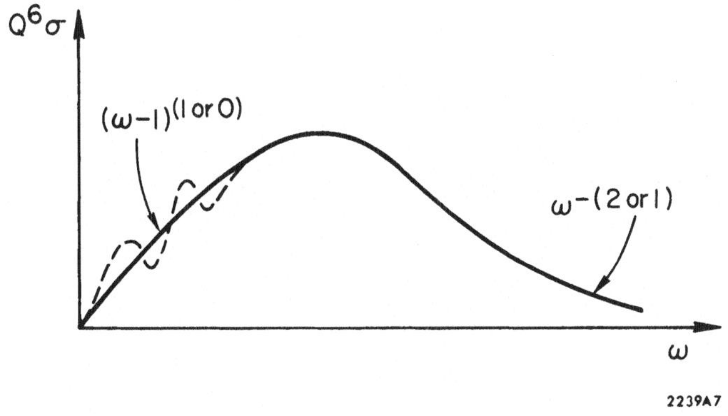

Fig. 5

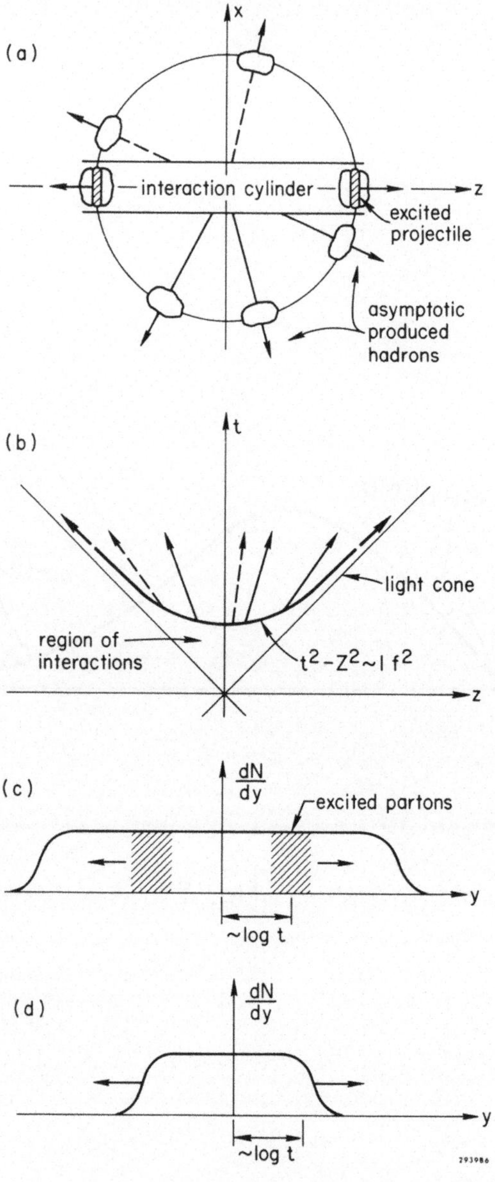
(a)
x
interaction cylinder
z
excited projectile
asymptotic produced hadrons
(b)
t
light cone
region of interactions
$t^2-Z^2\sim 1\,f^2$
z
(c)
dN/dy
excited partons
y
~log t
(d)
dN/dy
y
~log t
293986

Fig. 6

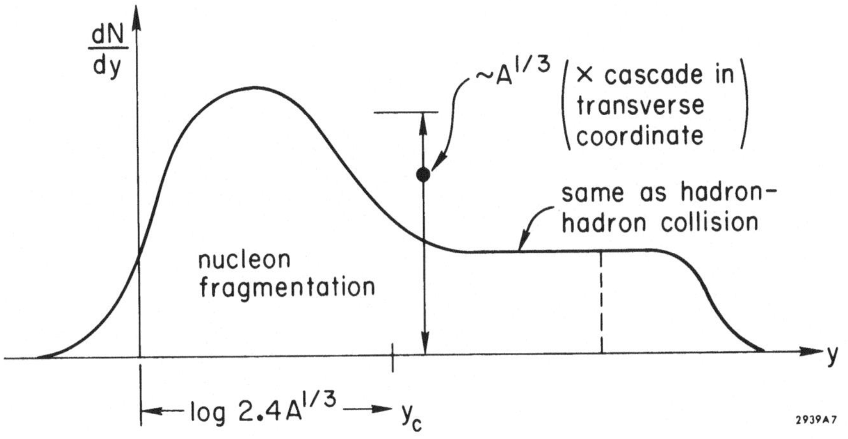
dN/dy
~A^{1/3}
(× cascade in transverse coordinate)
same as hadron-hadron collision
nucleon fragmentation
y
log 2.4A^{1/3}
y_c
2939A7

Fig. 7

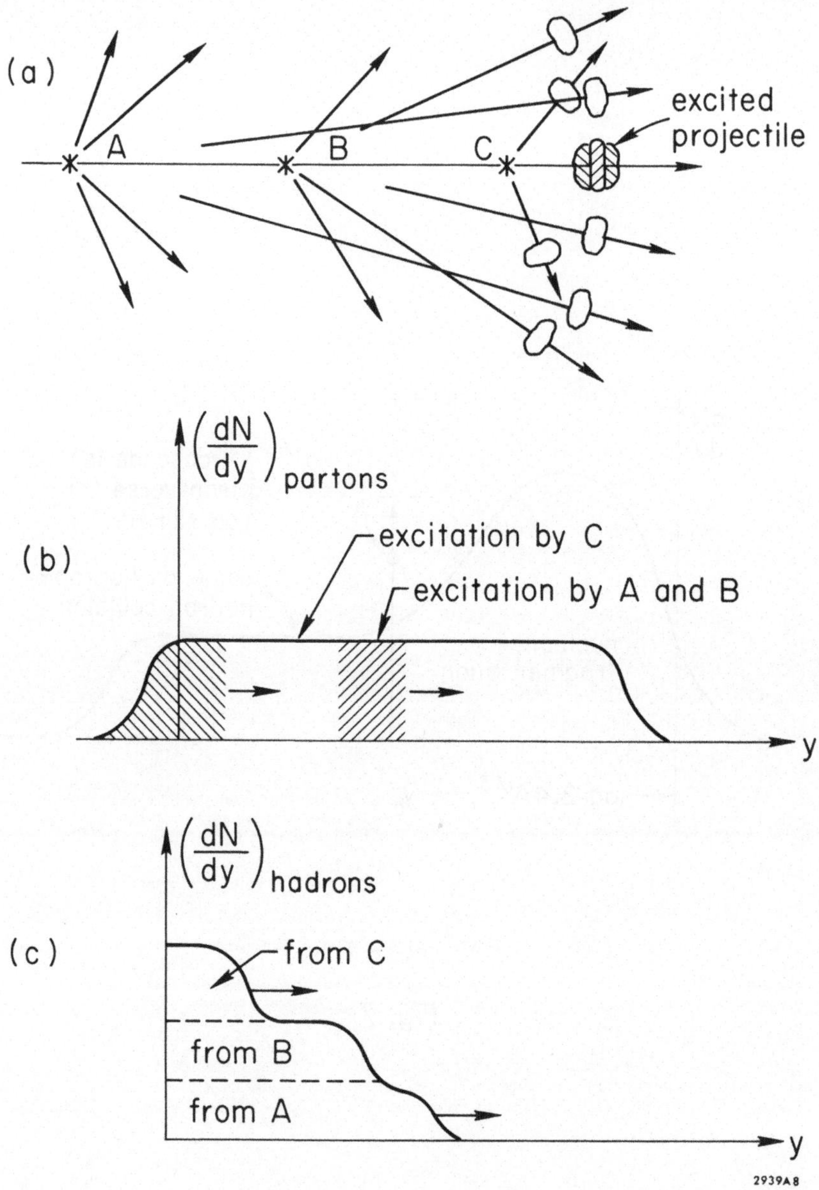

(a)
A
B
C
excited projectile
(b)
$\left(\frac{dN}{dy}\right)_{partons}$
excitation by C
excitation by A and B
y
(c)
$\left(\frac{dN}{dy}\right)_{hadrons}$
from C
from B
from A
y
2939A8

Fig. 8

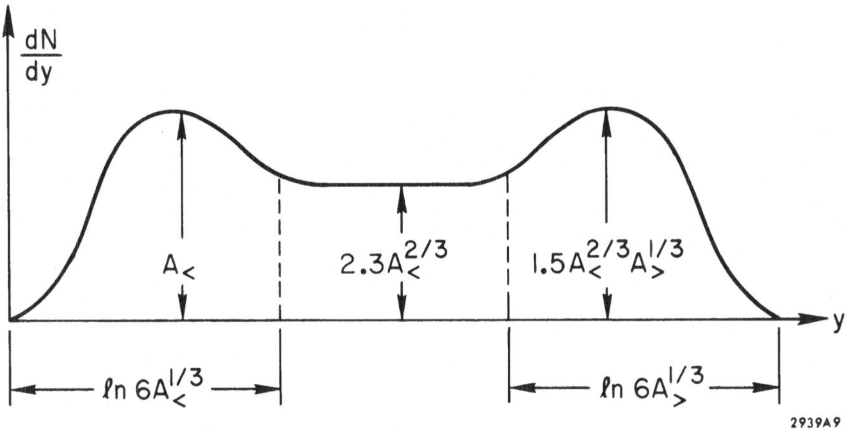

Fig. 9

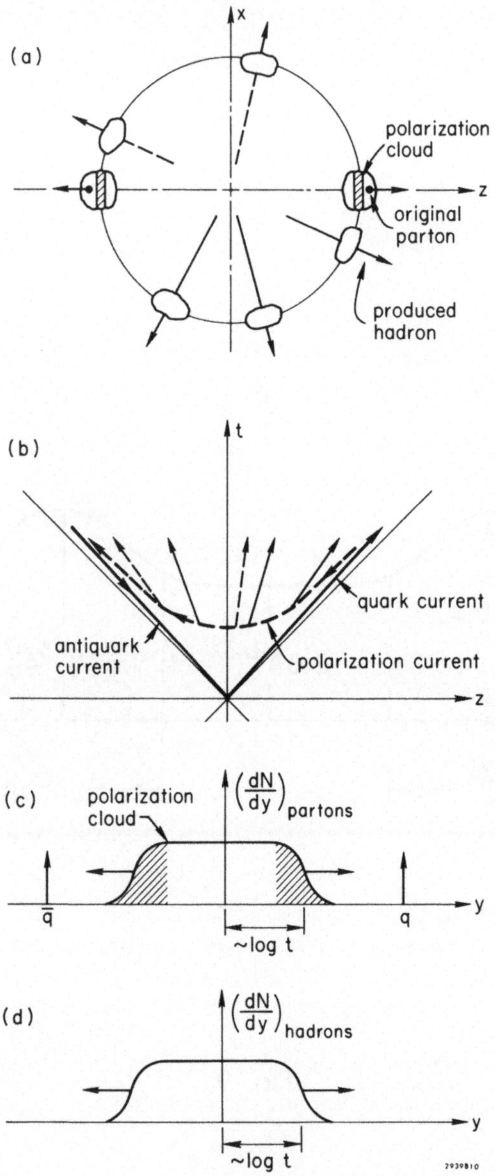

(a)
x
polarization cloud
z
original parton
produced hadron
(b)
t
quark current
antiquark current
polarization current
z
(c)
polarization cloud
$\left(\frac{dN}{dy}\right)_{partons}$
$\bar{q}$
q
y
~log t
(d)
$\left(\frac{dN}{dy}\right)_{hadrons}$
y
~log t
2939B10

Fig. 10

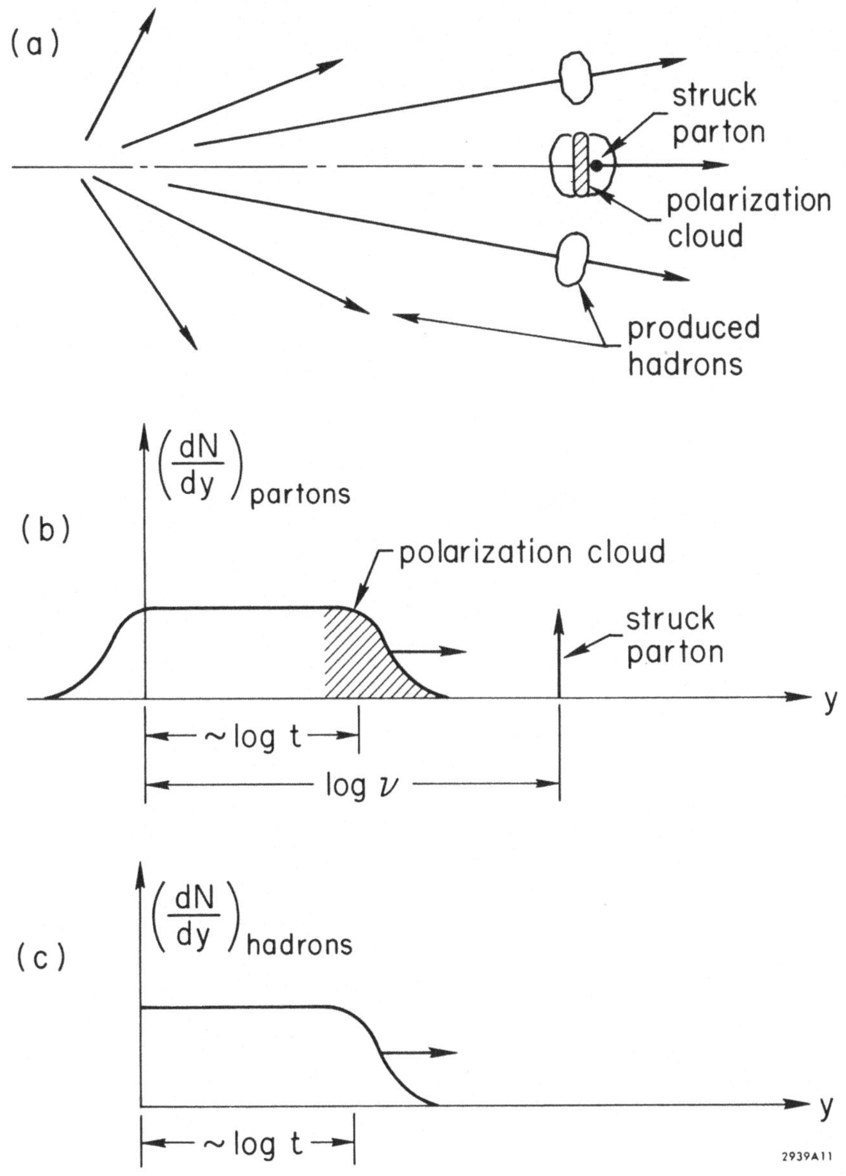
(a)
struck parton
polarization cloud
produced hadrons
(b)
(dN/dy) partons
polarization cloud
struck parton
y
~ log t
log ν
(c)
(dN/dy) hadrons
y
~ log t
2939A11

Fig. 11

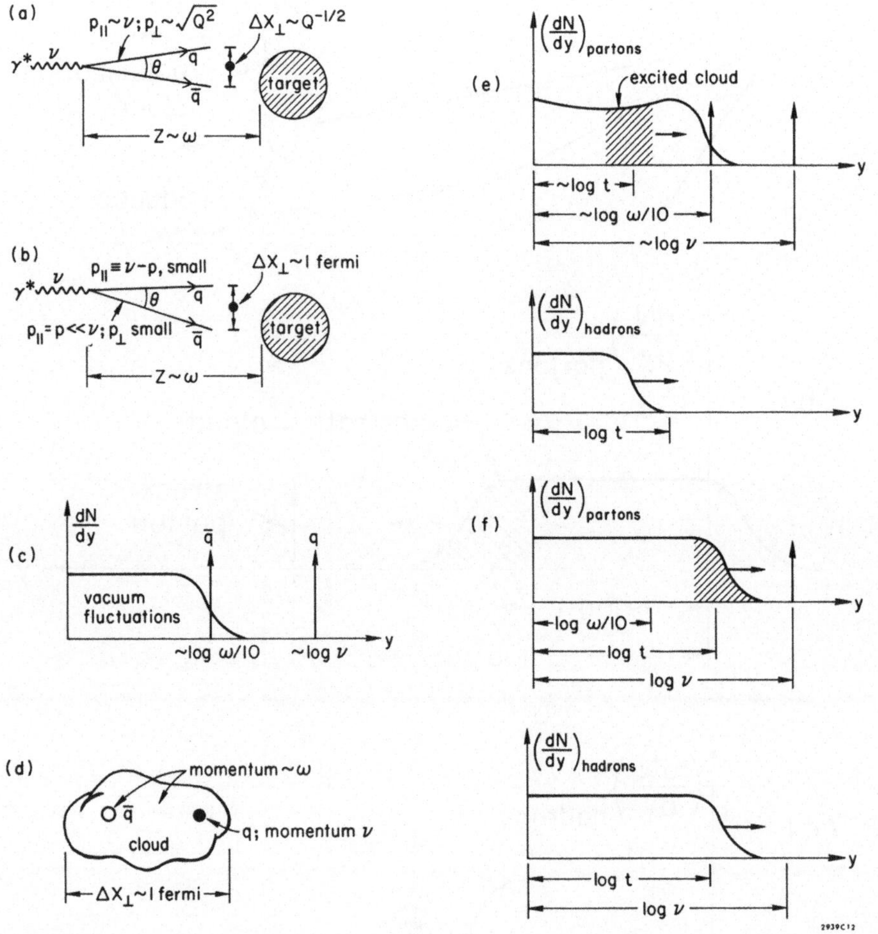
(a)
$p_{\|} \sim \nu ; p_{\perp} \sim \sqrt{Q^2}$
$\Delta X_{\perp} \sim Q^{-1/2}$
γ^*
ν
θ
q
$\bar{q}$
target
$Z \sim \omega$
(b)
$p_{\|} \equiv \nu - p$, small
$\Delta X_{\perp} \sim 1$ fermi
$p_{\|} = p \ll \nu ; p_{\perp}$ small
target
$Z \sim \omega$
(c)
dN/dy
vacuum fluctuations
~log ω/10
~log ν
y
(d)
momentum ~ω
cloud
q; momentum ν
$\Delta X_{\perp} \sim 1$ fermi
(e)
(dN/dy) partons
excited cloud
~log t
~log ω/10
~log ν
(dN/dy) hadrons
log t
(f)
(dN/dy) partons
log ω/10
log t
log ν
(dN/dy) hadrons
log t
log ν
2939C12

Fig. 12

DISPERSION THEORY OF NUCLEON FORM FACTORS+)

by

G. Höhler
Institut für Theoretische Kernphysik
Universität Karlsruhe, Germany

I. Introduction

The determination of the electromagnetic structure of the nucleon is a fundamental problem in Elementary Particle Physics. The structure is described by form factors which occur in the expression for the electromagnetic nucleon current. The simplest reaction induced by this current is electron-nucleon scattering, from which nucleon form factors can be derived easily as long as the one-photon exchange approximation is valid.

During the last 20 years experimentalists have spent a great effort in measuring ep-scattering and the reactions ed → ed and ed → enp, from which neutron form factors can be obtained. At present the best data have systematic errors of only a few per cent and the range of momentum transfer is as large as possible with the existing accelerators and spectrometers. Further progress will be slow, since it would be very difficult to reduce the systematic errors and there is no encouragement by spectacular theoretical predictions.

On the contrary, it is not even clear that a substantial improvement of the experimental accuracy would lead directly to a better knowledge of the nucleon form factors. The reason is that two-photon exchange contributions are expected on the level of 1 %. It would be difficult to separate the one-photon exchange part which contains the nucleon form factors.

Furthermore in the case of neutron form factors an important part of the uncertainty comes from the theoretical method used in the extraction of these form

+) The contributions of our group in this field have been supported by Bundesministerium für Forschung und Technologie.

factors from ed-scattering data. In fact, some of these data have been more useful for our knowledge of the two-nucleon system than for that of the neutron form factor.

Finally the present theory of radiative corrections is expected to be reliable only up to uncertainties of about one per cent.

From a theoretical point of view the reaction $eN \rightarrow eN$ is closely related to nucleon pair production in storage rings $e^+e^- \rightarrow N\bar{N}$ and to the inverse reaction $\bar{N}N \rightarrow e^+e^-$, since the amplitudes are connected by analytic continuation. For both reactions a considerable improvement of the present poor experimental information can be expected in the near future. The results will be of great interest not only for the investigation of resonant intermediate states but also for a combined analysis with the $eN \rightarrow eN$ data.

The present status of the theoretical analysis of eN scattering data is not satisfactory. A long time ago (1958-60) dispersion methods have been developed for the investigation of nucleon form factors. But most of the subsequent authors were content with strongly simplified versions, assuming that t-channel exchanges can be described completely by the exchange of vector mesons in narrow resonance approximations. The results for the parameters differ widely and there is no indication that the uncertainties decreased to the extent that data became more complete and more accurate.

It is the purpose of this lecture to review the status of the theoretical analysis of nucleon form factors. In particular I shall discuss the improvements following from a full exploitation of dispersion theory.

It will be seen that it is possible to obtain results for the spectral functions from which vector meson-nucleon coupling constants can be derived. These coupling constants are of interest for tests of nucleon models and for investigations of other reactions.

2. Determination of Nucleon Form Factors From Cross Sections

Aside from radiative corrections electron-nucleon scattering can be described by the following graphs (see Källén [1], Urban [2], Gourdin [3])

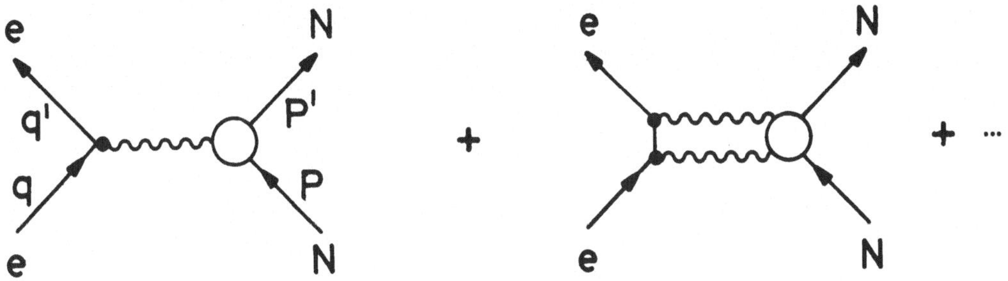

Fig.1: The lowest order graphs for eN scattering

One expects the first term to be dominating and the higher terms to be negligible because of the smallness of the electromagnetic coupling constant $\alpha = e^2/\hbar c \approx 1/137$.

We assume a point-like structure for the eeγ vertex, because ee $\rightarrow$ ee scattering experiments have shown no deviation from the predictions from QED [4]. The γNN and $\gamma\gamma$NN vertices have a structure due to strong interactions.

The γNN vertex is described by the most general expression for the electromagnetic current which is compatible with Lorentz invariance and charge conservation or charge conjugation (see Källen [1])

$$\langle p'|j_\mu(0)|p\rangle = e\,\bar{u}(p')\left\{\gamma_\mu F_1(t) + \frac{i}{2m}\sigma_{\mu\nu}(p-p')^\nu F_2(t)\right\} u(p) \qquad (2.1)$$

$$= e\,\bar{u}(p')\frac{1}{1-(t/4m^2)}\left\{\frac{P_\mu}{m}G_E(t) + \frac{1}{2m^2}(\gamma_\mu \not{P}\not{K} - \not{K}\not{P}\gamma_\mu)\,G_M(t)\right\} u(p). \qquad (2.2)$$

Notation:

$$P_\mu = \frac{1}{2}(p_\mu + p'_\mu); \quad K_\mu = \frac{1}{2}(q_\mu - q'_\mu); \quad \sigma_{\mu\nu} = \frac{i}{2}(\gamma_\mu\gamma_\nu - \gamma_\nu\gamma_\mu);$$

$$t = -Q^2 = (q_\mu - q'_\mu)^2 = -4E^2\sin^2\frac{\theta}{2}\left\{1 + \frac{2E}{m}\sin^2\frac{\theta}{2}\right\}^{-1}; \qquad (2.3)$$

E, θ = total energy and scattering angle of the electron in the laboratory system, m = nucleon mass, μ = pion mass.

Units for t: 1 $GeV^2 \triangleq 2.68\ fm^{-2} \triangleq 51.33\,ƛ_\pi^{-2}$ (natural units); $ƛ_\pi = \hbar/\mu c$

$F_1(t)$ and $F_2(t)$ are the Dirac and Pauli form factors. They are real-valued functions in the physical region ($t < 0$) and different for proton and neutron. Normalization:

$$F_{1p}(0) = 1 \; , \; F_{1n}(0) = 0 \; , \; F_{2p}(0) = g_p' = 1.793 \; , \; F_{2n}(0) = g_n = -1.913. \tag{2.4}$$

We have given in eq. (2.2) an alternative version of the expression for the current, which leads to the electric and magnetic Sachs form factors G_E and G_M

$$G_E(t) = F_1 - \tau F_2 \qquad F_1(t) = \frac{G_E + \tau G_M}{1+\tau} \; , \tag{2.5}$$

$$G_M(t) = F_1 + F_2 \qquad F_2(t) = \frac{G_M - G_E}{1+\tau} \; , \tag{2.6}$$

where

$$\tau = -t/4m^2 \; . \tag{2.7}$$

The Sachs form factors have been preferred in many papers for the following reasons [2]

i) They are related to t-channel helicity amplitudes. Therefore only their squares enter in the expression for the cross section.

ii) They allow a simple transition to the non-relativistic case, where they describe the distribution of charge and magnetic moment.

However, none of the sets of form factors is more fundamental than the other one and we shall see later that the Dirac and Pauli form factors are more suitable for describing the results of our form factor analysis.

Eqs. (2.5), (2.6) show that $F_i(t)$ have a pole at $t=4m^2$ unless G_E and G_M are restricted by the condition

$$G_E(4m^2) = G_M(4m^2) \; . \tag{2.8}$$

These poles cannot be excluded from general considerations, but they correspond to an unusual threshold behaviour of the reaction $e^+e^- \to \bar{N}N$. If eq. (2.8) is not fulfilled, one has D-waves together with S-waves at threshold ($t=4m^2$). If the differential cross section is calculated in the one-photon exchange approximation, one obtains the Rosenbluth formula

$$\frac{d\sigma}{d\Omega_{Lab}} = \frac{d\sigma}{d\Omega_{NS}} \left\{ \frac{G_E^2 + \tau G_M^2}{1+\tau} + 2\tau G_M^2 \tan^2\frac{\Theta}{2} \right\}, \qquad (2.9)$$

where the "no structure" cross section

$$\frac{d\sigma}{d\Omega_{NS}} = \frac{\alpha^2}{4E^2} \frac{\cos^2\frac{\Theta}{2}}{\sin^4\frac{\Theta}{2}} \left\{ 1 + \frac{2E}{m} \sin^2\frac{\Theta}{2} \right\}^{-1} \qquad (2.10)$$

is the Born approximation for scattering between a Dirac electron and a spinless point charge of proton mass m.

G_E^2 and G_M^2 can easily be determined from the parameters of the best straight line in a plot of $(d\sigma/d\Omega_{Lab})/(d\sigma/d\Omega_{NS})$ vs. $\tan^2(\Theta/2)$ at fixed Q^2 ("Rosenbluth plot"). Bartel et al. [5] preferred to plot $\cos^2(\Theta/2)\times$ $\times(d\sigma/d\Omega_{Lab})(d\sigma/d\Omega_{NS})$ vs. $\cos\Theta$ which, according to eq. (2.9), must also give a straight line and has the advantage that the physical range of the abscissa is finite.

One should notice that a deviation from linearity in these plots would indicate that the one-photon exchange approximation is not sufficient. However the linearity does not exclude certain two-photon exchange contributions (Gourdin [3]).

In general, neutron form factors have to be determined from ed-scattering, but the theory of the ed-system is not yet in good shape [5]. Although data for elastic ed-scattering are available up to $Q^2 = 6\ GeV^2$ (Ref. [6]), we shall use results for $2G_{Es} \equiv (G_{Ep} + G_{En})$ only up to $0.5\ GeV^2$. Results for G_{Mn} at small Q^2 have led to some difficulties which have been resolved in a recent paper by Bethe et al. (Ref. [7]). It is remarkable that even at $Q^2 < 0.02\ GeV^2$ ed → ed scattering data together with the slope of G_{En} at t = 0 give better information on the low energy parameters of the np system than that available from nuclear physics [8].

At $Q^2 \gtrsim 0.5\ GeV^2$ we have used neutron form factors determined from "quasi-elastic" ed scattering: ed $\rightarrow$ enp. The most reliable results follow from the "ratio method", in which the ratio of cross sections belonging to ep and en final states is measured at the quasielastic peak [5]. There are theoretical reasons to believe that this ratio agrees well with that of free ep and en cross sections. The results from other methods are more sensitive to features of the deuteron theory. In both elastic and inelastic ed-scattering the main problems are to understand the role of meson-exchange currents and of isobar admixtures [7,9,10].

A further important information has been obtained from scattering of thermal neutrons by atomic electrons, which allow to determine the slope dG_{En}/dQ^2 at $Q^2 = 0$. This was the first nucleon form factor experiment (Fermi and Marshall, 1947) and it is now one of the most accurate ones. The effect is, of course, very small and has to be separated from the large neutron-nucleus scattering. This can be done by using the fact that the two scattering amplitudes have a different dependence on scattering angle and neutron energy because of the different target size. It is interesting to consider the slope dF_{1n}/dQ^2 in addition to that of $G_{En}(Q^2)$

$$G'_{En}(0) = F'_{1n}(0) - \frac{F_{2n}(0)}{4m^2} , \tag{2.11}$$

because it turns out that the second term ("Foldy term") is strongly dominating, i.e. the charge distribution of the neutron is mainly an effect of the anomalous magnetic moment (Foldy [11]).

Up to now the reaction $e^+e^- \rightarrow \bar{p}p$ has been investigated only in a single experiment at Frascati [12] at $t = 4.4\ GeV^2$ and in the region of the $\psi(3095)$ resonance, the branching ratio being $0.21 \pm .04\%$(Ref. [13]). Very recently the first results for $\bar{p}p \rightarrow e^+e^-$ have been reported [14], only bounds have been available before.

Finally it should be noted that it is possible to obtain information on nucleon form factors from pion electroproduction ($ep \rightarrow eN\pi$) (see for instance Ref. [15]) and from the inverse reaction ($\pi^- p \rightarrow e^+e^- n$, Ref. [16]). However there are two difficulties:

i) One has to determine simultaneously the pion form factor,

ii) In contradistinction to the dispersion approach a further form factor appears in an approach based on current algebra and PCAC: the form factor $G_A(Q^2)$ of the axial part of the weak current.

The relation between the two approaches has been investigated in Refs. [17,18]. It seems to us that at present these methods do not improve our knowledge of the electromagnetic nucleon form factors, but it might be that they will help to determine the neutron form factors in the future. Furthermore we think that the errors of results presented for the pion form factor deserve further investigations as well as those of the approach of Ref. [18].

3. Two-Photon Exchange and Radiative Corrections

In a classical description the two-photon exchange graph is related to the polarizability of the nucleon, which follows from the deformation of its pion cloud in an external field. In quantum mechanics the same effect can be described by inserting intermediate states in that part of the graph which belongs to virtual Compton scattering, the most important intermediate state being the Δ(1232) nucleon isobar (Compare the situation in the case of scattering of electrons by atoms [19], where dispersion terms occur in addition to those containing the atomic form factor).

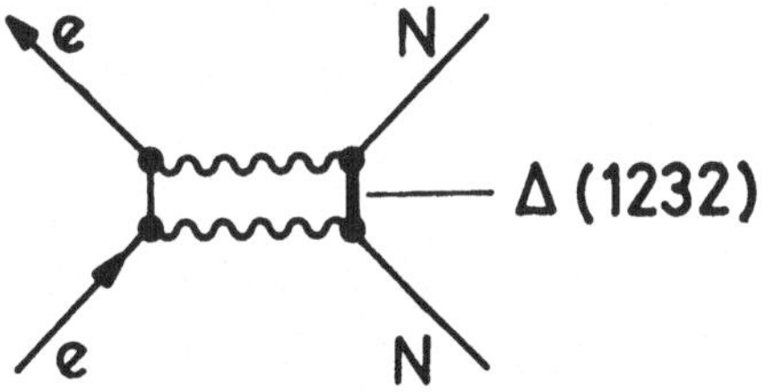

Fig. 2 Two photon exchange

A calculation of the two-photon exchange term in electron-nucleon scattering is difficult and cannot be performed at present in a quantitatively reliable way (Ref. [20] and papers quoted there). The estimates usually give corrections of the order of 1 % or less to $d\sigma/d\Omega$.

Recently it was claimed that the neutron polarizability contribution to the n-e scattering length is within the level of experimental precision [21], but Brown et al. [22] showed that the effect is negligible.

The electric proton polarizability can be obtained from data and from the forward dispersion relation for proton Compton scattering [23]. The result: $\approx 10^{-42}$ cm^3 indicates that the proton is rather rigid, since the classical estimate (radius)^3 gives the larger value 0.5 10^{-39} cm^3. - An attempt to estimate the neutron polarizability has been made from scattering of neutrons off uranium [24].

A direct experimental information on the magnitude of the two-photon exchange terms can be obtained in the following ways:

i) In the one-photon exchange approximation the amplitudes for $e^+p \to e^+p$ and $e^-p \to e^-p$ scattering differ only by a sign, so the cross sections are equal. The difference $\sigma_- - \sigma_+$ ($\sigma \equiv d\sigma/d\Omega$) is proportional to the interference term between the real one-photon exchange amplitude and the real part of the two-photon exchange amplitude. This is usually expressed by the formula

$$R = \frac{1}{2}\,\frac{\sigma_- - \sigma_+}{\sigma_- + \sigma_+} = \frac{\mathrm{Re}\,A_{2\gamma}}{A_{1\gamma}} \tag{3.1}$$

which is simplified because two independent amplitudes are needed for the description of ep scattering. (See Gourdin [5] for a more detailed treatment.) The experiments give [25]

$$R \lesssim 1\ \% \qquad 0.01 < Q^2 < 1\ \mathrm{GeV}^2$$

for

$$R \lesssim 2\ \% \qquad 1 < Q^2 < 4\ \mathrm{GeV}^2$$

ii) The imaginary part of the two-photon exchange amplitude can be measured in polarization experiments. One has to measure either the polarization parameter P which follows from the polarization of the recoil proton in the case of an unpolarized target or the asymmetry parameter A, for which a polarized target is needed. (A=P from time reversal invariance). The experimental results [25,26] show A=P=0 with an accuracy of (1-3) % in the range $0.2 < Q^2 < 1\ \mathrm{GeV}^2$ and 8 % at 2 GeV^2.

Radiative corrections have to be considered, because it is not possible to measure elastic ep-scattering alone. Since the energy resolution is finite, one always includes inelastic processes in which soft photons are emitted. In quantum electrodynamics this contribution diverges. But it turns out that the "infrared divergence" is cancelled, if one considers not only the emission of soft Bremsstrahlung photons but also the exchange of soft internal photons.

It can be expected that soft photons are well described in a classical treatment and this leads in fact to an understanding of the cancellation of the divergence. There remains a finite "radiative correction", which has to be subtracted from the data before the form factors are determined. This problem has been treated in the books of Akhiezer-Berestetskii [27] and Urban [2]. Other reviews and some original papers are given in Refs. [28-30]. There remain uncertainties of the order of 1.5 % (Ref. [31]). A test of the correction procedure has been performed in a recent DESY experiment [5].

4. Summary of the Experimental Results

4.1 ep → ep

The usual statement is that the data are approximately described by the relation

$$\underbrace{G_{Ep}(Q^2) \approx \frac{G_{Mp}(Q^2)}{g_p} \approx \frac{G_{Mn}(Q^2)}{g_n}}_{\text{Scaling Law}} \approx \underbrace{G_D(Q^2) \equiv \left(1 + \frac{Q^2}{Q_D^2}\right)^{-2}}_{\text{Dipole Law}} . \tag{4.1}$$

It is a generally accepted convention to take

$$Q_D^2 = 0.71 \text{ GeV}^2 \triangleq 18.2 \text{ fm}^{-2} \triangleq 36.4 \mu^2 . \tag{4.2}$$

We want to rewrite the Rosenbluth formula in such a way that a "modified Rosenbluth plot" shows immediately the deviations from the scaling law and the dipole law. For this purpose we introduce the following notation ($g_p = 2.793$)

$$\Delta_D(Q^2) \equiv \left[\frac{G_{Mp}/g_p}{G_D}\right]^2 - 1 , \quad \Delta_{Sc}(Q^2) \equiv \left[\frac{g_p G_{Ep}}{G_{Mp}}\right]^2 - 1 , \tag{4.3}$$

$$\Delta(Q^2,X) \equiv \frac{d\sigma/d\Omega_{lab}}{d\sigma/d\Omega_{DS}} - 1 = \frac{d\sigma_{lab}}{d\sigma_{NS}} \left\{ \frac{1+\tau}{1+g_p^2\tau} \; \frac{X(Q^2,\theta)}{G_D^2} \right\} - 1 , \tag{4.4}$$

$$X(Q^2,\theta) = \left\{ 1 + \frac{1+\tau}{1+g_p^2\tau} \, 2\tau g_p^2 \tan^2\frac{\theta}{2} \right\}^{-1} ; \quad \tau = \frac{Q^2}{4m^2} \tag{4.5}$$

In terms of these variables the modified Rosenbluth formula reads

$$\Delta(Q^2, X) = \Delta_D(Q^2) + \frac{(1+\Delta_D)\Delta_{Sc}}{1+g_P^2\tau} X(Q^2, \theta) . \qquad (4.6)$$

The cross section $d\sigma/d\Omega_{DS}$ in eq. (4.4) is obtained from the Rosenbluth formula (2.9) by inserting the dipole and scaling law, so it is known explicitly. Therefore $\Delta(Q^2,X)$ is given by the experimental data. It will be plotted in two different ways:

i) as a function of Q^2 for all X. For each Q^2 one obtains several points which belong to different scattering angles.

ii) As a function of X at fixed Q^2 ("modified Rosenbluth plot"). According to eq. (4.6) all experimental points must lie on a straight line for the same reason which leads to a straight line in the usual Rosenbluth plot. The advantage of the modified plot is that the intercept of the straight line at X=0 (corresponding to $\Theta = 180^\circ$) gives directly the violation of the dipole law and the slope is proportional to the violation of the scaling law.

Fig. 3 shows $\Delta(Q^2,X)$ vs Q^2 for all X. If the scaling law is valid, all points have to lie on a universal curve, i.e. $\Delta(Q^2,X)$ does not depend on the second variable X, deviations being only due to experimental errors. It is seen that this is true to a good approximation at small Q^2 ($\lesssim 1$ GeV^2). A quantitative determination of Δ_{Sc} from eq. (4.6), (Fig. 4a), shows that there are clear deviations from the scaling law at larger Q^2.

At fixed Q^2 (Fig. 4b) data at small and large X-values come frequently from different laboratories. Therefore the slope in Fig. 4b and Δ_{Sc} can easily be falsified by normalization errors. In some cases the data of a single laboratory differ more than statistically expected from a straight line (see the data of Hofstaedter et al., Phys. Rev. 142 (1966) 922 in Fig. 4b and Ref. [95], Fig. 3,4). Reliable data from the same experiment in a large X-interval have been determined in a recent DESY experiment [5].

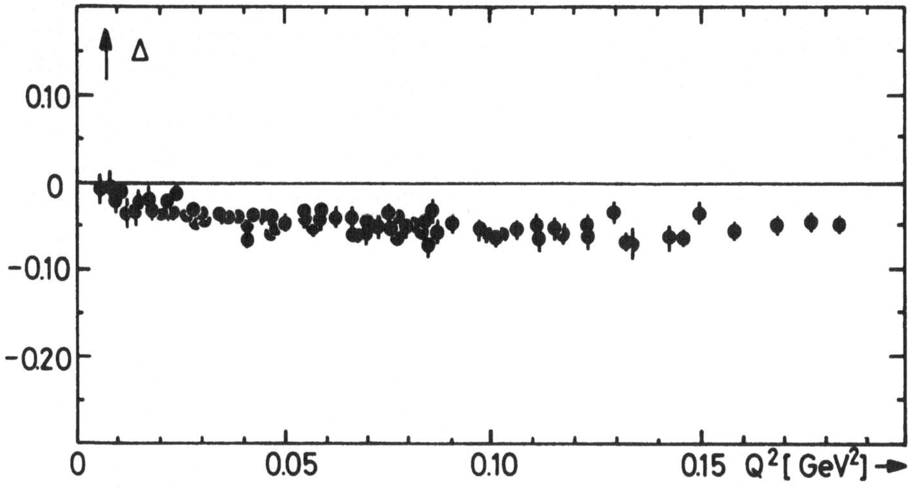

Fig. 3a

Fig. 3a. Δ vs. Q^2 at small Q^2. Data of the Mainz group (Ref. [94,95]).

Fig. 3b (next page). Δ vs. log Q^2 at larger Q^2. Data: | Bonn 1971, ┼ DESY 1973 (Ref. [5]), ⧫ SLAC 1973 (Ref. [31]), ⧫ SLAC 1975 (Atwood, thesis). The figures show only a selection of data. Plots containing all data will be included in the "Compilation" mentioned at the end of this article.

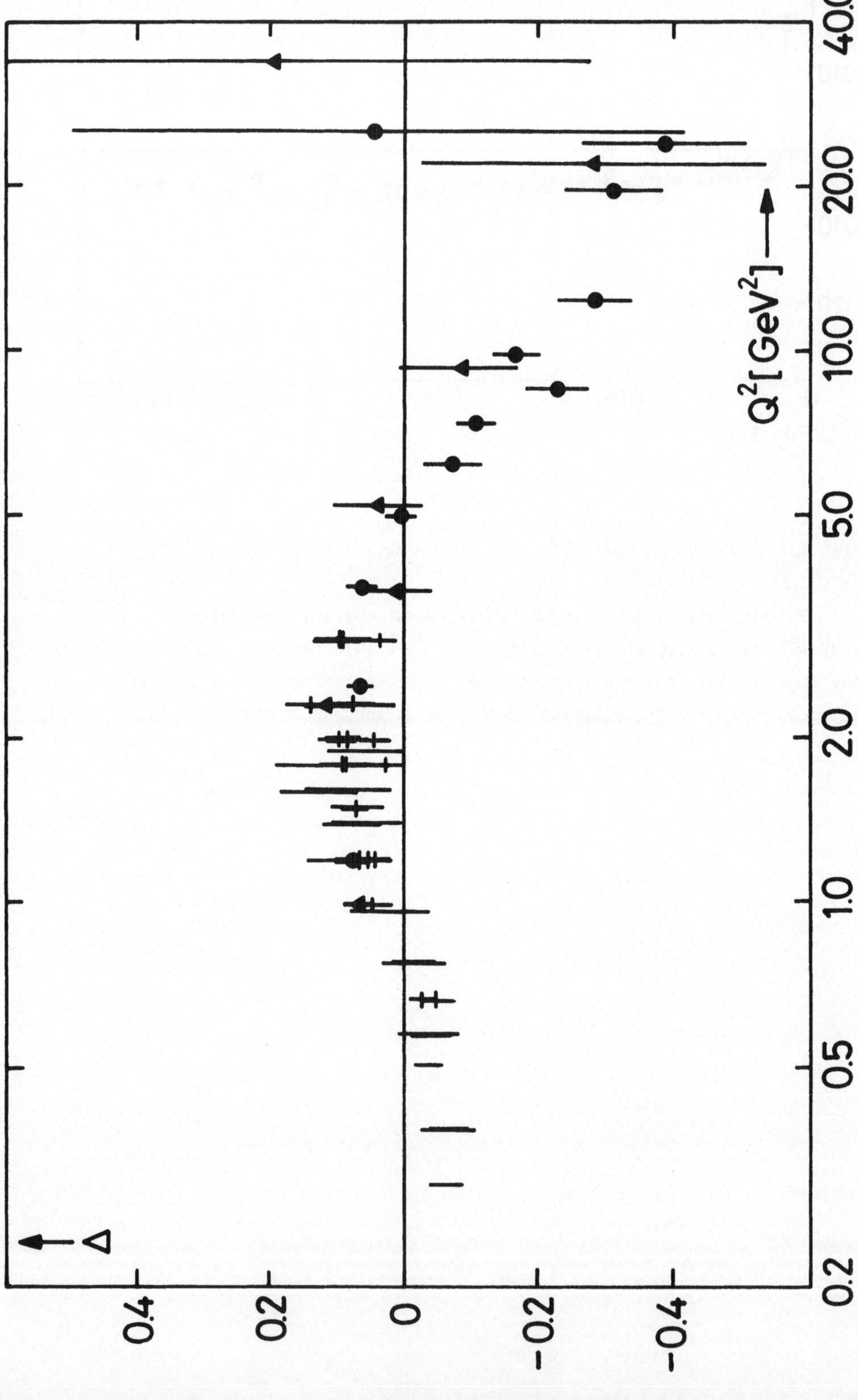

FIG. 3b

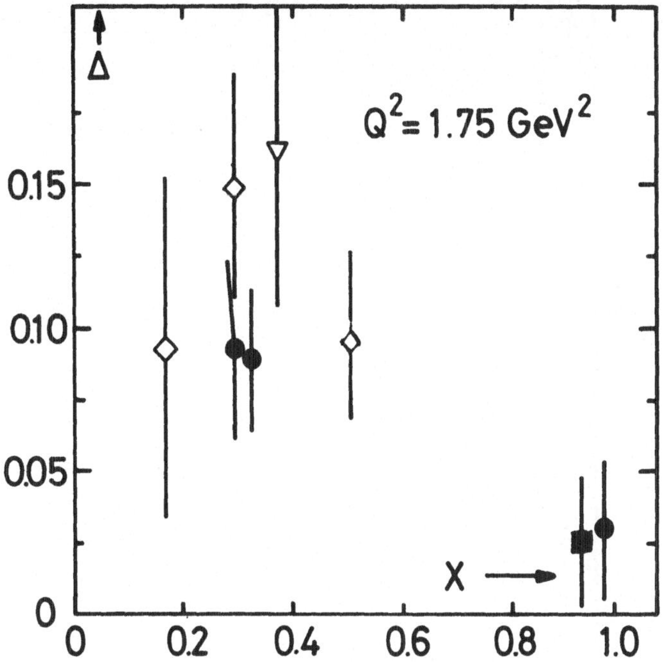

Fig.4a

Fig.4: Modified Rosenbluth Plots, Eq.(4.6).

a) $Q^2 = 1.75\ GeV^2$. Black circles: DESY[5].

b) $Q^2 = 0.78\ GeV^2$. Black triangles: Stanford data

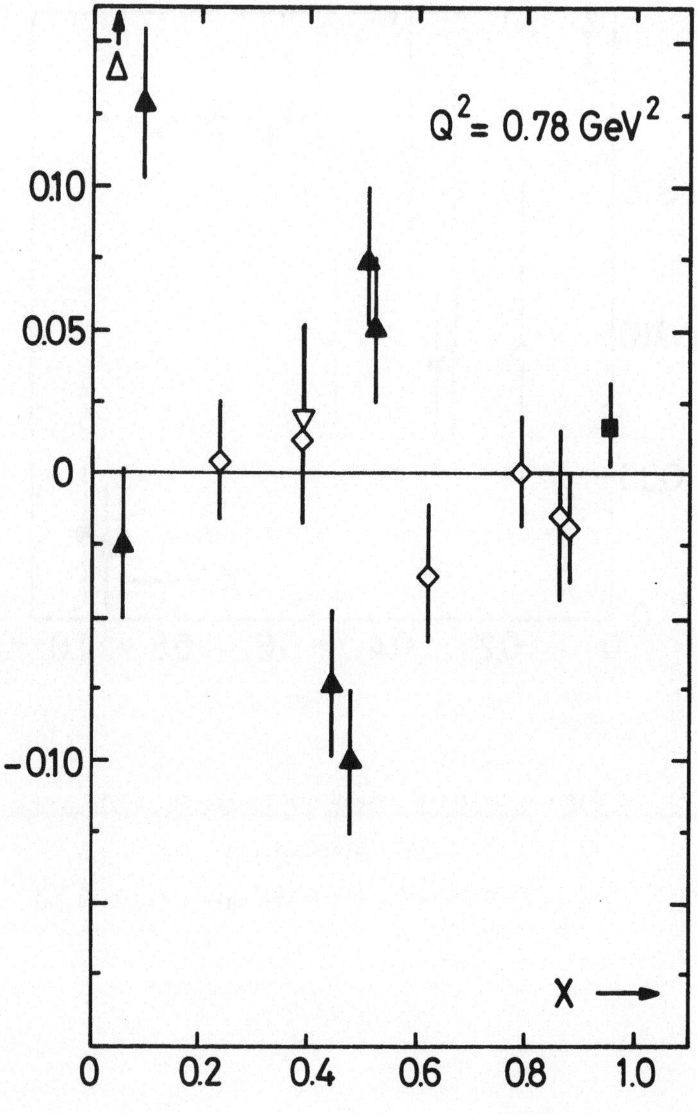

Fig.4b

Figs. 3,4 show that the deviations from the dipole law are of the order of a few per cent at small Q^2 (< 1 GeV^2) and less than about 20 % up to the largest Q^2-value where data are available (35 GeV^2).

4.2 Neutron Form Factors

The information on neutron form factors has recently been summarized by Bartel et al. [5], Felst [32] and Hanson et al. [33]. We restrict ourselves to new results from neutron-electron scattering on the slope of G_{En} at $Q^2 = 0$. +)

Some authors gave their result in terms of the neutron-electron scattering length and others preferred to describe the neutron-electron interaction by a square well potential of radius $r_e = e^2/m_e c^2 = 2.818$ fm and depth V_o (Fermi convention).

The relation between the different quantities reads

$$\left.\frac{dG_{En}}{dQ^2}\right|_{Q^2=0} = \frac{-V_0}{3e^2/\hbar c}\,\frac{r_e^3}{\hbar c} = -5.180\ 10^{-6}\ V_0\,[\mathrm{fm}^2]\ ;\ V_0 \text{ in eV}$$

$$= -b_{ne}\,\frac{\lambda\!\!\!^{-}{}_e^2}{2r_e}\,\frac{m_e}{m_p} = -14.41\ b_{ne}\,[\mathrm{fm}^2]\,;\ b_{ne} \text{ in fm}. \qquad (4.7)$$

$\lambda\!\!\!^{-}{}_e = \hbar/m_e c$, m_p = proton mass.

The most accurate experimental results are listed in Table 1.

+) I am much indebted to Dr. L. Koester for a discussion and private communications.

dG_{En}/dQ^2 at $Q^2 = 0$ 10^{-2} fm^2	dF_{1n}/dQ^2 at $Q^2 = 0$ 10^{-2} fm^2	$-b_{ne}$ 10^{-3} fm	$-V_o$ eV	Authors (Ref.[34])
2.25 ± 0.07	0.13 ± 0.07	1.56 ± 0.05	4340 ± 140	MELKONIAN
2.15 ± 0.08	0.03 ± 0.08	1.49 ± 0.05	4150 ± 150	Corrected by KOESTER +)
1.89 ± 0.04	-0.23 ± 0.04	1.30 ± 0.025	3630 ± 70	KROHN
2.23 ± 0.03	0.11 ± 0.03	1.55 ± 0.02	4300 ± 60	ALEXANDROV, Preliminary
1.99 ± 0.03	-0.13 ± 0.03	1.380 ± 0.017	3840 ± 50	KOESTER
2.115		1.468	4083	Foldy-term

Table 1. New results on the slope of G_{En} at $Q^2 = 0$. (1 fm$^2 \triangleq$ 25.68 GeV^{-2}).

It is not meaningful to calculate a "world average", since the results differ by many standard deviations and Alexandrov's value is preliminary. The very careful investigations of Koester et al. and Krohn and Ringo suggest strongly that the slope $F'_{1n}(0)$ is negative. At large Q^2(0.24 ... 0.50 GeV^2) the determination of F_{1n} from the elastic ed-scattering data of Galster et al.[35] gives also negative values ++).

+) Correction for Schwinger scattering and resonance contribution

++) There is a printing error in Fig. 13 of this paper (The dashed curve belongs to p = 4 and the dash-dotted one to p = 0). Presumably this is the reason why the dashed line in Fig. 43 of Gourdin's review[36] contradicts our statement.

5. Summary of Theoretical Methods for the Analysis of Nucleon Form Factors

If one attempts to classify theoretical papers on nucleon form factors, one can distinguish two groups:

i) Investigations based on models for extended nucleons (for instance quark models). The most important results follow from symmetry properties [37]. Static SU(6) predicts $g_p/g_n = -3/2$, which almost agrees with the experimental value: 1.459. Relativistic SU(6) predicts the scaling law for G_{Mn} and G_{Mp} and $G_{En} \equiv 0$. Since the second result is not satisfactory, several authors have tried to derive a more realistic behaviour (see for instance Refs. [38,39]). Up to now it is hard to see, how the semi-quantitative treatment can be improved to become reliable quantitatively.

ii) Investigations based on dispersion methods have a more modest aim. Nucleon form factors are analyzed together with other experimental data in order to obtain results on the mechanism which determines the structure of the nucleon. It will be seen that these methods are not just fits to the data. It can happen that experimental nucleon form factor data are incompatible with dispersion theory, if other reliable data are used as an additional input.

There are two different dispersion approaches:

i) Dispersion relations for the nucleon form factors as a function of $t = -Q^2$ (Refs. [40-42]). (Subtractions will be discussed later.)

$$\mathrm{Re}\, F_i(t) = \frac{1}{\pi} \int_{4\mu^2}^{\infty} \frac{\mathrm{Im}\, F_i(t)}{t'-t}\, dt' \tag{5.1}$$

ii) Dispersion relations for the nucleon form factors as a function of the mass W of the incoming nucleon at fixed t: $F_i(t,W)$, the outgoing nucleon being on its mass shell (Ref. [43])

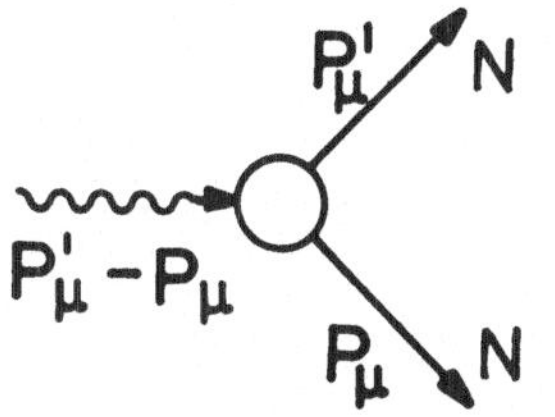

$$t = -Q^2 = (p'_\mu - p_\mu)^2$$

$$p_\mu^2 = W^2, \quad p'^2_\mu = m^2$$

Fig. 5 The γNN-vertex

Of course one considers finally the on-shell form factor $F_i(t,m^2)$. Its dispersion relations read [43]

$$\mathrm{Re}\, F_2(t,m) = \frac{1}{\pi}\int_{m+\mu}^{\infty} dW \left\{ \frac{\mathrm{Im}\, F_2(t,W)}{W-m} + \frac{\mathrm{Im}\, F_2(t,-W)}{W+m} \right\} ,$$

$$\mathrm{Re}\, F_3'(t,m) = \frac{1}{\pi}\int_{m+\mu}^{\infty} dW \left\{ \frac{\mathrm{Im}\, F_3(t,W)}{(W-m)^2} + \frac{\mathrm{Im}\, F_3(t,-W)}{(W+m)^2} \right\} . \tag{5.2}$$

The Dirac form factor F_1 is given in terms of F_3 by

$$F_1(t,m) = F_1(0,m) - t\, F_3'(t,m) . \tag{5.3}$$

The prime denotes differentiation with respect to W.

The main point in both dispersion approaches is the fact that Im F_i can be calculated from the extended unitarity relation, inserting intermediate states.

i) In the case of Im $F_i(t)$ only those intermediate states are admitted which have the quantum numbers of the photon and an invariant mass squared less than t:

2 pions	$t > 4\mu^2$	$\bar{K}K$	$t > 4m_K^2 \approx 50\mu^2$
3 pions	$t > 9\mu^2$	$\bar{N}N$	$t > 4m^2 \approx 180\mu^2$
4 pions	$t > 16\mu^2$	etc.	

It is useful to introduce the isoscalar and isovector combinations of form factors

$$F_{is} = \tfrac{1}{2}(F_{ip} + F_{in}) , \qquad F_{iv} = \tfrac{1}{2}(F_{ip} - F_{in}) , \tag{5.4}$$

because then the G-parity selection rule allows only an even number of pions in the intermediate state in the isovector case and an odd number in the isoscalar case. For example

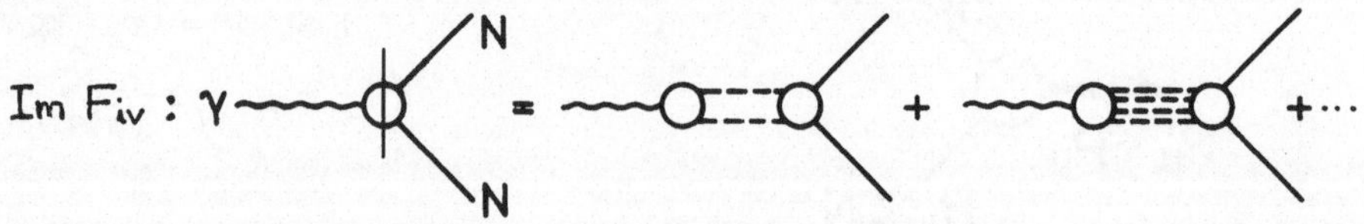

Fig. 6 Intermediate States in the isovector case

A large part of the contribution of the intermediate states to the unitarity relation can be approximated by the exchange of resonances, which have the quantum numbers of the photon (vector mesons).

isoscalar case	$\omega, \phi, \dots \psi$
isovector case	$\rho, \rho', \dots$

However it will be shown below that in contrast to the assumption of many authors a narrow resonance approximation is not sufficient in the case of ρ-exchange because of peculiar circumstances.

ii) The intermediate states in the unitarity relation for Im $F_i(t,W)$ at fixed t are obtained, if the vertex is cut sidewise ("Sidewise dispersion relation")

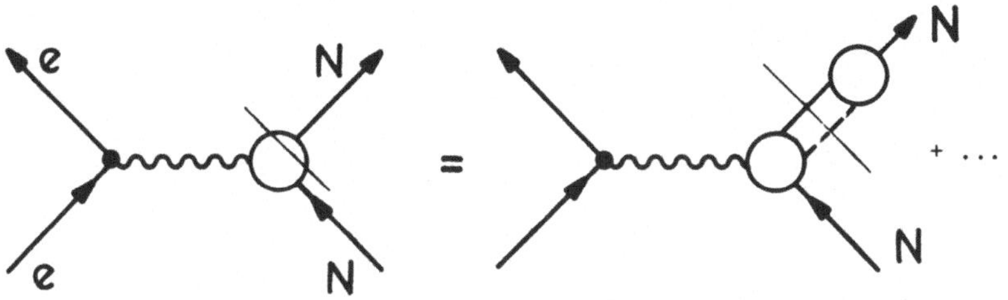

Fig.7: Intermediate states in sidewise dispersion relations.

The intermediate states have now the nucleon number 1 and $J^P = 1^{\pm}$ Since the one-nucleon state gives no contribution [43], the lowest state is the πN state. Fig. 7 shows that one factor in the unitarity relation is the amplitude for electroproduction of pions on nucleons ($e^- p \to e^- N\pi$). The second factor is the pionic form factor of the nucleon.

If the electroproduction amplitude is decomposed into multipoles, there are only contributions from final πN states $J^P = 1^{\pm}$: (E_{0+}, L_{0+}, M_{1-}, L_{1-}). It is remarkable that the list includes states of both parities (see Ref.[43]).

In order to determine the pionic form factor of the nucleon +) one can again write a dispersion relation and insert the lowest intermediate states into the unitarity relation. Since only $J^P = 1^{\pm}$ states are possible, one needs S and $P_{1/2}$ πN partial waves, which are not so well known at higher energies. Several

+) Other dispersion relations for the pionic form factor of the nucleon have been considered in connection with the Goldberger-Treiman relation. See Ref.[44]

interesting results have been derived from sidewise dispersion relations by Drell et al. [45] and others [46]. However it seems difficult to justify the neglect of πN intermediate states at higher energies [47] and of $\pi\pi N$ intermediate states (see Ref. [48] for an estimate). Therefore the predictive power is poorer than that of dispersion relations in t where, at small $|t|$, the dominator of the dispersion integral suppresses contributions from the exchange of higher vector mesons (see §8).

Sidewise dispersion relations have been proven from axiomatic field theory [43], whereas the proof of dispersion relations in t led to difficulties due to the unphysical region, similar to those in the cases of KN and NN scattering. However, this is no serious argument against phenomenological applications [40].

In the following we shall restrict ourselves to dispersion relations in t.

6. Dispersion Relations for $F_i(t)$, $G_E(t)$ and $G_M(t)$

The unsubtracted dispersion relation for the Dirac and Pauli form factors read ($t \equiv -Q^2$)

$$\mathrm{Re}\, F_i(t) = \frac{1}{\pi} \int_{4\mu^2}^{\infty} \frac{\mathrm{Im}\, F_i(t')}{t'-t}\, dt' . \tag{6.1}$$

For the isoscalar combination $\mathrm{Im}\, F_{is} \equiv 0$ below $9\mu^2$.
One can as well consider dispersion relations for G_E and G_M

$$\mathrm{Re}\, G_E(t) = \frac{1}{\pi} \int_{4\mu^2}^{\infty} \frac{\mathrm{Im}\, G_E(t')}{t'-t}\, dt' , \quad \mathrm{Re}\, G_M(t) = \frac{1}{\pi} \int_{4\mu^2}^{\infty} \frac{\mathrm{Im}\, G_M(t')}{t'-t}\, dt' . \tag{6.2}$$

Neither of them is more fundamental than the other one, but for special applications one of them might be more convenient. Furthermore, one should notice that a certain approximation (for instance ρ-dominance) gives different results if it is used in (6.1) and (6.2).

In order to discuss the large t behaviour, we start from the fact that the ep-scattering data up to 35 GeV^2 suggest

$$t^2 G_{Mp} \to \text{const} \quad \text{as} \quad -t \to \infty . \tag{6.3}$$

Therefore the integral in (6.2) converges and one even has a "superconvergence" sum rule

$$\int_{4\mu^2}^{\infty} \mathrm{Im}\, G_{Mp}(t)\, dt = 0 \quad . \tag{6.4}$$

Since according to (2.6): $M_{Mp} = F_{1p} + F_{2p}$, one has two possibilities:

i) Both F_{1p} and F_{2p} behave like G_{Mp} at large (-t),

ii) each of them has a slower decrease and the leading term cancels in the sum.

In order to distinguish these cases one needs G_{Ep}. Unfortunately the data do not allow us to determine its asymptotic behaviour. G_{Ep} is kinematically suppressed above 1 GeV^2 and therefore has rapidly increasing errors [+)].

The data show that G_{Ep} decreases more rapidly than G_{Mp} in the range 1 $GeV < Q^2 < 3$ GeV^2. Simplicity suggests either superconvergence also for G_{Ep} or a change of sign. Theoretically there is not even an argument to exclude a finite limit ("hard core") $G_{Ep} \to$ const as $-t \to \infty$. In this case a subtraction is required in the dispersion relation for G_{Ep}, but not in those for F_{1p} and F_{2p} [++)].

The discussion of superconvergence properties will be continued later.

As mentioned above an important feature of the dispersion approach is the possibility to determine Im $F_i(t)$ at $t > 4\mu^2$ from extended unitarity. In practice this can be done quantitatively only for two-pion intermediate states (Refs. [40,41]) which are important because of the existence of the ρ-meson resonance (Ref. [42]). For the other parts of the spectral functions unitarity is used only insofar as it suggests to assume a peak of Im F_i at t-values corresponding to the mass squared of the other vector meson resonances ($J^P = 1^-$).

+) The only way to determine G_{Ep} at large Q^2 is to use a polarized target and to measure the asymmetry on flipping its polarization which gives $G_{Ep}G_{Mp}$ (Ref. [49]).

++) The situation would be similar to that in πN forward scattering where a subtraction is required in C^+, but possibly not in A^+ (Ref. [50]). - See Ref. [87] for a discussion of hard cores.

The unitarity relation for two pion intermediate states reads [42]

$$\text{Im}\, F_{iv}(t) = \frac{q_t^3}{\sqrt{t}} F_\pi(t)\,\Gamma_i^*(t) \equiv \frac{q_t^3}{\sqrt{t}} |F_\pi|^2 J_i(t) \tag{6.5}$$

where $q_t^2 = \frac{1}{4}t - \mu^2$, $F_\pi(t)$ is the electromagnetic pion form factor, and $\Gamma_i(t)$ the I=J=1 $\pi\pi N\bar{N}$ partial wave amplitudes (i=1,2) [+)]. $J_i(t)$ is the numerator function in a N/D representation of $\Gamma_i(t)$

$$\Gamma_i(t) = \frac{J_i(t)}{D(t)} \equiv J_i(t)\,F_\pi(t) \tag{6.6}$$

and D is the D-function of the I=J=1 $\pi\pi \to \pi\pi$ partial wave. The above formula can be read from the graph

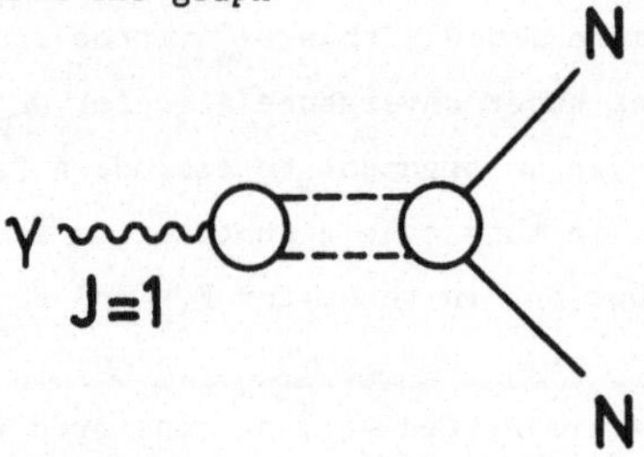

Fig. 8 Two pion intermediate states in the unitarity relation

In order to simplify the comparison with the literature, we also introduce $\pi\pi\, N\bar{N}$ partial waves $f_\pm^1(t)$ (defined as by Frazer-Fulco [42], $p_-^2 = m^2 - t/4$)

$$\Gamma_1(t) = \frac{m}{p_-^2}\left\{f_+^1 - \frac{t}{4\sqrt{2}\,m} f_-^1\right\}, \quad \Gamma_2(t) = \frac{m}{p_-^2}\left\{f_+^1 - \frac{m}{\sqrt{2}} f_-^1\right\} \tag{6.7}$$

which occur in the unitarity relations for the Sachs form factors

+) In order to have simple formulas we have chosen a notation slightly different from that of Frazer and Fulco [42] : $\Gamma_1 = -\Gamma_1^{FF}$, $\Gamma_2 = -2m\,\Gamma_2^{FF}$

$$\mathrm{Im}\, G_{EV}(t) = \frac{1}{m} \frac{q_t^3}{\sqrt{t}} |F_\pi|^2 J_+(t) \,, \qquad f_\pm^1(t) \equiv \frac{J_\pm(t)}{D(t)} \,. \qquad (6.8)$$

$$\mathrm{Im}\, G_{MV}(t) = \frac{1}{\sqrt{2}} \frac{q_t^3}{\sqrt{t}} |F_\pi|^2 J_-(t) \,,$$

Since $|F_\pi|^2$ follows directly from $e^+e^- \to \pi^+\pi^-$ experiments [+)], eq. (6.5) shows that the problem is to determine $J_i(t)$ for $t > 4\mu^2$ and to discuss the t-range, in which this relation can be applied in practice. It is valid exactly only for $t < 16\mu^2$.

Starting with the second point, it is useful to notice that the main correction is expected from the four-pion intermediate state.

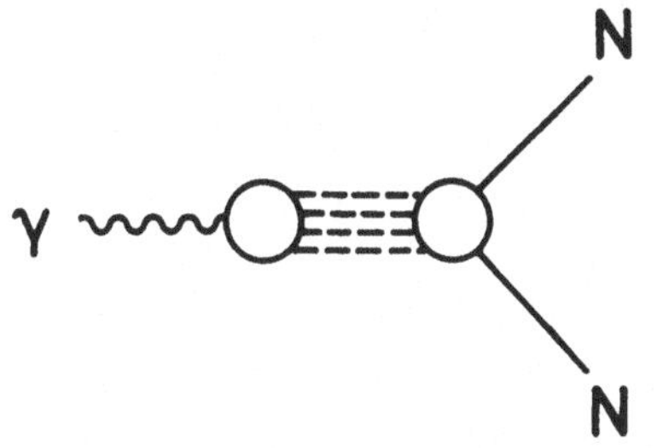

Fig. 9 4-pion states in the unitarity relation

$e^+e^- \to 4\pi$ experiments indicate that the cross section remains small below the $\omega\pi$ threshold [51)]. Furthermore phase shift analyses of $\pi\pi$-scattering have shown that in the I=J=1 state inelasticity is practically zero up to $t \approx 50\mu^2$ and probably not larger than 20 % up to $90\mu^2$ (Ref. [52)]). Therefore we shall use eq. (6.5) up to $t \approx 50\mu^2$.

The determination of $J_i(t)$ is technically somewhat complicated and therefore only a relatively small number of authors [53-63] has tried to improve the Frazer-Fulco method [42] or to exploit the better experimental information on the pion form factor and πN scattering amplitudes. There exists a much larger number of papers in which $\mathrm{Im}\, F_{iv}(t)$ at $t < 50\mu^2$ is represented by a delta function at $t = m_\rho^2$ or another simple ansatz, ignoring the possibility of calculating the spectral

---

+) A new $e^+e^- \to \pi^+\pi^-$ experiment of Quenzer et al. (Orsay preprint and contribution to the Stanford Conference 1975) at 484 MeV agrees well with the assumption made in [61].

function in this range from other data (see Refs.[31,32,64-66] and the papers quoted there. It will be shown that these simplified treatments are misleading, since the ρ-peak in Im $F_{iv}(t)$ has an unexpected shape due to a singularity just be below $t = 4\mu^2$.

The determination of $J_i(t)$ from the most recent πN amplitude analysis (Pietarinen[67,68]) has been discussed in detail in Ref.[61]. Other results are based on earlier πN phase shifts or even on an approximation of the πN amplitude by the nucleon and Δ(1232) isobar exchange terms only (Ref.[42,53,54]).

Our procedure consists of two steps:

i) Using fixed-t analyticity the invariant πN amplitudes are continued to the unphysical region between the s- and u-channels. A simple projection gives the I=J=1 $\pi\pi N\bar{N}$ partial waves $\Gamma_i(t)$ for $t < 0$. Taking an overall fit to the pion form factor data, one obtains from (6.6) $J_i(t)$ for $t < 0$.

ii) Next $J_i(t)$ is continued to $t > 4\mu^2$, using the fact that it is real up to $16\mu^2$ and, in practice, even up to about $50\mu^2$.

It was already pointed out by Frazer and Fulco[42] that $\Gamma_i(t)$ and $J_i(t)$ have a logarithmic singularity at $t = 4\mu^2 - \mu^4/m^2 \approx 3.98\,\mu^2$, which comes from the partial wave projection of the nucleon Born term in the πN scattering amplitude. If the dispersion relation (6.1) is used for the continuation of $F_i(t)$ to the 2nd sheet of the t-plane, it follows from (6.5) that $F_{iv}(t)$ has a logarithmic singularity at $t = 3.98\,\mu^2$ in the second sheet, just below threshold[+)].

This singularity is responsible for the rapid increase of $J_i(t)$ near $t = 4\mu^2$ in Fig. 10 which shows our result. If $J_i(t)$ were almost constant as usually assumed for an N-function, Im F_{iv} would have a shape similar to that of Im F_π or $|F_\pi|^2$, the region near threshold being suppressed because the two pions are in a p-wave state. In our result the left wing of the ρ-peak in Im F_i is strongly enhanced because of the increase of J_i at small $|t|$.

Although the existence of the nucleon singularity was discussed in the

+) See Oehme[69] for the continuation of dispersion integrals to the 2nd sheet.

Fig. 10 (next page) Our result for $J_i(t)$. Dashed line: non-Bornterm part.

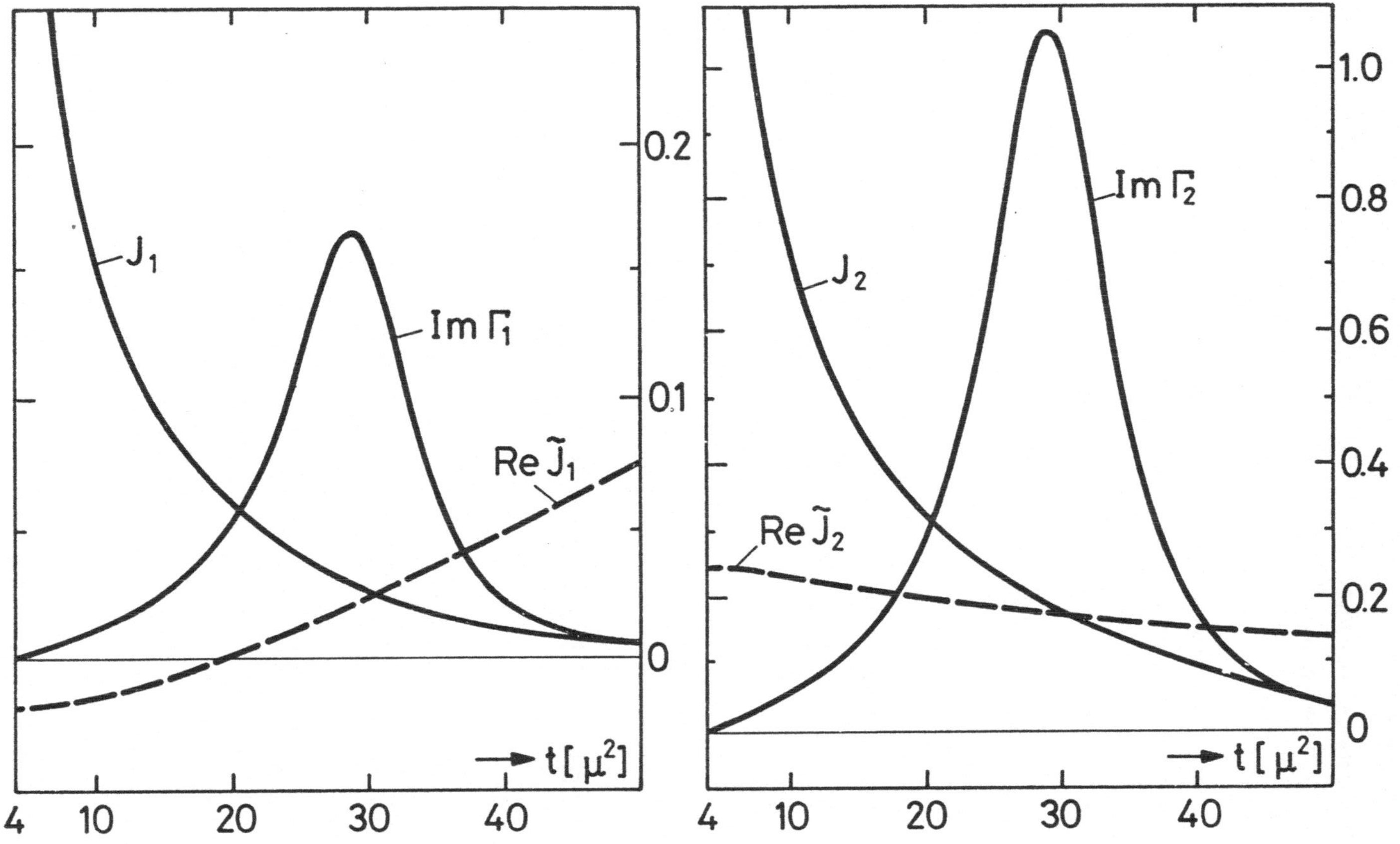
J1
Im Γ1
Re J̃1
0.2
0.1
0
→ t [μ2]
4
10
20
30
40
J2
Im Γ2
Re J̃2
1.0
0.8
0.6
0.4
0.2
0
→ t [μ2]
4
10
20
30
40

Fig. 10

very first papers on this subject [40-42], its influence has been ignored in many subsequent papers. We mentioned already the simple "vector-dominance" fits and some more elaborate investigations [65,66]. The consequence of unitarity were also ignored by Deo et al. [70] and Hammer et al. [71] (See our comment [72]).

Finally the nucleon singularity was ignored in a class of papers, in which the authors attempted to determine the spectral functions from form factor data by sophisticated methods of analytic continuation. (See the papers quoted in Pfister's review and Ref. [74]). There is no objection to that from a formal mathematical point of view, but it is clear that truncated expansions give a better result, if a known nearby singularity in the second sheet is removed.

The reliability of our prediction for Im $F_{iv}(t)$ is certainly good on the left wing of the ρ-peak. At the peak position we estimate the uncertainty to be of the order of 10 %. On the right wing there might be increasing corrections from the neglected contributions of four-pion states in the unitarity relation.

Attempts to estimate these corrections have been made by Brall and Rodenberg [56] and by Willrodt [63]. In our opinion the first mentioned paper is not tenable, because it disagrees with results derived from πN amplitudes. Willrodt's result will be discussed later.

7. Vector Dominance in Pion-Nucleon Scattering

Before applying the prediction for Im $F_i(t)$ to the analysis of nucleon form factors, we want to use our result for the $\pi\pi N\bar{N}$ partial waves for a discussion of "ρ-dominance" in pion-nucleon scattering. Historically this case has played an important role for the development of "vector dominance" ideas and furthermore it offers the best possibility for a quantitative study of this notion.

In 1960 Sakurai [75] pointed out that the difference of the I=1/2 and 3/2 πN S-wave scattering lengths is near to a result following from a simple-minded Feynman graph calculation for ρ-exchange

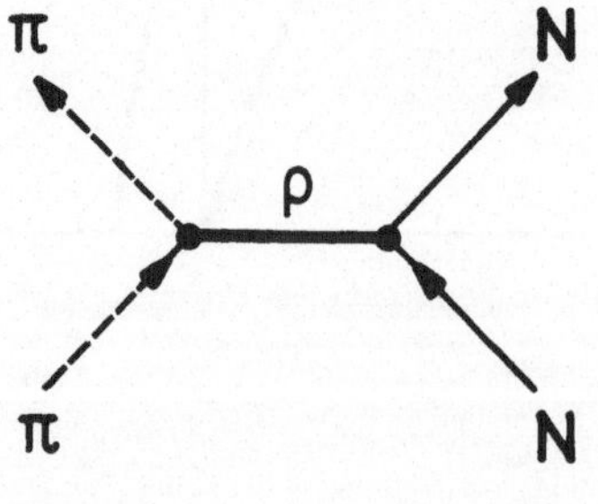

Fig. 11. Feynman graph for ρ-exchange

$$a_1 - a_3 = \frac{f_{\rho\pi\pi} f_1}{4\pi} \frac{3\mu}{m_\rho^2} \frac{1}{1+\frac{\mu}{m}} , \tag{7.1}$$

if one assumed ρ-dominance and universality $f_1 = f_{\rho\pi\pi} = f_\rho$.

In subsequent papers several authors considered derivations of this relation from current algebra or dispersion theory. However, in our opinion, none of these is acceptable.

Current Algebra. Eq. (7.1) is closely related to the Weinberg-Tomozawa version of the Adler-Weisberger relation [75]

$$a_1 - a_3 = \frac{3\mu}{4\pi f_\pi^2} \frac{1}{1+\frac{\mu}{m}} \; ; \; f_\mu = 0.93\mu . \tag{7.2}$$

If one assumes the KSFR relation [75]

$$f_\rho^2 f_\pi^2 = m_\rho^2 \tag{7.3}$$

and universality, Sakurai's conjecture follows. Unfortunately there exists no derivation of the KSFR relation from basic assumptions (See the discussion in Ref. [76] and the references given there).

Dispersion Theory

i) a_1-a_3 is the threshold value of the real part of the πN S-wave f_{0+}^- . This quantity has been studied in great detail by Hamilton and coworkers [53,77] using partial wave dispersion relations. Their result shows that about 50 % of a_1-a_3 can be attributed to that part of ρ-exchange which can be calculated ("long range part"). ρ-exchange has also a "short range part", but this cannot be separated from other contributions. Furthermore, in this approach a_1-a_3 receives a contribution from tensor coupling in contradistinction to (7.1). We conclude that (7.1) cannot be derived from partial wave dispersion relations.

ii) Sakurai [75] mentioned a possible derivation based on the Cini-Fubini approach, but a more detailed investigation showed that a reliable separation of the ρ-contribution is not possible [78].

iii) a_1-a_3 can be considered as the threshold value of the backward amplitude. Backward dispersion relations for the invariant πN amplitudes give a large ρ-exchange contribution from the left hand cut. But it differs from eq. (7.1) by a factor 1/2 and by a tensor coupling term [79]. - From a study of backward amplitudes alone it is not possible to separate the ρ-contribution from that of higher vector mesons (ρ',...) and from higher spin terms (g-meson). For this reason we made no attempt to determine the ρ-coupling constant in Ref. [79], pointing out that another method [55] is more reliable. Another difficulty was that the backward amplitudes as reconstructed from phase shifts were not quite compatible with the dispersion relations.

Nevertheless Banerjee et al. [80] calculated a ρ-coupling constant derived from our result, simply omitting that part of our figure which showed the discrepancy. Their treatment of ρ-exchange was used in Sakurai's Erice Lecture [75].

We conclude that up to now Sakurai's conjecture has not been derived from some basic assumptions. It has the same status as the KSFR relation.

As mentioned above Sakurai's formula (7.1) gives a reasonable numerical value for the amplitude at t=0. The question is, whether this kind of "ρ-dominance" can be generalized to a t-interval which includes $t=m_\rho^2$, where ρ-dominance is fulfilled trivially.

The best amplitude for this investigation is obviously the t-channel partial wave f_+^1, which belongs to the quantum numbers of the ρ-meson and approximately agrees up to a factor with a_1-a_3 at t=0.

Sakurai [75] has already mentioned that one has to consider the pv nucleon Born term in addition to ρ-exchange. But it turns out that this is not sufficient. N- and ρ-exchange do not even qualitatively describe the shape of the t-dependence of Re f_+^1 near t=0.

The difficulty is understood, if one considers either the partial wave dispersion relation for f_+^1 or consequences of the Ward identity: one has to add at least Δ(1232)-exchange, i.e. one has to take into account all nearby singularities (Ref. [76,78]).

In the case of the flip amplitude f_-^1, which approximately agrees with B^- up to a factor, ρ-exchange is not even dominant at t=0.

We conclude that vector dominance is not valid in the usual sense in πN scattering. Sakurai's conjecture has the character of a sum rule (Beg. [81]). It is valid only for Re f^1_+ and only at a single point: t=0, similar to the Adler-Weisberger relation.

ρNN Coupling Constants. We add a remark on the problem how to determine ρNN coupling constants. It is clear that these coupling constants must be derived from $f^1_\pm(t)$ together with information on the $\rho\pi\pi$-vertex. One possibility is to use the formalism of Gell-Mann and Zachariasen [82] and of Kroll, Lee and Zumino [83]. Using the Frazer-Fulco[42]-Gounaris-Sakurai[84] parametrization for the pion form factor we find

$$\mathrm{Re}\ \frac{f_i(t_\rho) f_\rho}{4\pi} = 3 t_\rho J_i(t_\rho) \ . \qquad (7.4)$$

Inserting our result for $J_i(t)$ one obtains

$$\mathrm{Re}\ \frac{f_1(t_\rho) f_\rho}{4\pi} = 2.4 \ , \qquad \frac{\mathrm{Re}\ f_2(t_\rho)}{\mathrm{Re}\ f_1(t_\rho)} = 6.6 \ . \qquad (7.5)$$

The value for f_ρ follows from $e^+e^- \rightarrow \pi^+\pi^-$ experiments, which have not yet a high accuracy. We have used $f_\rho^2/4\pi = 2.6 \pm 0.3$, but a different analysis of the same data leads to 2.26 ± 0.25 (Ref. [85]). Our value for $[\mathrm{Re}\ f_1\ (t_\rho)]^2/4\pi$ is 2.2 (Ref. [61]).

Other definitions of coupling constants and a more detailed discussion can be found in Ref. [61].

8. Nucleon Radii

In theoretical treatments of nucleon scattering against atoms or light nuclei one can start from a first Born approximation, assuming that the target is described by a given charge distribution $\rho(r)$. Then the scattering amplitude differs from that for scattering against a point charge only by a form factor $F(Q^2)$ which is the Fourier transform of $\rho(r)$

$$F(Q^2) = \iiint \rho(r)\, e^{i\vec{Q}\cdot\vec{r}} d^3x = \frac{4\pi}{Q} \int_0^\infty \rho(r) \sin(Qr)\, r\, dr \ . \qquad (8.1)$$

Q is the momentum transfer. The root-mean-square radius $\sqrt{\langle r^2 \rangle}$ of the charge distribution follows from

$$R^2 \equiv \langle r^2 \rangle = \int_0^\infty r^2 \rho(r) \, 4\pi r^2 dr \quad , \tag{8.2}$$

the total charge being normalized to 1.

The derivatives of (8.1) read

$$\left. \frac{d^n F(Q^2)}{d(Q^2)^n} \right|_{Q^2=0} = \frac{(-1)^n}{(2n+1)!} \iiint \rho(r) r^{2n} d^3x \quad . \tag{8.3}$$

In particular the first derivative is related to R^2

$$R^2 = - \left. \frac{6 \, dF/dQ^2}{F(Q^2)} \right|_{Q^2=0} \quad . \tag{8.4}$$

A table of form factors and the related charge distributions is given in Ref. [86].

Our treatment of nucleon form factors is based on the more general relativistic ansatz (2.1), which includes spin, recoil and a second form factor for the magnetic moment. Formally, one can calculate from F_i and from G_E, G_M, distributions of charge and magnetic moment densities, using the inverse of (8.1). However these quantities have a simple physical meaning only in the non-relativistic limit (See Ref. [87], a more recent discussion is found in Ref. [88]).

In the following we shall <u>define</u> the radii R_i, R_E, R_M by the derivative (8.4) of the form factors F_i, G_E, G_M at $Q^2 = 0$.

Taking the derivative in the dispersion integrals (6.1), (6.2), one obtains for instance

$$R_{Ev}^2 = \frac{6}{\pi} \int_{4\mu^2}^\infty \frac{\mathrm{Im}\, G_{Ev}(t)}{t^2} dt \quad . \tag{8.5}$$

Our discussion in § 6 has shown that Im G_{Ev}, eq. (6.8), can be calculated in the range $4\mu^2 < t < 50\mu^2$ from πN scattering amplitudes and $F_\pi(t)$. It turns out that the shape of the integrand in (8.5) differs qualitatively from that assumed in a vector dominance model ("VDM", narrow peak at $t=m_\rho^2$), the steep increase at $t=4\mu^2$ being due to the enhancement by the nucleon singularity (Fig. 12). As a consequence the radius R_{Ev} is considerably larger than the VDM estimate.

It is interesting to discuss the "tail" of the charge distribution far outside the nucleon. Classically it can be described by the higher moments which, according to (8.3), follow from the higher derivatives of the form factors. If these higher derivatives at $Q^2=0$ are taken in the dispersion relation, the integrand becomes Im G_{Ev}/t^{n+1}, i.e. the region near $t=4\mu^2$ is even more enhanced and the ρ-peak is suppressed as n increases. Since the region near $t=4\mu^2$ is dominated by the nucleon exchange graph

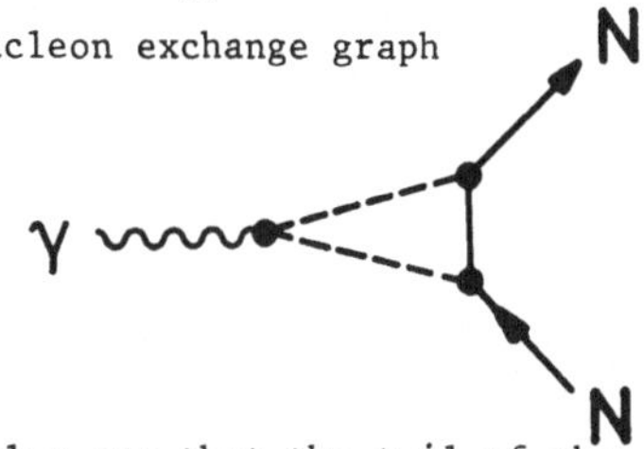

Fig. 13. Lowest order perturbation theory graph for the nucleon form factor.

one can also say that the tail of the charge distribution is determined by (uncorrelated) two-pion-exchange in the t-channel, which agrees with the lowest order of perturbation theory.

We conclude that the proton charge distribution at large distances is well understood and that the "halo" proposed in Ref. [89] is not acceptable.

Fig. 12 shows also the remarkable fact that the shapes of Im $2F_{1v}(t)$ and Im $F_\pi(t)$ are much different (our normalization: $2F_{1v}(0) = F_\pi(0) = 1$). Therefore the shapes of $2F_{1v}(t)$ and $F_\pi(t)$ in the physical region ($t< 0$) must be different too. Several authors assumed that the equality $F_\pi = 2F_{1v}$ is valid [90] or attempted to derive it [91].

In the analysis of experiments in which both form factors occur [16,92], one should not assume $F_\pi(t) \equiv 2F_{1v}(t)$ and $G_{nE} \equiv 0$, as it is sometimes done, but rather use the results of the analysis of nucleon form factors.

In some papers one finds the remark that the equality of $F_\pi(t)$ and

Fig. 12 (next page). Prediction for the integrand of (8.5)

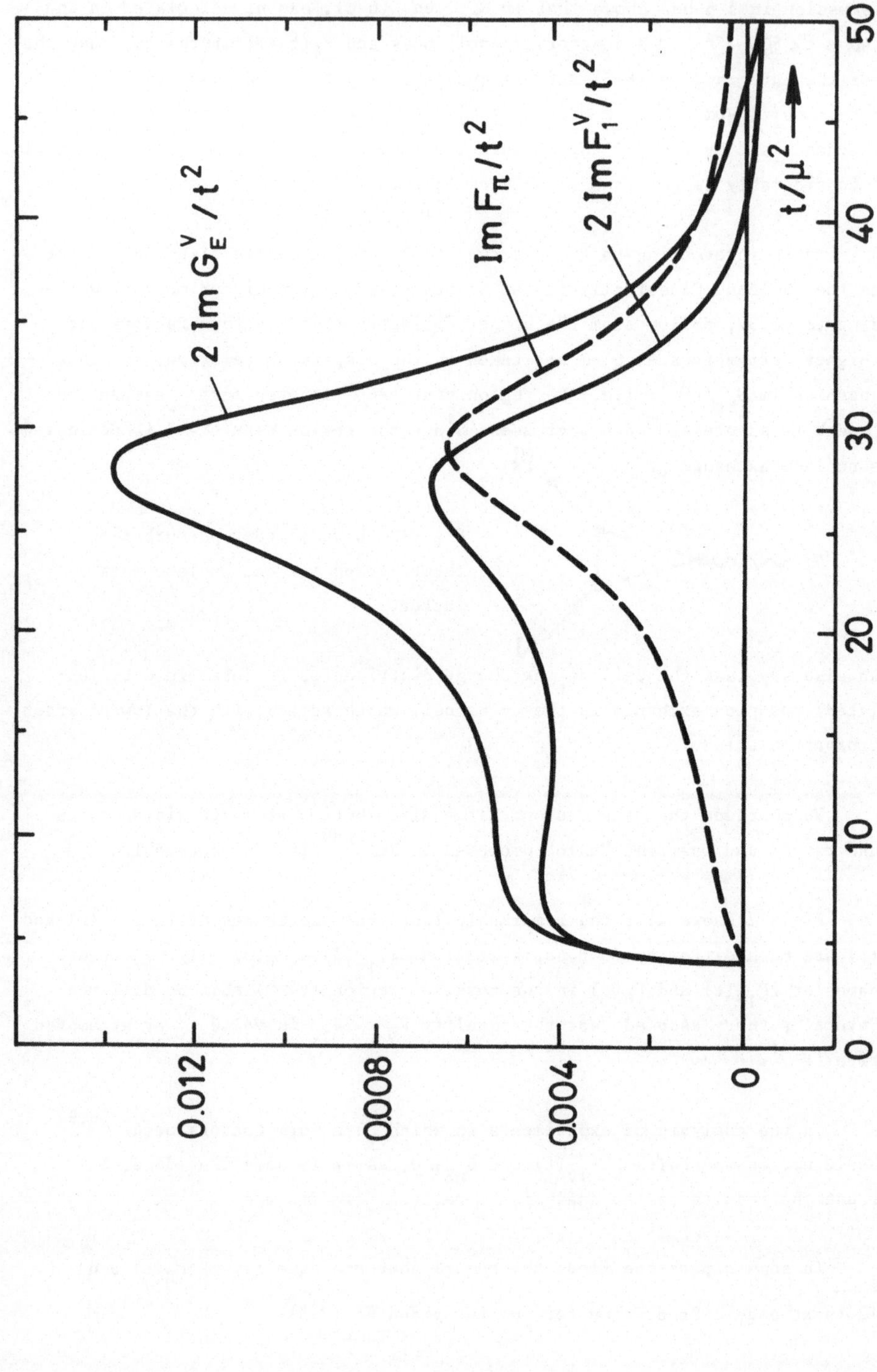

0.012
0.008
0.004
0
0
10
20
30
40
50
$2\,\mathrm{Im}\,G_E^V/t^2$
$\mathrm{Im}\,F_\pi/t^2$
$2\,\mathrm{Im}\,F_1^V/t^2$
$t/\mu^2 \longrightarrow$

Fig. 12

$2F_{1v}(t)$ follows from gauge invariance. However this is not true (See Behrends et al.[93] +))

Eq. (8.5) can be used for an estimate of the nucleon radius by neglecting the contribution to the integral from the range $t > 50\mu^2$. - For an accurate determination of the radius we start from the subtracted dispersion relation, taking the known part of the dispersion integral to the left hand side

$$t\,d(t) \equiv G_{Ev}(t) - G_{Ev}(0) - \frac{t}{\pi}\int_{4\mu^2}^{50\mu^2} \frac{\operatorname{Im} G_{Ev}(t')}{t'(t'-t)}\,dt' = \frac{t}{\pi}\int_{50\mu^2}^{\infty} \frac{\operatorname{Im} G_{Ev}(t')}{t'(t'-t)}\,dt' \qquad (8.6)$$

and inserting experimental values for the form factors $G_{Ev}(t)$. The unknown part of the integral (8.5) can now be accurately determined by extrapolating the experimental points for the "discrepancy" $d(t)$ to $t=0$.

This extrapolation is model-independent, since for all physically acceptable assumptions for the spectral function $\operatorname{Im} G_{Ev}$ on the distant cut $t > 50\mu^2$ the discrepancy function is slowly variable near $t=0$ and can be represented for instance by an effective pole.

An equivalent method has been used in Ref. [60], where

$$g(t) \equiv G_{Ev}(t) - \frac{1}{\pi}\int_{4\mu^2}^{50\mu^2} \frac{\operatorname{Im} G_{Ev}(t')}{t'-t}\,dt' = \frac{1}{\pi}\int_{50\mu^2}^{\infty} \frac{\operatorname{Im} G_{Ev}(t')}{t'-t}\,dt' \qquad (8.7)$$

has been considered and $g'(0)$ is the quantity which gives the unknown contribution in (8.5).

Fig. 14 shows that $1/g(t)$ is approximately linear in a fairly wide t-range. The straight line belongs to an effective pole

$$g(t) = \frac{-0.42}{1 - t/85\mu^2} \; . \qquad (8.8)$$

+) I am grateful to P. Stichel for a discussion of this point.

It is interesting to notice that g(0) = - 0.42 is comparable with $G_{Ev}(0) = 0.5$. This means that beyond $50\mu^2$ Im G_{Ev} can be roughly described by a large negative dip at $t \approx 85\mu^2$, corresponding to a mass of 1.3 GeV for the exchanged higher vector meson. g'(0) contributes about 15 % to R^2_{Ev}.

In addition to a slight systematic deviation from a straight line Fig. 14 shows irregular deviations which must be due to experimental errors, since g(t) has only a distant cut. It cannot have wiggles at $t<0$ unless the spectral function Im G_{Ev} is very strange.

We conclude that the isovector radii can be determined in a model-independent way from the prediction for the spectral functions and experimental form factor data. One could ask why one should not prefer a direct determination of the slope from form factor data, using an empirical fit [94,95]. The answer is that the slope is sensitive to normalization errors of the data. Because of the theoretical input our extrapolation to t=0 has considerably less freedom than an empirical fit. In fact it turns out that consistency with the dispersion relation can only be achieved, if some data are renormalized within the errors estimated by the authors.

9. t-Channel Exchanges in Electron-Nucleon Scattering

It is our aim to analyse electron-nucleon scattering data in such a way that one obtains information on the mechanism which is responsible for the structure of the nucleon. In a straightforward application of the dispersion approach the first step would be to determine the spectral functions Im $F_i(t)$ from experimental form factor data. Then a discussion of the unitarity relation would show the mechanism which governs the distribution of charge and magnetic moment densities within the nucleon. In the simplest case the shape of the form factors would be determined by the exchange of vector mesons.

The first step is a problem of analytic continuation. Unfortunately the information on Im $F_i(t)$ which can be derived from the existing form factor data is rather poor [73,74,94]. As mentioned in § 1 there is no hope for a substantial improvement during the next decade, because there are experimental difficulties in reaching a higher accuracy of the normalization; even more serious are the problems with the separation of two-photon exchange terms and radiative corrections.

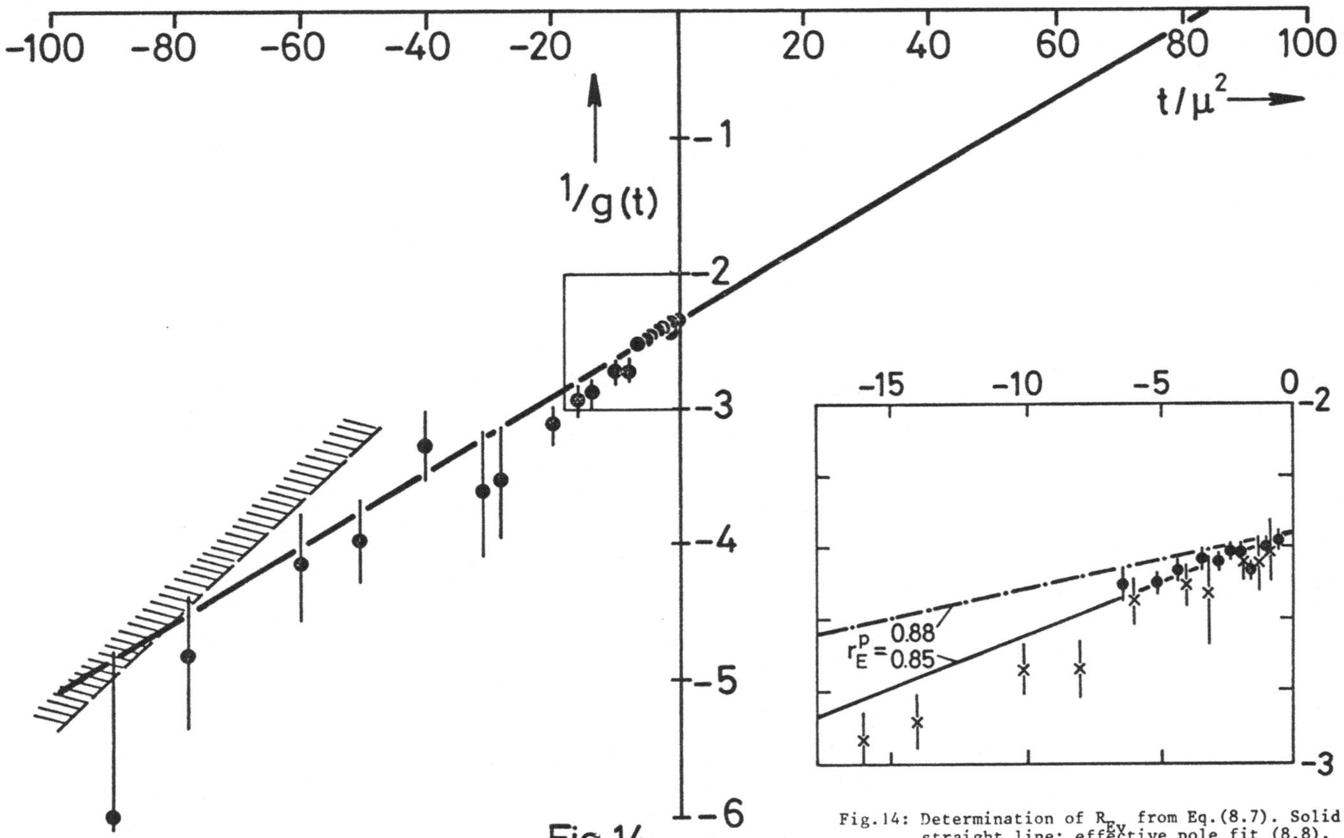

Fig.14

Fig.14: Determination of R_{Ev} from Eq.(8.7). Solid straight line: effective pole fit (8.8). Hatched region: $G_{Ev} < 0$. Inset: Sensitivity to R_{Ep} (fm-units) if the experimental value for $G'_{En}(0)$ is used.

For this reason it is necessary to use a further input already in the first step: the prediction for the isovector spectral functions below $50\mu^2$ which has been derived from other experimental data (§ 6).

We add the assumption that there is no appreciable contribution to the isoscalar spectral function below $t=m_\omega^2$. Unfortunately there exist no data for $e^+e^- \rightarrow \pi^+\pi^0\pi^-$ in this range. Our assumption would be supported, if this reaction had a small cross section.

Since the ω-resonance is narrow (width 10 MeV), the ω-exchange contribution to Im F_{is} can very well be approximated by a delta function which gives a pole term in $F_{is}(t)$.

The problem is now to determine the residues of the ω-pole and Im F_i at larger t. It is clear that there is no chance to obtain a detailed result for the shape of Im F_i, since F_i at $t<0$ is known only within error bars in a limited t-range at a discrete set of points. Furthermore the possibility of resolving structures in Im F_i will worsen rapidly as t increases.

A direct application of the dispersion relation in the isovector case gives an integral over Im F_{iv} from $50\mu^2$ to infinity ("discrepancy function"), from which one can derive a few "moments" of Im F_{iv}. In the isoscalar case the situation is more complicated, because one has to treat the ω-pole residues as adjustable parameters.

For a discussion of the unitarity relation the moments are not transparent and one would like to see curves for Im F_i. Of course there are many curves which are compatible with the discrepancy function. The selection should be guided by physical arguments rather than by a somewhat vague probabilistic reasoning for an "optimal" choice.

In the following we shall discuss a pole term expansion which is reasonable, if Im F_i above $50\mu^2$ is dominated by the exchange of vector-mesons.

Of course this is not sure and it could as well be that Im F_i shows broad structures. Then other expansions are preferable which are usually written in terms of a variable z obtained from a suitable conformal mapping of the t-plane. Methods of this type have recently been proposed by Cutkosky [74] and Pietarinen (eq. (10) in Ref. [95]).

Our ansatz reads

$$F_{iv}(t) = F_i^{\rho}(t) + \frac{a_i(\rho')}{t_{\rho'} - t} + \frac{a_i(\rho'')}{t_{\rho''} - t} + \cdots \tag{9.1}$$

$$F_{is}(t) = \frac{a_i(\omega)}{t_\omega - t} + \frac{a_i(\omega')}{t_{\omega'} - t} + \frac{a_i(\omega'')}{t_{\omega''} - t} + \cdots \tag{9.2}$$

The ρ-contribution is taken from an evaluation of the dispersion integral (6.1) from $4\mu^2$ to $50\mu^2$, using our result for Im F_i (§ 6). Instead of working with a numerical table, it is more convenient to use an empirical parametrization [61]

$$2F_1^{\rho}(t) = \frac{0.955 + 0.090\,(1 - t/0.355)^{-2}}{1 - t/0.536}\,, \quad 2F_2^{\rho}(t) = \frac{5.335 + 0.962\,(1 - t/0.268)^{-1}}{1 - t/0.603} \tag{9.3}$$

The notation ρ', ρ'' etc. refers to effective poles. A possible relation to meson resonances will be discussed later.

The number of parameters is reduced by 4, because the values of $F_{is}(0)$ and $F_{iv}(0)$ are known. Two more parameters are practically fixed by the accurate information on $G'_{En}(0)$ and on $G_{Ep}(Q^2)$ at small Q^2. One further parameter is fixed by the superconvergence condition $Q^2 G_{Mp} \to 0$ for $Q^2 \to \infty$ which is strongly suggested by the SLAC data up to $Q^2 = 35$ GeV^2.

It can happen that a fit to a sum of pole terms (9.1), (9.2) gives a rather unstable result, in which neighboring poles have large residua of opposite signs. Since a wildly oscillating spectral function is doubtful, we have imposed stabilizing conditions as $|a_i(\omega)| + |a_i(\omega')| + |a_i(\omega'')| + \ldots = M$, where M is a given number. M is chosen in such a way that a further increase does not improve the fit significantly.

Some authors determined the parameters by fits to form factors which were calculated from Rosenbluth plots. We prefer to fit the ep-scattering cross sections together with neutron form factors,+) because this simplifies the treatment of renormalization factors of the experimental data and one does not have to worry about

+) Note added in proof. In the final calculation we have used en-scattering cross sections as derived from ed-scattering data and our fit to the ep-cross sections.

correlations of the errors of proton form factors.

For a given number of pole terms the main uncertainty results from the normalization errors of the cross section data. It turns out that the renormalization factor of data at small Q^2 is well determined, because the fit must go to known values $F_i(0)$ at $Q^2 = 0$. For instance all of our fits suggest a renormalization of the Mainz data [94,95] by about + 2 % (N = 1.02), which is compatible with the systematic error given by the authors. A possible reason is Greenhut's two-photon exchange correction [20] which has the right sign. The magnitude is somewhat smaller but one should notice that Greenhut's result has a considerable uncertainty. - For $Q^2 \geq 1$ GeV^2 the restrictions of our ansatz are not sufficient for a unique determination of the renormalization factors and therefore we decided to assume a given renormalization factor N_D=1000 for the DESY data of Ref.[5].

Since our analysis [62] is not yet completed, I shall give only a summary of the main features of the preliminary results.

ω-exchange contribution

The result for $a_1(\omega)$ is nearly the same in all of our fits: $(33 \pm 3)\mu^2$. According to the VDM the residue $a_i(\omega)$ is equal to the product of the γ-ω and ωNN coupling constants (i=1: vector coupling, i=2 tensor coupling)

$$a_i(\omega) = \frac{t_\omega}{f_\omega} g_i(\omega NN) , \tag{9.4}$$

where f_ω follows from the $\omega \to e^+e^-$ decay width (α = 1/137)

$$\frac{f_\omega^2}{4\pi} = \frac{\alpha^2}{3} \frac{m_\omega}{\Gamma(\omega \to e^+e^-)} . \tag{9.5}$$

Inserting the value $\Gamma(\omega \to e^+e^-) = (0.76 \pm .17)$ keV of the Particle Data Group (1974), one finds $f_\omega^2/4\pi = 18.3 \pm 4$ (Benaksas et al. [85] give a smaller error). This value gives

$$\frac{1}{4\pi} g_1^2(\omega NN) = 20 \pm 5 \tag{9.6}$$

for the vector coupling constant. The tensor coupling constant $g_2(\omega NN)$ comes out at least 5 times smaller and not even the sign is well determined at present. The ωNN coupling constants play an important role in nuclear physics, where they are determined from NN scattering data. Published values [96] for $g_1^2(\omega NN)/4\pi$ lie in the range 4 ... 20 and some authors give small errors. However one should notice that in this case the exchange is not limited to vector mesons, so it is much more difficult to separate the different contributions. We think that at least our ρNN coupling constants (7.5) are more reliable and that this will be true also for the ωNN coupling constants as soon as a new $e^+e^- \rightarrow \pi^+\pi^-\pi^0$ experiment gives a more accurate value for f_ω.

Our results are at variance with SU(6) relations which have been used by some authors as a constraint in determinations of coupling constants from NN scattering, assuming a mixing angle ($\tan\Theta = 1/\sqrt{2}$) which gives vanishing ϕNN coupling (Ref. 97)

$$\frac{g_1^2(\omega NN)}{4\pi} = \frac{9}{4}\,\frac{f_1^2(\rho NN)}{4\pi}\,, \quad \frac{g_2(\omega NN)}{g_1(\omega NN)} = \frac{1}{5}\,\frac{f_2(\rho NN)}{f_1(\rho NN)}\,. \tag{9.7}$$

If one inserts our result for the ρNN couplings, $f_i(\rho NN)$, Eq.(7.4), is given in Sakurai's normalization. It is equal to $2g_i(\rho NN)$, if $g_i(\rho NN)$ is defined in analogy to $g_i(\omega NN)$, Eq.(9.4), one finds 5 and 1.3 for the right hand sides. Both values are not compatible with our results for the ωNN couplings.

Another possible source of information on the ωNN coupling is π^0-photoproduction on protons. Unfortunately the analysis is even more difficult. The high energy data show a contribution from reggeized ω-exchange[98], but the extrapolation from the physical region to the ω-pole is very uncertain. (Cf. the difficulty in extrapolating to the ρ-pole in πN charge-exchange scattering).

b) $\rho'(1250)$-Exchange

Since G_{Ep}, G_{Mp} and G_{Mn} are approximately described by a dipole fit, the main feature of the spectral functions must be a bump-dip structure. The bumps can be identified with ρ- and ω-exchange, so the problem is to study the location and the physical origin of the dips.

In the isovector case we find in general a large dip of Im F_{2v} in the vicinity of $t = 80\mu^2$, corresponding to the exchange of a particle $\rho'(1250)$ which was predicted from the Veneziano model[99] as the first daughter of $\rho(770)$. Unfortunately the Table of the Particle Data Group shows no compelling evidence for a resonance at this mass.

There is no evidence for a $\rho'(1250)$ resonance in the I=J=1 phase shift of $\pi\pi$-scattering[52] and no significant structure in the reactions $e^+e^- \to \pi^+\pi^-$ (Ref.[100]) and $e^+e^+ \to 2\pi^+ + 2\pi^-$ (Refs.[51,101]). However a structure has been reported in $e^+e^- \to \pi^+\pi^-\pi^o\pi^o$ (Ref.[51]) which could be due to a $\omega\pi^o$ state not far above the $\omega\pi^o$ threshold (0.92 GeV, $t = 43.2\mu^2$) and near to the $J^{PC} = 1^{+-}$ state B(1235). (See also Ref.[102]). It will be difficult to understand, how such a small effect can lead to a dip in Im F_{2v} comparable with that caused by ρ-exchange.

It is remarkable that our result for Im F_{1v} does not show a bump-dip structure. Beyond the ρ-bump we find another bump at $t \approx 80\mu^2$ followed by a dip at larger t. The magnitude of the 2nd bump and of the dip are not well determined. This structure of the spectral function is related to the fact that $F_{1v}(Q^2)$ decreases more slowly than other form factors and is near to $F(Q^2)$ (cf. Sakurai's "universality").

If one constructs Im G_{Ev} and Im G_{Mv}, one finds a bump-dip structure in both cases because of the large dip in Im F_{2v}.

A first attempt to evaluate the unitarity relation above the ρ-region, inserting amplitudes for $\pi\pi \to N\bar{N}$, $\pi^o\omega \to N\bar{N}$ and $\rho\epsilon \to N\bar{N}$ has been made by Willrodt[63]. He finds that the $\omega\pi^o$ and $\rho\epsilon$ contributions are small except for the $\omega\pi^o$ contribution to the dip of IM F_{2v}.

ω'(1250) and ϕ(1020)-Exchange

Our fit describes the dipole-like behaviour of $F_{1s}(Q^2)$ by a peak of Im F_{1s} at $t = t_\omega$ and a dip which again is located roughly in the region around $80\mu^2$.

Most of the earlier authors assumed a pole at $t = t_\phi = 53.4\mu^2$ and obtained a residue comparable with that of ω-exchange. This is not satisfactory, since in general one finds reasonable results from a quark model, in which ϕ is described by strange quarks only ($\bar{s}s$) and then its coupling to the nucleon must be strongly suppressed. (See for instance Ref.[103]).

For this reason we treated the masses of all contributions beyond the ω as adjustable parameters. Of course our effective pole in the region near to $80\mu^2$ can belong to a broad dip which includes a small contribution from ϕ-exchange.[+)]

+) See "note added in proof" at the end.

The table of the Particle Data Group has no entry for a $\omega'(1250)$. The $e^+e^- \to \pi^+\pi^-\pi^0$ cross section in this region is even smaller (Ref. [104]) than for $e^+e^- \to \pi^+\pi^-\pi^0\pi^0$. $e^+e^- \to 5\pi$ has not yet been investigated.

The form factor $F_{2s}(Q^2)$ is the small difference between two large quantities, since the anomalous moments of proton and neutron and their distributions are almost equal in magnitude and have opposite signs. In the spectral function not only the ω-contribution is strongly suppressed but also the structure at $80\mu^2$ (a dip) is smaller than in the other cases. To our knowledge a good explanation for the suppression (Cf. the failure of eq. (9.7)) is not known. - Both G_{Es} and G_{Ms} have a dipole-like decrease and therefore one observes a pronounced bump-dip structure in Im G_{Es} and Im G_{Ms}, as expected from the dominance of Im F_{1s} over Im F_{2s}. The smallness of Im F_{2s} is also the reason for an approximate validity of the "isoscaling law" $G_{Es}(Q^2)/G_{Es}(0) = G_{Ms}(Q^2)/G_{Ms}(0)$ (Ref. [105]).

If the isoscaling law were exactly valid for the isovector form factors, we could apply it to the unitarity relations for Im G_{Ev} and Im G_{Mv} and derive

$$f_+^1(t) = \frac{G_{Ev}(0)}{G_{Mv}(0)} \frac{m}{\sqrt{2}} f_-^1(t) = 1.010\, f_-^1(t)\mu \; . \tag{9.8}$$

Our results for the $\pi\pi N\bar{N}$ partial waves show that this relation is valid up to 20 % in the range $4\mu^2 < t < 30\mu^2$, although f_+^1 and f_-^1 are completely different below $t = 4\mu^2$. - Of course the isoscaling law cannot be exact because of (2.8).

Exchange of Higher Vector Mesons

In order to obtain good fits, it is necessary to admit a third pole in both the isoscalar and isovector case. Its position (1.5 - 2 GeV) and residue are not well determined separately. Therefore it is possible to impose superconvergence conditions on several form factors without worsening the fit.

It is not yet known whether dispersion integrals are almost saturated by contributions from $t < 4\ \mathrm{GeV}^2$ or not, since the information from $e^+e^- \to N\bar{N}$ is still poor. If superconvergence integrals exist but are far from saturation in the above range, they are of little practical interest.

Our result indicates a slow convergence of the dispersion integral for $F_{1n}(Q^2)$. As discussed in § 4.2, F_{1n} vanishes and has an almost vanishing derivative at $Q^2 = 0$. Im F_{1n} starts with a broad ρ-dip, on which a narrow ω-peak is superimposed. In the dispersion integral for $F_{1n}(0)$ only half of the large ω-contribution is cancelled by the ρ-contribution and it is still far from saturation, if we include the 1250 MeV mass region. This means that explanations of the peculiar behaviour of F_{1n} from symmetry assumptions and vector dominance models will be very difficult, aside from the fact that the ρ cannot be treated in the narrow resonance approximation.

As a final remark we want to point out that for theoretical investigations Dirac and Pauli form factors have a clear advantage over Sachs form factors. The spectral functions Im $G_{Es,v}$ and Im $G_{Ms,v}$ are roughly similar, whereas Im $F_{is,v}$ show qualitative differences and therefore offer better possibilities to test the predictions of a model.

10. Conclusions

i) In analyses of electromagnetic nucleon form factors one should take the isovector spectral functions Im F_{iv} at $t \approx m_\rho^2$ and below as predicted by the unitarity relation, inserting πN amplitudes and the pion form factor (Frazer-Fulco method [42]). Many authors used instead a narrow resonance or Breit-Wigner approximation without noticing that this simplification led to considerable errors, for instance in determinations of nucleon radii.

ii) Dispersion relations in t have some predictive power. If Im F_{iv} is determined as described in i), it can happen that the experimental form factors are not compatible with the dispersion relation and a reasonable behaviour of Im F_{iv} above the ρ-peak.

iii) Isovector nucleon radii can be determined in a model-independent way.

iv) It is possible to determine the magnitude of the ω-contribution.

v) The physical origin of the success of the dipole fit is the fact that the spectral functions Im $G_{Es,v}$ and Im $G_{Ms,v}$ have a peak from ρ- or ω-exchange and a (possibly broad) dip of comparable magnitude at $t \approx 80\mu^2$, corresponding to a mass of 1250 MeV of the exchanged particle. For a good fit it is necessary to assume a third bump or dip at a higher t. - The mechanism for the dips in the vicinity of $80\mu^2$ is not known.

vi) For a comparison of the spectral functions with models Dirac and Pauli form factors are more suitable than Sachs form factors.

vii) ρ-dominance is not valid in πN scattering amplitudes between $t = 0$ and $t = m_\rho^2$.

The final result of our analysis [62] of the nucleon form factors will be available soon. In collaboration with the Mainz group we are also preparing a "Compilation of Electron-Nucleon Scattering Data", which includes modified Rosenbluth plots, tables of form factors and our proposal for the renormalization of data.

Acknowledgements: I am grateful to E. Borie and to I. Sabba-Stefanescu for reading the manuscript and for many discussions. It is a pleasure to thank S. Ciulli, R. Felst, P.G.O. Freund, J. Hamilton, J.L. Petersen, E. Pietarinen, G. Simon and V.H. Walther for interesting comments.

Note added in proof (Jan. 1976).

At present our best value for the ωNN vector coupling constant is slightly larger than Eq.(9.6): $g_1^2(\omega NN)/4\pi = 24 \pm 6$.

The final conclusion on ϕ-exchange is different from that given in the text. If our values for the ρ- and ω vector coupling constants are inserted into the SU(3) relation, one obtains a prediction for the ϕNN vector coupling which is well compatible with the large dip in Im F_{1s}. Therefore I think that there is evidence for a fairly large ϕNN vector coupling, i.e. for a violation of Zweig's rule. Details are given in Ref. [106].

References

1. G. Källén, Elementarteilchenphysik, 2nd ed., Bibliograph. Institut Mannheim
2. P. Urban, Topics in Applied Quantum Electrodynamics, Springer Verlag Wien, 1970
3. M. Gourdin, Diffusion des Electrons de Haute Energie, Manon et Cie., Paris 1966
4. K. Strauch, Proceedings of the 6th International Symposium on Electron and Photon Interactions at High Energies (Bonn). Ed. H. Rollnik and W. Pfeil, 1974
5. W. Bartel et al., Nucl. Phys. B58 (1973) 429
6. R.G. Arnold et al., SLAC-PUB-1596, June 1975
7. R.E. Rand et al., Phys. Rev. D8 (1973) 3229
8. R.W. Berard et al., Phys. Lett. 47B (1973) 355
   C.R. Schumacher and H.S. Bethe, Cornell Report LNS 180, unpublished
9. J. Hockert et al., Nucl. Phys. A 217 (1973) 14
10. W. Fabian, H. Arenhövel and H.G. Miller, Z. Physik 271 (1974) 93
11. L.L. Foldy, Rev. Mod. Phys. 30 (1958) 471
12. M. Castellano et al., Nuovo Cim. 14A (1973) 1
13. A.M. Boyarski et al., SLAC-PUB-1599 (June 1975) and Proceedings of the International Conf. for High Energy Physics, Palermo (1975)
14. G. Bassompierre et al. (CERN). Paper submitted to the Palermo Conference 1975
15. R.L. Crawford, Nucl. Phys. B 28 (1971) 573
    A. Del Guerra et al., Daresbury Laboratory DL/P 242 (August 75)
16. S.F. Berezhnev et al., Soviet J. Nucl. Phys. 18 (1974) 53
    A. Bietti and S. Petrarca, Nuovo Cim. 22A (1974) 595 and Lett. Nuovo Cim. 13 (1975) 539
17. G. von Gehlen, Nucl. Phys. B 20 (1970) 173; Springer Tracts in Modern Physics 59 (1971) 164 and Proceedings of the 6th International Symposium on Electron and Photon Interactions (Bonn 1973), p. 117
18. N. Dombey and B.J. Read, Nucl. Phys. B 60 (1973) 65
19. B.H. Bransden, Atomic Collision Theory, Chapter 5.3 Benjamin, New York 1970
    J. Sucher and J. Soffer, Phys. Rev. 161 (1967) 1664

20. G.K. Greenhut, Phys. Rev. 184 (1969) 1860
D. Drechsel, Dispersion Corrections. Talk given at the Advanced Institute on Electron Scattering and Nuclear Structure. Cagliari (Italy), Sept. 1970. Unpublished.
U. Günther and R. Rodenberg, Nuovo Cim. 2A (1971) 25

21. J. Bernabeu and T.E.O. Ericson, CERN TH-1525, 1972

22. L.S. Brown et al., Phys. Lett. 42B (1972) 111

23. G.C. Fox and D.Z. Freedman, Phys. Rev. 182 (1969) 1628
J. Bernabeu, T.E.O. Ericson and C. Ferro Fontan, Phys. Lett. 49B (1974) 381
P. Barenov et al., Phys. Lett. 52B (1974) 122

24. G.V. Anikin and I.I. Kotukhov, Soviet J. Nucl. Phys. 14 (1972) 152

25. B. Bartoli, F. Felicetti and V. Silvestrini, Rivista del Nuov. Cim. 2 (1972) 241

26. H.C. Kirkmann et al., Phys. Lett. 32B (1970) 519

27. A.I. Akhiezer and V.B. Berestetskii, Quantum Electrodynamics, Interscience 1965

28. Y.S. Tsai, Phys. Rev. 122 (1961) 1898 and SLAC-PUB 848 (Jan. 1971)
L.W. Mo and Y.S. Tsai, Rev. Mod. Phys. 41 (1969) 205

29. D.R. Yennie, S. Frautschi and H. Suura, Annals of Physics 13 (1961) 379
N. Meister and D.R. Yennie, Phys. Rev. 130 (1963) 1210

30. L.C. Maximon, Rev. Mod. Phys. 41 (1969) 193

31. P.N. Kirk et al., Phys. Rev. D 8 (1973) 63

32. R. Felst, DESY 73/56 (unpublished)

33. K.M. Hanson et al., Phys. Rev. D 8 (1973) 753

34. E. Melkonian et al., Phys. Rev. 114 (1959) 1571
V.E. Krohn and G.E. Ringo, Phys. Rev. D 8 (1973) 1305
Yu. A. Alexandrov et al., Dubna preprint P3-7745 (1974)
L. Koester, W. Nistler and W. Waschkowski, T.U. Munich, preprint (Oct.1975)

35. S. Galster et al., Nucl. Phys. B 32 (1971) 221

36. M. Gourdin, Physics Reports 11C (1974) 29

37. A. Pais, Rev. Mod. Phys. 38 (1966) 215
B. Sakita, Advances in Particle Physics, Vol. 1, P. 247, 279
Ed. R.L. Cool and R.E. Marshak, Interscience 1968
P.G.O. Freund and R. Oehme, Phys. Rev. Lett. 14 (1965) 1085

38. A. Le Yaouanc et al. , Phys. Rev. D12 (1975) 2137 and preprint

39. A. Barut, O.D. Corrigan and H. Kleinert, Phys. Rev. Lett. 20 (1968) 167
A. Barut, Springer Tracts in Modern Physics 50 (1969) 1

40. G.F. Chew, R. Karplus, S. Gasiorowicz and F. Zachariasen, Phys. Rev. 110 (1958) 265

41. P. Federbush, M.L. Goldberger and S.B. Treiman, Phys. Rev. 112 (1958) 642

42. W.R. Frazer and J.R. Fulco, Phys. Rev. 117 (1960) 1603

43. A.M. Bincer, Phys. Rev. 118 (1960) 855

44. H. Pagels, Physics Reports 16 (1975) 219
H.F. Jones and M.D. Scadron, Phys. Rev. D 11 (1975) 174

45. S.D. Drell and H.R. Pagels, Phys. Rev. 140 (1965) B 397
S.D. Drell and D.J. Silverman, Phys. Rev. Lett. 20 (1968) 1325

46. B.B. Deo and L.P. Singh, Phys. Rev. D 10 (1974) 308
R.E. Bluvstein et al., Nucl. Phys. B64 (1973) 407

47. A. Love and R.G. Moorhouse, Nucl. Phys. B9 (1969) 577

48. P.S. Lee, G.L. Shaw and D. Silverman, Phys. Rev. D10 (1974) 2251

49. N. Dombey, Rev. Mod. Phys. 41 (1969) 236

50. G. Höhler and R. Strauss, Z. Physik 232 (1970) 205

51. M. Conversi et al., Phys. Lett. 52B (1974) 493

52. J.L. Basdevant, C.D. Froggatt and J.L. Petersen, Nucl. Phys. B72 (1974) 413. C.D. Frogatt and J.L. Petersen, Nordita 75/2
W. Männer, CERN preprint and contribution to the 17th Int. Conf. on high energy physics, London 1974, B. Hyams et al., Nucl.Phys. B100 (1975) 205

53. J.S. Ball and D.Y. Wong, Phys. Rev. Lett. 6 (1961) 29
J. Hamilton et al., Phys. Rev. 128 (1962) 1881
L.L.J. Vick, Nuovo Cim. 31 (1964) 643
N.G. Antoniou and J.E. Bowcock, Phys. Rev. 159 (1967) 1257

54. S. Furuichi et al., Progr. Theor. Phys. 38 (1967) 636 and 42 (1969) 744, 861

55. G. Höhler, R. Strauss and R. Wunder, Karlsruhe preprint, submitted to the Int. Conf. on High Energy Physics, Vienna 1968 and
R. Strauss, thesis, University of Karlsruhe (1968)

56. U. Brall and R. Rodenberg, Nuovo Cim. 8A (1972) 381

57. H. Nielsen, Nucl. Phys. B33 (1971) 152
H. Nielsen and G.C. Oades, Nucl. Phys. B49 (1972) 586 and unpublished work (1975)

58. G.N. Epstein and B.H.J. Kellar, Phys. Rev. D 10 (1974) 2169

59. G.E. Bohannon and P. Signell, Phys. Rev. D 10 (1974) 815

60. G. Höhler and E. Pietarinen, Phys. Lett. 53B (1975) 471

61. G. Höhler and E. Pietarinen, Nucl. Phys. B95 (1975) 210

62. G. Höhler, E. Pietarinen, I. Sabba-Stefanescu, F. Borkowski, G.G. Simon, V.H. Walther and R.D. Wendling, TKP 76/1, Univ. of Karlsruhe

63. J. Willrodt, Diplomarbeit and Thesis, University of Hamburg (1975)

64. N. Zovko, Fortschritte der Physik 23 (1975) 185

65. F. Iachello, A.D. Jackson and A. Lande, Phys. Lett. 43B (1973) 191

66. S. Mehrotra and M. Roos, preprint, University of Helsinki (August 1975), submitted to the 1975 International Symposium on Lepton and Photon Interactions at High Energies (Stanford)

67. E. Pietarinen, Karlsruhe preprint TKP 15/75, to be published in the Proceedings of the Palermo Conference 1975

68. E. Pietarinen, Karlsruhe preprint 4/75 and 5/75, to be published

69. R. Oehme in "Lectures on High Energy Physics", Ed. B. Jaksic, Gordon and Breach 1965

70. B.B. Deo and M.P. Parida, Phys. Rev. D8 (1973) 2939

71. C.L. Hammer et al., Phys. Rev. D9 (1974) 158

72. G. Höhler, H.D. Kiehlmann and W. Schmidt, Phys. Rev. D11 (1975) 2667

73. H. Pfister, Fortschritte der Physik 19 (1971) 1

74. S.C. Cheung, Carnegie-Mellon preprint COO-3066-1 (1971)

75. J.J. Sakurai, Current and Mesons. University of Chicago Press 1969; Invited paper presented at the Canadian Institute of Particle Physics, Summer School 1972 (UCLA/72/TEP/62 and 63)
Erice Lecture Notes 1971 (UCLA/TEP/139)

76. G. Höhler and P. Stichel, Z. Physik 245 (1971) 387

77. H. Nielsen, J. Lyng-Petersen and E. Pietarinen, Nucl. Phys. B 22 (1970) 525
J. Hamilton and J. Lyng-Petersen, Nucl. Phys. B 29 (1971) 51

78. G. Höhler, J. Baacke and F. Steiner, Z. Physik 214 (1968) 381

79. J. Engels, G. Höhler and B. Petersen, Nucl. Phys. B 15 (1970) 365

80. J. Banerjee et al., Nuovo Cim. 66 (1970) 475

81. M.A. Beg, Phys. Rev. Lett. 19 (1967) 767

82. M. Gell-Mann and F. Zachariasen, Phys. Rev. 124 (1961) 953

83. N.M. Kroll, T.D. Lee and B. Zumino, Phys. Rev. 157 (1967) 1376

84. G.J. Gounaris, Phys. Rev. 181 (1969) 2066

85. D. Benaksas et al., Phys. Lett. 39B (1972) 289

86. R. Hofstadter, Rev. Mod. Phys. 28 (1956) 214

87. R.G. Sachs, Phys. Rev. 126 (1962) 2256 (Appendix)

88. N.B. Skachkov, Dubna preprint E2-8007 (1974)

89. R.C. Barret et al., Phys. Rev. 166 (1966) 1589

90. G. Shaw, Phys. Lett. 39B (1972) 255
F.M. Renard, Phys. Lett. 47B (1973) 361

91. K. Fujikawa et al., preprint DAMTP 73/17, Cambridge, England

92. C.N. Brown et al., Phys. Rev. D8 (1973) 92

93. F.A. Behrends and G.B. West, Phys. Rev. 188 (1969) 2538

94. F. Borkowski et al., Nucl. Phys. A 222 (1974) 269

95. F. Borkowski, G.G. Simon, V.H. Walther and R.D. Wendling, Nucl. Phys. B 93 (1975) 461

96. H. Pilkuhn et al., Nucl. Phys. B65 (1973) 460

97. M.M. Nagels, T.A. Rijken and J.J. de Swart, Annals of Physics (N.Y.) 79 (1973) 338

98. A. Donnachie, Rapporteur's talk presented at the International Symposium on Lepton and Hadron Interactions at High Energies, Stanford 1975

99. G. Veneziano, Nuovo Cim. 57A (1968) 190
J.A. Shapiro, Phys. Rev. 179 (1969) 1345
C. Lovelace, Phys. Lett. 28B (1968) 264

100. M. Bernardini et al., Phys. Lett. 46B (1973) 261

101. F. Ceradini et al., Phys. Lett. 43B (1973) 341
M. Bernardini et al., Phys. Lett. 53B (1974) 384

102. P. Frenkiel et al., Nucl. Phys. B47 (1972) 61

103. R.D. Feynman, Photon Hadron Interactions, Benjamin 1972

104. C. Bacci et al., Phys. Lett 44B (1973) 533

105. C.R. Schumacher and I.M. Engle, Argonne preprint ANL/HEP 7032

106. G. Höhler and H. Genz, TKP 76/2, University of Karlsruhe.

THE DISCREET CHARM OF THE NEW PARTICLES+

A. De Rújula

The Physics Laboratories

Harvard University

Cambridge, Mass. 02138

FOREWORD

I discuss the charmonium interpretation of the newly discovered resonances, not as an isolated topic, but as something that fits naturally into the standard model of elementary particles. I emphasize the consistency of the overall picture, in particular as a discription of e^+e^- annihilation. I try to asses our degree of understanding of the model in a fieldtheoretic framework. Chromodynamics quark alchemy, flavor counting, asymptotics, gamma ray spectroscopy and the search for charm are amoung the topics discussed. In an attempt to reach a hypothetical readership of non-experts, I sacrifice the customary bon ton to make many "trivialities" and "well known facts" explicit. In an attempt to guide the reader with very specific interests, I subdivide the material ad nauseam.

I. HISTORY: A DISTILLATION OF RUMOUR

What used to be vices
have become fashions. (Seneca)

The discovery of strangeness, a new quantum number conserved by the strong interaction, and the study of the regularities of the hadron spectrum, eventually led to the acceptance of SU(3) as a classification scheme for hadrons and an approximate symmetry of their interactions. In 1963 and '64 the idea that nature

+ Work partly supported by the NSF under grant MPS 75-20427.

may not have stopped at the SU(3) level started creeping into peoples' minds. Teplitz and Tarjanne[1] were first to point out the possible existence of another quantum number conserved by the strong interactons that along with strangeness and isospin would build up SU(4) as an approximate hadron symmetry. The analogy between four leptons (e, ν_e, μ, ν_μ) and four quarks (p, n, the charmed p' and λ), including the form of the weak interaction current, was first drawn by Hara[2]. Bjorken and Glashow[3] independently drew this analogy and gave the new quantum number the name that would eventually catch: charm. Many others can be quoted among the early exponents of these or closely related ideas[4]. The philosophy behind these proposals was "why not?", simple but not entirely compelling.

The coronation of charm as a rather unique and compelling necessity within an attractive framework (the field theory of weak decays in the quark model) had to wait for the 1970 work of Glashow, Iliopoulos and Maiani[5]. These authors proved that the introduction of a weak hadron current

$$J_\mu = \bar{p}\gamma_\mu(1+\gamma_5)[n\cos\theta_C + \lambda\sin\theta_C] + \bar{p}'\gamma_\mu(1+\gamma_5)[-n\sin\theta_C + \lambda\cos\theta_C] \qquad (1)$$

involving the charmed p', was a satisfactory and elegant way of suppressing strangeness changing neutral currents. The "GIM" mechanism, they proved, would work both in cutoff nonrenormalizable Fermi-type theories and in the gauge theories that at the time were only suspected be to renormalizable[6]. The motivation for charm, they pointed out, was particularly strong in gauge theories: without charm, strangeness changing neutral currents would generally occur to leading order in the weak interactions.

In an SU(4) scheme many new particles are predicted, some of them with charm, others with no charm. The family of neutral nonstrange vector mesons (ρ, ω, ϕ), in particular, should be extended to include an extra particle. The possibility of this extra vector meson being very heavy and very narrow had been entertained by theoreticians before the 1974 November Revolution. In their timely review of the search for charm, Gaillard, Lee and Rosner[7] predicted a total width 2 MeV for a hypothetical state of mass 2 GeV (for the actual mass they would have predicted $\Gamma = 3$ MeV, an overestimate of "only" a factor of 50). Appelquist and Politzer[8], on the other hand, were considering a "Coulombic" picture in which the quark-quark binding forces would have become so weak that one could expect, not only an extremely narrow resonance, but a rapid succession of them with a hydrogen-like mass spectrum. Nature chose to sit halfway between these theoretical expectations.

In November 1974 a new very heavy meson was simultaneously discovered at Brookhaven and SLAC. As a fair solution to the problem of whether to call it J or ψ,

I will in what follows, with the help of a coin, randomly give it one or the other name. The excitement that the codiscovery produced needs no comment. The theoretical literature was immediately saturated with different interpretations of ψ. In These lecture notes I only discuss the status of the "charmonium" picture in which the new mesons are interpreted as bound states of a heavy quark and its antiquark. Even in the best of circumstances, it may take years to establish whether or not the hypothetical quark constitutents are "charmed" in the technical sense of carrying the quantum number specifically used in the GIM scheme[5] to tackle weak interaction problems. While the charm of charmonium is an entirely open issue, I will argue that the "onium" of charmonium, one year after the discovery, is in a very satisfactory shape. "Coloured" schemes are reviewed by Prof. Stech elsewhere in these proceedings.

Could J be interpreted in November 1974 as a bound state of a charmed quark and its antiquark: a hadron? At first sight the answer was definetly negative, there being a seemingly deadly problem. This is dramatized in Figure (1), where I have plotted the masses of all established particles versus their lifetimes. The names of the "stable" particles are written at their positions in this plot. The open circles are meson resonances, the crosses baryon resonances. Also indicated are J and ψ'. They are seen to last a thousand times longer than a "normal" hadron with such a large mass. It is clear that without rather specific dynamical reasons (or at least excuses) the charmonium hypothesis could not be entertained. The width (or lifetime) measurement was in fact more suggestive of a semiweak decay. In Figure (1) I have also plotted mass versus lifetime as naively estimated in a quark model for a neutral intermediate vector boson with conventional semiweak couplings. The weak boson interpretation was soon "supported" by the spurious detection of a forward-backward asymmetry at Frascati, and somewhat discredited by the discovery of ψ'.

The interpretation of J as orthocharmonium immediately led to several trivial predictions that turned out to be correct: the quantum numbers should be $J^{PC} = 1^{--}$, G-parity negative, SU(3) singlet, SU(2) scalar. The widths into electron and muon pairs should be equal. Radial excitations with the same quantum numbers should exist, the lightest of them with a chance of being narrow and a large branching ratio into $\psi\pi\pi$[9]. There should be a threshold in e^+e^- annihilation at $W \sim 3.9$ GeV[9], where pairs of mesons with opposite charm would start being produced (At the time of this writing we are still waiting for this last bit of information to be confirmed experimentally.)

II. THE LEPTONIC AND HADRONIC WIDTHS OF ψ

As I have stated in the historical introduction the crucial problem in the charmonium approach is the understanding of total widths. I will argue in this section that the leptonic width of ψ is "normal" and its hadronic width is fascinating, but not too abnormal. In subsequent sections I will try to elaborate on our present degree of real understanding of this problem.

Suppose a vector meson annihilates into e^+e^- (or $\mu^+\mu^-$) via a photon as in Figure (2a). Assume conventional quark charges (2/3, 2/3, -1/3, -1/3 for p', p, n, λ respectively). Assume ϕ to be purely made of $\lambda\bar{\lambda}$, ψ of $p'\bar{p}'$ and ρ^0 and ω to contain only light quarks ($\bar{p}p \pm \bar{n}n$). (This "purity" assumption is crucial, I postpone its elaborate justification till Chapters IV and V). From the above assumptions, it follows that if J, ϕ, ω and ρ all had equal masses, their widths into e^+e^- would be in the ratio 8 : 2 : 1 : 9. Experimentally they are found to be in the ratio (7.3 ±1.0) : (2.0 ±1.2) : (1.2 ±0.2) : (9.4 ±1.4) in agreement with the naive expectation. The argument is a little fallacious but not enough to stop us from concluding that the e^+e^- width of ψ is in the right ball park. In the real world with big mass differences it is not clear what combination of widths and masses should be used in the comparison between the different mesons. In a nonrelativistic bound state approximation, for example, the electronic widths of J and ϕ would be

$$\Gamma_e = 4\alpha^2 Q^2 |\psi(0)|^2 / m^2 \tag{2}$$

where m and Q are the quark mass and charge and $\psi(0)$ is the wave function at the origin. For what combination of $|\psi(0)|$ and m, if any, should we assume SU(4) symmetry? No answer is possible without detailed dynamics.

I now turn to the study of hadronic widths of vector mesons and, having convinced myself that the leptonic widths are normal, I will consistently take the ratio of hadronic to leptonic widths. I expect this to reduce the model dependence of my considerations. To treat all ratios on equal footing, I will also "normalize" the leptonic widths to equal quark charge. Thus, consider the following ratios of widths, as determined from experiment:

$$\gamma_\rho \equiv \frac{\Gamma(\rho \to 2\pi)}{\{\Gamma(\rho \to e^+e^-)/9\}} = (2.1 \pm 0.3)10^5 \tag{3a}$$

$$\gamma_\phi \equiv \frac{\Gamma(\phi \to 3\pi)}{\{\Gamma(\phi \to e^+e^-)/2\}} = 990 \pm 110 \tag{3b}$$

$$\gamma_J \equiv \frac{\Gamma(J \to \text{hadrons})}{\{\Gamma(J \to e^+e^-)/\gamma\}} = 98 \pm 26 \tag{3c}$$

The right hand side of (3a) is of order $1/\alpha^2$, as one would expect for the ratio of a strong process to a second order electromagnetic process. The right hand side of (3b) is much smaller. In a quark model where ϕ is assumed to be dominantly a $\lambda\bar{\lambda}$ bound state, there is an excuse for this (Zweig-Iizuka's rule[10]). In the decay of ϕ into pions (states containing no λ quarks) the strange quarks must meet and annihilate. Contrary-wise, in the decay of ρ into pions no such quark annihilation is necessary. Since different processes may have different rates, $\gamma_\rho >> \gamma_\phi$ is not surprising. Stated this way Zweig-Iizuka's rule is somewhat trivial, it is only when other "allowed" and "forbidden" processes are compared that the rule becomes nontrivial. J is supposed to be too light to decay into a pair of oppositely charmed particles. Thus, if it is a rather pure $p'\bar{p}'$ bound state, its constitutents must meet and annihilate in the decay. It follows that γ_J need not resemble γ_ρ, but must be of the same order as γ_ϕ. This is observed to be the case and follows from the rule that similar processes must have similar rates (I have been unable to trace down the adequate reference on this).

I will justify the purity of ϕ as a $\lambda\bar{\lambda}$ state and of ψ as a $p'\bar{p}'$ state in the "standard" model of quarks and gluons. This will allow me to be more ambitious in trying to understand the relation between γ_ψ and γ_ϕ and, more important, to make predictions for other states of the charmonium family.

III. THE STANDARD MODEL[11] OF NEARLY EVERYTHING

Born of years of strugle with reality, the standard model of nearly everything (i.e. everything but CP violation and renormalizable gravity) is very specific and predictive. The building material blocks of the model are the four known leptons and twelve quarks q, coming in four flavours (p, p', n and λ), each in a color - SU(3) triplet[12]. The rules and interactions are the following:

i) Hadrons are color singlets, either $q\bar{q}$ (mesons) or qqq (baryons). This rule is incompletely understood, but it solves in one stroke the quark statistics problem and reproduces the gigantic success of the quark model in hadron spectroscopy.

ii) Quarks interact with photons as pointlike Dirac particles with fractional charges.

iii) The weak interactions are analogous to the electromagnetic: they couple the GIM current of Eq (1) to a heavy intermediate vector boson. Weak and electromagnetic interactions are unifiable à la Weinberg-Salam[13].

iv) "Chromodynamics", the "strong" interaction of the standard model is also analogous to electrodynamics and to the weak interactions. It is the coupling of an octet of currents $\bar{\psi}\gamma_\mu\lambda^a\psi$ to an octet of massless (yet impossible to isolate) colored gluons G^a_μ. In what follows the quark-gluon strong "gauge" coupling constant is denoted g_s. The strong gauge group commutes with the weak and electromagnetic gauge group[11]. Conventional hadrodynamics is a Van der Waals leftover of chromodynamics. The usual strong interactions are secondary, not fundamental and, I am almost tempted to say, not more interesting than chemistry. The observed SU(4), SU(3) and SU(2) breaking is a consequence of mass differences between differently flavored quarks. Only this breaking respects the renormalizability of the model. What produces this mass differences is tomorrow's spectroscopical challenge. By construction and demonstration, the "strong" interactions of the standard model are asymptotically free[14]. They are also conjectured to suffer infrared slavery.

<u>Asymptotic freedom</u>, in layman's words, is the statement that the coupling "constant" g_s effectively decreases at short distances or large momenta Q. Somewhat less imprecisely, in suitably chosen and unfortunately uncommon processes it is possible to apply the machinery of the renormalization group to organize perturbation theory in terms of a Q^2 dependent $\bar{g}_s$ such that no large logarithms ($\ln Q^2$) pop up and threaten the convergence of the perturbation expansion. The "usual" logarithms (old friends of anybody who has witnessed a one-loop integration) are under control and only surface as a slow logarithmic decrease of g_s. At short distances the theory effectively becomes quasi-free field theory. This is the basis for our understanding of Bjorken scaling: at higher and higher Q^2 the photon sees charged quarks progressively stripped of their strong interaction dressings. The standard model predicts calculable corrections to exact scaling. Scaling deviations not unlike the predicted ones have actually been observed in high energy muon scattering.

<u>Infrared slavery</u> is the converse notion: at large distances or smaller momenta the strong coupling increases, perhaps such as to provide permanent quark confinement. For a large coupling, perturbation theory breaks down and it is necessary to invoke other approximations or techniques. In the "approximation" that space and time are discrete (i.e. in a gauge theory in a lattice) it has been proved that the standard model displays infrared slavery[15]. It is found, for example, that the energy of interaction between two static distant quarks increases linearly with

their separation. This provides an excuse for the use of linear potentials in charmonium calculations, a point to which I come back in Chapter IX.

IV. THE MASSES AND QUARK CONTENT OF CONVENTIONAL MESONS IN THE STANDARD MODEL

To argue that the width of J is not unthinkable for a hadron we necessarily had to assume that it is a very pure state of charmed quarks, with no significant admixture of lighter quarks through which to decay fast. We found an analogy in ϕ , assumed to be a rather pure $\lambda\bar{\lambda}$ state. It would be entirely unsatisfactory to have to make these assumptions with no justification. Fortunately, in the context of the standard model, the assumptions are not arbitrary, but find support in the systematics of the spectrum of light mesons[9]. The $J^P = 1^-$ vector mesons satisfy the equal spacing rules

$$\rho = \omega \;[770 \text{ MeV} \sim 784 \text{ MeV}] \tag{4a}$$

$$2K^* - \rho = \phi \;[1014 \text{ MeV} \sim 1019 \text{ MeV}] \tag{4b}$$

where particle names stand for particle masses. If SU(3) is only broken by differences in quark masses, to first order in these differences the masses of the 1^- multiplet will equal a universal constant term plus a term linear in the masses of the constitutent quarks. Mass differences are smaller than particle masses and this is a satisfactory first order perturbation of the SU(3) symmetric limit. The eigenvalues of the mass matrix are dictated by quark content, ϕ is a pure $\lambda\bar{\lambda}$ state, ω and ρ are made of light quarks (of roughly equal mass) and the equal spacing rules follow automatically. I have tacitly assumed that there is no interaction capable of annihilating a $\lambda\bar{\lambda}$ quark to recreate a light quark pair, mixing ϕ and ω and de-diagonalizing the mass matrix. I shall challenge this assumption soon.

If one extends the previous analysis to the $J^P = 0^-$ pseudoscalars the result is total disaster:

$$\pi = \eta \;[140 \text{ MeV} = 548 \text{ MeV !}] \tag{5a}$$

$$2K - \pi = \eta' \;[854 \text{ MeV} = 958 \text{ MeV}] \tag{5b}$$

Something must provide for nondiagonal terms in the mass matrix, as labelled by quark content. In the standard model interactions capable of doing the job exist: the annihilation of, say, a $\lambda\bar{\lambda}$ pair into two[16] or more gluons that rematerialize as a light quark pair as in Fig. (3a). The mass difference between π and η is considerable and the nondiagonal terms must be relatively large, making the whole perturbative approach in quark mass differences entirely unreliable. But, back to the 1^- mesons, terms that de-diagonalize the mass matrix are also present: annihilation via three or more gluons as in Fig. (3b). The contribution of these interactions to the vector meson masses must be small for the equal spacing rules not to be fortuitous. It would be overoptimistic to invoke asymptotic freedom and precocious scaling to argue for a small quark-gluon coupling at the mass of the 1^- multiplet relative to the 0^- multiplet. Fortunately, there is an important kinematical effect working in the right direction: three body phase space is much less favorable than two body phase space and one does not need an unbelievably small coupling to make the three gluon annihilation diagram much smaller than the two-gluon one. Thus, we qualitatively understand the systematics of the vector mesons and we understand why we do not understand the pseudoscalars.

The above considerations become predictive (rather than merely entertaining and consistent) when one considers the nonet of tensor mesons with $J^P = 2^+$. In a quark model, these are P wave bound states and, although quark annihilation may proceed via two gluons, it should be somewhat damped by the angular momentum barrier. Thus equal spacing should be good. It is:

$$A_2 = f \quad [1310 \text{ MeV} = 1270 \text{ MeV}] \tag{6a}$$

$$2K_A^* - A_2 = f' \quad [1530 \text{ MeV} = 1516 \text{ MeV}] \tag{6b}$$

The preceeding discussion has an immediate implication: ϕ decay into nonstrange hadrons must be damped. The light quark content of the ϕ must be small for the equal spacing rules to be so good. In the annihilation of its strange constitutents into gluons that subsequently evolve into pions we encounter again the necessarily small three gluon process. In conclusion, goodness of linear spacing and smallness of $\Gamma(\phi \to 3\pi)$ are aspects of the same underlying dynamics.

V. THE PHANTOM OF ASYMPTOTIC FREEDOM: γ_J VERSUS γ_ϕ

A trivial repetition of the arguments of the preceding chapter implies that J, yet another vector meson, must be a rather pure $p'\bar{p}'$ state. In the standard model the "impurities" in J and the annihilation of its constitutents into conventional hadrons are again governed by three (or more) gluon diagrams. More ambitiously one can surpass the hand-waving stage and compute the hadronic width of J. This is done in the following manner: trade hadron dynamics by the underlying chromodynamics and "forget" (as one does in other successful quark parton model calculations) the effect of the interactions that transmogrify the fundamental fields into hadrons. The lowest order amplitude for J decay is then $p'\bar{p}' \to 3$ gluons. The decay rate to lowest order is as in Fig. 3d. In an attempt to minimize the model dependence, take the ratio of calculated strong and electromagnetic decays. Define $\alpha = e^2/4\pi$ and its strong analog $\alpha_s = g_s^2/4\pi$. Following the instructions, compute γ_J to find

$$\gamma_J/8 \equiv \frac{\Gamma(J \to \text{hadrons})}{\Gamma(J \to e^+e^-)} \sim \frac{5(\pi^2-9)\{\alpha_s(J)\}^3}{18\pi\alpha^2} \qquad (7)$$

The formula would be correct to leading order in α and α_s in a nonrelativistic bound state approximation, provided the radius of the state is much larger than the inverse of the heavy quark mass. From a formula analogous to Eq (7) and the observed γ_ϕ I obtain $\alpha_s(\phi) \simeq .5$. Optimistically assume that ϕ is a sufficiently nonrelativistic bound state, that 1019 MeV is large enough for asymptotic freedom to have set in, and that $\alpha_s = .5$ is small enough for higher order corrections not to matter. Then one can use the functional form of $\alpha_s(M^2)$, explicit within the model, to predict $\alpha_s(J)$ from $\alpha_s(\phi)$. The result is

$$\alpha_s(\psi) \sim \frac{\alpha_s(\phi)}{1 + \frac{25}{12\pi}\alpha_s(\phi)\ln\frac{m^2(\psi)}{m^2(\phi)}} \sim .28 \qquad (8)$$

which, inserted back into Eq (7) predicts $\gamma_J \simeq 250$, to be compared with the observed value $\gamma_J = 98 \pm 26$. This is considerable improvement relative to the naive identification $\gamma_J \sim \gamma_\phi$ which would have been in error by one order of magnitude. Values of α_s $(Q = M(\psi))$ in this neighborhood are also found is studies of scaling deviations (see Section X). Many ifs entered the derivation and many buts can be raised. I cannot conclude that we have numerically explained the width of J from the observed widths of ϕ or that we have seen asymptotic freedom in operation. The picture, though, is nice and consistent. I will proceed with farther predictions after a section on the ifs and buts.

VI. HOW NAIVE IS THE NAIVE APPROACH?

Before I embark in the task of deriving predictions of the charmonium approach in the standard model, I pause to assess our degree of understanding of the simplistic considerations of previous chapters. As far as possible (not very far) I stick to a West Coast definition of understanding: "We understand a proposition if it follows from general principles or a justifiable use of perturbative field theory".

To what degree can we ascertain that three-gluon annihilation is a fair description of J decay into hadrons? Some problems and the solutions to some of them are:

i) Bound state singulatiries should not affect the perturbative computation of γ_ψ[17]. An annihilation diagram of a $p'\bar{p}'$ pair into light quarks and gluons can always be factorized as in Figure 4. The B amplitude is by definition two-heavy-quark irreducible. Eventually small quantities like $M(\psi) - 2\, m(p')$ do not make B singular. The sum $\Sigma|B|^2$ of all diagrams to a specified order of perturbation is finite. These points have been checked to several orders of perturbation theory[17]. The A amplitude contains Coulomb and Yang-Mills singularities but drops from ratios like γ_J. Thus, bound-state problems are rendered less severe in the perturbative computation of ratios of widths.

ii) The mass M of the particle may be used as the argument of the running coupling constant g_s in the calculation of the annihilation of its quark constitutents into gluons. In principle any argument Q^2 in $g_s(Q^2)$ is as good as any other in the perturbative calculation of any process at an arbitrary energy. Choosing $g_s(Q^2)$ to compute a process characterized by the scale of momentum Q is just a devise to avert evil (i.e. large logarithms). What has been checked to several orders of perturbation is that, in the calculation of $\Sigma|B|^2$ to a given order from diagrams containing N gluons or gluons plus light quarks, no small momenta Q/N naturally appear[17].

iii) To be confident in the application of short-distance ideas to the annihilation of ψ's constitutent quarks one may check that in a reasonable bound state model charmed quark masses are large compared with the inverse of the "size" of the bound state. This has been done by the Cornell group[18] in a nonrelativistic picture where binding is provided by a linear plus coulombic potential. The result is that the mass of the quark is roughly five times larger than the inverse of the classical turning point of the lowest lying bound state. Whether a similarly satisfactory result would be obtained for the ϕ meson is doubtful.

iv) For J to be narrow it must be kinematically forbidden to decay into charmed pairs. Otherwise diagrams of order in g_s lower than the three gluon diagram would have nonzero absorptive parts and would contribute to decay (into charmed particles). The condition $m_J < 2\, m$, where m is the mass of the lightest charmed particle, is met by all estimates of charmed masses of which I am aware. In a naive quark model approach, for example, the charmed nonstrange pseudoscalar meson D is lightest and $m_D \sim 1.85$ GeV [19].

v) I have tacitly neglected the strong interaction fudge factor: the effect of the amplitude for gluons to transfigurate into conventional hadrons, as in Figure 5a. This is only justified by tradition and success. In quark parton model calculations of relations between electron and neutrino structure functions (as in Figure 5c) or of the total e^+e^- annihilation cross section at energies below the threshold for new physics (Fig. 5b) the amplitude for the evolution of the fundamental fields (in this case quarks) into hadrons is also neglected, with satisfactory results.

vi) Last but not least, I have assumed that surprises undectable in perturbation theory do not spoil the relevance of the three gluon picture. It may be that when the charmed quarks in J are far apart their strong interactions, which by then are really strong (infrared slavery), mediate the production of a light quark pair. The heavy quarks subsequently annihilate and the light quarks sneak out as conventional hadrons, as in Figure 6b. If this process is important, all considerations based on the short distance three gluon amplitude of Figure 6a are wrong. What picture (short or large distance annihilation) is closer to reality may be resolved in the study of paracharmonium.

VII. PARACHARMONIUM: THE CORNERSTONE OF CHROMODYNAMIC GLUON COUNTING

If ψ is orthocharmonium, the lowest lying S wave bound state of a charmed quark-antiquark pair in the triplet spin state; paracharmonium, the $J^P = 0^-$ singlet state should also exist and be nearby in mass. We interpret the state of mass 2.75 GeV discovered at DESY as the predicted paracharmonium.

The chromodynamic ideas discussed in previous chapters can be put to simple and stringent test in the study of paracharmonium decays. Should three gluon annihilation with a relatively small g_s be the dominant process for orthocharmonium decay, paracharmonium, a pseudoscalar, would dominantly decay via two gluons, as in Fig. 3c. The ratio of total hadronic widths of para and orthocharmonium (two versus three gluon decays) can be computed in the standard model. To leading order in $\alpha_s = g_s^2/4\pi$, the result is[20]:

$$\frac{\Gamma_P}{\Gamma_0} \equiv \frac{\Gamma(\text{Para} \to \text{hadrons})}{(\text{Ortho} \to \text{hadrons})} \sim \frac{27\pi\alpha_s^{-1}}{5(\pi^2-9)} \sim 70 \quad . \qquad (9)$$

The dependence of α_s on mass in the region M(Para) to M(Ortho) is very small, in Eq (9) I have used the orthocharmonium result of Eq (8). Thus paracharmonium is expected to have a hadron width $\Gamma_P \sim 4.1 \pm 1.2$ MeV, roughly two orders of magnitude larger than orthocharmonium. The difference is mainly due to how favorable two-body phase space is, relative to three-body phase space. The large ratio ~70 is a weird and very specific prediction of the standard model, supplemented by the short distance annihilation ansatz. Should experiment confirm Eq (9) we would have a successful understanding of hadron dynamics in the new domain of gluon counting. On the other extreme, long distance effects could dominate Ortho and Para decay, as in Figure 6b. In this case we do not expect spin effects to be very relevant: Γ_P/Γ_O should be roughly unity. Values of Γ_P/Γ_O well outside the range 1 to 70 would be rather puzzling. How close the experimental result is to one of the extremes measures how good the long or short distances ansatzs are.

A very neat measurement of α_s at the mass of paracharmonium would be provided by the ratio

$$\frac{\Gamma(\text{Para} \to 2\gamma)}{\Gamma(\text{Para} \to \text{all})} \equiv \frac{\Gamma(\text{Para} \to 2\gamma)}{\Gamma(\text{Para} \to 2 \text{ gluons})} = \frac{2\alpha^2}{\alpha_s^2} \sim 0.3\,\% \tag{10}$$

where I have assumed short distance two gluon dominance. All kinematical factors drop from this ratio, that could only be modified by the strong interaction fudge factor that we always "forget". At the time of this writing I am not aware of a measurement of the ratio in Eq (10). Preliminary experimental indications are consistent with a ratio as small as the predicted ~.3 %. We know, with rather pathetic statistics that $\Gamma(\text{Para} \to 2\gamma) \sim \Gamma(\text{Para} \to p\bar{p})$. We also know that $\Gamma(\text{Ortho} \to \bar{p}p)/\Gamma(\text{Ortho} \to \text{all}) = (.21 \pm .04)\ \%$. Theoretically we expect the branching ratio into $\bar{p}p$ to be quite similar for Ortho and Para. From the above considerations and numbers, $\Gamma(\text{Para} \to 2\gamma)/\Gamma(\text{Para} \to \text{all}) \sim .2\ \%$, in agreement with Eq (10).

VIII. CHARMONIUM LEVELS: THE RESURRECTION OF GAMMA RAY SPECTROSCOPY

Facts are stubborn things. (J. Elliot)

The interpretation of ψ' as an S wave radial excitation of the two quark system in ψ quite trivially leads to the prediction of a full spectrum of charmonium states[18,20,21]. The analogous spectrum of conventional mesons, interpretable as light quark-antiquark bound states is well known, and displayed in Fig. (7). The Ortho ground state is ρ, its radial excitation ρ'. The Para ground state is π, π' has not been seen. There are P wave states with positive charge conjugation ($J^{PC} = 2^{++}, 1^{++}, 0^{++}$) and negative charge conjugation (1^{+-}). All of them are established, with the time-dependent exception of A_1 (1^{++}). The P wave states lie halfway between ρ and ρ', as they would in a linear oscillator quark-quark potential. Figure 8 shows the predicted spectrum of narrow charmonia. (The states anticipated to be below charm threshold). I have added to a Figure from Appelquist et al.[20], a subset of the flourishing vocabulary of charmonia. The P wave states were predicted at 3.5 GeV. In a harmonic oscillator potential they would be halfway between $\psi'(3.7)$ and $\psi(3.1)$. In a Coulomb potential they would sit up by $\psi'(3.7)$. For a linear potential (hinted by gauge theories on a lattice) one expects[20] or computes[18,22] an intermediate position. A harmonic oscillator potential for $(\text{mass})^2$ gives a similar result. To predict the splitting between P states or the Ortho-Para splittings it is necessary to guess the quark-quark spin-orbit, tensor and spin-spin forces. (More on this subject in Chapter XIII).

The novel and stimulating feature of the charmonium states is that they must be significantly produced in the γ ray decays of ψ' or J. This is because the hadron widths of these mesons are very small, while their electromagnetic interactions are normal.

The possible magnetic and electric dipole transitions, expected to be dominant, are indicated in Figure 8. The naive nonrelativistic quark model estimate of the transition rates is

$$\Gamma(M_1) \sim \frac{16}{27}\,\alpha\,\frac{k^3}{m_{p'}^2}\,I \tag{11a}$$

$$\Gamma(E_1) \sim \frac{16}{27}\,\alpha\,k^3 R^2\,\tilde{I} \tag{11b}$$

where k is the photon momentum and R is a measure of the size of the deca ying state. The factors I and $\tilde{I}$ are the square of wave function overlaps and are bound to be smaller than unity. In transitions like Ortho II $\to$ Para I + γ; $I \ll 1$ because of the orthogonality of the radial wave functions. If only because of the naiveté of the nonrelativistic quark model, it goes without saying that Eqs (11a,b) are not hoped to be excellent estimates[23].

The γ rays were first seen at DESY[24] in the cascade transition $\psi' \to J\gamma\gamma$ with the intermediate state at a mass of ~3.5 GeV (assuming the first emitted photon to be the less energetic one) and perhaps also 3.41 GeV (2 events, same assumption). The intermediate states where baptized P_C, apparently meaning positive charge conjugation, not P-wave charmonia. The discovery was confirmed at SLAC, where γ ray decays of ψ' into states at 3.53 and 3.41 GeV followed by hadronic decays of the latter have been observed[25]. These states have been rebaptized χ[26].

The observed Γ ray transition rates from ψ' down to the P-wave states have turned out to be one order of magnitude smaller than estimated in Eq (11b). A possible excuse for this (naively assuming that we need one) is the predicted proximity of $\psi'(3.7)$ to charm threshold. As a consequence of this propincuity ψ' would have a large probability of consisting of virtual charmed meson pairs. This would make its wave function "more orthogonal" to the P-wave state wave functions than one would estimate in a simple potential model. Predictions that do not depend on details of wave functions should be better satisfied. An example is the nonrelativistic statistical rule

$$\frac{\Gamma(\psi' \to 2^{++}\gamma(k_2))}{5k_2^3} \sim \frac{\Gamma(\psi' \to 1^{++}\gamma(k_1))}{3k_1^3} \sim \frac{\Gamma(\psi' \to 0^{++}\gamma(k_0))}{k_0^3} \tag{12}$$

which should be good in the limit of E_1 dominance and negligible tensor and spin-orbit forces in the P wave multiplet[23].

In the charmonium picture the P-wave states are expected to have γ ray decays into J that compete or even dominate over direct decays into hadrons[20]. Hadron decays of 2^{++} and 0^{++} may proceed via two gluons but are suppressed by the angular momentum barrier. Yang's rule forbids the decay of 1^{++} into two 1^- gluons, and may further suppress the coupling of 1^{++} to hadrons[27]. Preliminary experimental indication is that the photonic branching ratios of the P-wave states are indeed very large[24,25]. In a model where dipole transitions are dominant, it is possible to predict the distribution in angle θ of γ rays relative to the e^+e^- beam direction in ψ' decay to the P-wave states. It is also possible to compute the correlation in the angle ϕ between the two γ rays in cascade decays via an intermediate P state. In the approximation where recoil is neglected, the results are[18]:

$$\begin{array}{llcl} \psi' \to & 2^{++} + \gamma & : & 1 + (\cos^2\theta)/13 \\ & \hookrightarrow J + \gamma & : & 1 \\ \psi' \to & 1^{++} + \gamma & : & 1 - (\cos^2\theta)/3 \\ & \hookrightarrow J + \gamma & : & 1 + (\cos^2\phi)/13 \\ \psi' \to & 0^{++} + \gamma & : & 1 + \cos^2\theta \\ & \hookrightarrow J + \gamma & : & 1 + 69(\cos^2\phi)/377 \end{array} \tag{13}$$

At present we do not have a measurement of the branching ratio for $\psi \to$ Paracharmonium I(2.75) + γ. However, an experimental estimate of some convoluted branching ratios exists[24,25]:

$$\{B_1\}\{B_2\} \equiv \left\{\frac{\Gamma(\psi \to \mathrm{Para}\,\gamma)}{\Gamma(\psi \to \mathrm{all})}\right\}\left\{\frac{\Gamma(\mathrm{Para} \to \gamma\gamma)}{\Gamma(\mathrm{Para} \to \mathrm{all})}\right\} \sim 2\cdot 10^{-4} \tag{14a}$$

$$\{R_1\}\{R_2\} \equiv \left\{\frac{\Gamma(\psi \to \text{Para}\,\gamma)}{\Gamma(\psi \to \eta\gamma)}\right\}\left\{\frac{\Gamma(\text{Para} \to \gamma\gamma)/\Gamma(\text{Para} \to \text{all})}{\Gamma(\eta \to 2\gamma)/\Gamma(\eta \to \text{all})}\right\} \sim 1 \tag{14b}$$

We have estimated B_2 in Eq (14a) to be ~.3 % (see Eq (10)). If this is correct B_1 ~ 6 %, an order of magnitude smaller than one would naively expect B_1 to be (see Eq (11a)) and a factor of ~3 smaller than some less naive estimates[28]. Should this bother us, we ought to remember that the M_1 transition rate Eq (11a) is computed with the assumption that quarks are Dirac fermions with no anomolous magnetic moment. For onshell photons this is not a prediction of the standard model, and similar estimates for light mesons (i.e. $K \to K\gamma$) are also found to be in error by a factor ~3.

R_2 in Eq (14b) can be computed from the measured decays of η and the predicted ratio B_2 of para decays. The result is R_2 ~ .01. It follows that R_1 ~ 100. This large number is not surprising since $\psi \to \eta\gamma$ is Zweig-Iizuka and SU(3) forbidden.

IX. THE e^+e^- ANNIHILATION CROSS SECTION IN THE THRESHOLD REGION

Following tradition, define

$$R(W) = \frac{\sigma(e^+e^- \to \text{hadrons})}{\sigma(e^+e^- \to \mu^+\mu^-)} \tag{15}$$

where W is the e^+e^- invariant mass. The measured[25] values of R in the region 2.4 GeV < W < 7.8 GeV are shown in Fig. (9). Fig. (10) is a blow up of part of Fig. (9), showing the transition region between two approximately constant regimes in R(W). In both figures the narrow resonances at 3.1 and 3.7 GeV have been removed.

The shape of R shows striking qualitative agreement with naive predictions of the charmonium approach. There is a threshold somewhere in the region 3.6 - 3.9 GeV where new action starts. Above this energy the final states should contain considerable production of charmed particle pairs: all of the R excess above the "background" R ~ 2.5 should be charmed. Advocates of nonrelativistic linear potential models[18,22] had predicted S wave radial excitations of J beyond ψ' with masses ~4.2, 4.6, 5, ... GeV. These states should have typical hadron widths, since they are expected to lie above charm threshold. Increasing phase space would make the heavier states too wide to be visible as bumps in R. Two and perhaps three peaks are visible in Fig. (10). It is not trivial to make a detailed "Who is Who" correspondence between the observed structures and the states predicted by the model. Three reasons are:

i) With every new set of improved data the structure in this region seems to become richter and richer.

ii) We expect several two-body thresholds to open up at these energies. In a quark model with charm there are, among others, charmed nonstrange pseudoscalar and vector mesons (D and D^*, each in an isotopic doublet with charges 0 and +1) and charmed strange pseudoscalars and vectors (F and F^*, with charge +1). Their level structure, estimated in a naive quark model calculation[19], is shown in Fig. (11). Thus, there are at least six thresholds in the e^+e^- channel:

$$\begin{array}{ll} D\bar{D} & 3.6 \text{ GeV} < W < 3.72 \text{ GeV} \\ D^*\bar{D} & 3.73 \text{ GeV} < W < 3.85 \text{ GeV} \\ D^*\bar{D}^* & 3.86 \text{ GeV} < W < 3.98 \text{ GeV} \\ F\bar{F} & W \sim 3.95 \text{ GeV} \\ F^*\bar{F} & W \sim 4.04 \text{ GeV} \\ F^*\bar{F}^* & W \sim 4.12 \text{ GeV} \end{array} \qquad (16)$$

I do not expect the above predictions of mass differences to be accurate to better than a factor of 3, but the conclusion that a complicated set of thresholds should be crossed between 3.6 and 5 GeV is unavoidable.

iii) Tensor forces will inevitable mix S and D levels of charmonium[18]. This implies that the "predominantly D" levels will not have vanishing wave functions at the origin. Hence, they will couple a little to the one photon channel and be produced in e^+e^- collisions. D states in a nonrelativistic linear potential of charmonium have masses[18,22] ~3.75, 4.3, ... GeV.

In conclusion, we expect the transition between the two "asymptopias" so far seen in R to be enormously rich and complicated. This has not stopped Kogut and Susskind[29] from developing a successful and very interesting attempt to understand the gross features of the transition region. These authors base their work on an analogy with the Born-Oppenheimer approximation scheme to the study of the hydrogen molecule. This problem has two very different time scales associated with the slow motion of the heavy nuclei, and the fast motion of the electrons, that may swiftly adapt to the instanteous nuclear coordinates. As a consequence of the two characteristic times, it is useful to factorize from the molecule's wave function, the would-be-wave-function of the electrons in the presence of static protons. This eventually leads to a perturbation approach in $(m_e/m_p)^{1/2}$, which in the case of H_2 can be pushed to a precision as high as one may need in practice. Kogut and Susskind trade the protons for a heavy charmed quark-antiquark pair and the electrons for gluons and light quark pairs. The large mass of the charmed quarks, they argue, justifies the collective treatment of the other degrees of freedom of the standard model as a potential between the quasi-static heavy quarks. Explicit calculations can be done in a power series expansion on the not very small parameter $(m_p/m_{p'})^{1/2}$. Fig. (12a) shows the form of the potential V(r) chosen by Kogut and Susskind. Their V is V-shaped. At short distances V is a one-gluon exchange Coulomb potential with logarithmic corrections dictated by asymptotic freedom. At intermediate distances V(r) is linear in r, a result that is borrowed from the study of the theory on a space-time lattice. At large distances between the heavy quarks there is a lot of energy accumulated in the "string" (the gluon and light quark pair fields extending between the oppositely colored heavy quarks). As r increases it becomes energetically favorable to break the string into a pair of light quarks. Each light quark associates with a heavy quark, the net result is two oppositely charmed mesons interacting only in a screened, exponentially decreasing manner. It is possible in this potential to have two narrow bound states (J and ψ') below charm threshold, followed by a quasi-bound state ψ'' above threshold. The parameters in the potential can be estimated from the masses and leptonic widths of the narrow resonances. Kogut and Susskind placed charm threshold at W ~ 4 GeV and computed R in their approach. Their results are shown in Fig. (12b), along with the data published in January 1975[30]. It may not be superfluous to point out that the prediction predated the data. Even in comparison with the much more structured data of Fig. (10) it is fair to say that the predicted shape of R is generally correct. One could not expect more in view of the complications discussed earlier in this section. A nice feature of the Kogut-Susskind calculation of R is that it connects smoothly above W ~ 5 GeV with the "asymptotic" predictions of the theory.

In the standard model asymptotia is a land within our reach. I turn in the next section to the discussion of the "asymptotic" predictions for R.

X. FINDING FANCY FLAVORS COUNTING COLORED QUARKS

In a model with nonstrongly interacting spin 1/2 pointlike hadron constitutents (in which even the vocabulary makes no sense), the value of R is predicted to be

$$R = \sum_i Q_i^2 \tag{17}$$

where Q_i are the constitutent quark charges and I have assumed quark masses to be negligible relative to the photon "mass" W. It is possible but rather antiaesthetic to fix the number of colors to be different from three[31]. With three colors, one may argue that R is a measure of the number of flavors. Below charm threshold and with the three conventional flavors (p, n,λ) active, R = 2; not very different from the observed value R ~ 2.5. Above charm threshold R = 10/3, very different from the observed value R ~ 5.5. Thus one may conclude, with charming naiveté, that some extra fancy flavor(s) beyond charm have actually been tasted at SLAC.

Alas, the free field theory considerations of the previous paragraph are not very compelling. On top of that heavy leptons whose decays would contribute to the experimentally defined R may have been found[32].

In asymptotically free field theories like the standard model the strong interactions are turned off as the photon mass increases and the naive prediction (17) is reached, on the average, at infinite energy. Unfortunately this, like all asymptotic statements, is in itself entirely useless. But fortunately the theory is much more powerful and may be used to predict how the limit is reached from finite energies, with inclusion of nonvanishing strong interactions and quark masses. Unfortunately the theoretical predictions are firm and simple in the spacelike domain, where complications due to bound states and multiquark thresholds, among others, do not arise. Fortunately, the timelike e^+e^- annihilation data can be converted via a dispersion relation for the hadronic contribution to the photon vacuum polarization tensor Π into the spacelike information with which the theoretical predictions may be directly compared[33]. Explicitly

$$\begin{aligned} D(Q) &= Q^2 \int_{4m_\pi^2}^{\infty} \frac{R(W)\,dW^2}{(Q^2+W^2)^2} \\ &= W^2 \left. \frac{d\Pi(W^2)}{dW^2} \right|_{(W^2=-Q^2)} \times \text{(known factors)} \end{aligned} \tag{18}$$

The theoretical prediction for D(Q) in the limit Q >> quark masses, is known to be[34]:

$$D(Q) = \sum_i Q_i^2 \left[1 + \frac{\alpha_s(Q)}{\pi} + \cdots \right] \tag{19}$$

The result is based on the calculation of the diagrams of Fig. (13): one quark loop without or with one gluon corrections. At large Q or small α_s, for which perturbation theory may make sense, α_s depends on only one scale parameter Λ, not specified by the theory. In the standard model, for example

$$\alpha_s(Q) \simeq \frac{\alpha_s(M)}{1 + \frac{25}{12\pi}\alpha_s(M) \ln \frac{Q^2}{M^2}} \equiv \frac{12\pi}{25}\left[\ln \frac{Q^2}{\Lambda^2}\right]^{-1} \tag{20}$$

An obvious feature of the data in Figs. (9) and (10) is the threshold. Clearly a mass scale (represented in the standard model by the heavy quark mass) plays a nontrivial, indeed essential role in the region experimentally explored so far. When continued via a dispersion relation to an spacelike domain with similar $|W| \sim |Q|$, the data do also display a characteristic threshold. Thus, the mass of the heavy quark (or quarks) cannot be ignored. The formulas corresponding to Eqs (19,20) with nonzero quark masses are unfortunately too complicated to display here, and I refer the eager reader to the original literature[35]. We have compared the data, continued into the spacelike domain, with several models. The results are[35]:

i) The "old" model with nothing but three light quark flavors is untenable. This is so even if one or two species of heavy leptons are produced at SLAC energies. Upper and lower limits to the continuation of the data to the spacelike domain are shown in Fig. (14). The limits correspond to the unmeasured R(W > 7.8 GeV) staying at the observed plateau level or jumping immediately to the model-dependent theoretical limit. Predictions, labelled by Λ in GeV, are also shown.

ii) The standard model is acceptable <u>only if</u> one or two new species of heavy leptons are being produced. For only one type of heavy lepton, the agreement between theory and experiment is marginal, in that it requires a large $\Lambda \sim 1.1$ GeV. For such large values of Λ the early scaling observed in electro-production becomes rather difficult to understand. Very roughly a value of $\Lambda \sim 1$ GeV corresponds to scaling being approximately true (to withing 30 %) at $Q^2 = 5$ GeV^2. Predictions and "data" are compared in Fig. (15), where m is the heavy quark mass in GeV.

iii) Six-quark models, of the kind that became fashionable in the spring of '75[36,37] are acceptable provided no heavy leptons are produced. They are also acceptable with heavy leptons if one or two of the heavy quarks are too heavy for their effects to be observable at SLAC energies. Theory and experiment are compared in Fig. (16).

iv) The conclusions for five-quark models[38], not surprisingly, lie between four and six quark conclusions. Excellent fits can be obtained with these models. With one heavy lepton and a transcharmanian quark charge -1/3, for example $\Lambda \sim .5$ GeV, a very reasonable value.

It is possible to have a field theorist look at the data from a closer standpoint than the spacelike domain. A dispersion relation like Eq. (18) can be used to analytically extend R to a complex value of Q^2, far enough from the timelike positive real axis for the resonances to be averaged, yet close enough for the general threshold structure to survive. For small enough α_s the theory can also get to this complex value of Q^2. This optimal compromise has been invented and analyzed by Poggio, Quinn and Weinberg[39]. Their conclusions are similar to the ones presented earlier in this chapter, but their sensitivity to quark-mass parameters and the position of thresholds is far better. In particular an optimist may see in their "experimental" curves a second threshold at $W^2 \sim 25$ GeV.

XI. SEARCH FOR CHARM IN e^+e^- ANNIHILATION

In this section I discuss the bump hunt, the kaon yield as a function of electron energy, and the role of parity violating observables in the search and study of charm.

Bumps in the invariant mass of a subset of two or three final hadrons in e^+e^- annihilation have not been seen so far[40]. This may mean that charmed particles have large branching ratios into states containing undetected neutrals or more than three charged particles. It is not inconceivable for example, that D's decay abundantly into a $K^*\rho$, in which case two or three body bumps will not be seen. With the presently available detectors K/π discrimination is difficult and the invariant mass search has a large artificial "statistical" background due to the assignment by fiat of a kaon identity (and mass) to one and each of the charged tracks.

Just above charm threshold and probably all the way to the peak at 4.1 GeV, charmed particles will be produced in pairs ($D\overline{D}$, $\overline{D}D^*$ etc.). A pair-produced particle has a fixed momentum, the momentum of its decay products peaks at a fixed value. A search for lab momentum peaks lacks the artificial K identification background and may be more successful than the invariant mass search[41]. The momentum of the decay products of D may peak at two different values, corresponding to $\overline{D}D$ and $\overline{D}^*D$ production: D^* may be found even if it always decays electromagnetically ($D^* \to D\gamma$). These remarks apply to F and F^* as well.

In the standard model with the GIM[5] weak current of Eq (1), particles with conventionally defined positive (negative) charm are expected to decay predominantly, both semileptonically and nonleptonically, into states of strangeness -1 (+1). During the winter of 1975, critics of the charmonium picture found support in the non observation of the predicted increase in the number of kaons per event above charm threshold. An uninvited customer, the heavy lepton supposedly responsible for the μ-e events[32] ($e^+e^- \to \mu^+e^-$ (or μ^-e^+) + missing energy) came to the rescue of the naive charm approach. Relative to the uncharmed background, heavy lepton events are expected to have small charged multiplicities and to contribute few kaons per event . The charged multiplicity and the number of kaons per event do not appreciably change with energy as the threshold is crossed. Nature may have made the perverse choice of having two close thresholds (charm and heavy leptons) contribute in opposite directions to the average number of kaons and charged tracks[42].

The study of parity violation in e^+e^- annihilation may play an important role in the search for charm (or in its study, once invariant mass peaks are observed). It may also help disentangle effects due to neutral currents, heavy leptons and charmed particles. On the 4.1 GeV bump $R \sim 6$, compared to $R \sim 2.5$ below the thresholds: 60 % of the events are "new physics". In the flat-R regime above 5 GeV, $R \sim 5.5$, again more than half of the events are new physics. Supposedly the new physics is charm and heavy lepton production. The produced charmed particles may cascade via photon or pion emission to lighter charmed particles. These, and the heavy leptons, will decay weakly. Thus about half of the high energy events contain a weak step: parity violating (P-odd) observables may take large values. Some P-odd observables, like the longitudinal polarisation of the outgoing muons in μ-e events, are expected to have nearly maximal values, but are not easy to measure. Other P-odd observables are extremely easy to measure, but are expected to have smaller expectation values. I end this section with a discussion of an example of the latter type of observable[41].

Consider the process $e^+e^- \to k_1k_2k_3 + \ldots$, where k_i are mesons, labelled by their momenta. The observable $B \equiv (\vec{k}_1 \times \vec{k}_2) \cdot \vec{k}_3$ is P-odd. Under time reversal B changes sign but, in the presence of strong interactions, time reversal invariance does not require $\langle B \rangle$ to vanish. If two mesons in the definition of B are identical, their momenta must be ordered: $|\vec{k}_1|>|\vec{k}_2|$. Consider the asymmetry $b \equiv (N_+-N_-)/N_++N_-)$, where $N_+(N_-)$ are the number of events where B is positive (negative). The measurement of b is not demanding in momentum resolution. A nonzero b is also evidence for parity violation. The largest contributions to b should come from three sources:

i) The interference between the electromagnetic and weak (neutral current) e^+e^- annihilation amplitudes.

ii) The weak decays of charmed mesons into more than three particles.

iii) The decays of heavy leptons into more than three particles.

The various origins of a nonzero b are distinguishable. If due to charm or heavy leptons, b will have a W threshold. If due to charm, it should "resonate" on the 4.1 resonance, if due to heavy leptons, it must decrease on the 4.1 resonance. On a charmed invariant mass bump, values of $b(K^+\pi^-\pi^-)$ or $b(K_s\pi^+\pi^+)$ up to $\sim 1/2$ are theoretically possible. With no bumps detected and no K/π identification, b may drop to the level of a percent.

Some day, I believe, invariant mass or lab momentum bumps will be discovered. Bumps are evidence for a new approximately conserved quantum number but, is it color, peculiarness[43], charm...[44]? Charmed particles decay via parity violating interations, unlike the many other specific candidates for a new quantum number[4,43,44]. The search for parity violation in e^+e^- annihilation is interesting in its own right; the day that invariant mass bumps are observed, the search for parity violation becomes mandatory.

XII. HADRON SPECTROSCOPY: QUARK ALCHEMY[19]

The standard model is a quark model and it incorporates the well known successful classification of the "old" hadrons in multiplets of SU(6) $\supset$ flavor SU(3) $\otimes$ spin SU(2). The idea that quarks interact via the exchange of flavor -SU(3)-singlet massless vector gluons (all adjectives essential!), complemented by hints from lattice gauge theories, has been successfully used to develop a qualitative unterstanding of "fine" and "hyperfine" splittings among members of SU(6) multiplets. In gauge theories on a lattice, the spin-spin force between two distant static quarks decreases exponentially with distance, while the spin-independent force remains constant. At short distances, on the other hand, the quark-gluon coupling constant becomes small and one-gluon exchange dominates. Thus, interquark spin-spin interactions leading to "hyperfine" splittings may be due to just one-gluon exchange. One may argue along similar lines for spin-orbit "fine" couplings.

The Σ^o-Λ mass difference is a good example where the idea works that quark spin-spin interactions are dominated by one-gluon exchange. Σ,Λ, along with N, N^* (3/2) and others belong to a 56-plet of SU(6). Let their SU(6) wave functions be identical in the limit where the p, n and λ quark masses are equal. In the study of the mass operator describing the "56" baryons we will depart from full symmetry and do perturbation theory in quark mass differences. (The renormalizability and respectability of the standard model is only maintained by a breaking of flavor - SU(3) via quark masses: the form of SU(3) breakdown is rather unique). Massless vector-gluon exchange is well known from electrodynamics to give rise, to leading order in the coupling, to a Fermi interaction potential energy

$$V^F \sim \delta^3(r_{12}) \left\{ \frac{1}{m_1^2} + \frac{1}{m_2^2} + \frac{16\,\vec{S}_1\cdot\vec{S}_2}{3\,m_1 m_2} \right\} \tag{21}$$

where the notation is self-explanatory. The N^*-N mass difference will arise from the fact that in N^* the quark spins add up to S=3/2, while in N they add up to S=1/2: The total gluon-exchange energy will be different in N^* and N . The Λ-N mass difference is even more trivial: Λ contains a heavy λ-quark, while N does not. But Σ^0 and Λ have the same spin and quark content (p n λ), what produces their considerable mass difference? The answer is simple: in Σ^0 the p and n quarks must be in a triplet spin state, because they are in an isospin 1 state. In Λ both spin and isospin are zero. Thus the spin-spin correlations and interactions of Eq (21) are different within the Λ and the Σ. The Σ-Λ mass difference may thus be related to the N^*-N mass difference with a result that agrees with experiment in magnitude and sign[19].

From the one gluon exchange dominance of fine and hyperfine mass splittings, several old and new successful mass formulae for baryons and mesons can be derived. Armed with this understanding of the mass systematics of conventional hadrons, one cannot resist the temptation to try and predict how the spectrum of charmed hadrons is going to look like. An example, the spectrum of charmed vector and pseudoscalar mesons, is given in Fig. (11). Twice the mass of the lightest pseudoscalar ($2\,m_D \sim 3.7$ GeV) is in rough agreement with the e^+e^- threshold position of Fig. (10). Another example, the spectrum of nonstrange singly charmed baryons, is given in Fig. (17). The characteristics of one observed $\Delta S = -\Delta Q$ event[45] are also indicated. The event is compatible with the production of a doubly charged Σ_c of mass ~2420 MeV, followed by its strong pion emission decay into $\Lambda_c(2260)$ and the weak decay of the latter into $\Lambda\pi^+\pi^+\pi^-$. In this interpretation the Σ_c-Λ_c mass difference is the predicted 160 MeV. It may also be that the prediction is not so correct and the produced particle is Σ_c^*. The Σ_c triplet would then have a mass ~2370 MeV. It would be reassuring to have other events of this type.

The general structure of the charmed particle mass spectra, based on the one-gluon exchange mechanism, has a much better chance to be correct than the actual numerical values of mass differences. These rely heavily on the unjustified use of SU(8) symmetry for the wave functions. In fact, as I discuss in the next chapter, the prediction of the mass difference of Ortho and Paracharmonium failed miserably.

XIII. FINE AND HYPERFINE SPLITTINGS OF CHARMONIUM LEVELS

In the one-gluon exchange approximation[19], if $\alpha_s|\psi(o)|^2$ is the same for ρ as for ψ, and to first order in mass differences, the hyperfine mass splitting Δ of the ground state charmonium is predicted to be

$$\Delta = M(\text{Orthocharmonium}) - M(\text{Paracharmonium})$$

$$\sim (m_\rho - m_\pi)\left(\frac{m_\rho}{m_\psi}\right)^2 \sim 30 \text{ MeV} \qquad (22)$$

This prediction is wrong by one order of magnitude, if the state at 2.75 GeV seen at DESY is Paracharmonium. This remarkable failure is not very surprising, since the perturbation theory in mass differences makes no sense for such large mass splittings. One may doctor Eq. (22) by "correcting" for the error incurred in assuming SU(8) symmetry of the wave functions at the origin $\psi(o)$. The leptonic widths of the vector mesons and the expectation values of one-gluon exchange spin-spin interactions are both proportional to $|\psi(o)|^2$. Fudging with this fact, I get

$$\Delta = (m_\rho - m_\pi)\,\frac{9}{8}\,\frac{\Gamma(\psi \to e^+e^-)}{\Gamma(\rho \to e^+e^-)} \sim 400 \text{ MeV} \qquad (23)$$

where I have assumed $\alpha_s(\rho) = \alpha_s(\psi)$ for the "exchange" α_s, for lack of a better guess. (The exchange α_s need not be equal to the "annihilation" α_s about which asymptotic freedom may have a say.) The difference between Eqs. (22) and (23) may mean than the SU(8) wave function symmetry was not a good guess, while the one-gluon exchange idea may still be viable. Support for this point of view is found in the work of the MIT bag group[46]. The bag model with the bag constant modified to fit the ψ'-J mass difference gives a nice value for $\Delta \sim 280$ MeV[46], if hyperfine splittings are again attributed to one-gluon exchange. Advocates of linear potential calculations, also using the one-gluon exchange ansatz, obtain values of $\Delta \sim 50 - 60$ MeV[47]. The moral seems to be that Δ is very model dependent. The splittings between P-wave levels were also underestimated by a factor of ~10 in the naive approach[19].

A completely different point of view on fine and hyperfine splittings is the one taken by Schnitzer and collaborators[48]. These authors argue that the spin-spin coupling may not be a contact term, like in the one-gluon exchange model, but may be proportional to $\nabla^2 V$, with V the "usual" linear plus Coulomb potential. It is possible in this approach to obtain values of Δ in the range 40 - 80 MeV, but once again the observed value is too large. Faced with this fact Schnitzer considers the possibility that $\psi''(4.2)$ rather than $\psi'(3.7)$ be the first radial excitation of $\psi(3.1)$. In this case a satisfactory result $\Delta \sim 330$ MeV is obtained. The narrow peak $\psi'(3.7)$ would have to be a bound state of quarks with a fifth flavor. The decay $\psi' \to \psi\pi\pi$ would be due to mixing.

XIV. PARACHARMONIUM II

Paracharmonium II, the 0^- comrade of ψ' should also exist and be nearby in mass. The mass difference between ψ' and Para II in the one-gluon exchange model is

$$\Delta' \equiv (\text{Ortho II} - \text{Para II}) \sim (\text{Ortho I} - \text{Para I}) \frac{\Gamma(\psi' \to e^+e^-)}{\Gamma(\psi \to e^+e^-)} \sim 160 \text{ MeV} \tag{24}$$

Should this be a correct estimate, Para II would lie suspiciously close in mass to $\chi(3.53)$. This prediction circumvents the problems discussed in the previous chapter, but may be rendered useless by a large admixture of charmed meson pairs in $\psi'(3.7)$. The smaller hyperfine splitting between the excited charmonium states diminishes the transition rate $\psi' \to$ PARA II + γ relative to $\psi \to$ PARA I + γ by a factor $\sim(\Delta'/\Delta)^3$ ~ 0.1. Moreover PARA II has a smaller $\gamma\gamma$ branching ratio than PARA I because of the extra hadronic decay PARA II $\to$ PARA I + 2 π, and its extra electromagnetic decays into ORTHO I and the 1^{+-} state at ~ 3.5 GeV. Thus PARA II may be somewhat more difficult to find than PARA I, at least in the 3 γ decays of ψ'. A nice signature for the detection of PARA II may be

$$\begin{aligned}
\psi'(3.7) &\to \text{Para II} + \gamma \\
&\qquad \hookrightarrow \text{Para I} + \pi^+\pi^- \ (\text{or } \pi^0\pi^0) \\
&\qquad\qquad \hookrightarrow \gamma\gamma \ (\text{or } \bar{p}p)
\end{aligned}$$

Its cascade γ ray decays into ORTHO I and the 1^{+-} state would also be rather spectacular.

Let the decays of ORTHO II and PARA II into hadrons not containing charmonia be called "direct". Direct decays proceed via quark annihilation. In the short distance perturbative approach discussed in Sections IV to VII, direct decays of ORTHO II (PARA II) proceed via three (two) gluons. The ratio of direct decay widths has a better chance to be an overestimate than the corresponding one for the ground state levels, because of the larger interquark distances in the radially excited state.

XV. THE DECAY $\psi' \to \psi\pi\pi$

We cannot explain the magnitude of the decay width $\psi' \to \psi\pi\pi$. The observed width is ~ 110 keV. This is very small, relative to "typical" hadron decays, say $\rho' \to \rho\pi\pi$, whose width is ~ 400 MeV. Vox populi explains the small width by pointing out that the decay, depicted in Fig. (18), is Zweig-forbidden. The field-theoretic artillery that flanks the standard model is enough to convince some that Zweig's rule is understood in its application to total hadronic decay rates of very heavy mesons, as discussed ad infinitum in chapters IV to VII. But our understanding of exclusive channels at the same fundamental level is meager, if not null. To my knowledge, nobody has proved that one can even start talking about asymptotic freedom with the diagram of Fig. (18) in sight. With the definition of "understanding" given in Chapter VI, we do not understand why $\psi' \to J\pi\pi$ is so small. Perhaps α_s is not large all the way to rather long distances or small momenta: infrared slavery comes late.

XVI. FOUR QUARKS OR MORE?

If only one kind of heavy lepton is produced at SLAC energies, and if the overall normalisation of the data is taken very seriously at face value, the agreement with the standard four-flavor model is only marginal[35]. The smeared out data may show a second threshold at $W \sim 5$ GeV[39]. (These facts were discussed in Chapter X). The most naive predictions of radiative transitions between charmonium states are wrong by as much as an order of magnitude (Chapter VIII). Understanding $\psi' \to \psi\pi\pi$ is not a trivial task (Chapter XIV). The present situation in high energy neutrino physics seems to be more complicated than we expect in the standard model[49]. In current times, when the introduction of a new quantum number is treated with iconoclastic irreverence, an obvious way to tackle the problems just mentioned is to introduce an extra quark or two. With the extra parameters and degrees of freedom and a finite (sometimes large) degree of ingenuity, the problems may be solved.

If charged heavy leptons have been seen[32] and they unimaginatively come with their own neutrino counterparts, two new quarks are needed to restore lepton-hadron symmetry. But this symmetry is rather ill-defined except in theories where lepton number is a fourth color (lilac[50]).

Six-quark models have a strong theoretical appeal (witness the number of authors in Reference 37). They are the basis of the construction of superunifiable theories of the "vector" type, where weak neutral currents are parity conserving and the parity violation and Cabibbo angle of the charged currents are low energy phenomena, due to the flavor asymmetries of the quark mass matrix[37]. Thus, there are theoretical justifications for as many as six flavors. Unfortunately (in my opinion), none of the models proposed so far has the appealing uniqueness of truth. Barring accidental mass degeneracies between the heavy quarks, all the models naively predict at least one and perhaps as many as four new very narrow resonances in e^+e^- annihilation. To predict "only" a new resonance one may associate ψ and the first threshold at $W \sim 3.8$ GeV with one type of quark, ψ' and the wooly cusp at 5 GeV (or the 4.4 structure) with a second type of quark.

XVIII. CONCLUSIONS AND OUTLOOK

The charmonium scheme fits naturally into our overall understanding of hadron dynamics and adequately describes all the hard facts about the new resonances. Most of the observed features of e^+e^- annihilation were anticipated by the theory. The model no doubt requires further test and development, but it is not unlikely that its essential ingredients will survive as part of the truth.

The predicted bound states of a quark-antiquark pair and the particles that are being discovered in the radiative decays of ψ' and J could hardly be more similar. If only because of this, it is rather difficult, in my opinion, not to be convinced of the truth of this aspect of the model. But, the observation of these levels only establishes the "onium" of charmonium. The crucial test of the rest of the scheme is the discovery of charmed particles, of course. The obervation of narrow invariant mass peaks in the final states in e^+e^- annihilation will be rather welcome, but will not completely settle the issue. The detection of parity violation in the decay of these yet to be observed peaks will be an important step in eliminating alternatives to charm. To settle the point of whether charmed particles do the job for which they were invented (have weak couplings à la GIM) will require a careful study of their weak decays, particularly a comparison between their semileptonic strange and nonstrange decay modes. This may not sound like an easy task, but we believe that charmed particles are being made in e^+e^- collisions by the thousands. If this is so, it is impossible not to expect that our ingenious experimental colleagues, armed with such a copious source, will not provide us with all the answers (and some new problems). If seen in neutrino physics, charmed particles will automatically come with some of the crucial weak interaction information.

The study of charmonium decay may give us precious information on aspects of chromodynamics in which gluons play a central role. This will complement the knowledge on quark interactions gathered in scaling experiments. The width of paracharmonium and its branching ratio into $\gamma\gamma$ are key measurements in this respect.

Heavy leptons may have been seen. Some soft evidence for more than four quark flavors exists. There is no doubt that these topics will occupy theorists and experimentalists alike for some time to come.

After a year like last year, it is impossible to list all the people whom I should thank for illuminating discussions. I have learned most of the topics I discuss in collaborating with Tom Appelquist, Howard Georgi, Sheldon Glashow, David Politzer and R. Shankar.

REFERENCES

1) V. Teplitz and P. Tarjanne, Phys. Rev. Letters 11 (1963) 447.

2) Y. Hara, Phys. Rev. 134 (1964) B701.

3) J.D. Bjorken and S.L. Glashow, Phys. Lett. 11 (1964) 255.

4) D. Amati et al., Nuovo Cimento 34 (1964) 1732, Phys Lett. 11 (1964) 190; L.B. Okun' ibid 12 (1964) 250; Z. Maki and Ohnuki, Prog. Theor. Phys. 32 (1964) 144; M. Nauenberg (unpublished).

5) S.L. Glashow, H. Iliopoulos and L. Maiani, Phys. Rev. D2 (1970) 1285.

6) For a history of gauge theories, see M. Veltman in Proceedings of the Sixth International Symposium on Electron and Photon Interactions at High Energies, Bonn 1973. North-Holland Pub.

7) Mary K. Gaillard, Benjamin W. Lee and Jonathan L. Rosner, Revs. Mod. Phys. 47, (1975) 277.

8) T.W. Appelquist and H.D. Politzer, Phys. Rev. Letters 34 (1975) 43.

9) A. De Rújula and S.L. Glashow, Phys. Rev. Letters 34 (1975) 46.

10) G. Zweig, unpublished (1964); Iizuka, Supplement to Prog. Theor. Phys. 37 - 38 (1966) 21.

11) The standard model cannot be attributed to a single person or group. For references and a review, see for instance Steven Weinberg, Rev. Mod. Phys. 46 (1974) 255.

12) In essence, color was invented by O.W. Greenberg, Phys. Rev. Letters 13 (1964) 598. For recent developments and other references, see W. Bardeen. H. Fritzch and M. Gell-Mann, in Scale and Conformal Symmetry in Hadron Physics, edited by R. Gatto (Wiley, New York 1973).

13) S.L. Glashow, Nucl. Phys. 22 (1961) 579; S. Weinberg, Phys. Rev. Letters 19 (196) 1264; A. Salam, in Elementary Particle Theory, edited by V. Svartholm (Almquist and Wiksell, Stockholm, Sweden 1968), p. 367.

14) H.D. Politzer, Phys. Rev. Letters 30 (1973) 1346; D.J. Gross and F. Wilczek, Phys. Rev. Letters 30 (1973) 1343.

15) K. Wilson, Phys. Rev. D10 (1974) 2445; J. Kogut and L. Susskind, Phys. Rev. D9 (1974) 3501.

16) The minimal number of gluons into which a $J^P = 0^-$ (1^-) state can annihilate (Two[three]), follows from arguments analogous to the ones used in para- (ortho)- positrorium decay into gamma rays.

17) T.W. Appelquist and H.D. Politzer, to be published in Phys. Rev. D.

18) E. Eichten, K. Gottfried, T. Kinoshita, J. Kogut, K.D. Lane and T.M. Yan, Phys. Rev. Letters 34 (1975) 369.

19) A. De Rújula, H. Georgi and S.L. Glashow, Phys. Rev. D12 (1975) 147.

20) T.W. Appelquist, A. De Rújula, S.L. Glashow and H.D. Politzer, Phys. Rev. Letters 34 (1975) 365.

21) C.G. Callan, R.L. Kingsley, S.B. Treiman, F. Wilczek and A. Zee, Phys. Rev. Letters 34 (1975) 52.

22) B. Harrington, S.Y. Park and A. Yildiz, Phys. Rev. Letters 34 (1975) 168.

23) For a discussion of gamma ray charmonium spectroscopy in models where more than one multipole contributes, see G. Karl, S. Meshkov and J.L. Rosen (unpublished).

24) W. Braunschweig et al., DESY 75/20, July 1975; B. Wiik, Proceedings of the International Symposium on Lepton and Photon Interactions at High Energies; J. Heinthe, ibid.

25) Talks by R. Schwitters and C.J. Feldman in the Proceedings quoted in Ref. 24.

26) The number of names is increasing much faster than the number of particles. A possible advantageous outcome will be the eventual reconsideration of all the nomenclature and the birth of a consistent one.

27) I am indebted to H. Schnitzer for bringing this point to my attention. The argument was developed by K. Lane.

28) For a nonrelativistic calculation, see J.M. Borenstein and R. Shankar, Phys. Rev. Letters 34 (1975) 619. For a relativistic calculation, see G. Feinberg and J. Sucher (unpublished).

29) J. Kogut and L. Susskind, Phys. Rev. Letters 34 (1975) 767 and Cornell Preprint CLNS-303 (1975). See also R. Barbieri, R. Gatto, R. Kogerler and Z. Kunszt, CERN preprint TH-2025.

30) J.E. Augustin et al., Phys. Rev. Letters 34 (1975) 764.

31) The $\pi^o \to \gamma\gamma$ decay rate is a measure of the number of colors. It agrees with there being three colors. S. Adler, Phys. Rev. 177 (1969) 2426.

32) M.L. Perl et al., SLAC-PUB-1626 (or LBL-4228), August 1975.

33) S.L. Adler, Phys. Rev. D10 (1974) 3714.

34) T.W. Appelquist and H. Georgi, Phys. Rev. D8 (1973) 4000; A. Zee, Phys. Rev. D8 (1973) 4038.

35) A. De Rújula and H. Georgi, Harvard Preprint.

36) In the context of the new particles, models with more than four quarks were resurrected by M. Barnett, Phys. Rev. Letters 34 (1975) 41, Phys. Rev. D11 (1975) 3246 and FERMILAB-Conf. 75/71 THY.

37) A partial list includes A. De Rújula, H. Georgi and S. L. Glashow; Phys. Rev. Letters 35, (1975) 69, and Harvard Preprint; F.A. Wilczek,

cont. 37) A. Zee, R.L. Kingsley and S.B. Treiman, FERMILAB-Pub-75/44-THY, H. Fritzch, M. Gell-Mann and P. Minkowski, Cal.Tech.Preprint CALT-68-503, S. Pakvasa, W.A. Simmons, and S.F. Tuan, Phys.Rev.Lett. 35, 702 (1975), G. Branco, T. Hagiwara and R.N. Mohapatra, CCN4-HEP 75/8 and COO-223B-84 (1975)

38) M. Barnett, reference 36; F. Wilczek, unpublished.

39) E. Poggio, Helen Quinn and S. Weinberg, in preparation.

40) A.M. Bogarski et al. Phys. Rev. Letters 35 (1975) 195.

41) A. De Rûjula, S.L. Glashow and R. Sahnkar, Harvard Preprint.

42) H. Harari in Proceedings of the International Conference on Lepton Physics at High Energies, Stanford 1975.

43) Peculiarness (sic), is the name proposed by Amati et al., reference 4.

44) The introduction of an entirely new quantum number to describe ψ was first proposed by H.T. Nieh, T.T. Wu and C.N. Yang, Phys. Rev. Letters 34 (1975) 49.

45) L.G. Cazzoli et al., Phys. Rev. 34 1975) 1125.

46) R. Jaffe, private communication.

47) R. Barbieri, R. Gatto, R. Kogerler and Z. Kunszt, CERN preprint.

48) H.J. Schnitzer, Brandeis Preprint, J.S. Kang and H.J. Schnitzer, Brandeis University preprint and Phys. Rev. D, August 1975.

49) A recent discussion can be found in A. Pais and S.B. Treiman, Brookhaven preprint.

50) J.C. Pati and A. Salam, these proceedings.

FIGURE CAPTIONS

Fig. 1 Mass versus lifetime of all established particles.

Fig. 2 Annihilation of a vector meson

(a) $\psi \rightarrow e^+e^-$,

(b) $\phi \rightarrow \pi\pi\pi$,

(c) $\rho \rightarrow \pi\pi$.

Fig. 3 Gluon counting.

(a) is a two gluon nondiagonal annihilation interaction within a pseudoscalar meson.

(b) is a three gluon annihilation within a vector meson.

(c) Rate of paracharmonium decay.

(d) Rate of orthocharmonium decay.

Fig. 4 Separation of the amplitude for two heavy quark annihilation.

Fig. 5 The strong interaction fudge factor

(a) in ψ decay,

(b) in e^+e^- annihilation,

(c) in deep inelastic electron scattering.

Fig. 6 (a) Short distance three gluon decay of charmonium into ordinary hadrons.

(b) Long distance many gluon decay.

Fig. 7 Conventional mesons.

Fig. 8 The new narrow resonances.

Fig. 9 The ratio R for photon energies ~2.4 to 8 GeV.

Fig. 10 The ratio R in the transition region ~3 to 5 GeV. This is a blow up of part of Fig. 9.

Fig. 11 Spectrum of charmed pseudoscalar and vector mesons.

Fig. 12 (a) Kogut-Susskind potential between heavy quarks.

(b) Predicted form of R(W) above charm threshold.

Fig. 13 One and two loop contributions to the vacuum polarisation tensor.

Fig. 14 D as a function of a spacelike Q, in the old model. Each shaded area corresponds to an assumed number of heavy leptons.

Fig. 15 D as a function of a spacelike Q, in the standard model. With two heavy leptons assumed the allowed domain shrinks to a line.

Fig. 16 D as a function of a spacelike Q, in a six-quark model.

Fig. 17 Spectrum of non-strange singly charmed baryons.

Fig. 18 The decay $\psi' \rightarrow \psi\pi\pi$.

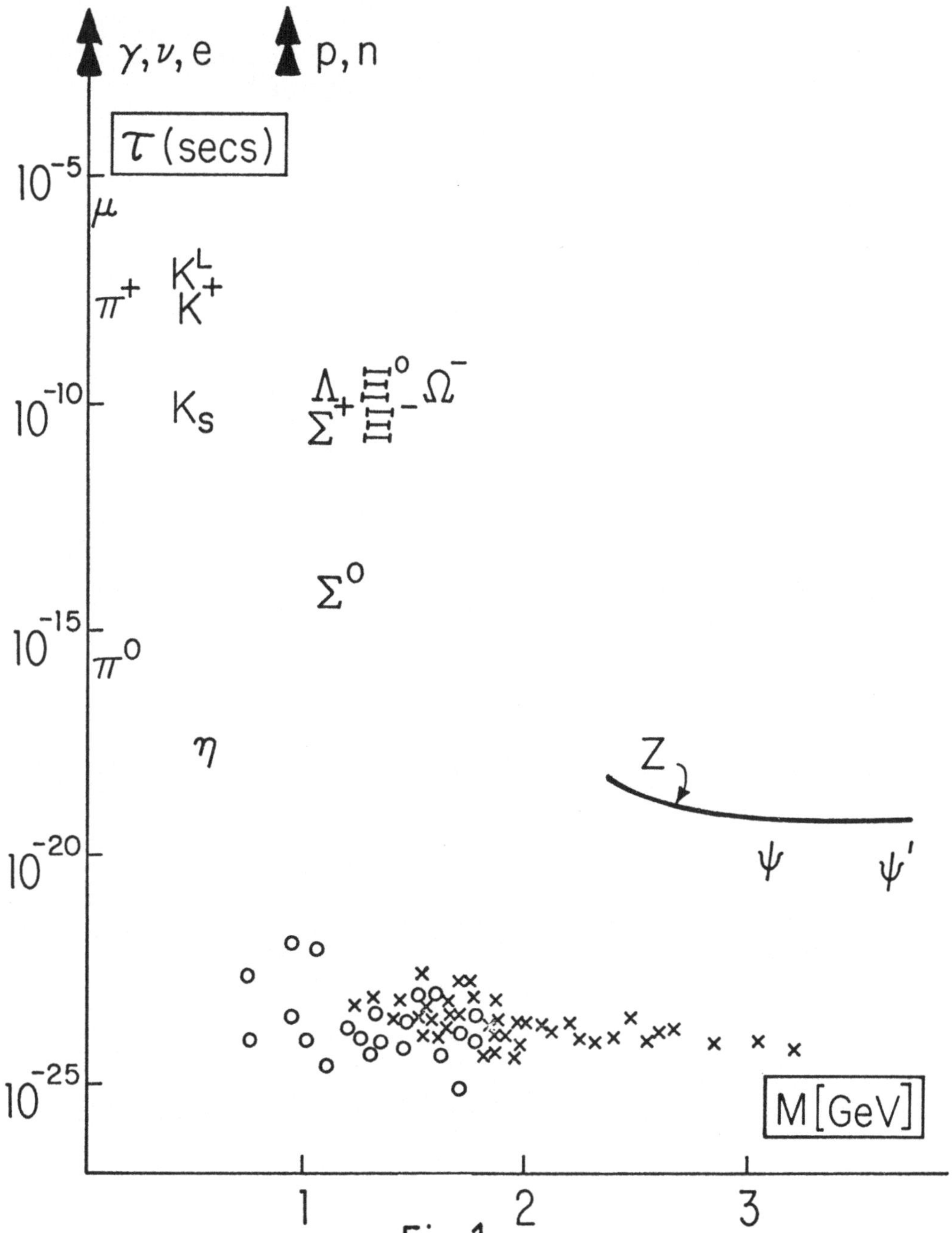
γ, ν, e
p, n
τ (secs)
10^-5
μ
π^+
K^L
K^+
10^-10
K_S
Λ
Σ^+
Ξ^0
Ξ^-
Ω^-
Σ^0
10^-15
π^0
η
Z
10^-20
ψ
ψ'
10^-25
M [GeV]
1
2
3

Fig. 1

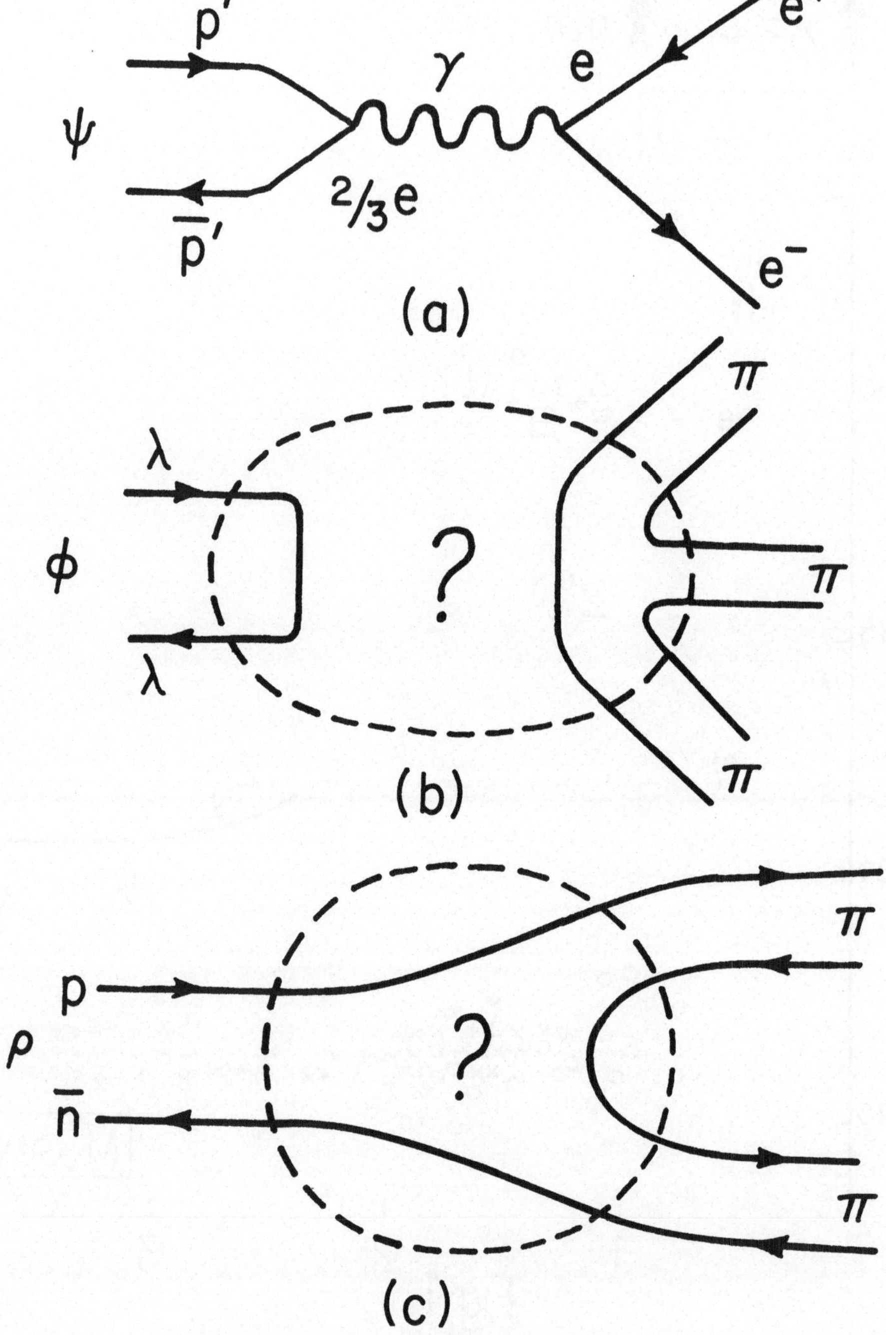

Fig. 2

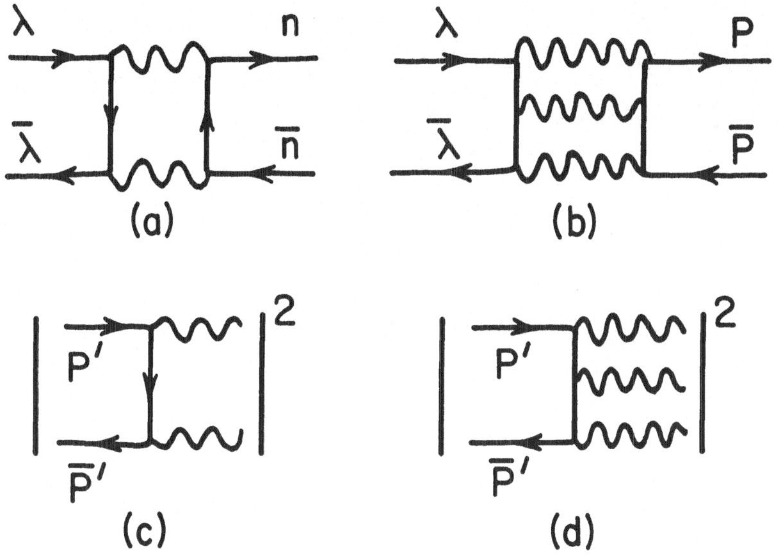
λ
n
$\bar{\lambda}$
$\bar{n}$
(a)
λ
P
$\bar{\lambda}$
$\bar{P}$
(b)
P′
$\bar{P}'$
2
(c)
P′
$\bar{P}'$
2
(d)

Fig. 3

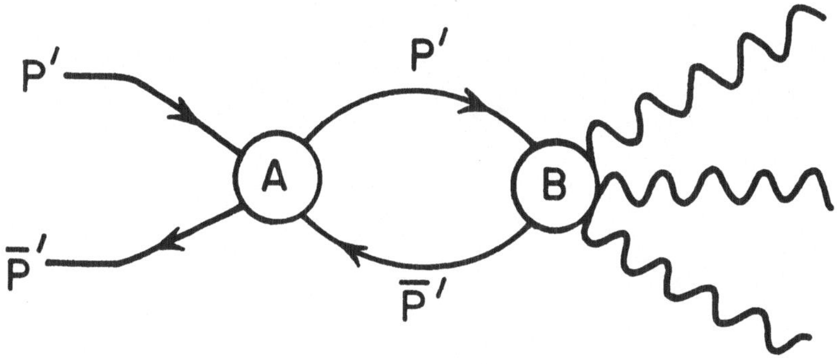
P′
P′
A
B
$\bar{P}'$
$\bar{P}'$

Fig. 4

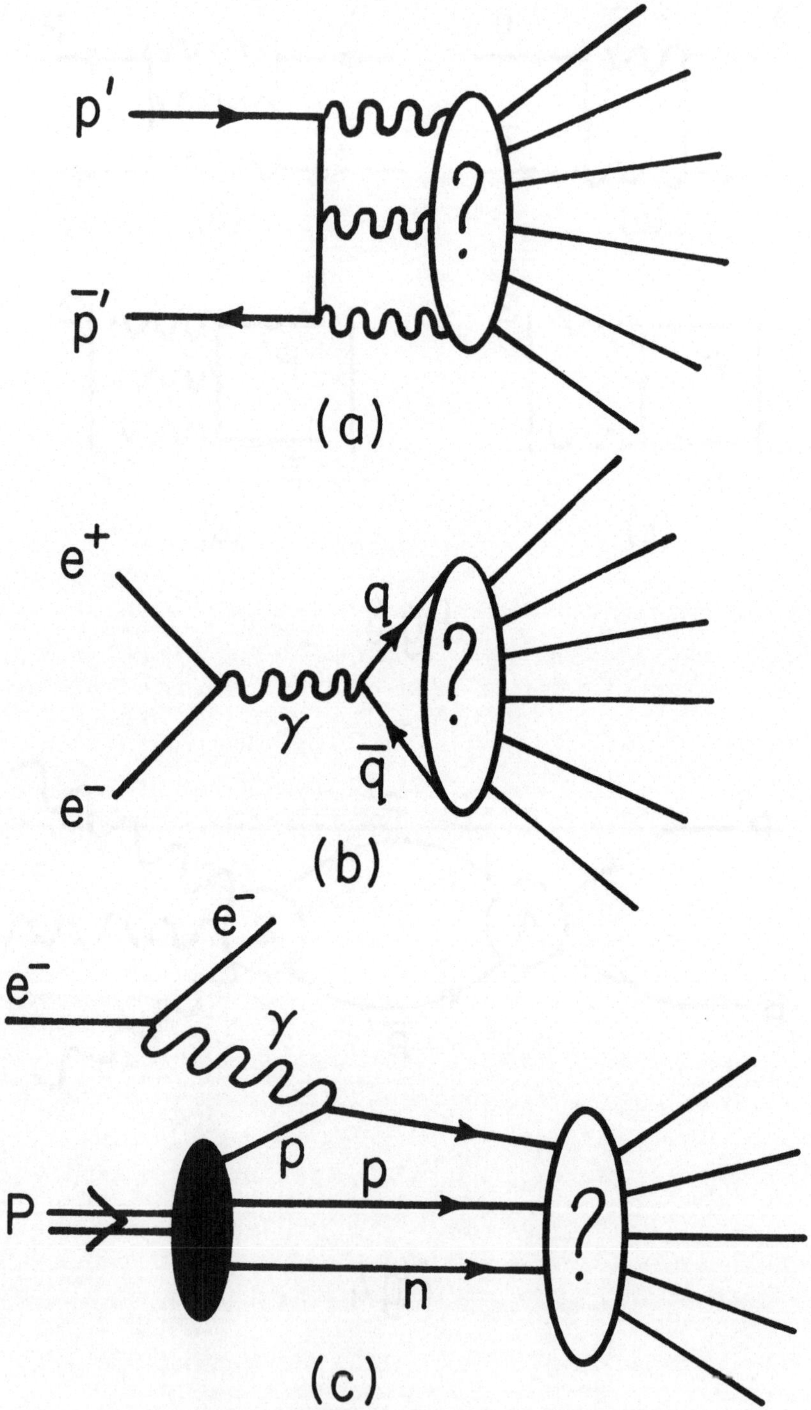

Fig.5

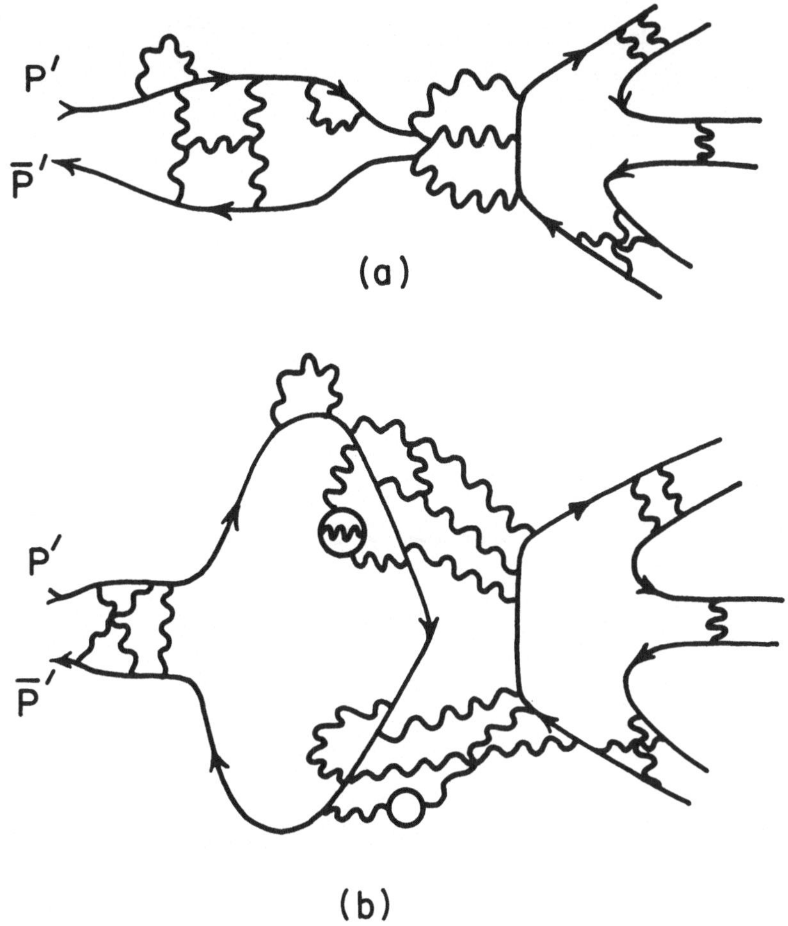

Fig.6

SPECTRUM OF LIGHT (p,n) QUARK-ANTIQUARK BOUND STATES

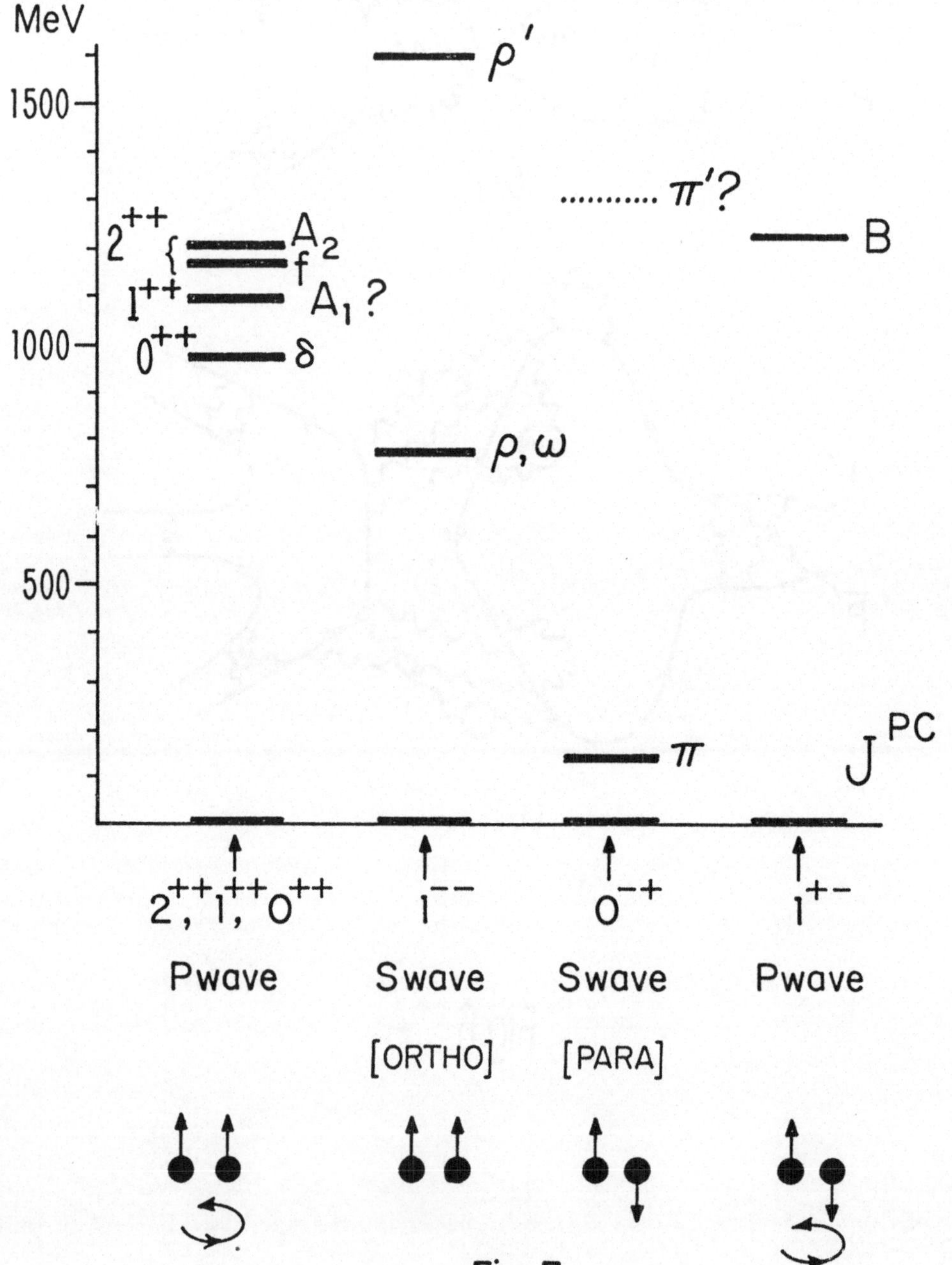

Fig. 7

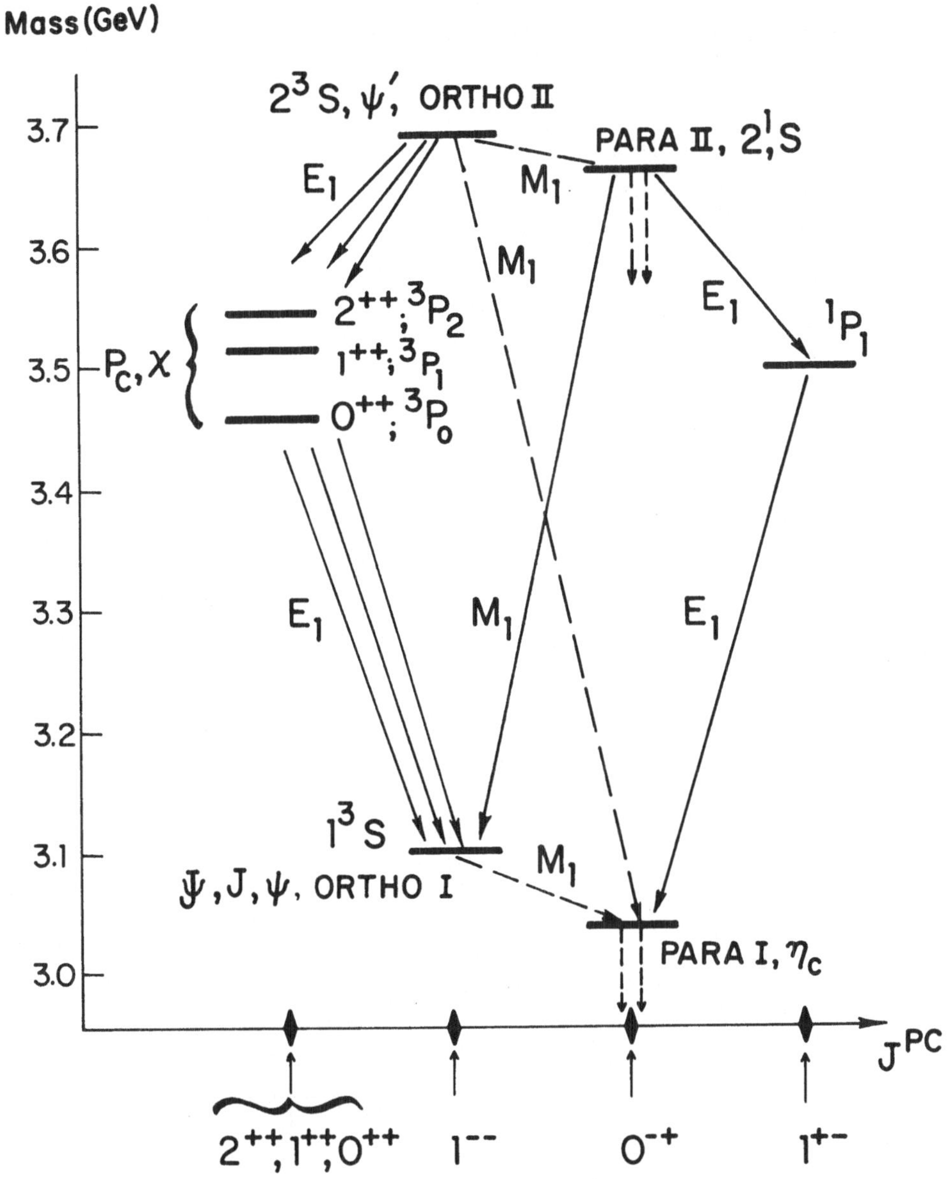

THE VOCABULARY AND SPECTRUM OF CHARMONIUM

Fig.8

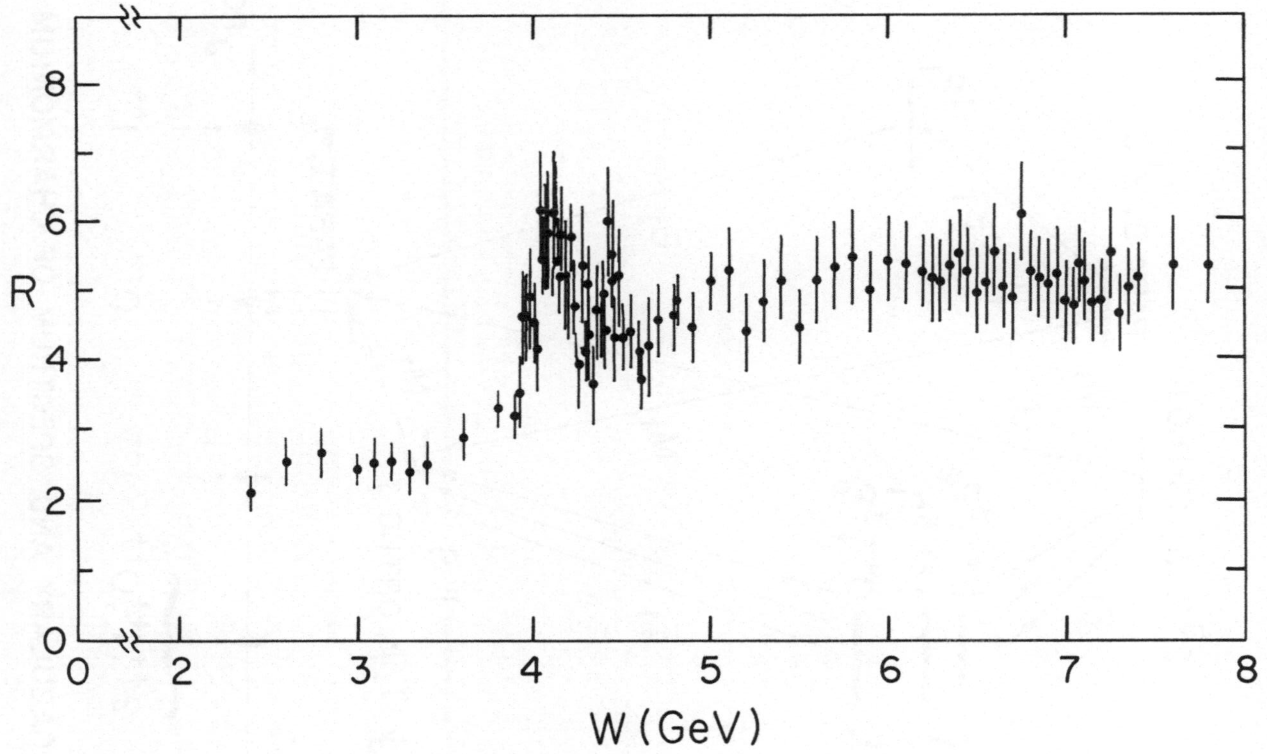

Fig. 9

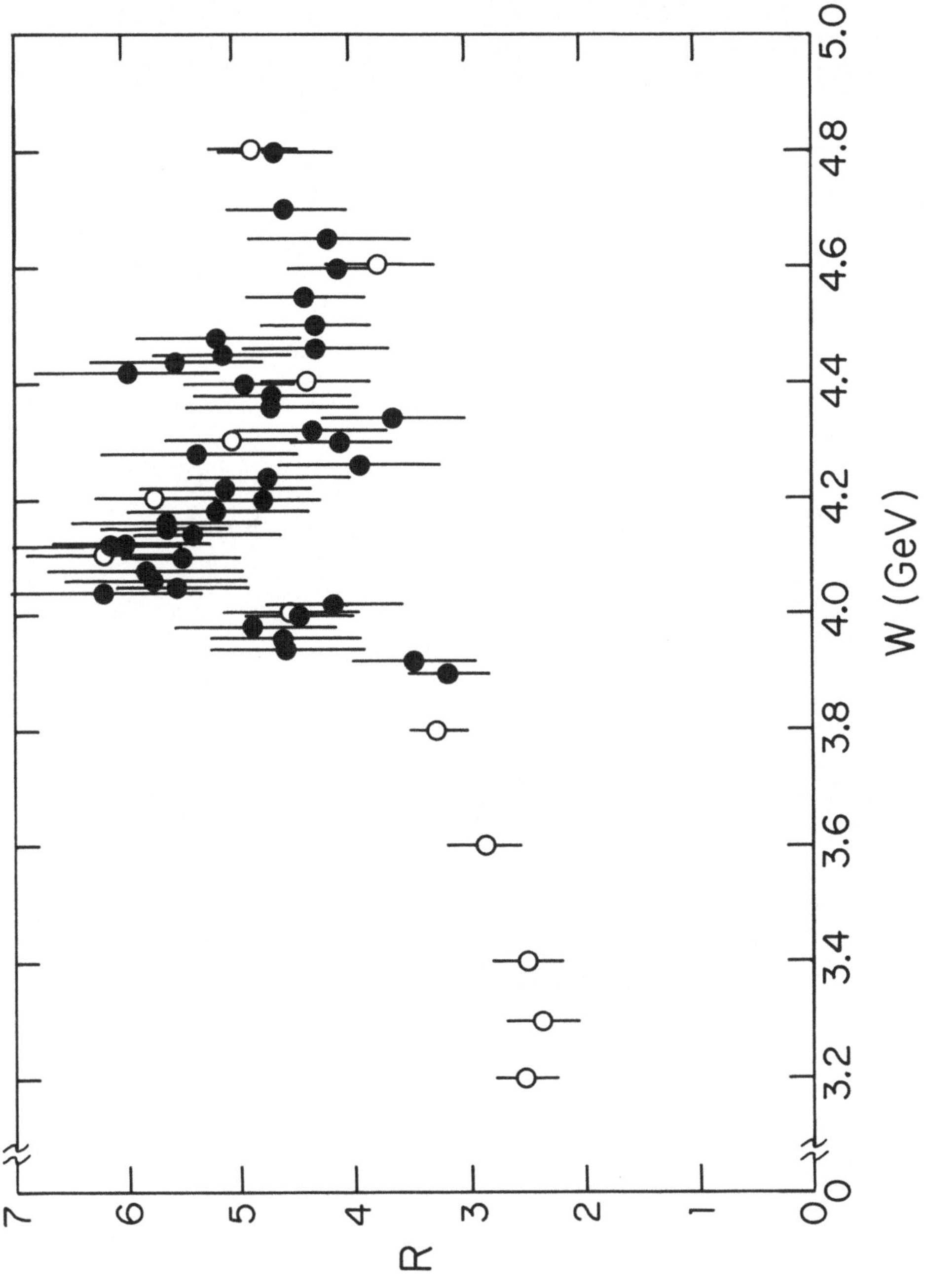

Fig.10

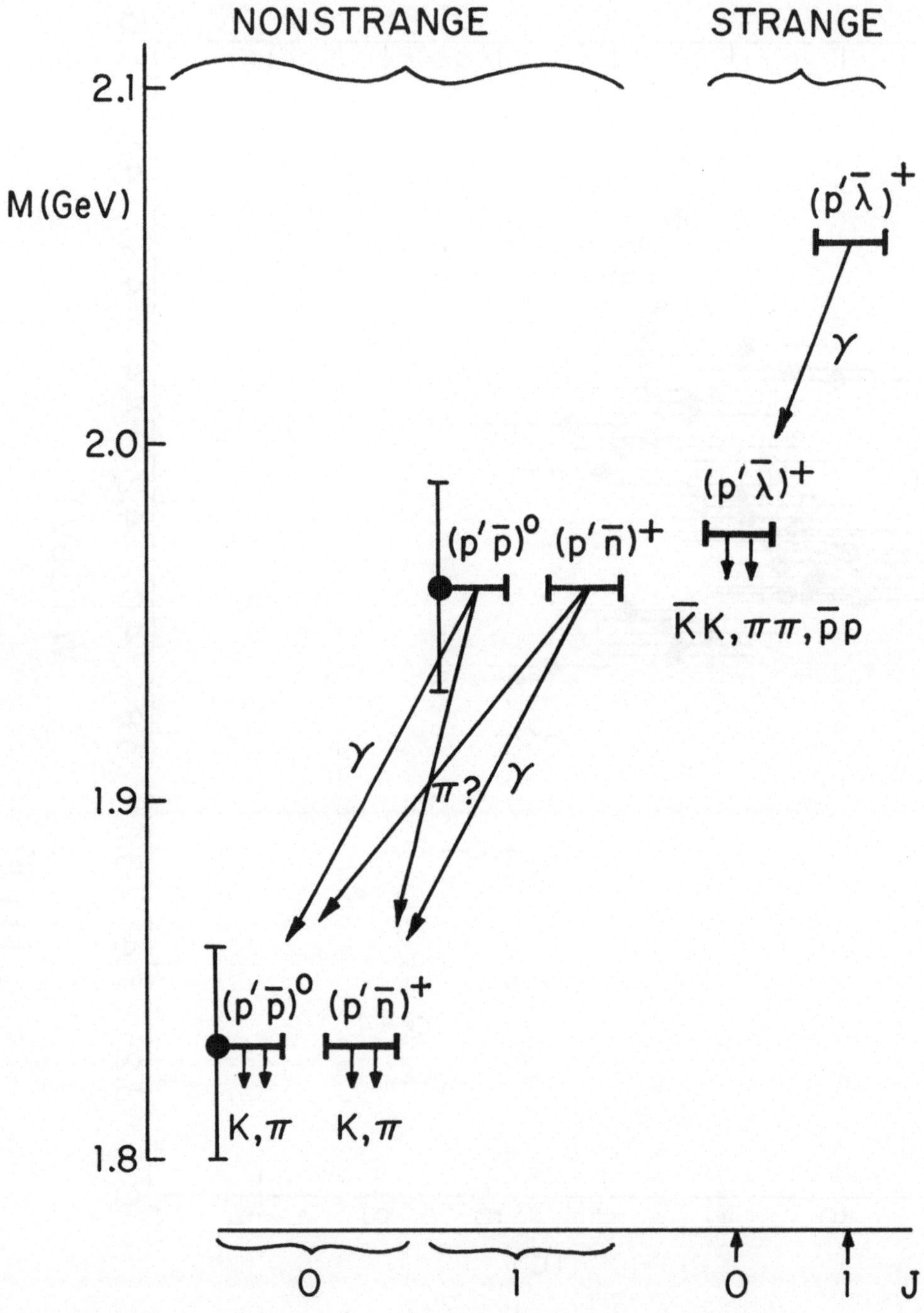

THE MASSES OF CHARMED VECTOR AND PSEUDOSCALAR MESONS

Fig. 11

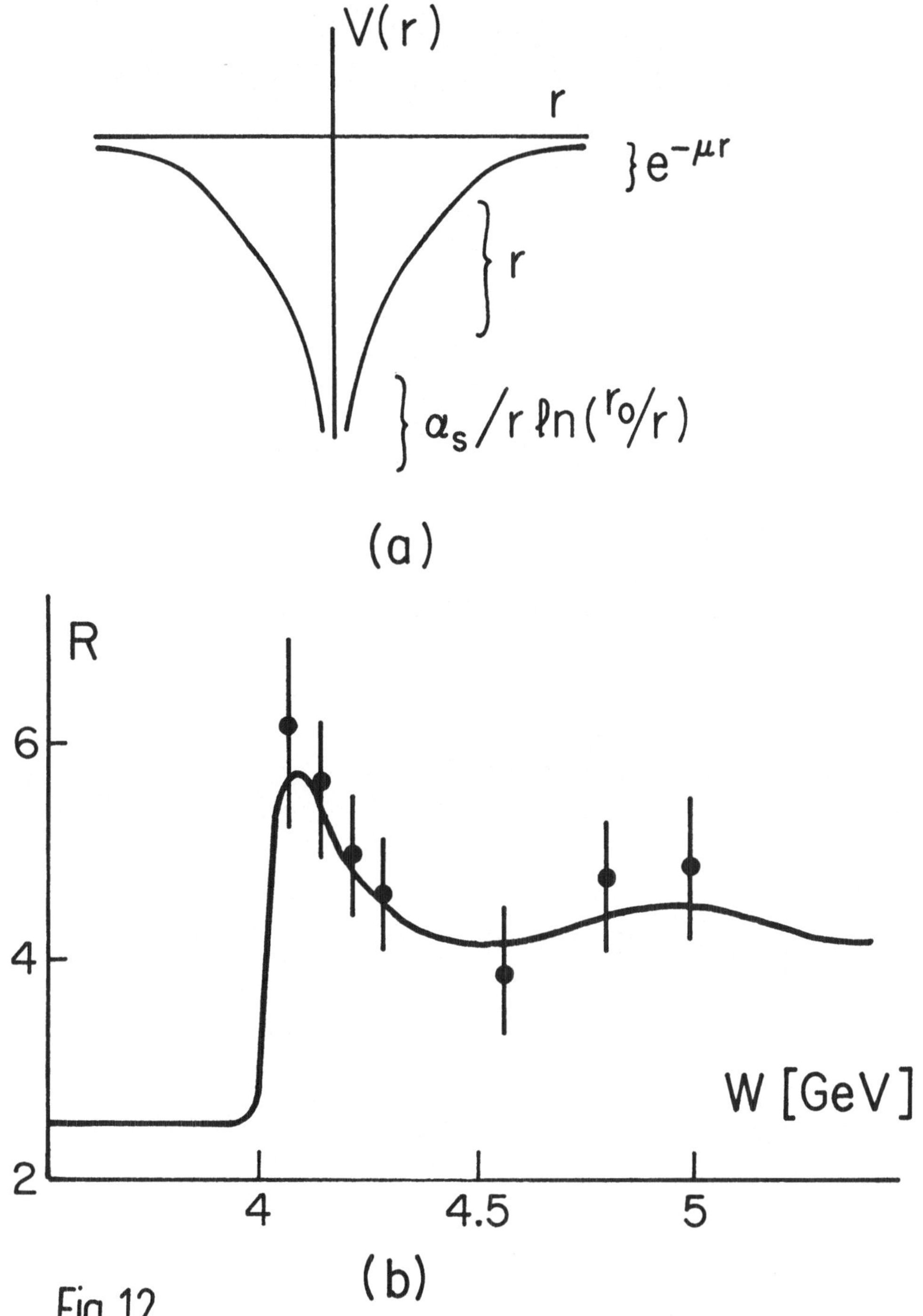

Fig.12

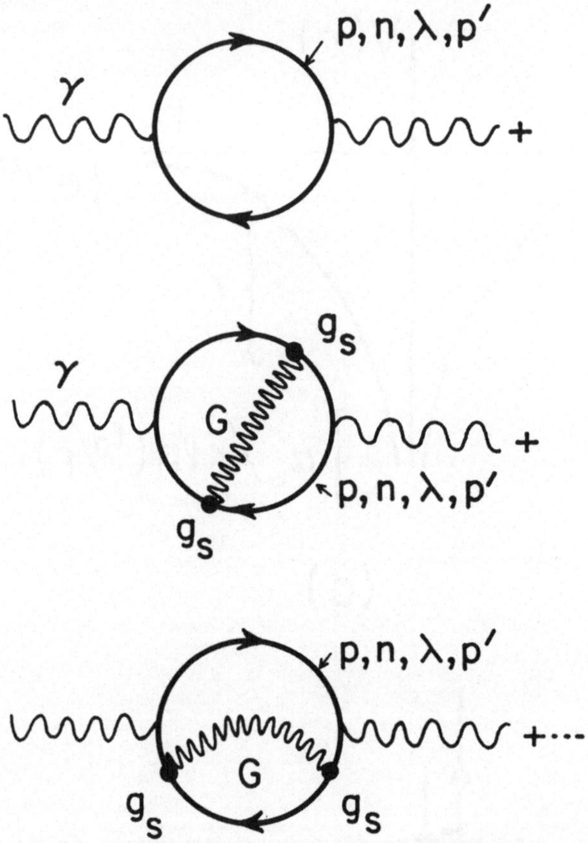

Fig. 13

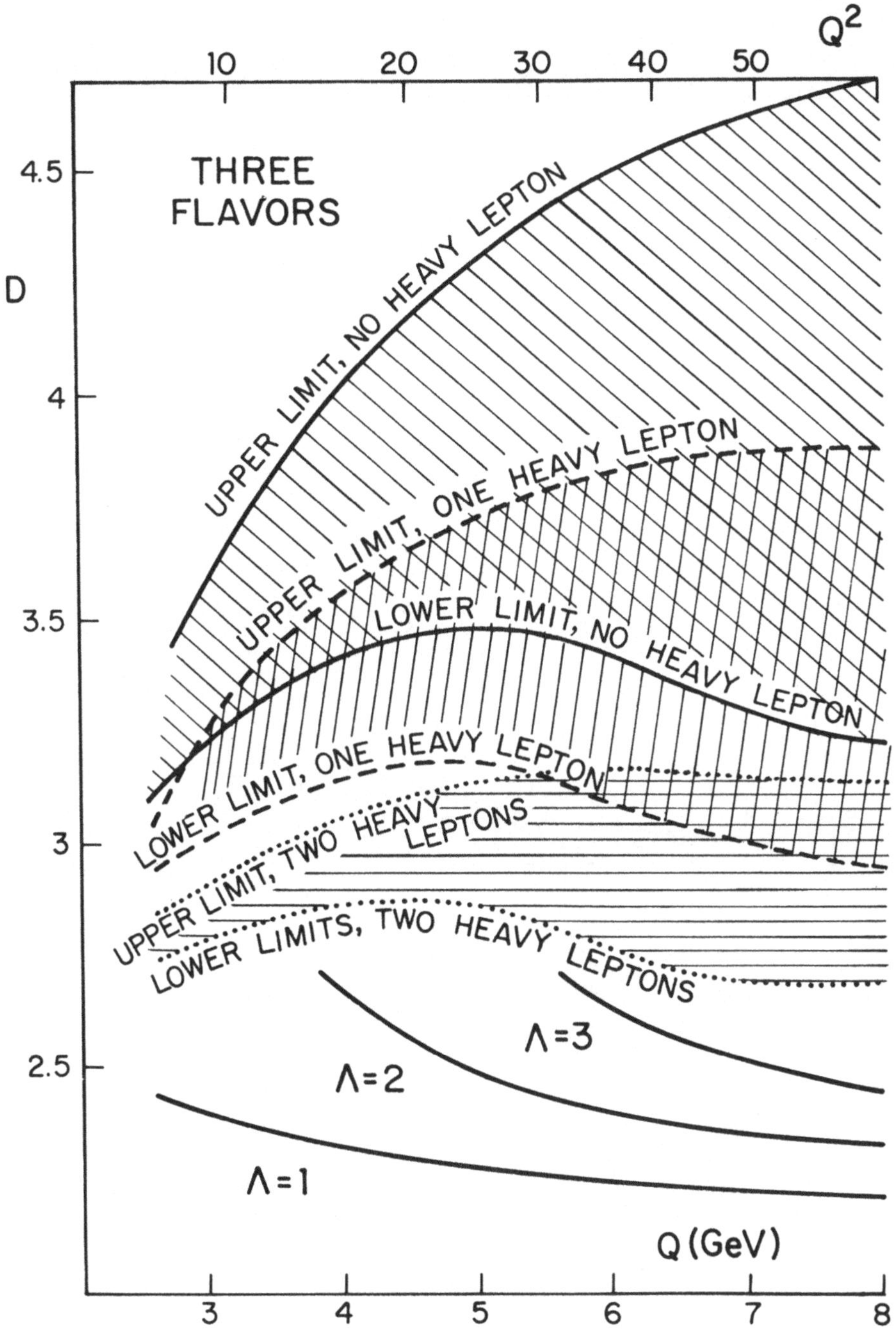
Q²
10
20
30
40
50
THREE
FLAVORS
D
4.5
4
3.5
3
2.5
UPPER LIMIT, NO HEAVY LEPTON
UPPER LIMIT, ONE HEAVY LEPTON
LOWER LIMIT, NO HEAVY LEPTON
LOWER LIMIT, ONE HEAVY LEPTON
UPPER LIMIT, TWO HEAVY LEPTONS
LOWER LIMITS, TWO HEAVY LEPTONS
Λ=3
Λ=2
Λ=1
Q (GeV)
3
4
5
6
7
8

Fig. 14

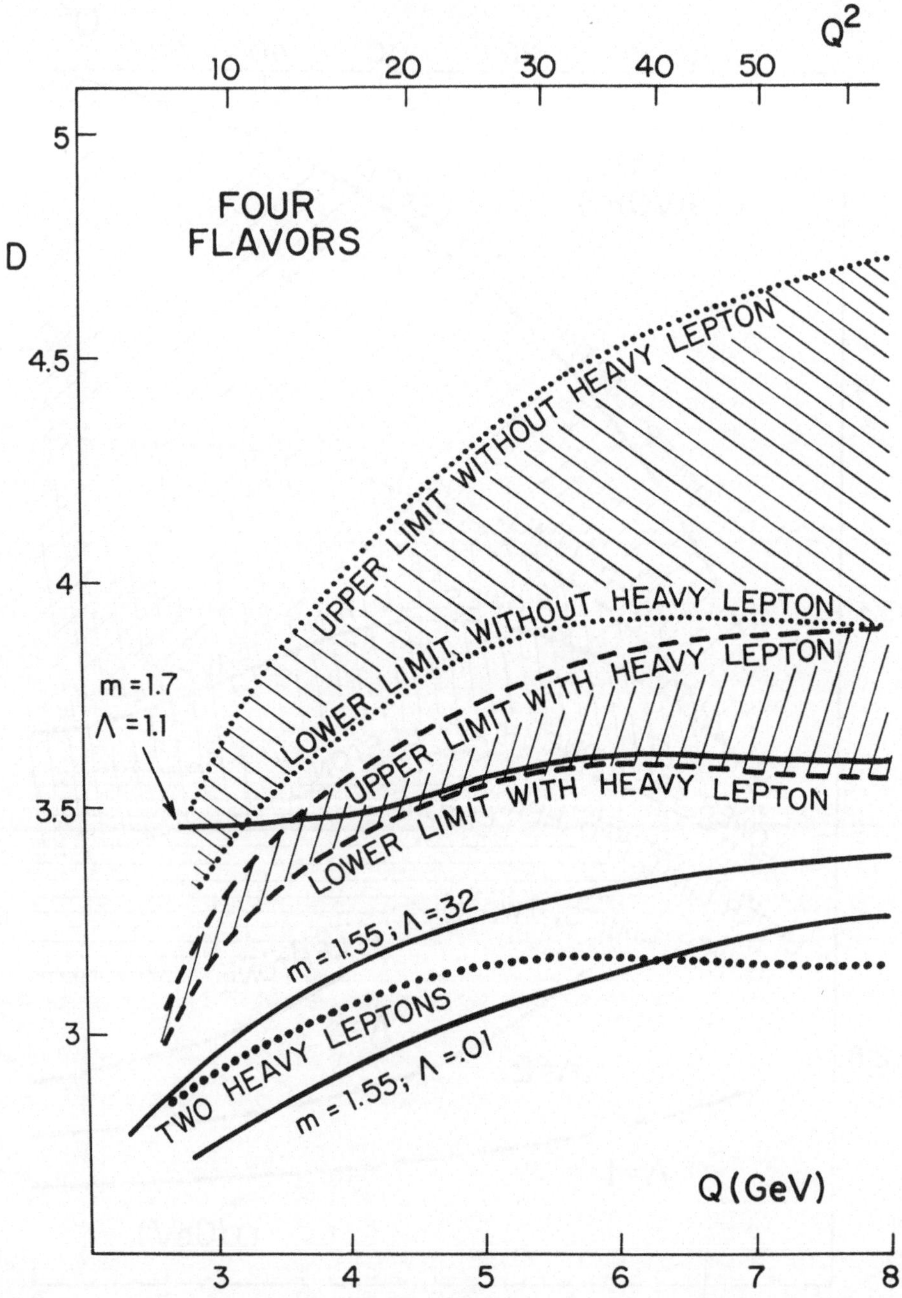
Q^2
10
20
30
40
50
5
4.5
4
3.5
3
D
FOUR FLAVORS
UPPER LIMIT WITHOUT HEAVY LEPTON
LOWER LIMIT WITHOUT HEAVY LEPTON
UPPER LIMIT WITH HEAVY LEPTON
LOWER LIMIT WITH HEAVY LEPTON
m = 1.7
Λ = 1.1
m = 1.55 ; Λ = .32
TWO HEAVY LEPTONS
m = 1.55 ; Λ = .01
Q (GeV)
3
4
5
6
7
8

Fig. 15

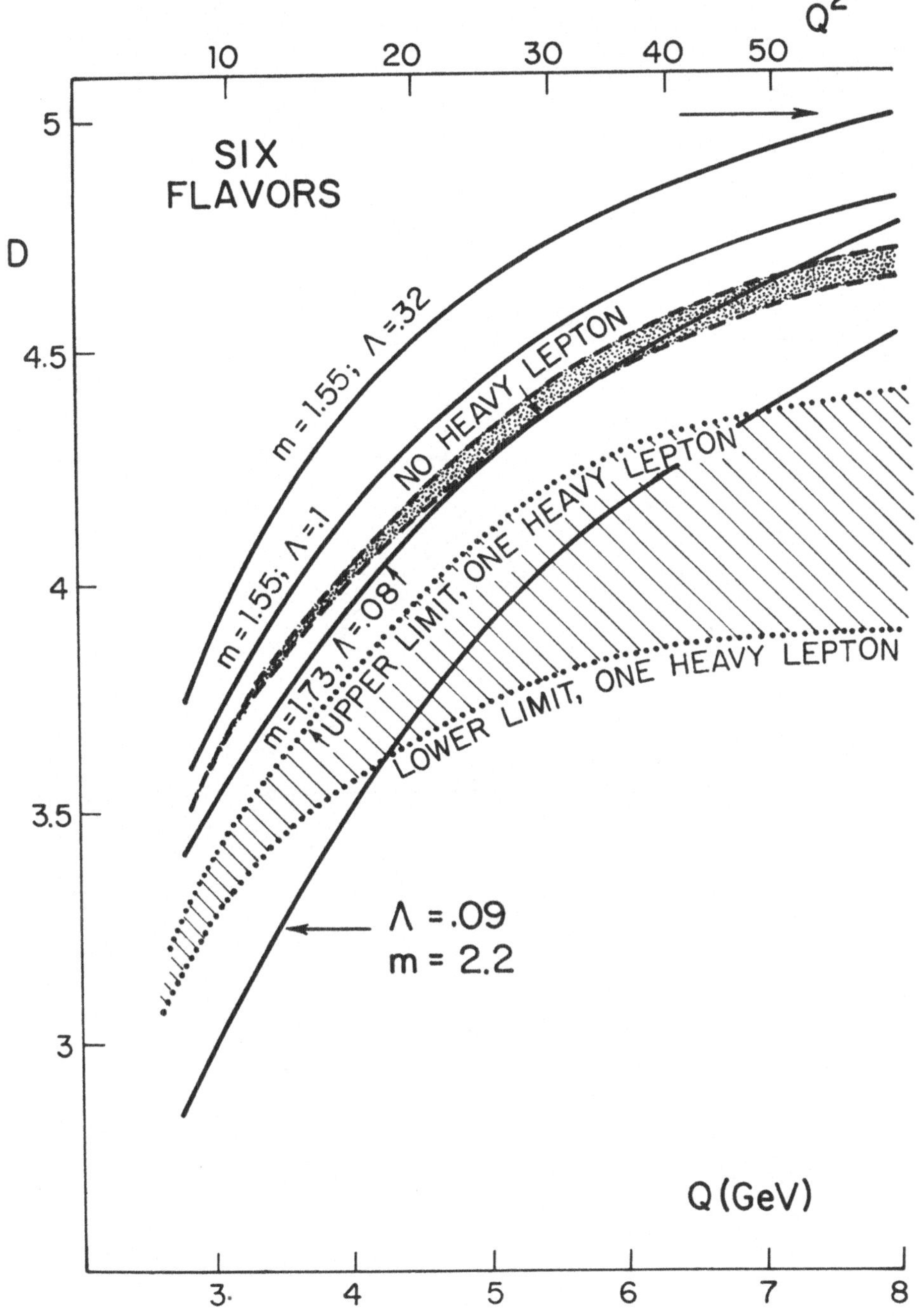
SIX
FLAVORS
D
Q²
10
20
30
40
50
5
4.5
4
3.5
3
m = 1.55; Λ = .32
m = 1.55; Λ = .1
NO HEAVY LEPTON
m = 1.73, Λ = .081
UPPER LIMIT, ONE HEAVY LEPTON
LOWER LIMIT, ONE HEAVY LEPTON
Λ = .09
m = 2.2
Q (GeV)
3.
4
5
6
7
8

Fig. 16

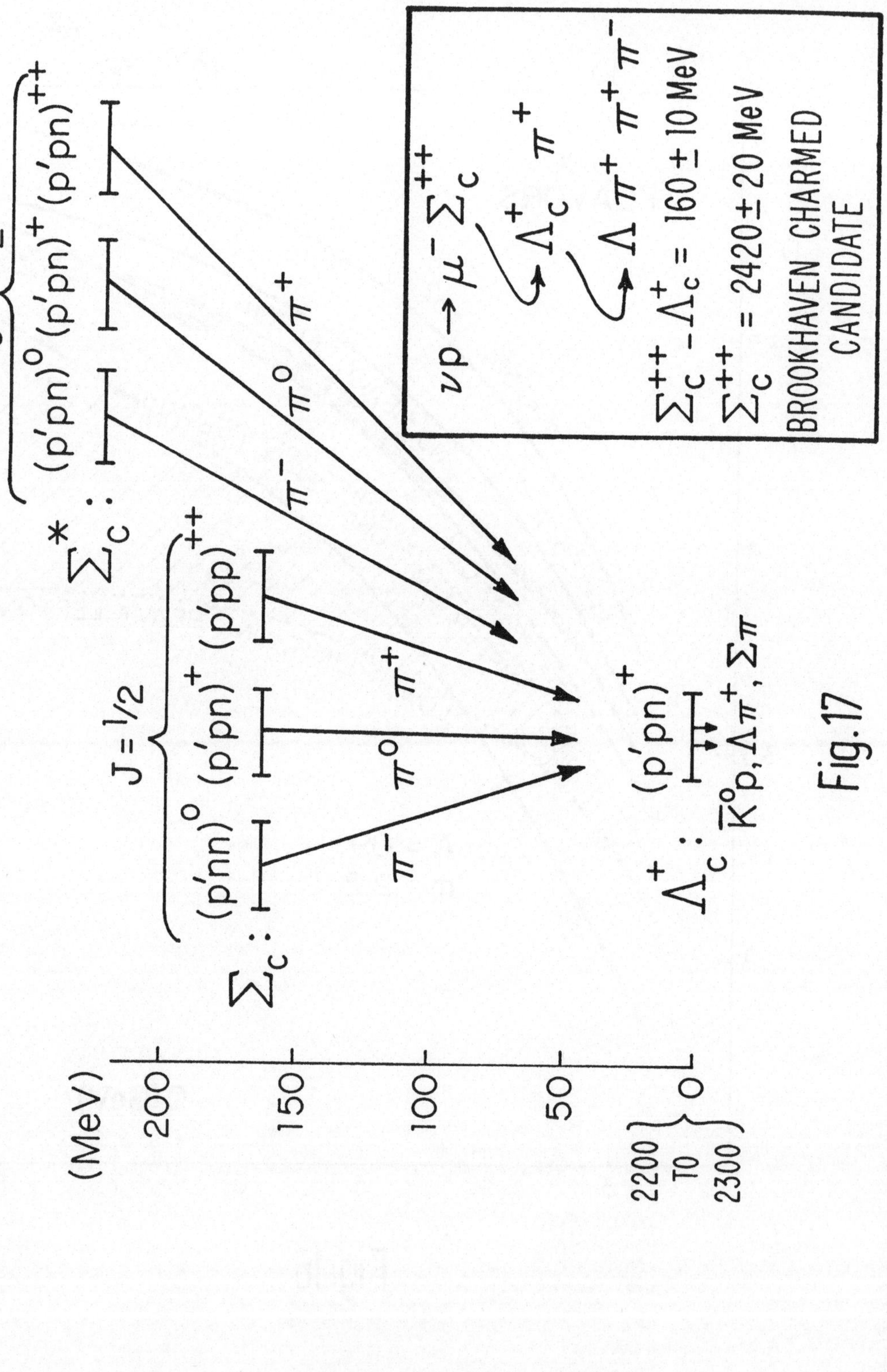
(MeV)
200
150
100
50
0
2200 TO 2300
Σ_c :
$J = 1/2$
$(p'nn)^0$ $(p'pn)^+$ $(p'pp)^{++}$
π^- π^0 π^+
Σ_c^* :
$J = 3/2$
$(p'pn)^0$ $(p'pn)^+$ $(p'pn)^{++}$
π^- π^0 π^+
Λ_c^+ : $(p'pn)^+$
$\bar{K}^0 p, \Lambda\pi^+, \Sigma\pi$
$\nu p \to \mu^- \Sigma_c^{++}$
$\Lambda_c^+ \pi^+$
$\Lambda \pi^+ \pi^+ \pi^-$
$\Sigma_c^{++} - \Lambda_c^+ = 160 \pm 10$ MeV
$\Sigma_c^{++} = 2420 \pm 20$ MeV
BROOKHAVEN CHARMED CANDIDATE

Fig.17

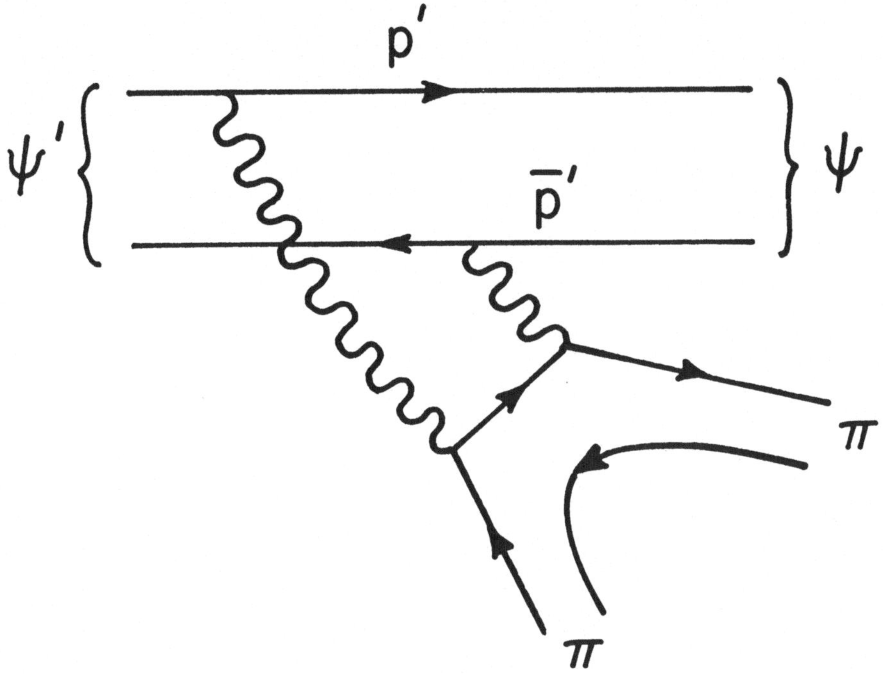

Fig. 18

NEUTRAL CURRENTS WITHOUT GAUGE THEORY PREJUDICES

by

J.J. Sakurai
CERN, Geneva

1. Introduction

Most theoretical lectures - and even some experimental lectures - on neutral currents begin by pointing out that neutral currents are needed because otherwise the cross-section for

$$\nu + \bar{\nu} \rightarrow W^+ + W^- \tag{1.1}$$

would go like s. You will find that my lectures are very different.

The main thing I would like to emphasize throughout these lectures is that neutral current phenomena should be studied in their own right. The discovery of neutral currents by the Gargamelle Collaboration [1)] in 1973 is a great milestone in the development of high-energy physics, quite comparable to the discovery of CP violation [2)] in 1964 or that of ψ/J particles [3)] in 1974. Neutral current phenomena are interesting regardless of whether they may be helpful in constructing renormalized gauge theories that unify weak and electromagnetic interactions. From 1932, when Fermi [4)] proposed a four-Fermion interaction, until 1973, there was no evidence for an interaction of the form $(\bar{\nu}\nu)$ $(\bar{p}p)$ with strength comparable to that of $(\bar{e}\nu)$ $(\bar{p}n)$. It is gratifying that after forty years of weak interaction physics with neutrinos something qualitatively new is finally happening.

Another reason for de-emphasizing particular theoretical approaches to neutral currents is that theoretical prejudices are often short-lived. It is possible that a renormalizable model of weak interactions looks attractive only because we do not know anything better at this present moment. A lesson from strong interaction physics may be helpful here. When I started graduate study in physics in the middle fifties, my superiors were telling me that the γ_5 theory was the only sensible theory of strong interaction physics because it is renormalizable like quantum electro-dynamics. We now know that a dogmatic attitude of that kind had harmful effects on the subsequent course of strong interaction physics; it discouraged people from examining effective Lagrangians based on the derivative coupling which reproduce many of the consequences of the PCAC hypothesis; it discouraged people from speculating on the existence of vector mesons which are at least as fundamental as pions; it discouraged people from studying seriously composite structure of hadrons, etc. When I hear that asymptotic freedom or quantum chromodynamics is the only correct theory of strong interactions, I am reminded of what was said about the γ_5 theory twenty years ago.

We cannot deny that the theorists' enthusiasm for unified gauge models stimulated a series of painstaking experimental investigations which ultimately led to the discovery of neutral current phenomena. At the same time, however, we should not forget that the existence of neutral currents per se does not provide strong evidence in favour of unified gauge models. We often read in news magazines, etc., that the discovery of neutral currents was an important step towards unifying weak and electromagnetic forces. But let us recall: before people started talking about unification we had two constants, e^2 and G. Now, after the neutral-current discovery we have three constants, e^2, m_W and $\sin^2\Theta_W$. Of course, confirmation of the detailed predictions of the Weinberg-Salam model [5], e.g.

$$m_Z \simeq \frac{74\ \text{GeV}}{\sin 2\Theta_W} \tag{1.2}$$

with Θ_W determined from $\bar{\nu}_\mu e$ scattering would be truly impressive. However, until then we should be open-minded on various alternative possibilities.

2. New Particles or old Neutrinos?

It is now widely accepted that the existence of neutral current effects is firmly established by at least six independent experiments [6-10]. We may, however, argue that the reactions we are talking about involve invisible particles coming in and invisible particles going out. How do we know that the incoming and outgoing particles are neutrinos?

We can now give a fairly good argument for the hypothesis that the incoming particle is an ordinary neutrino from π (K) decay. That the observed neutral current effects are not due to known hadrons such as neutrons and K_L mesons has been extensively discussed in all the experimental papers on neutral currents; I have nothing to add here. It may appear a little more difficult to rule out the possibility that the incoming particles are some yet unknown "background particles" that penetrate several hundred metres of earthen shielding, but the fact that the measured neutral-to-charged-current ratios are independent within errors of the various manners in which the neutrino beam is prepared makes this exotic possibility rather remote.

Coming now to the nature of the outgoing particles, let us first recall that the neutral-to-charged-current ratios are similar within a factor of ~ 2 between CERN energies [1] ($E_\nu \sim$ 2-6 GeV typically) and Fermilab energies [6,8] ($E \sim$ $\sim$ 10-200 GeV typically); so there does not seem to be an obvious threshold effect. Furthermore, the observed hadron energy distributions look similar between the neutral and charged current events. All this suggests, but does not conclusively prove, that the outgoing particle is also a neutrino.

As a specific proposal, let us examine the possibility that the outgoing particle is a heavy lepton of some sort:

$$\begin{aligned} \nu + N &\to L^{\pm,0} + \text{hadrons} \;, \\ L^{\pm,0} &\to \text{hadrons} + \nu(\bar{\nu}) \;. \end{aligned} \tag{2.1}$$

The main difficulty with this idea is that no threshold dependence of any kind has been observed. In particular the fact that neutral current effects have also been reported at Argonne where the neutrino spectrum peaks at $\sim$500 MeV shows that the hypothetical heavy lepton mass must be very low. It is of interest to mention in this connection that the ANL Bubble Chamber Group has looked for a possible peaking in the invariant mass distribution of the $\pi\nu$ system in

$$\nu + n \to \nu + \pi^- + p \;, \tag{2.2}$$

which is a zero-constraint reaction provided the incoming and outgoing invisible particles are assumed to be massless neutrinos. If we had a heavy lepton reaction

$$\nu + n \to L^- + p \tag{2.3}$$

with L^- subsequently decaying into $\nu + \pi^-$, a sharp peak in the $\nu\pi^-$ invariant mass distribution would be expected. Experimentally no peak has been observed [11]. Note also that it is not possible to explain away the pure-leptonic neutral-current events ($\bar{\nu}_\mu e$ scattering) observed by the Gargamelle Collaboration [12] in this manner.

The possibility that the outgoing particle is a neutral heavy lepton of low mass - sufficiently low to forbid its semileptonic decay - may be more difficult to rule out particularly if its main decay mode turns out to be

$$L^0 \rightarrow \nu + \bar{\nu} + \nu \quad . \tag{2.4}$$

If the L^0 mass is close to zero, this proposal is practically indistinguishable from a model in which the outgoing particle is a new kind of neutrino, a case I will treat later.

Even though the neutral-current events observed by the Gargamelle Collaboration are unlikely to be due to exotic particles such as heavy leptons, it is conceivable that some of the neutral-current events at Fermilab energies receive contaminations from exotic particle production. One specific suggestion which was alive until about a month ago went as follows. As discussed in Dr. von Krogh's lecture, the HPW (Harvard-Pennsylvania-Wisconsin) Collaboration [13] has reported an anomaly in the y-distribution in the charged-current reactions, interpretable as some kind of threshold effect due to new particle production. The same group also has claimed anomalous dimuon production [14]. Suppose we have a neutral heavy lepton of a few GeV produced in neutrino-nucleon collisions

$$\nu + N \rightarrow L^0 + \text{hadrons} \quad . \tag{2.5}$$

The heavy lepton L^0 may subsequently decay as

$$L^0 \rightarrow \mu^- + \text{hadrons} \quad , \tag{2.6a}$$

$$L^0 \rightarrow \mu^+ + \mu^- + \nu \quad , \tag{2.6b}$$

$$L^0 \rightarrow \nu + \text{hadrons} \quad , \tag{2.6c}$$

$$L^0 \rightarrow \nu + \bar{\nu} + \nu \quad . \tag{2.6d}$$

The production process (2.5) followed by (2.6a) may explain the charged-current y-distribution anomaly; if (2.5) is followed by (2.6b) we may have an explanation of the dimuon anomaly; if (2.5) is followed by (2.6c) and (2.6d) we expect neutral-current-like signals with no muons in the final state. It has to be pointed out, however, that a more recent study of the dimuon events by the HPW Collaboration does not favour the neutral-heavy-lepton hypothesis as an explanation of the anomaly; this is because there is a large difference in the μ^+ and μ^- spectrum which, according to Pais and Treiman [15)], is impossible for heavy lepton decays. Nevertheless, it is still possible that the production of even more exotic particles may contaminate the neutral-current sample.

We now turn to the question of whether the final neutrino involved in the neutral-current reactions is the same as the initial neutrino. Sehgal [16)] has pointed out that it is, in principle, possible to settle this question of neutrino identity by looking at

$$\bar{\nu}_e + e^- \rightarrow \bar{\nu}' + e^- \quad , \tag{2.7}$$

which is now being studied at reactor energies at Savannah River by Reines and collaborators [17)]. First, note that the reaction occurs with $\bar{\nu}' = \bar{\nu}_e$ even in the old fashioned charge-current V-A theory, as is clear from Fig. 1(a) where the current appears in the s-channel. The presence of the neutral-current interaction adds a t-channel diagram, as shown in Fig. 1(b). The question is whether the final neutral particle in Fig. 1(b) is the same as the $\bar{\nu}_e$ appearing in the final state of Fig. 1(a). With the most general combination of vector and axial-vector interactions, it is possible to show that the energy distribution of the final electron is given by

$$\frac{d\sigma}{dE_e} = (2G^2 m_e/\pi)\left[A + B\left(1 - \frac{E_e}{E_\nu}\right)^2 - C\, m_e \frac{E_e}{E_\nu^2}\right] . \tag{2.8}$$

If we just had the usual V-A charged-current interaction, we would have

$$A = 0 \quad , \quad B = 1 \quad , \quad C = 0 . \tag{2.9}$$

This, incidentally, can be shown to be a consequence of the rule which says that relativistic leptons (antileptons) are left-handed (right-handed) in the V-A theory. The presence of the neutral-current diagram Fig. 1(b) in general makes B different from 1 and, in addition, makes A (opposite chirality term) finite. However, this is not the whole story. The helicity rule is not exact for a finite mass particle; as a result, there is a term that represents interference between interaction of opposite chirality, the C term of (2.8), proportional to the electron mass. If the final $\bar{\nu}'$ in (2.7) is identical to the usual $\bar{\nu}_e$, then we expect

$$C^2 = AB. \tag{2.10}$$

On the other hand, if $\bar{\nu}'$ in (2.7) is distinct from $\bar{\nu}_e$, the interference is reduced - Fig. 1(a) and Fig. 1(b) cannot interfere - and as a result we obtain

$$C^2 = A(B-1). \tag{2.11}$$

The question of neutrino identity can also be examined in semileptonic reactions. Consider

$$\nu + I \rightarrow \nu' + F \ , \tag{2.12a}$$

$$\bar{\nu} + I \rightarrow \bar{\nu}' + F \ , \tag{2.12b}$$

where I and F are some hadronic states. If the $\nu'(\bar{\nu}')$ in the final state is identical to the $\nu(\bar{\nu})$ in the initial state, then the effective Lagrangian must look like $i\bar{\nu}\gamma_\lambda(1+\gamma_5)\nu J_\lambda$ where the space and time components of J_λ are required to be Hermitian by the CPT theorem or equivalently by the Hermiticity of the effective Lagrangian. As a result, the hadronic parts of the matrix elements for the two reactions must be the same, $\langle F|J_\lambda|I\rangle$ in both cases; however, there is a difference in the cross-section because of VA interference which changes sign as we go from the neutrino reaction to the antineutrino reaction. On the other hand, with $\nu' \neq \nu$, it is in principle possible to consider a non-Hermitian neutral current even though in the usual quark model simple diagonal currents we can construct are automatically Hermitian. (I will say more about non-diagonal currents in Section 4.) So it is logically conceivable for the hadronic matrix element for the neutrino reaction given by $\langle F | J_\lambda | I\rangle$ to be different from that for the antineutrino reaction given by $\langle F | J_\lambda^\dagger | I\rangle$.

We can test all this by examining whether the cross-sections for (2.12a) and (2.12b) are the same when we specialize to kinematical configurations where VA interference is required to vanish. This can be accomplished for exclusive reactions by going to $q^2 \rightarrow 0$ for W (final hadronic invariant mass) fixed, or $E_\nu \rightarrow \infty$ with W fixed, and for inclusive reactions by letting $v \equiv q^2/2mE_\nu \rightarrow 0$ or $y \equiv 1 - E'_\nu/E_\nu \rightarrow 0$. The Hermiticity of the effective Lagrangian is as sacred as unitarity or probability conservation. Therefore, if the cross-sections for (2.12a) and (2.12b) turned out to be different in the kinematical configurations mentioned above, we would be driven to the inescapable conclusion that the final neutrino is different from the initial neutrino. This argument is due to Wolfenstein [18].

Before concluding this section I would like to discuss the possibility that the observed neutral current effects might be explained by attributing anomalous properties to the "old"neutrino. For a long time it has been recognized that neutral-current-like effects would be generated if the neutrino had a large electromagnetic radius [19]. To see this, consider the interaction between a neutrino and hadrons via one-photon exchange. Since the total electric charge of the neutrino is zero, the Dirac form factor of the neutrino must start like q^2; this just cancels the $1/q^2$-dependence arising from the photon propagator. As a result, the net effect simulates the q^2-dependence of the usual current-current interaction.

The main defect of this proposal [20] is that with almost any reasonable estimate for the neutrino charge radius, the effect is much too small to explain the observed neutral-current effects. Diagrams such as Fig. 2 responsible for the electromagnetic structure of the neutrino have logarithmic mass singularities as $m_\mu \to 0$; so just on dimensional grounds we expect the derivative of the Dirac form factor at $q^2 = 0$ to be of the form

$$\left.\frac{dF}{dq^2}\right|_{q^2=0} \simeq G \ln(\Lambda^2/m_\mu^2) \tag{2.13}$$

where Λ may be identified with the cut-off energy in divergent models and the W boson mass in convergent models. More detailed calculations carried out by several authors [19,21] confirm the simple result given above within numerical factors like $1/12\pi^2\sqrt{2}$.

We know experimentally that the effective strength of the observed neutral-current interactions is similar to that of the charged-current interactions. On the other hand the charge radius effect given by (2.13) would imply a neutral-current cross-section of order $\alpha^2[\ln(\Lambda/m_\mu^2)]^2$ times the charged-current cross-section. Even if Λ is taken to be as large as 100 GeV, this factor is only 0.01. It is therefore unlikely that a normal mechanism for generating a neutrino charge radius is sufficient to explain the observed strength of the neutral-current interactions. We may, of course, just postulate, in an ad hoc manner, a large charge radius for the neutrino without violating any of the well-established physical principles [20].

As far as practical consequences on neutrino-induced reactions are concerned, this proposal yields the same predictions as the "electromagnetic current model", i.e. a model in which the hadronic and the charged-leptonic part of the weak neutral current are postulated to be identical to the electromagnetic current [22]. The neutral current is pure vector and the final states in neutrino-induced semi-leptonic neutral current reactions must be the same as in analogous electroproduction. We do not expect, however, deviations from the usual QED predictions in processes like electron-positron annihilation into muon pairs of the kind to be discussed in Section 8.

3. Space-Time Properties - VA Orthodoxy

In this section I assume that the neutral-current interaction is effectively of the current-current form where the current is made up of a linear combination of vector and axial vector covariants. The more heretical possibility that the muonless neutrino interactions involve S, P and/or T covariants will be treated in the next section.

For pedagogical purposes it is best to begin by reviewing deep inelastic phenomenology for charged-current processes [23]. Consider

$$\begin{aligned} \nu_\mu + N &\rightarrow \mu^- + \text{hadrons} \,, \\ \bar{\nu}_\mu + N &\rightarrow \mu^+ + \text{hadrons} \,, \end{aligned} \tag{3.1}$$

where N may stand for proton or neutron. We introduce the usual kinematical variables -- $q^2 = (p_\nu - p_\mu)^2$ (> 0 for space-like momentum transfer in my metric), the energy transfer in the target rest frame and E the laboratory neutrino energy. In addition we use two dimensionless variables

$$x \equiv q^2/2m_N\nu \quad , \qquad y \equiv \frac{\nu}{E} \,. \tag{3.2}$$

Note that x and y both range from 0 to 1. Ignoring the lepton mass the double differential cross-section can be written as

$$\begin{aligned} \frac{d\sigma}{dx\,dy} &= \left(G^2 m_N E \frac{1}{\pi}\right)\left[x y^2 F_1(x) + (1-y) F_2(x) \mp x y \left(1-\tfrac{1}{2}y\right) F_3(x)\right] \\ &= \left(G^2 m_N E \frac{1}{\pi}\right) F_2(x) \left[\begin{Bmatrix} L(x) \\ R(x) \end{Bmatrix} + (1-y)^2 \begin{Bmatrix} R(x) \\ L(x) \end{Bmatrix} + (1-y)\, S(x)\right] \end{aligned} \tag{3.3}$$

$$\left(L(x) + R(x) + S(x) = 1 \right)$$

where the upper (lower) sign refers to the $\nu(\bar{\nu})$ reaction. In writing down (3.3) Bjorken's scaling hypothesis [24] has been assumed. However, even if we relax scaling, the only change needed is to let $F_{1,2,3}$, L, R, and S be functions of q^2 as well as of x. Upon integrating over x and y, we obtain a linearly rising cross-section provided the structure functions are q^2-independent.

Experimentally [25], to accuracies of 10 % to 20 %, the following relations [26] are satisfied for a major part of the kinematical region:

$$2xF_1 = F_2 \; , \qquad -F_3 = 2F_1 \tag{3.4}$$

or, equivalently,

$$R = 0 \quad , \qquad S = 0 \; . \tag{3.5}$$

This means that the y-dependence is flat for the ν reaction and $(1 - y)^2$ for the $\bar{\nu}$ reaction. The significance of this remarkable experimental fact may be appreciated by computing the cross-section for neutrino scattering on quarks and antiquarks of the u (proton-like) and d (neùtron-like) type according to the usual V-A rule:

$$\begin{aligned} \frac{d\sigma}{dy}\begin{pmatrix} \nu + d \to \mu^- + u \\ \bar{\nu} + \bar{d} \to \mu^+ + \bar{u} \end{pmatrix} &= 2G^2 m_N E/\pi \; , \\ \frac{d\sigma}{dy}\begin{pmatrix} \bar{\nu} + u \to \mu^+ + d \\ \nu + \bar{u} \to \mu^- + \bar{d} \end{pmatrix} &= (2G^2 m_N E/\pi)(1-y)^2 \, . \end{aligned} \tag{3.6}$$

The observed y-dependence of deep inelastic ν and $\bar{\nu}$ scattering is just what we expect from scattering on quarks, not on antiquarks. So, regardless of whether or not you like quark partons, the simple description based on scattering off point-like valence quarks within the nucleon provides a convenient framework for summarizing the basic experimental facts to accuracies of 10 % to 20 %. The famous one-to-three ratio for the $\bar{\nu}$ to ν cross-section also follows in a simple manner:

$$\int_0^1 (1-y)^2 dy \Big/ \int_0^1 dy = 1/3 \; . \tag{3.7}$$

Coming back to the subject of semileptonic neutral-current interactions, we can take advantage of what we learned in studying charged-current deep inelastic phenomenology. We assume that the quark parton description also works for

$$\nu + N \rightarrow \nu + \text{hadrons} , \qquad (3.8a)$$

$$\bar{\nu} + N \rightarrow \bar{\nu} + \text{hadrons} \qquad (3.8b)$$

and proceed to infer the space-time structure of the current from the y-distribution as follows:

$$\begin{aligned} &\text{V-A} &&: \quad 1 \text{ for } \nu \ , \ (1-y)^2 \text{ for } \bar{\nu} \ , \\ &\text{pure V or pure A} &&: \quad \tfrac{1}{2}\left[1+(1-y)^2\right] \text{ for both } \nu \text{ and } \bar{\nu} \ , \qquad (3.9) \\ &\text{V + A} &&: \quad (1-y)^2 \text{ for } \nu \ , \ 1 \text{ for } \bar{\nu} \ . \end{aligned}$$

Pathological cases where the simple conclusion above does not follow will be discussed later. The ratio of $\bar{\nu}$ to ν total cross-section can be obtained by straightforward integration. Notice in particular that both the differential and total neutral-current cross-sections must be equal between ν and $\bar{\nu}$ in the pure V or pure A case.

Experimentally it is not easy to measure the y-distribution. At Fermilab energies y, defined as ν/E, is roughly equal to E_{had}/E, and E_{had} can be inferred from the total energy deposited in the calorimeter. The main difficulty is that the incident neutrino energy E is not known; to solve for E, we must know, in addition to E_{had}, the direction of the hadronic debris. Nevertheless, knowing the incident neutrino flux and assuming Bjorken's scaling, it is possible to predict the E_{had} distribution for each of the three cases listed above. Such an analysis has been made by the Caltech Group [8)] whose ν beam is dichromatic. The main conclusion is that the data rule out V+A, do not favour V-A, and that the best fit lies somewhere between V-A and pure V/pure A. For the relevant figures consult Dr. von Krogh's lectures.

Let us now look at the cross-section ratios. According to the published data of the Gargamelle Collaboration [1,27)] we have

$$R_\nu \equiv \frac{\sigma(\nu + N \to \nu + \text{hadrons})}{\sigma(\nu + N \to \mu^- + \text{hadrons})} = 0.22 \pm 0.03 \quad ,$$

$$R_{\bar\nu} \equiv \frac{\sigma(\bar\nu + N \to \bar\nu + \text{hadrons})}{\sigma(\bar\nu + N \to \mu^+ + \text{hadrons})} = 0.43 \pm 0.12 \qquad (3.10)$$

with $E_{had} > 1$ GeV. With this energy cut the $\bar\nu$ to ν cross-section ratio for the charged-current reactions is even smaller than 1/3, perhaps something like 0.26. Hence the published Gargamelle data rule out the pure V/pure A hypothesis by about two standard deviations. More recently, however, a new preliminary number for $R_{\bar\nu}$ with three times the previous statistics was reported by Morfín [28] at the SLAC Conference:

$$R_{\bar\nu} = 0.55 \pm 0.07 \ . \qquad (3.11)$$

If we combine this result with R_ν given by (3.10), we see that the ν and $\bar\nu$ cross-sections for the neutral-current reactions are closer to being equal. So there may still be some hope for pure V (or pure A) enthusiasts.

As for R_ν and $R_{\bar\nu}$ at higher energies, the published data of the HPW Collaboration [6] give

$$R_\nu = 0.11 \pm 0.05 \ ,$$
$$R_{\bar\nu} = 0.32 \pm 0.09 \ , \qquad (3.12)$$

which is in good agreement with the pure V/pure A hypothesis. However, more recently it was reported that their new measurement shows a considerably higher value of R_ν [29].

Let us be more quantitative and attempt to obtain the vector and axial-vector coupling constants for semileptonic neutral-current processes. This can be done in a straightforward manner in models - subsequently referred to as pure isovector models - where the vector and axial vector pieces of the neutral current are pure isovector and belong to the same isospin triplets as the vector and axial-vector pieces of the charged current. The relevant interaction is given by

$$\mathcal{L} = \frac{G}{\sqrt{2}}\Big\{\big[i\bar{\mu}\gamma_\lambda(1+\gamma_5)\nu\,(j_\lambda^{1+i2} + j_{5\lambda}^{1+i2}) + h.c.\big]$$

$$+\, i\bar{\nu}\gamma_\lambda(1+\gamma_5)\nu\,(v_3 j_\lambda^{3} + a_3 j_{5\lambda}^{3})\Big\} \tag{3.13}$$

where v_3 and a_3 are the vector and axial-vector coupling constants. The cosine of the Cabibbo angle has been set equal to 1 for simplicity, not a bad approximation. The quark model normalization for the currents I am using is

$$j_\lambda^{1+i2} = i\bar{u}\gamma_\lambda d \;, \quad j_\lambda^{3} = \frac{i}{2}(\bar{u}\gamma_\lambda u - \bar{d}\gamma_\lambda d) \tag{3.14}$$

and similarly for $j_{5\lambda}^{1+i2}$ and $j_{5\lambda}^{3}$ with $\gamma_\lambda \to \gamma_\lambda\gamma_5$. Isospin invariance leads to

$$\begin{aligned} \sigma(\nu\to\nu) &= \tfrac{1}{2}(v_3^2 V + a_3^2 A + v_3 a_3 I), \\ \sigma(\bar{\nu}\to\bar{\nu}) &= \tfrac{1}{2}(v_3^2 V + a_3^2 A - v_3 a_3 I), \\ \sigma(\nu\to\mu^-) &= V + A + I, \\ \sigma(\bar{\nu}\to\mu^+) &= V + A - I, \end{aligned} \tag{3.15}$$

where $\sigma(\nu\to\nu)$, etc. stands for the cross-section for the inclusive reaction (3.8a), etc., after averaging over equal numbers of target protons and neutrons, and V, A and I stand for the vector, axial-vector and VA interference contributions to the charged-current cross-sections.

In obtaining (3.15) only isospin invariance has been used. We can now invoke chiral symmetry or the quark-parton condition to set

$$V = A \;, \tag{3.16}$$

which actually can be justified on experimental grounds because the charged-current data indicate that VA interference is essentially maximal with $R(x) \simeq 0$ [cf. (3.5)]. It is now possible to solve for $v_3^2+a_3^2$ and v_3a_3 as follows:

$$\frac{1}{4}(v_3^2 + a_3^2) = \frac{\sigma(\nu\to\nu) + \sigma(\bar\nu\to\bar\nu)}{\sigma(\nu\to\mu^-) + \sigma(\bar\nu\to\mu^+)} \;, \qquad (3.17a)$$

$$\frac{1}{2} v_3 a_3 = \frac{\sigma(\nu\to\nu) - \sigma(\bar\nu\to\bar\nu)}{\sigma(\nu\to\mu^-) - \sigma(\bar\nu\to\mu^+)} \;, \qquad (3.17b)$$

which are generalizations of results first obtained by Paschos and Wolfenstein [30].

Using these equations several authors have proposed a graphical method [31] for displaying the constraints imposed by the experimental data on a v_3-a_3 coupling constant plane. Equation (3.17a) defines a circle while Eq. (3.17b) defines hyperbolae; so we look for the intersections of the circle and the hyperbolae. I have made such a plot using the central values of R_ν and $R_{\bar\nu}$ reported by the Gargamelle Collaboration; for $R_{\bar\nu}$ the more recent preliminary value (3.11) is used. The effect of the energy cut, which emphasizes higher values of y, is taken into account by setting the charged-current antineutrino to neutrino cross-section ratio to 0.26 (rather than 1/3). The result is shown on Fig. 3. The intersection points are close to the v_3 and a_3 axes, which shows that the interaction is predominantly (if not purely) vector or predominantly axial-vector. Our currents are normalized so that v_3 and a_3, are equal for V minus A; the fact that the intersection points are not in the second or fourth quadrant implies that $V+\alpha A$ with $\alpha > 0$ is unlikely. Note also that either v_3^2 or a_3^2 is close to unity.

Let us now turn to models with pure isoscalar neutral currents. Here the situation is a little less straightforward because the charged and neutral currents no longer belong to the same isospin multiplet. It becomes necessary to rely on quark parton models to deduce the relevant coupling constants. As will be discussed in details in Section 5 there are at least two - and possibly three if you believe in charm - kinds of isoscalars, the kind that goes like hypercharge [the isoscalar member of the SU(3) octet to which the charged current belongs] and the kind that goes like baryonic charge SU(3) singlet ; in the quark model the difference is just $\bar{s}s$. Fortunately in discussing deep-inelastic scattering it appears legitimate to assume that strangeness-bearing quarks or antiquarks can be ignored. With this assumption we need not distinguish hypercharge-like currents from baryon-like currents; we simply write the relevant effective Lagrangian as

$$\mathcal{L} = \frac{G}{\sqrt{2}} i \bar\nu \gamma_\lambda (1+\gamma_5) \nu (v_s j_\lambda^s + a_s j_{5\lambda}^s) \qquad (3.18)$$

where my normalization is such that in the quark-parton model

$$j_\lambda^s = \frac{i}{3}(\bar{u}\gamma_\lambda u + \bar{d}\gamma d + \ldots\ldots),$$

$$j_{5\lambda}^s = \frac{i}{3}(\bar{u}\gamma_\lambda\gamma_5 u + \bar{d}\gamma_\lambda\gamma_5 d + \cdots) \qquad (3.19)$$

with ... standing for currents made up of s- and c-type quarks, which we need not specify in the naive quark-parton model based on non-strange valence quarks for the nucleon. With this normalization the physical nucleon (not a quark) has one unit of isoscalar charge.

According to the naive quark-parton model without quark-antiquark seas, the charged-current inclusive distributions measure the distribution function $d_p(x) = u_n(x)$ on proton targets and $d_n(x) = u_p(x)$ on neutron targets where $d_p(x)$, etc., stand for the distribution function for the down quark d within the proton, etc. So, when averaged over proton and neutron targets, the charged current reaction measures the sum

$$d_p(x) + u_p(x) = u_n(x) + d_n(x) \,. \qquad (3.20)$$

But precisely this combination appears when the isoscalar neutral current probes the nucleon. Thus even though the charged and neutral currents do not belong to the same isospin multiplet, we can still express the coupling constants v_s and a_s in terms of the observable cross-sections just as in (3.17a) and (3.17b) [32]

$$\frac{1}{9}(v_s^2 + a_s^2) = \frac{\sigma(\nu\to\nu) + \sigma(\bar\nu\to\bar\nu)}{\sigma(\nu\to\mu^-) + \sigma(\bar\nu\to\mu^+)} \,, \qquad (3.21a)$$

$$\frac{2}{9} v_s a_s = \frac{\sigma(\nu\to\nu) - \sigma(\bar\nu\to\bar\nu)}{\sigma(\nu\to\mu^-) - \sigma(\bar\nu\to\mu^+)} \,. \qquad (3.21b)$$

In a model in which the neutral current contains both isovector and isoscalar, we simply add the left-hand sides of (3.17a) and (3.21a); of (3.17b) and (3.21b). Isoscalar-isovector interference is absent because we average over proton and neutron targets.

I have tried to present a general framework for discussing the deep inelastic reactions without commitment to particular models. If we so desire, we can, of course, easily specialize to some specific model. As an example, consider the one-parameter Weinberg-Salam model [5], in which the coupling constants are given by

$$v_3 = 1 - 2\sin^2\Theta_W \,, \qquad a_3 = 1 \,,$$
$$v_s = -\sin^2\Theta_W \,, \qquad a_s = 0 \,. \tag{3.22}$$

Ignoring the isoscalar contributions for a moment and assuming the usual one-to-three ratio for the $\bar{\nu}$-to-ν charged-current cross-sections, we can readily derive from (3.17) and (3.22) [33]

$$R_\nu = \frac{1}{2} - \sin^2\Theta_W + \frac{2}{3}\sin^4\Theta_W \,,$$
$$R_{\bar{\nu}} = \frac{1}{2} - \sin^2\Theta_W + 2\sin^4\Theta_W \,. \tag{3.23}$$

The inclusion of the isoscalar contributions only slightly modifies (3.23) as follows [34]:

$$R_\nu = \frac{1}{2} - \sin^2\Theta_W + \frac{20}{27}\sin^4\Theta_W \,,$$
$$R_{\bar{\nu}} = \frac{1}{2} - \sin^2\Theta_W + \frac{20}{9}\sin^4\Theta_W \,. \tag{3.24}$$

This pair of equations yields a one-parameter nose-like curve on an R_ν - $R_{\bar{\nu}}$ plane, with which many of you may be familiar.

I emphasized earlier that the pure V or pure A hypothesis demands that the cross-sections for ν and $\bar{\nu}$ be equal. However, the converse is not necessarily correct. For example, if for some reason the quark-parton description, which is so successful in the charged-current case, fails in the neutral-current case, the VA interference structure function F_3 may vanish even with both V and A contributing. This happens in a pure diffraction model. Even within the framework of quark-parton models the ν-$\bar{\nu}$ cross-section difference vanishes as long as the coupling constants satisfy [32]

$$v_3 a_3 + \frac{4}{9} v_s a_s = 0 \,. \tag{3.25}$$

This condition is met, for example, in a model where the u-type quark is coupled left-handed but the d-type quark is coupled right-handed. It has been pointed out that such a pathological model predicts different inclusive pion distributions from pure V/pure A models [35], but I will not elaborate on this point any further. While we are considering pathological models, it may be mentioned that a large departure from the quark-parton expectations is possible if gluons, which are dormant in the charged-current interactions, play important roles [36]. Since the gluons are presumably spin-one elementary bosons, there may be large deviations from the scaling predictions due to the $p_\mu p_\nu$ term of the propagator with neutrino cross-sections going like E^2 or E^3. Needless to say, it is very important to examine carefully whether R_ν and $R_{\bar{\nu}}$ are indeed energy independent.

Coming back to the more straightforward models, let us suppose that VA interference is shown to be small or absent. Because of the quadratic ambiguity inherent in (3.17) we will still be left with the important task of deciding whether vector or axial-vector is the dominant piece. The history of weak interaction physics has taught us that low-energy beta decay and capture processes have been quite informative in deciphering the space-time properties of the charged-current interactions. Along this line let us first consider the weak analog of the photodisintegration of the deuteron

$$\bar{\nu}_e + D \rightarrow \bar{\nu}_e + n + p \tag{3.26}$$

at reactor energies. This reaction is essentially forbidden for vector-type interactions because the relevant Fermi matrix element vanishes by the orthogonality of the initial and final wave functions. In the axial vector case we have an allowed Gamow-Teller transition from the deuteron state $^3S_1(I = 0)$ to $^1S_0(I = 1)$, which is possible if the axial vector current contains an isovector piece. So this reaction at low (reactor) energies isolates the a_3 term (isovector axial) of the neutral current [37]. A preliminary feasibility study of this process was made by Gurr, Reines and Sobel [38]. The present status can be summarized by saying that

$$a_3^2 = -0.9 \pm 2.2 \quad , \tag{3.27}$$

which is consistent with both a pure isoscalar model ($a_3 = 0$) and the Weinberg-Salam model ($a_3 = 1$). It is expected, however, that the error will eventually come down to the level of $\pm$ 0.2.

There are other low-energy nuclear transitions that isolate the isovector axial piece. Walecka and collaborators [39] have considered

$$\bar{\nu} + {}^{7}\mathrm{Li}\,(3/2^-) \rightarrow \bar{\nu} + {}^{7}\mathrm{Li}^{*}(1/2^-) \qquad (3.28)$$

where the excited state is to be identified by looking at its subsequent γ decay. The cross-section can be estimated from the isospin-related electron capture process, ${}^{7}\mathrm{Be}\,(3/2^-) \rightarrow {}^{7}\mathrm{Li}^{*}(1/2^-)$. They conclude that with a typical Savannah River flux of $2 \times 10^{13} \bar{\nu}_e/\mathrm{cm}^2$ sec., one may observe four γ-ray transitions per day for one kilogram of Li.

Another method for detecting axial vector takes advantage of Adler's theorem for forward neutrino scattering. In 1964 Adler [40] showed that in any elastic neutrino reaction

$$\nu + N \rightarrow \mu^-(e^-) + X \,, \qquad (3.29)$$

if we choose kinematical configurations in which the incident neutrino and the outgoing charged lepton move in the same direction, then the cross-section is proportional to $|\langle X|\partial_\lambda J_\lambda|N\rangle|^2$ provided the lepton mass could be ignored. Applying this to the neutral current case, as first proposed by Pais and Treiman [41], we expect no contribution from the vector piece because simple vector currents we can consider, using quark fields, etc., are necessarily conserved. So, if there is a finite cross-section for these parallel configurations, the most natural conclusion we can draw is that the neutral currents contain axial vector. Furthermore, if the isovector axial current is an isospin transform of the charged axial current, we can appeal to PCAC and relate the isovector axial contributions to the cross-section for

$$\pi^0 + N \rightarrow X \,. \qquad (3.30)$$

However, to perform this test quantitatively we must know the coupling constant a_3, one of the quantities we are trying to determine.

From the examples discussed so far, it appears difficult to think of reactions that isolate isoscalar axial vector. Particular importance is attached to this component because it is a genuinely new piece which makes no appearance in the charged-current interactions, nor in the electromagnetic interactions. Pais and Treiman [41] have pointed out that a careful study of elastic neutrino-deuteron scattering

$$\nu + D \rightarrow \nu + D \quad ,$$
$$\bar{\nu} + D \rightarrow \bar{\nu} + D \qquad (3.31)$$

may reveal the presence of isoscalar axial. First, let us note that these reactions proceed only via the isoscalar (I = 0) piece of the neutral currents. We examine whether the neutrino cross-section is the same as the antineutrino cross-section; any difference must be due to VA interference, which is possible only if the neutral currents contain both isoscalar axial and isoscalar vector. A practical difficulty with this proposal is that the VA interference term (W_3) vanishes as q^2 goes to zero while, because of the rapidly falling deuteron form factor, the cross-sections for these reactions are expected to be large only when q^2 is small.

It is also possible to settle the question of V vs. A, of isoscalar vs. isovector, by looking at diffractively produced vector and/or axial vector meson states, or by studying coherent elastic scattering processes off nuclei which isolate isoscalar vector. I will discuss these in Section 6.

4. Space-Time Properties - SPT Heresy

Before proceeding further with a more systematic discussion of the isospin and other internal properties of the neutral currents, I would like to discuss the heretical possibility, first entertained by Rosen, Kayser and others [42], that the observed neutrino-induced muonless events are due to "neutral density" interactions of the scalar, pseudoscalar and/or tensor (SPT) type. As is well known, the covariants $\bar{\nu}\nu$, $\bar{\nu}\gamma_5\nu$ and $\bar{\nu}\sigma_{\lambda\tau}\nu$ connect neutrino states of opposite helicity. In accelerator and reactor neutrino physics, the incident neutrino (antineutrino) is always left-handed (right-handed). So with S, P and T the outgoing neutrino (antineutrino) is right-handed (left-handed), in violation with the two-component condition for the neutrino. However, we may take the point of view that the success of the two-component neutrino theory is due to a special property of the charged-current interactions, not due to any intrinsic property of the neutrino. A priori there is nothing inconsistent with the idea that the neutrinos are actually ambidextrous but happen to use only their left hands when participating in the charged-current interactions.

For orientation purposes I begin by considering

$$\nu_\mu + e^- \rightarrow \nu_\mu + e^- \quad ,$$
$$\bar{\nu}_\mu + e^- \rightarrow \bar{\nu}_\mu + e^- \quad ; \qquad (4.1)$$

which are forbidden by the usual charged-current coupling. Ignoring the electron mass, we define the scaling variable y as

$$y \equiv E_e/E_\nu = E_e/E_e^{(max)} \tag{4.2}$$

where E_e stands for the electron recoil energy in the laboratory system. The variable y is related to the scattering angle in the centre-of-mass system by

$$\tfrac{1}{2}(1 + \cos\Theta_{c.m.}) = 1 - y \quad . \tag{4.3}$$

Note that y = 0 and y = 1 correspond to forward and backward scattering, respectively.

As is familiar from deep-inelastic phenomenology with quark partons, we have the following y distributions for $\nu_\mu e$ scattering

$$\begin{aligned} V-A &: \quad 1 \quad , \\ V+A &: \quad (1-y)^2 \ . \end{aligned} \tag{4.4}$$

For $\bar{\nu}_\mu e$ scattering interchange V-A with V+A. If the interaction is between pure V-A and pure V+A, we just take an appropriate linear combination of 1 and $(1 - y)^2$ with positive-definite coefficients. When written in this way, there is no interference term because states of different helicities cannot interfere. The main point to be observed here is that the y-distribution either is flat or falls with increasing y.

With S, P and/or T, quite different y-distributions are possible [42,43]:

$$\begin{aligned} S, P &: \quad y^2 \\ T &: \quad (1 - \tfrac{1}{2}y)^2 \\ \text{ST or PT interference} &: \quad y(1 - \tfrac{1}{2}y) \ . \end{aligned} \tag{4.5}$$

So, if we observe a y-distribution that rises with increasing y, then we will have unambiguous evidence for the presence of S and/or P. On the other hand, a falling y-distribution cannot be taken as proof of the VA combination.

If we now consider semileptonic inclusive reactions, there are as many as eight structure functions even with time reversal invariance. The double differential cross-section in the SPT case, taken from a paper by Pakvasa and Rajasekaran [44], reads as follows:

$$\frac{d\sigma}{dx\,dy} = (G^2 m_N E/\pi)\{xy^2(F_{SS}+F_{PP}) + 4\,[x(2-y)^2 F_{TT}^{(1)} + 4(1-y)F_{TT}^{(2)} - 2x^2y^2 F_{TT}^{(3)} + 2xy^2 F_{TT}^{(4)}] \pm 4xy(2-y)(F_{ST}+F_{PT})\} \tag{4.6}$$

where the structure functions depend on x and possibly on q^2 as well. To this we may add (3.3); there is no interference between SPT and VA because of opposite helicities in the final state. With such a complicated quadratic function of y, we can simulate any y-dependence produced by the VA case. Notice in particular that the ν-$\bar{\nu}$ difference is given by $y(1 - 1/2y)$ in both the SPT case and the VA case; there is a simple explanation for this in terms of s-u crossing [45]. So we have what is known as the "Confusion Theorem" [42]:

> "For any admixture of V and A interactions there is a corresponding admixture of S, P and T interaction, which yields the same ν and the same $\bar{\nu}$ cross-section".

Nevertheless there are certain y-dependences that cannot be simulated by VA interactions. A rising y-distribution would be impossible for VA unless we also have a scaling violation, e.g. due to a new particle threshold effect. On the other hand, a y-distribution going like y^2 is characteristic of S and/or P. More generally, if we have a linear combination of 1 and $(1 - y)^2$ as in spin-1/2 constituent models with VA, then we must have

$$A + C + \bar{C} = 0 \tag{4.7}$$

where A, C and $\bar{C}$ are coefficients that characterize the y-distribution, as defined below:

$$\begin{aligned} &A + By + Cy^2 && \text{for } \nu \quad , \\ &A + \bar{B}y + \bar{C}y^2 && \text{for } \nu \quad . \end{aligned} \tag{4.8}$$

A violation of (4.7) would imply, within the framework of quark-parton models, that some S, P and/or T must be present [44]. It is also worth mentioning that the range of R, the $\bar{\nu}$-to-ν cross-section ratio, is wider in the SPT case [44,46],

$$R_{min} = \frac{\sqrt{7}-2}{\sqrt{7}+2} \simeq 0.139 \; , \; R_{max} = \frac{\sqrt{7}+2}{\sqrt{7}-2} \simeq 7.20 \; . \tag{4.9}$$

So, if R falls outside the interval, 1/3 to 3, then we have evidence for SPT. The data mentioned in the previous section, however, show that this is an academic point.

It should be clear from the foregoing discussion that while it is relatively easy to test the pure S or P case, it is difficult to rule out, on the basis of the y-distribution, an arbitrary combination of S, P and T. Experimentally, already at the 1974 Downington Conference [47] it was noted that the hadronic energy distribution in the neutral current data of the HPW Collaboration [6] appeared to be inconsistent with a y-distribution rising like y^2. More recently, the Caltech Group [8] presented at the 1975 Paris Conference a hadron energy distribution that disagrees with the pure S and/or P hypothesis.

From a theoretical point of view, while it is easy to construct a simple intermediate boson theory that yields S or P type interactions, we cannot so easily build a boson model that leads to an effective tensor interaction as well. In the tensor case we may have to resort to a complicated Fierz-type transformation with an intermediate boson carrying lepton number as well as baryon number. So we have a natural theoretical bias against SPT combinations.

In exclusive channels such as elastic νp scattering the s-dependence at fixed q^2 is dependent on the space-time structure. The S and/or P case corresponds to spin-zero exchange between neutrinos and hadrons while the V, A and/or T case corresponds to spin-one exchange. So, as $s \to \infty$ with q^2 fixed, we must have

$$\begin{aligned} \frac{d\sigma}{dq^2} &\sim \text{const.} \; , \qquad && \text{V, A and/or T} \; ; \\ \frac{d\sigma}{dq^2} &\sim 1/s^2 \; , \qquad && \text{S and/or P} \; . \end{aligned} \tag{4.10}$$

Explicit model calculations with ST combinations show that ealstic νp scattering is expected to exhibit a bizarre energy dependence [48].

Low-energy neutrino physics with definite nuclear states may settle the question of S vs. V, or, of T vs. A, as emphasized by Kayser et al. [42]. For example, in elastic neutrino scattering off spin-zero nucleus the centre-of-mass angular distribution is $1 + \cos\theta$ for the vector case and $1 - \cos\theta$ for the scalar case; this can be seen to be a simple consequence of the helicity rule (no flip for V, flip for S) and angular momentum conservation. In a pure Gamow-Teller transition

$$\nu + {}^{12}C(0^+) \rightarrow \nu + {}^{12}C^*(1^+) \tag{4.11}$$

the helicity rule leads to angular distributions shown in Table 1.

Table 1

Angular distribution in inelastic νC scattering (4.11)

	A	T
$m_f = 0$	$1 + \cos\theta$	$1 - \cos\theta$
$m_f = -1$	$1 - \cos\theta$	$1 + \cos\theta$

The different angular distributions in the scattering process lead to different γ-ray spectra in the subsequent radiative decay of the excited nucleus

$$ {}^{12}C^*(1^+) \rightarrow {}^{12}C(0^+) + \gamma \ , \tag{4.12}$$

which should be measurable [42].

With a pseudoscalar density interaction we expect

$$\pi^0 \rightarrow \nu + \bar{\nu} \ , \tag{4.13}$$

which in the usual two-component neutrino theory is forbidden by the helicity rule. The branching ratio for this process relative to $\gamma\gamma$ may be estimated on dimensional grounds as follows [42]

$$(G m_\pi^2)^2/\alpha^2 \simeq 4 \cdot 10^{-8} \ . \tag{4.14}$$

This unusual decay mode might be looked for in

$$K^+ \to \pi^+ + \pi^0 \quad \hookrightarrow \nu + \bar{\nu} \quad . \tag{4.15}$$

There are a couple of miscellaneous remarks to be made before closing the subject of SPT heresy (once and for all!). Suppose we attempt to construct a parity-violating interaction that contains both S and P using the usual quark fields. We might try, for instance

$$\bar{\nu}\nu\,(g_s \bar{q} q + i g_P \bar{q}\gamma_5 q) \qquad (g_s, g_P \text{ real}) \tag{4.16}$$

where q may stand for u, d or s, and the factor i is required by Hermiticity. But this interaction violates time reversal invariance. So, in the SP case parity violation appears to imply time reversal violation [49]. This conclusion, however, is not quite correct if we are willing to tolerate non-diagonal covariants. For example, we may consider, in place of (4.16),

$$\bar{\nu}\nu\,[g_s(\bar{q}_a q_a + \bar{q}_b q_b) + g_P(\bar{q}_a\gamma_5 q_b - \bar{q}_b\gamma_5 q_a)\;, \qquad (g_s, g_P \text{ real}) \tag{4.17}$$

without violating time reversal invariance. The price we have to pay is that we must introduce q_a and q_b which have the same charge and strangeness. Experimental tests for the presence of such non-diagonal covariants can be found in a paper by Kingsley et al. [50].

Another peculiar feature of the SPT case, discussed by Pakvasa and Rajasekaran [44], has to do with dimensionality and scale invariance. As long as we have spin-1/2 constituents, the S, P, T covariants - $\bar{q}q$, $\bar{q}\gamma_5 q$, $\bar{q}\sigma_{\lambda\tau}q$ - have the same dimension as the vector and axial-vector covariants, $\bar{q}\gamma_\lambda q$ and $\bar{q}\gamma_\lambda\gamma_5 q$. As a result, point-like couplings imply scale invariance à la Bjorken even in the SPT case. However, if we admit spin-zero constituents the situation is very different. Even though the vector covariant constructed out of non-Hermitian spin-zero fields, $\partial_\lambda\phi^+\phi - \phi^+\partial_\lambda\phi$ has the same dimension, viz. 3, as $\bar{q}\gamma_\lambda q$, the scalar covariant $\phi^+\phi$ has dimension 2 while the antisymmetric tensor covariant, $\partial_\lambda\phi^+\partial_\tau\phi - \partial_\tau\phi^+\partial_\lambda\phi$, has dimension 4. As a result, point-like couplings do not imply a linearly rising cross-section. For spin-zero partons with a tensor coupling it is actually more natural to expect a cross-section rising like E^2.

5. Isospin and SU(3) Properties

I now turn to a detailed discussion of the internal symmetry properties of the hadronic part of the neutral currents. It is a well-established experimental fact that the strangeness-changing neutral current is either absent or highly suppressed, e.g. [51)]

$$\frac{\Gamma(K^+ \to \pi^+ + e^+ + e^-)}{\Gamma(K^+ \to \text{all})} = (2.3 \pm 0.8) \cdot 10^{-7} \quad . \tag{5.1}$$

We are therefore dealing with a current whose quantum numbers are Q = 0, S = 0, B = 0. The Gell-Mann/Nishijima rule then requires $I_3 = 0$. This means that the following possibilities are open:

$$I = 0, 1, 2, \cdots \tag{5.2}$$

We are prejudiced against I = 2 or higher for two reasons. First, in simple quark models it is impossible to construct a bilinear current with $I \geq 2$. Second, there does not seem to be evidence for an isotensor current in the electromagnetic interactions [52)], nor in the charged-current weak interactions [53)]. In any case, the possible existence of an $I \geq 2$ piece in the neutral currents can be checked by examining whether the following "quadrangular relation" and its permutations are violated [54)]:

$$\sqrt{\sigma(\nu p \to \nu \pi^+ n)} \leq \sqrt{\sigma(\nu n \to \nu \pi^- p)} + \sqrt{2\sigma(\nu p \to \nu \pi^0 p)} + \sqrt{2\sigma(\nu n \to \nu \pi^0 n)} . \tag{5.3}$$

In the following we assume that the hadronic part of the neutral currents is isovector (I = 1) and/or isoscalar (I = 0).

It has been recognized for some time that the cleanest way to determine the isospin of the neutral current is to study the cross-section ratios in single-pion production [55)]

$$\begin{aligned} \nu + p &\to \nu + \pi^0 + p \,, \\ \nu + p &\to \nu + \pi^+ + n \,, \\ \nu + n &\to \nu + \pi^0 + n \,, \\ \nu + n &\to \nu + \pi^- + p \,. \end{aligned} \tag{5.4}$$

If the current is pure isoscalar, only I = 1/2 πN final states are possible. So we expect

$$\sigma(\pi^0 p):\sigma(\pi^+ n):\sigma(\pi^0 n):\sigma(\pi^- p) = 1:2:1:2 . \tag{5.5}$$

The equalities

$$\sigma(\pi^0 p) = \sigma(\pi^0 n) \quad , \quad \sigma(\pi^+ n) = \sigma(\pi^- p) \tag{5.6}$$

also follow in the pure isovector case. So violations of (5.6) would imply that both isoscalar and isovector pieces must be present. If we work in the $W \simeq 1236$ MeV region, it may be reasonable to assume that the reactions (5.4) are dominated by $\Delta(1236)$ provided, of course, that the current is predominantly isovector. We then expect

$$\sigma(\pi^0 p):\sigma(\pi^+ n):\sigma(\pi^0 n):\sigma(\pi^- p) = 2:1:2:1 , \tag{5.7}$$

in sharp contrast with (5.5).

Experimentally the ANL Bubble Chamber Group [7)] has reported

$$\frac{\sigma(\nu p \to \nu \pi^0 p)}{\sigma(\nu p \to \nu \pi^+ n) + \sigma(\nu n \to \nu \pi^- p)} = 1.7 \pm 1.1 . \tag{5.8}$$

Since this ratio should be 1/4 for I = 1/2 πN and 1 for I = 3/2 πN, it appears that the isovector hypothesis is favoured. However, this conclusion is not universally accepted because the cross-section for $\nu \pi^0 p$ used in the numerator of (5.8) is several times larger than an upper limit obtained in the old CERN propane-chamber experiment [56)] and also disagrees with more recent $\pi^0 N$ data of the Columbia-Illinois-Rockefeller Collaboration [9)] and the Gargamelle Collaboration [57)] unless nuclear charge-exchange corrections in these experiments are unreasonably large.

A more direct method for detecting the presence of isovector is to examine the π^0p (or π^-p) mass distributions to see whether $\Delta(1236)$ is excited. If the current is pure isoscalar, Δ production without extra π would, of course, be forbidden. Recently the Columbia-Illinois-Rockefeller Collaboration [9)] working at Brookhaven has compared the π^0p mass distributions in

$$\nu + p \to \nu + \pi^0 + p , \tag{5.9a}$$

$$\nu + n \to \mu^- + \pi^0 + p , \tag{5.9b}$$

where aluminium plates are used as targets. The results are shown in Fig. 4. A clear signal for $\Delta(1236)$ is visible in the charged-current data. In contrast the $\pi^0 p$ mass distribution in the neutral-current data is inconclusive as regards the presence or absence of Δ. The best thing we can say is that you may draw your own conclusion. The $\pi^0 p$ mass distributions studied by the Gargamelle Collaboration [57)], shown in Dr. von Krogh's lectures, are also inconclusive.

There are other exclusive channels such as N(1530), ηp, ρp, Ap, ωp, etc., which are relevant to the question of isospin determination. Some of these channels will be discussed in Section 6.

I would now like to mention some general results that follow from simple isospin considerations. In pure isovector models where the neutral current and the charged current are related by isospin rotations [see (3.13)], we can derive clean relations which may be of some practical interest. I already remarked in Section 3 that for deep-inelastic scattering on targets composed of equal numbers of protons and neutrons the neutral-to-charged-current ratio R is given by

$$
\begin{aligned}
R(X) &\equiv \frac{\sigma(\nu p \to \nu X^+) + \sigma(\nu n \to \nu X^0)}{\sigma(\nu p \to \mu^- X^{++}) + \sigma(\nu n \to \mu^- X)} \\
&= \frac{1}{2} \, \frac{v_3^2 V + a_3^2 A + v_3 a_3 I}{V + A + I} \, .
\end{aligned}
\tag{5.10}
$$

This formula actually holds outside the deep-inelastic region for any X <u>as long as we sum over all possible charge configurations</u>; it does not matter here whether the isospin of X is 1/2, 3/2, or mixed. On the other hand, if we select isospin-1/2 final states, we get

$$
\begin{aligned}
R(N^*_{1/2}) &\equiv \frac{\sigma(\nu p \to \nu N^{*+}_{1/2}) + \sigma(\nu n \to \nu N^{*0}_{1/2})}{2\sigma(\nu n \to \mu^- N^{*+}_{1/2})} \\
&= \frac{1}{4} \, \frac{v_3^2 V + a_3^2 A + v_3 a_3 I}{V + A + I} \, ,
\end{aligned}
\tag{5.11}
$$

and, if we isolate isospin-3/2 final states of Q = 0 and +1 so that a direct comparison may be made with (5.11), then the analogous ratio is given by

$$R(\Delta^{+,0}) \equiv \frac{\sigma(\nu p \to \nu\Delta^+) + \sigma(\nu n \to \nu\Delta^0)}{2\sigma(\nu n \to \mu^-\Delta^+)}$$

$$= \frac{v_3^2 V + a_3^2 A + v_3 a_3 I}{V + A + I} \quad . \tag{5.12}$$

In (5.11) and (5.12) $N^*_{1/2}$ and Δ may stand for any isospin-1/2 and isospin-3/2 final system, not necessarily a nucleon isobar of definite spin-parity. Note that there is a factor of four difference between (5.11) and (5.12). These formulas will become important when we discuss Δ and $N^*_{1/2}$ excitation in the next section.

There are very simple isospin tests for pion inclusive reactions

$$\nu + N \longrightarrow \nu + \pi^{\pm,0} + X \tag{5.13}$$

where, as usual, N stands for the nucleon averaged over equal numbers of protons and neutrons. For a pure isoscalar current we obviously expect

$$\sigma(\pi^+) : \sigma(\pi^-) : \sigma(\pi^0) = 1:1:1 \ . \tag{5.14}$$

Departure from (5.14) is measured by the combination

$$\sigma(\pi^+) + \sigma(\pi^-) - 2\sigma(\pi^0) \tag{5.15}$$

which isolates the isovector contribution. Isoscalar-isovector interference can be studied by looking at [58)]

$$A_\nu \equiv \sigma(\nu N \to \nu\pi^+ X) - \sigma(\nu N \to \nu\pi^- X) \tag{5.16}$$

and

$$A_{\bar\nu} \equiv \sigma(\bar\nu N \to \bar\nu\pi^+ X) - \sigma(\bar\nu N \to \bar\nu\pi^- X) \quad . \tag{5.17}$$

The sum and the difference of A_ν and $A_{\bar\nu}$ are proportional to products of isoscalar and isovector constants as follows:

$$A_\nu + A_{\bar\nu} \propto v_3 v_s \ , \ a_3 a_s \ ,$$
$$A_\nu - A_{\bar\nu} \propto v_3 a_s \ , \ v_s a_3 \ . \tag{5.18}$$

Experimental data relevant to these isospin tests are still scarce but at the 1975 SLAC Symposium Morfin mentioned a preliminary Gargamelle value for the $\pi^+ : \pi^- : \pi^0$ ratio for $p_\pi > 1$ GeV/c which appears to be more like 1 : 1.1 : 1.8 than 1 : 1 : 1. So models based on pure isoscalar currents will be in grave difficulty if further data confirm this preliminary ratio.

Neutrino experiments with hydrogen and deuterium bubble chambers, of the kind now in progress at Brookhaven and at Fermilab, will give separate data on proton and neutron targets. If the neutral current is isospin pure, i.e. pure I = 1 or pure I = 0, we, of course, expect

$$\sigma(\nu p \to \nu X^+) = \sigma(\nu n \to \nu X^0) \ ,$$
$$\sigma(\bar\nu p \to \bar\nu X^+) = \sigma(\bar\nu n \to \bar\nu X^0) \tag{5.19}$$

where X may stand for any final state (inclusive or exclusive). Violations of (5.19) would provide conclusive evidence that both isovector and isoscalar currents are present.

I now turn to the SU(3) and other higher symmetry properties of the neutral currents. The first important point to be examined is whether the isoscalar piece, if present, belongs to the same SU(3) octet as the charged current, or is something new. In the quark model language we may inquire whether the current goes like the hypercharge current (and/or its axial analog),

$$\tfrac{1}{3}(\bar u u + \bar d d) - \tfrac{2}{3}\bar s s \ , \quad (\gamma_\lambda \text{ or } \gamma_\lambda\gamma_5 \text{ omitted}) \tag{5.20}$$

or the baryon current (and/or its axial analog) which, of course, is an SU(3) singlet:

$$\tfrac{1}{3}(\bar u u + \bar d d + \bar s s) . \quad (\gamma_\lambda \text{ or } \gamma_\lambda\gamma_5 \text{ omitted}) \tag{5.21}$$

The difference between (5.20) and (5.21) is $\bar{s}s$, just a ϕ -like piece. To complicate the situation, charm enthusiasts may speculate how $\bar{c}c$, a ψ -like piece, may enter.

Quark diagram ("Zweig rule") considerations [59] show that

$$\text{current} + N \rightarrow N^* \tag{5.22}$$

is insensitive to the presence or absence of a ϕ -like component in the current. Invoking duality, we see that two-body reactions with ordinary Regge exchange are also insensitive to the presence or absence of a ϕ -like component; this is consistent with the experimental observation that the total ϕ N cross-section deduced from ϕ-meson photoproduction is flat, as expected from pure Pomeron exchange. We therefore expect that neither resonance excitations nor non-Pomeron exchange two-body reactions throw light on the SU(3) properties of the isoscalar neutral current. Deep inelastic scattering does not seem to be helpful either because the success of the naive quark-parton model indicates that the nucleon is made up mostly of non-strange valence quarks; indeed, it was because of this that we could derive in Section 3 the expressions for v_s and a_s in terms of the observable cross-section cf. (3.21) without specifying the SU(3) properties of the isoscalar current.

To discriminate (5.20) from (5.21) we must rely on Pomeron exchange reactions that violate the Zweig rule [60]. One possibility is to compare diffractive ϕ and ω production; this will be discussed in the next section.

In a model with a large isoscalar vector part it is of interest to consider the isoscalar analog of the CVC hypothesis where we compare the isoscalar part of

$$e^- + p \rightarrow e^- + N^{*+}_{1/2} \tag{5.23}$$

with the isoscalar vector part of

$$\nu + p \rightarrow \nu + N^{*+}_{1/2} \tag{5.24}$$

Under the same kinematical conditions, we must have [60]

$$\sigma^{(\text{isoscalar vector})}(\nu p \rightarrow \nu N^{*+}_{1/2}) = \frac{G^2}{2\pi^2\alpha^2} v_s^2 q^4 \sigma^{(\text{isoscalar})}(e^- p \rightarrow e^- N^{*+}_{1/2}) \,. \tag{5.25}$$

If the isoscalar neutral current is proportional to the isoscalar electromagnetic current, this relation is self-evident. The non-trivial point I wish to emphasize is that even if the isoscalar vector part goes like the baryon current rather than the hypercharge current, or even if it has some extra piece involving charmed quarks, we will expect (5.25) to hold for a large class of I = 1/2 states whose production mechanism does not violate the Zweig rule. I have tried to look for practical applications of (5.25) in N^* production; unfortunately the isospin properties of resonance electroproduction in the second and third resonance region are too poorly known to provide meaningful estimates on N^* production by the neutral-current interactions.

I would like to finish this section by examining the U-spin properties of the neutral currents. To motivate our discussion let us recall that in the quark model language the charged-current interactions make use of Cabibbo-rotated quark fields

$$\begin{aligned} d_C &= d\cos\Theta_C + s\sin\Theta_C \quad , \\ s_C &= s\cos\Theta_C - d\sin\Theta_C \quad . \end{aligned} \tag{5.26}$$

Cabibbo's proposal [61)] to use d_C rather than d may be justified by saying that in the absence of the medium strong interactions that pick up the "eighth" direction, it would be just as plausible to use the (u,d_C,s_C) set as the original (u,d,s) set.

In the neutral-current interactions we may wonder whether nature takes advantage of (u,d_C,s_C) or (u,d,s). One possibility which we may contemplate is that the neutral-current interactions are such that it should not matter which set we use. Since the transformation that carries (u,d,s) into (u,d_C,s_C) is a U-spin rotation, this hypothesis amounts to requiring invariance under U-spin rotations.

Quite generally we have two U-spin invariants. The first goes like

$$\bar{u}u + \bar{d}d + \bar{s}s \quad , \tag{5.27}$$

which, being a unitary singlet, is invariant under Cabibbo-type rotations of any kind; the vector current with this property is just the baryon current. We can also construct

$$\tfrac{2}{3}\bar{u}u - \tfrac{1}{3}\bar{d}d - \tfrac{1}{3}\bar{s}s \quad , \tag{5.28}$$

which has the same transformation properties as the electromagnetic current, a well-known U-spin scalar. So the most general neutral current invariant under U-spin rotations must behave under SU(3) transformations like a linear combination of the baryon current and the electromagnetic current. It is amusing that the requirement of U-spin invariance automatically rules out a strangeness-changing neutral current. Historically the importance of invariance under Cabibbo-type rotations was first stressed within the context of a model in which the hadronic part of the neutral currents is postulated to be the baryon current [55]. Subsequently Mathur, Okubo and Kim [62] derived a host of relations that follow from U-spin invariance, e.g.

$$\sqrt{\sigma(\nu p \to \nu \pi^0 p)} \leq \sqrt{2\sigma(\nu p \to \nu K^0 \Sigma^+)} + \sqrt{\sigma(\nu p \to \nu K^+ \Sigma^0)}\ , \tag{5.29}$$

and emphasized, among other things, that the necessary and sufficient condition for the absence of strangeness-changing neutral currents is that the neutral currents transform like a U-spin scalar.

Even though the transformation properties under U-spin rotations are not emphasized in the usual discussion of gauge theory models, we can analyse the neutral currents in those models using the U-spin language. For example, let us look at the famous Glashow-Iliopoulos-Maiani construction [63] where the neutral current is made up of a linear combination of

$$(\bar{u}u - \bar{d}d) + (\bar{c}c - \bar{s}s) \tag{5.30}$$

and the electromagnetic current. It is amusing that this is automatically U-spin invariant with the understanding that the newly added fourth quark is unaffected by U-spin rotations.

6. Exclusive Semileptonic Reactions

One of the most important neutral current reactions yet to be studied is elastic νp scattering

$$\nu + p \to \nu + p\ , \tag{6.1}$$

which is as fundamental as neutron beta decay. At $q^2 = 0$ various neutral-current models give definite and unambiguous predictions for the neutral-to-charged current ratio as follows:

$$R_{el}\Big|_{q^2=0} \equiv \left. \frac{\frac{d\sigma}{dq^2}(\nu p \to \nu p)}{\frac{d\sigma}{dq^2}(\nu n \to \mu^- p)} \right|_{q^2=0}$$

$$= \begin{cases} \dfrac{v_3{}^2 + (1.25)^2 a_3{}^2}{4[1+(1.25)^2]} , & \text{pure isovector} ; \\ \dfrac{(1 - 4\sin^2\theta_W)^2 + (1.25)^2}{4[1+(1.25)^2]} , & \text{Weinberg-Salam} ; \\ \dfrac{v_S{}^2}{1+(1.25)^2} , & \text{isoscalar vector} . \end{cases} \tag{6.2}$$

I do not give here predictions for models with isoscalar axial currents because the matrix element of the isoscalar axial current ("8th" or "9th") between the physical nucleons is model-dependent. Those who are interested in this subject may consult a paper by Adler and collaborators [64].

Numerically with values of the coupling constants consistent with the deep inelastic data we obtain for $R_{el}|_{q^2=0}$ 0.11 to 0.18 in pure isovector models - the lower value (higher value) corresponding to the V(A) dominant solution - 0.15 to 0.18 in the Weinberg-Salam model, and a much larger value, 0.7-1.0, in the isoscalar vector ("baryon current") model.

As we go away from $q^2 = 0$, the ratio becomes dependent on both q^2 and the incident neutrino energy E. This is because form factor dampings do not affect the numerator and the denominator in the same manner, and, what is worse, VA interference effects (absent at $q^2 = 0$) are strongly dependent on E and also affect the numerator and the denominator differently. This is unfortunate. The experimentalists are forced to make q^2 cuts because very small q^2 events are (i) difficult to detect and (ii) quite often severely contaminated by neutron background due to elastic np scattering. As a result, it is very dangerous to identify the measured ratio with the $q^2 = 0$ predictions given by (6.2). For example, the pure isoscalar vector model that predicts $R_{el} \simeq 0.7$ at $q^2 = 0$ gives $R_{el} \simeq 0.17$ with $q^2 > 0.4$ GeV^2, E = 2 GeV. Sample calculations that illustrate this point are presented in Fig. 5 [65].

Experimentally there are limits on elastic νp scattering quoted by the ANL Bubble Chamber Group [7] and also by the old CERN Propane Chamber Group [66], but they are not sufficiently stringent to provide critical tests of the various models. An experiment specifically designed for elastic νp scattering is now in progress at Brookhaven by the HPW Collaboration.

Let us now turn to single-pion production near threshold. This is a good place to apply soft-pion techniques based on PCAC and current algebra. Arguments developed in the sixties lead us to believe that the single-pion production matrix element is the sum of a commutator term and a nucleon pole term. The relevant commutator is between the neutral current and the axial charge of the same isospin component as the emitted pion; it vanishes for $\pi^{\pm,0}$ production with an isoscalar neutral current and also for π^0 production even with an isovector current. The nucleon pole term is obtained by attaching a pion line to the external nucleon line of the elastic νp scattering matrix element using the gradient coupling prescription. At threshold there is a particularly simple relation between single-pion production and elastic scattering, first derived by Adler [67], applicable to any model in which the commutator term vanishes:

$$\frac{1}{|\vec{q}_\pi|}\left.\frac{d\sigma(\nu N\to\nu\pi N)}{dq^2 dW}\right|_{\text{thresh.}} = \begin{Bmatrix}1\\2\end{Bmatrix}\frac{G^2_{\pi NN}}{4\pi}\,\frac{q^2}{4m_N^2\pi}\,\cdot \tag{6.3}$$

$$\cdot\,\frac{(1+q^2/4m_N^2)}{(1+q^2/2m_N^2)^2}\,\frac{d\sigma}{dq^2}(\nu p\to\nu p)\quad\text{for}\ \begin{Bmatrix}\pi^0\\ \pi^\pm\end{Bmatrix}\,.$$

where $\vec{q}_\pi$ is to be evaluated in the πN rest system and $G^2_{\pi NN}/4\pi \approx 14$. Initial attempts to apply this formula to preliminary ANL data on single-pion production caused a great deal of excitement because the measured cross-section near threshold was an order of magnitude larger than is implied by the elastic scattering limit for any combination of V and A. This stimulated a tremendous amount of theoretical activities; some of the ablest young men and women at the Institute for Advanced Study were drafted to Adler's Army to make detailed examinations of various alternative possibilities such as SPT combinations [48] and second-class VA currents [68]. The preliminary ANL data which stimulated these activities, however, have subsequently been withdrawn.

There have been many theoretical attempts to estimate the neutral-to-charged current ratio in $\Delta(1236)$ production. Even though most authors investigate this process within the framework of the Weinberg-Salam model, it is just as easy to generalize their results for other models with predominantly isovector currents. The basic equation is (5.12), applicable even if v_s and/or a_s are non-vanishing.

The quantities V, A and I can be estimated from static model calculations or more sophisticated dispersion-theoretic calculations, which adequately account for charged-current Δ production and Δ electroproduction. One of the earliest calculations along this line was performed by B.W. Lee [69] who obtained

$$V : A : I = 0.263 : 0.202 : 0.235 \tag{6.4}$$

at E = 1 GeV. If we use the values of the coupling constants indicated by the deep-inelastic data, we obtain

$$R(\Delta^{+,0}) = 0.4 - 0.5 \quad . \tag{6.5}$$

The Weinberg-Salam model prediction for the ratio is also in this range with the currently accepted value of $\sin^2\theta_W$.

Experimentally the quantity that is most easily determined is R_0' defined by

$$R_0' \equiv \frac{\sigma(\nu + "N" \rightarrow \nu + "N" + \pi^0)}{2\sigma(\nu + "N" \rightarrow \mu^- + "N" + \pi^0)} \quad , \tag{6.6}$$

where the symbol "N" reminds us that the initial nucleon is in a complex nucleus and that the final nucleon may be mixed up among nuclear disintegration products. The most recent experimental numbers for this ratio are

$$R_0' = \begin{cases} 0.17 \pm 0.04 \quad , \text{Columbia-Illinois-Rockefeller} \quad , \\ 0.07 - 0.22 \quad , \text{Gargamelle} \quad . \end{cases} \tag{6.7}$$

If we had pure Δ production and nuclear-charge exchange corrections could be ignored, then R_0' would be equal to $R(\Delta^{+,0})$ defined by (5.12). Unfortunately the charged-current data on single-pion production indicate that there are sizeable I = 1/2 contributions even in the $\Delta(1236)$ region; this is inferred from the fact that the $\pi^+p : \pi^0p : \pi^+n$ ratios do not quite satisfy the famous 9 : 2 : 1 ratios expected for pure I = 3/2 [53]. Second, there are theoretical and experimental reasons to believe that nuclear charge-exchange effects, first emphasized by Perkins [70], are, in fact, important. I will not go into many of the detailed calculations [71] that have been performed to attack these two problems. The concensus of the experts is that when these two effects are taken into account, there is no serious discrepancy between the theoretically expected ratio (6.5) and the experimental ratio (6.7).

All this may turn out to be an idle exercise if Δ is not seen. As mentioned in the previous section, the experimental $\pi^0 p$ mass distribution is still inconclusive as regards the presence or absence of a Δ signal. A Δ peak shows up strongly in both charged-current single-pion production and single-pion electroproduction, and an equally strong Δ signal must be present in the neutral-current case as long as the current is predominantly isovector. Paschos [31] has recently examined the effect of I = 1/2 contaminations on the Δ shape by taking into account the I = 1/2 amplitude deducted from the charged-current data. His main conclusion is that the I = 1/2 contaminations do not seriously alter the Δ resonance peak. In fact, simple isospin considerations based on (5.11) and (5.12) show that the distortions in the $\pi^0 p$ mass distribution due to the I = 1/2 background should be less serious in the neutral-current data. Failure to observe a conspicuous Δ peak in the $\pi^0 p$ (or $\pi^- p$) mass distribution would constitute a serious argument against models based on predominantly isovector currents.

If we now increase the invariant hadronic mass, many I = 1/2 resonances are expected, according to the Particle Data Table, in the 1400-1700 range. Our theoretical understanding of resonance excitation in this mass range is on much less secure grounds. However, I believe that it is safe to assume: (i) VA interference continues to be constructive in the charged-current neutrino interaction, and (ii) V and A are again of the same order of magnitude. These features are strongly supported by the fact that the famous one-to-three ratio for the $\bar{\nu}$ -to- ν cross-section ratio holds even at low energies, which may be understood as just another manifestation of duality of the Bloom-Gilman type [72]. Quark model calculations also support these conclusions; for instance, in Ravndal's work [73] the right-handed cross-section σ_R is essentially negligible for every $N^*_{1/2}$ resonance.

With this view in mind let us look once again at (5.11). We can compute $R(N^*_{1/2})$ in pure isovector models assuming that V = A and $0 \leq I/V \leq 2$, as required by positivity. With values of v_3 and a_3 consistent with the deep inelastic data, a typical number we get for $R(N^*_{1/2})$ is between 0.1 and 0.15. In the Weinberg-Salam model we must add contributions due to the isoscalar electromagnetic current, but even in this case it is highly unlikely that $R(N^*_{1/2})$ exceeds 1/4. This simple argument [60] is supported by much more elaborate model calculations recently made by Adler and collaborators [74].

A comparison of (5.11) and (5.12) reveals that if the relative ratios of V, A and I are unchanged throughout, then the neutral-to-charged current ratio for $\Delta^{+,0}$ or for $N^{*+,0}$ must decrease by a factor of 4 as we go from the I = 3/2 dominance region to the I = 1/2 dominance region. In reality the change would not be so dramatic; after all we expect some I = 1/2 background in the Δ(1236) region; some I = 3/2 background in the 1400-1700 region. However, it would be interesting to examine the neutral-to-charged current ratio for $\pi^0 N$ as a function of the invariant mass W to see if there is any noticeable decrease in the ratio as we go away from the Δ(1236) region. Other channels of particular interest are ηN and $K\Lambda$, which are necessarily in the I = 1/2 state.

We now turn to diffractive channels, the importance of which was first stressed in the charged-current case by Piketty and Stodolsky [75]. As is well known, ρ^0 , ω and ϕ show up very clearly in photoproduction and also, to some extent, in low-q^2 electroproduction. Moreover, these vector meson reactions are known to exhibit all the features of diffractive processes with Pomeron exchange even though some one-pion-exchange contribution is still present in ω production with $\nu \lesssim 6$ GeV. We therefore expect, in low-q^2 neutral-current reactions, diffractive production of vector and/or axial-vector mesons of the same quantum numbers as the current - ρ^0 for isovector vector, ω and ϕ for isoscalar vector, A_1 (if it exists) for isovector axial-vector, etc. Indeed this may be the only practical way to determine whether V or A dominates in the semileptonic neutral-current interactions.

A comparison of diffractive ω and ϕ production is relevant to the question of whether the isoscalar vector current (if present) goes like the hypercharge current or like the baryon current [55]. If the isoscalar vector current is proportional to the hypercharge current, then we, of course, expect that the ω-to-ϕ ratio in diffractive vector meson production is the same as the corresponding ratio in the electroproduction reactions. In the baryon current case we expect that this is given by the electroproduction ratio times 4, which is the fourth power of the cotangent of the $\omega\phi$ mixing angle, assumed to be "ideal". (It is hoped here that the symmetry violating effects cancel out because we are comparing the ratios.) Unfortunately, away from $q^2 = 0$, the existing data on the electroproduction of ω and ϕ are not sufficiently accurate to enable us to predict with firmness the percentage of ω and ϕ expected in neutral-current reactions.

It is possible to make more detailed estimates on the q^2-dependence, etc., of these diffractive channels using vector (and/or axial vector) meson dominance. Some model calculations along this line were carried out by Gaillard, Jackson and Nanopoulos [76]. Those who are interested in this may consult Dr. Nanopoulos' lecture notes which contain figures on the fractions of the diffractively produced states expected at various energies. In the Salam-Weinberg model the A_1^o is the most important diffractive channel comprising several per cent of the total cross-section at $E \simeq 10$ GeV. It goes without saying that before we start seeing these marvellous diffractive states in the neutral-current data, we must first observe ρ^+ and A_1^+ in the charged-current data.

One of the most interesting possibilities opened up by the discovery of neutral current phenomena is that neutrinos may now be scattered coherently off nuclei with an A-dependence going like A^2, where A stands for the atomic number. Consider, for simplicity, target nuclei made up of equal numbers of protons and neutrons. The isovector vector contribution, being proportional to v_3 for the proton and $-v_3$ for the neutron, has no net effect as far as coherence processes are concerned. However, the isoscalar vector contribution is proportional to $+v_s$ for both the proton and the neutron; therefore we expect a coherent cross-section proportional to $v_s^2A^2$. The axial piece does not contribute because the spin orientations of the individual nucleons inside the target nucleus do not yield any coherence effect. So the coherent cross-section is proportional to $v_s^2A^2$, and such scattering processes, in principle, give the most direct means of isolating the isoscalar vector piece of the neutral current.

Quantitative considerations along this line were first worked out by Freedman [77)] in the Fall of 1973 immediately after he attended a seminar I gave at NAL on the baryon current model of weak neutral currents. The differential cross-section for

$$\nu + A \rightarrow \nu + A \tag{6.8}$$

is given by

$$\frac{d\sigma}{dq^2} = \frac{G^2}{2\pi} v_s^2 A^2 e^{-2bq^2} \left[1 - q^2 \frac{2ME + M^2}{4M^2E^2}\right] \tag{6.9}$$

for a nuclear target of mass M with a nuclear density form factor assumed to be of the form Ae^{-bq^2}. The quantity inside the square bracket just shows that there is a complicated way of writing $(1 + \cos\Theta_{cm})/2$.

In planning experiments relevant to coherent scattering there are some practical considerations that must be kept in mind. First, q^2 is related to the kinetic energy E_{KE} of the recoil nucleus in the laboratory system via $q^2 = 2M\,E_{KE}$; so it is difficult to observe extremely low-q^2 events, say, $\sqrt{q^2} < 100$ MeV/c. For high q^2 the form factor drops off sharply; this, of course, reflects the fact that if we attempt to explore the target nucleus using a current whose "de Broglie wavelength" is much shorter than the nuclear radius, $R = \sqrt{6b}$, then the target is likely to break up. As a numerical example, if we take a carbon nucleus of radius $R = 2.42 \times 10^{-13}$ cm, the cross-section for $\sqrt{q^2}$ between 100 and 200 MeV/c is $v_s^2 \times 11.2 \times 10^{-39}$ cm^2 at E = 200 MeV. We may recall here that v_s^2 is in the neighbourhood of 0.1-0.2 in the Weinberg-Salam model [see (3.22)] and about 2 in models with a pure isoscalar vector current.

At extreme low energies, i.e. in the MeV range, the nucleon density form factor can be regarded as constant, and we get

$$\left(\frac{d\sigma}{d\Omega}\right)_{c.m.} = v_s^2 G^2 A^2 E^2 (1 + \cos\Theta) \quad . \tag{6.10}$$

Note that backward scattering $\Theta = \pi$ is forbidden by angular momentum conservation - no helicity change in the VA interaction; if the interaction were scalar, we would still expect coherent scattering but with an angular distribution going like $1 - \cos\Theta$. Numerically, with a 10 MeV ν beam on A = 50 nuclei, we get a total coherent cross-section of $v_s^2 \times 3.7 \times 10^{-39}$ cm^2.

Coherent neutrino scattering has important implications in an unexpected domain. Prior to the discovery of neutral currents there were many attempts to estimate neutrino cross-sections in stellar matter. First, there are low-energy charged-current processes due to inverse beta decay with cross-sections estimated to be $\sim 10^{-43 \pm 1}$ cm^2 [78]. Second, there is νe scattering which, for an electron gas of $kT \simeq 2$ MeV, gives a cross-section in the neighbourhood of typically 2×10^{-42} cm^2 [79]. Let us compare these numbers with what we expect from coherent scattering. With v_s^2 of order 0.1 to 1 the cross-section implied by (6.10) is two to three orders of magnitude larger. As a result, stellar matter may become opaque to neutrinos at lower than conventional density.

All this may have interesting applications on the question of stellar collapse and supernova explosion. The importance of neutrino flow in gravitational collapse of a star was recognized as early as 1964 by Hoyle and Fowler [80]. Subsequently many calculations [81] were made to see whether in stellar collapse the neutrinos that escape from the central region where free nucleons exist can interact sufficiently strongly with the outer region of high atomic weight (Fe-Ni) so that the stellar envelope may be blown off with a neutron star left behind. What is needed is a very high cross-section off heavy nuclei; at the same time the cross-section off free individual nucleons should not be too high if the neutrinos are to excape from the inner core. Pre-Gargamelle mechanisms did not appear to be adequate.

The coherent effect discussed by Freedman [77] altered the situation dramatically. If neutrinos get scattered coherently by heavy nuclei, an explosion of the stellar envelope may become more probable. Detailed calculations by Wilson and others [82] show that, with $v_s^2 \simeq 0.1-0.2$ suggested by the Weinberg-Salam model, a supernova explosion is <u>not</u> possible, but if v_s^2 is as large as 1 or 2 as in a model with a predominantly isoscalar vector current, a violent explosion is comfortably predicted. For quantitative estimates we must consider the effect of non-isoscalar pieces because the effect is sensitive to the ratio of the neutrino-heavy-nucleus cross-section to the neutrino-neutron cross-section. I am unable to judge the reliability of such calculations; to do justice to this subject we need an entire lecture course devoted to the problem of neutrino flow and gravitational collapse, of the kind given by Wilson [83] at the 1975 Varenna Summer School. But I cannot help get impressed by "unity of science" when I realize that our exploration into short distances is relevant after all to our understanding of large-scale phenomena such as supernova explosion.

7. Neutrino-Electron Scattering

The standard current-current structure of weak interactions implies that semileptonic neutral-current processes will most likely be accompanied by pure-leptonic neutral-current processes. Indeed, the first hint of the existence of neutral currents came from one event, discovered by the Gargamelle Collaboration [12], interpretable as $\bar{\nu}_\mu e^-$ scattering.

In discussing pure-leptonic neutral-current reactions let us first note that there are four ν- and $\bar{\nu}$-induced reactions of interest:

$$\nu_e + e^- \rightarrow \nu_e + e^- \quad , \tag{7.1a}$$

$$\bar{\nu}_e + e^- \rightarrow \bar{\nu}_e + e^- \quad , \tag{7.1b}$$

$$\nu_\mu + e^- \rightarrow \nu_\mu + e^- \quad , \tag{7.1c}$$

$$\bar{\nu}_\mu + e^- \rightarrow \bar{\nu}_\mu + e^- \quad . \tag{7.1d}$$

Of these the last two reactions are allowed only by the neutral-current interactions while the first two reactions are allowed even in the old V-A charged-current theory with the charged current appearing in the s- and u-channels. So observation of the last two reactions can be taken as evidence for the existence of neutral currents whereas to establish neutral-current effects in the first two reactions detailed studies of the rate and spectrum shape are necessary.

As of this summer institute, the Gargamelle Collaboration [12)] has three events which are most likely due to the elastic $\bar{\nu}_\mu e^-$ scattering reaction (7.1d). The cross-section implied by these three events is

$$0.01 \cdot 10^{-41}\ \mathrm{cm}^2/\mathrm{GeV} \leq \sigma_{obs} / E_{\bar{\nu}} \leq 0.14 \cdot 10^{-41}\ \mathrm{cm}^2/\mathrm{GeV} \tag{7.2}$$

before signal loss corrections. The corrections needed because of signal loss are somewhat model-dependent, but, if interpreted within the framework of the Weinberg-Salam model with $\sin^2\Theta_W < 0.4$, the observed number of events is reported to correspond to

$$\sigma/E_{\bar{\nu}} = (0.13 \pm 0.08) \cdot 10^{-41}\ \mathrm{cm}^2/\mathrm{GeV}\ . \tag{7.3}$$

As for $\nu_\mu e$ scattering, no event has been found, leading to a 90 % confidence limit of

$$\sigma/E_\nu < 0.26 \cdot 10^{-41}\ \mathrm{cm}^2/\mathrm{GeV}\ . \tag{7.4}$$

The most important conclusion we can draw from the observed $\bar{\nu}_\mu e$ cross-section is that the effective strength of the $\bar{\nu}_\mu e$ interaction is not too different from that of the semileptonic inclusive neutral-current interactions. An $\bar{\nu}_\mu e$ cross-section of 0.1×10^{-41} cm^2 x $E_{\bar{\nu}}$ ($E_{\bar{\nu}}$ in GeV) corresponds to a cross-section of $0.06\ G^2 s/\pi$ while, for the hadronic final states, the observed neutral-current inclusive cross-sections for ν and $\bar{\nu}$ are typically in the range $(0.04\text{-}0.06) G^2 s/\pi$ per nucleon. So, independently of any detailed theory, we can conclude that the coupling of $\bar{e}e$ to $\bar{\nu}_\mu \nu_\mu$ is not too different in strength from that of $\bar{q}q$ to $\bar{\nu}_\mu \nu_\mu$.

To be quantitative, I write down the most general Lagrangian relevant to the four processes induced by left- (right-)handed $\nu(\bar{\nu})$ in charge-retention order as follows:

$$\mathcal{L} = -\frac{G}{\sqrt{2}} \left[\bar{\nu}_\mu \gamma_\lambda (1+\gamma_5) \nu_\mu\, \bar{e} \gamma_\lambda (g_V + g_A \gamma_5) e \right.$$
$$\left. + \bar{\nu}_e \gamma_\lambda (1+\gamma_5) \nu_e\, \bar{e} \gamma_\lambda (G_V + G_A \gamma_5) e \right]\ . \tag{7.5}$$

If we invoke μe universality, we must have

$$G_V = 1 + g_V$$
$$G_A = 1 + g_A \qquad (7.6)$$

The physical origin for this shift in the coupling constants can be traced to the fact that the reactions (7.1c) and (7.1d) can occur even within the framework of the old charged-current V-A theory with no neutral current.

Given the Lagrangian (7.5), we can compute the cross-sections for the four reactions (7.1a)-(7.1d):

$$\frac{d\sigma}{dy} = (2G^2 m E_\nu/\pi)\left[A + B(1-y)^2 - C\,\frac{my}{E_\nu}\right] \qquad (7.7)$$

where y stands, as usual, for E_e/E , and the coefficients A, B and C are given in Table 2, taken from a paper by t'Hooft [84].

Table 2

Spectrum coefficients in neutrino-electron scattering

	A	B	C
$\nu_e e^-$	$1/4\ (G_V + G_A)^2$	$1/4\ (G_V - G_A)^2$	$1/4\ (G_V^2 - G_A^2)$
$\bar\nu_e e^-$	$1/4\ (G_V - G_A)^2$	$1/4\ (G_V + G_A)^2$	$1/4\ (G_V^2 - G_A^2)$
$\nu_\mu e^-$	$1/4\ (g_V + g_A)^2$	$1/4\ (g_V - g_A)^2$	$1/4\ (g_V^2 - g_A^2)$
$\bar\nu_\mu e^-$	$1/4\ (g_V - g_A)^2$	$1/4\ (g_V + g_A)^2$	$1/4\ (g_V^2 - g_A^2)$

It is important to note that a careful study of the recoil electron spectrum can distinguish various coupling schemes. To illustrate this point let us consider $\bar\nu_e e^-$ scattering (7.1b). In the absence of the neutral-current interaction we expect, with the approximation $m_e \ll E_e$, the usual $(1-y)^2$ distribution characteristic of the V-A interaction. The presence of the neutral-current interaction significantly alters the y-distribution as well as the absolute rate. In particular the $y \simeq 1$ region now gets populated unless the neutral-current interaction is pure V-A. This point is of practical importance in analyzing the data of Reines and collaborators [17], who are studying this reaction using reactor antineutrinos whose typical energies are in the range 1 MeV-6 MeV. Reactor-off background in this experiment is known to be particularly high for $E_e \lesssim 3.5$ MeV; so for practical reasons they look at electrons of energies between 3.5 MeV and 5 MeV. As a result the experiment is very sensitive to the high-y region. Unfortunately this also means that the $\bar{\nu}$ spectrum, which falls very steeply near the maximum energy, must be accurately known if we are to draw a meaningful conclusion on the coupling schemes.

Another point to be noted in connection with (7.5)-(7.7) is that, with universality, the four processes are characterized by only two coupling constants. As a result, there are relations among the spectrum coefficients and total cross-sections, e.g.

$$A(\bar{\nu}_e e^-) = A(\bar{\nu}_\mu e^-) \quad , \quad B(\nu_e e^-) = B(\nu_\mu e^-) \quad ,$$
$$\sigma(\nu_e e^-) - \sigma(\nu_\mu e^-) = 3\left[\sigma(\bar{\nu}_e e^-) - \sigma(\bar{\nu}_\mu e^-)\right] . \tag{7.8}$$

More formulas of this kind are found in papers by Sehgal [85].

I have remarked earlier that the coupling of $\bar{e}e$ to $\bar{\nu}_\mu \nu_\mu$ is not grossly different from that of $\bar{q}q$ to $\bar{\nu}_\mu \nu_\mu$. For a more quantitative formulation of universality it is necessary to specify the group structure of the leptonic and hadronic currents, $SU(2) \otimes U(1)$ in the Weinberg-Salam model, etc. Definite predictions can then be made on the coupling constants g_V and g_A. In the Weinberg-Salam model we have [5,84]

$$g_V = -\tfrac{1}{2} + 2\sin^2\Theta_W \quad ,$$
$$g_A = -\tfrac{1}{2} . \tag{7.9}$$

As another example, we may consider the "fermion-current model", which is a natural extension of the baryon-current model discussed earlier. Here we postulate an idiotically simple fermion-current self-interaction as follows [55]:

$$\mathcal{L} = \frac{G\lambda}{\sqrt{2}} j_\lambda^{(F)} j_\lambda^{(F)} \tag{7.10}$$

where

$$j_\lambda^{(F)} = i(\bar{e}\gamma_\lambda e + \bar{\mu}\gamma_\lambda\mu + \bar{\nu}_e\gamma_\lambda\nu_e + \bar{\nu}_\mu\gamma_\lambda\nu_\mu$$

$$+ \bar{u}\gamma_\lambda u + \bar{d}\gamma_\lambda d + \bar{s}\gamma_\lambda s + \cdots) \quad . \tag{7.11}$$

In this model the basic group structure is just that of U(1) and the νe scattering coupling constants are given by

$$g_V = \lambda \quad , \quad g_A = 0 \quad . \tag{7.12}$$

The model is "universal" in the sense that the coupling constant λ is, in turn, related to the semileptonic constant v_s as follows [86]:

$$v_s = 3\lambda \quad . \tag{7.13}$$

The best way to compare the experiments with the theoretical predictions is to represent the constraints imposed by the data on a g_V-g_A plane [87]. Such a plot is made in Fig. 6. The experimental upper limits on (7.1b) and (7.1c) imply that the allowed domain must lie inside the ellipses denoted by $\bar{\nu}_e e^-$ (outer ellipse) and $\nu_\mu e^-$. If we take the three Gargamelle events for $\bar{\nu}_\mu e^-$ scattering as positive evidence for this scattering process, then even the interior of the inner ellipse, as well as the exterior of the outer ellipse, denoted by $\bar{\nu}_\mu e^-$ is excluded. The allowed domain is getting smaller now, but better data are needed to discriminate the various models.

8. Neutral-Current Effects without Neutrinos

In this section I would like to discuss neutral-current effects not induced by neutrinos.

The most spectacular process that belongs to this category is the direct-channel formation of an intermediate boson, Z, in electron-positron collisions. In November 1974, when the first ψ (or J) particle was discovered [3], many people speculated that this particle might be the mediator of the neutral-current interactions. The rich spectrum of new particles subsequently found in the mass range 2.8-4.4 GeV as well as diffractive photoproduction of $\psi(3.1)$ indicate that $\psi(3.1)$ is most likely a hadron. Nevertheless the techniques developed at that time can be applied mutatis mutandis to a weak boson which might show up in the future at higher mass.

The effective Lagrangian for Z decay into an electron pair can be taken to be

$$\mathcal{L} = i\bar{e}\gamma_\lambda(a+b\gamma_5)eZ_\lambda \quad . \tag{8.1}$$

We can easily compute the partial decay width as follows:

$$\Gamma(Z\to e^+e^-) = \tfrac{1}{3}\left[(a^2+b^2)/4\pi\right]m_Z \quad . \tag{8.2}$$

Now, if the coupling strengths of Z to $\bar{\nu}\nu$, $\bar{q}q$ and $\bar{\ell}\ell$ $(\ell = e,\mu)$ are all comparable, then we must have

$$a^2+b^2 \simeq m_Z^2 G \quad . \tag{8.3}$$

This means that the partial decay width into an electron-positron pair is proportional to the <u>cube</u> of the boson mass, which could have been guessed from dimensional considerations.

To be specific let us look at the prediction for this width in the fermion-current model [88]:

$$\Gamma(Z\to e^+e^-) = (\lambda/6\pi)\frac{G}{\sqrt{2}}\, m_Z^3 \quad . \tag{8.4}$$

If we put the numbers in, we get

$$\Gamma(Z\to e^+e^-) \simeq 6\,\mathrm{keV}\left(\frac{m_Z}{m_\psi}\right)^3 \tag{8.5}$$

where m_ψ = 3.1 GeV. Recall now that, for any spin-1 resonance, the area under the colliding-beam total cross-section curve is related to the partial decay width into electron pairs as follows:

$$\int_{Z\,\mathrm{peak}} \sigma(e^+e^-\to \mathrm{all})\,dE_{c.m.} = 6\pi^2\,\Gamma(Z\to e^+e^-)/m_Z^2 \quad . \tag{8.6}$$

$$(E_{c.m.} \equiv \sqrt{s})$$

Note that the area under the Z resonance curve increases linearly with the boson mass. So if there indeed exists a neutral weak boson whose mass can be reached by an e^+e^- facility, we are assured of a peak at least as spectacular as the $\Psi(3.1)$ peak. Conversely, the empirical fact that the narrow boson search of the SLAC-LBL Collaboration [89] at SPEAR II did not uncover a peak comparable to $\Psi(3.1)$ and $\Psi(3.7)$ implies that there is no neutral weak boson with $m \lesssim 7.6$ GeV. If the Z boson coupling to lepton-pairs violates parity, spectacular parity interference effects are predicted in the Z-boson region [90].

Coming back for a moment to neutrino-induced reactions, let us examine how the existence of a low-mass Z boson affects the x- and y-distributions. Suppose we expect a differential distribution $d\sigma/dxdy$ that goes like F(x,y) in the limit $m_Z \to \infty$. The effect of the Z boson is to modify the distribution as follows:

$$F(x,y) \to F(x,y)/(1+2m_N E xy/m_Z^2)^2 , \tag{8.7}$$

For example, the y-distribution for the pure V-A spin-1/2 parton model is no longer flat; instead it falls with increasing y.

Neutral-current effects in electron-positron annihilations into muon pairs

$$e^+ + e^- \to \mu^+ + \mu^- \tag{8.8}$$

can be detected even if the boson mass turns out to be much bigger than the total centre-of-mass energy of the colliding beam, or even if the boson had an infinite mass. Since there is a great deal of interest here at DESY for a higher energy colliding-beam apparatus, I treat this reaction in some detail. Possible neutral-current effects in this reaction were discussed by Cabibbo and Gatto [91] in 1961, long before neutral currents became fashionable.

First, let us agree on the notation convention. Assuming again μe universality, we take our effective Lagrangian to be [92]

$$\begin{aligned} \mathcal{L} = -\frac{G}{\sqrt{2}} [& h_{VV}(\bar{e}\gamma_\lambda e + \bar{\mu}\gamma_\lambda\mu)(\bar{e}\gamma_\lambda e + \bar{\mu}\gamma_\lambda\mu) \\ & + 2h_{VA}(\bar{e}\gamma_\lambda e + \bar{\mu}\gamma_\lambda\mu)(\bar{e}\gamma_\lambda\gamma_5 e + \bar{\mu}\gamma_\lambda\gamma_5\mu) \\ & + h_{AA}(\bar{e}\gamma_\lambda\gamma_5 e + \bar{\mu}\gamma_\lambda\gamma_5\mu)(\bar{e}\gamma_\lambda\gamma_5 e + \bar{\mu}\gamma_\lambda\gamma_5\mu)] . \end{aligned} \tag{8.9}$$

It is assumed here that the centre-of-mass energy $\sqrt{s}$ is much lower than the Z mass. Otherwise just multiply the whole expression by a Z-boson propagator factor

$$m_Z^2/(m_Z^2 - s - i m_Z \Gamma_Z) \quad . \tag{8.10}$$

If the interaction arises from the exchange of a single Z boson, then we must have

$$h_{VA}^{\ 2} = h_{VV} h_{AA} \quad . \tag{8.11}$$

However, if there are many intermediate bosons, this condition is, in general, not fulfilled. Notice in particular that the question of parity non-conservation - $h_{VA} = 0$ or $\neq 0$ - is logically independent of that of the coexistence of V and A. As an extreme example, consider a model in which the vector and axial vector interactions are mediated by different weak bosons. It is then possible to have parity conservation - $h_{VA} = 0$ - even though both h_{VV} and h_{AA} are non-vanishing. So it is desirable to devise experiments that test each of the three terms separately.

Various models give definite predictions for h_{VV}, h_{VA} and h_{AA}. For example, in the Weinberg-Salam model we have

$$\begin{aligned} h_{VV} &= \tfrac{1}{4}(1 - \sin^2\theta_W)^2 \ , \\ h_{VA} &= \tfrac{1}{4}(1 - \sin^2\theta_W) \ , \\ h_{AA} &= \tfrac{1}{4} \ . \end{aligned} \tag{8.12}$$

In contrast, in the fermion-current model we get

$$h_{VV} = \lambda \quad , \quad h_{VA} = h_{AA} = 0 \quad . \tag{8.13}$$

The ratio of the weak to the electromagnetic amplitudes goes like $(G/\alpha)s$, so at sufficiently high energies the weak interactions become quite comparable. However, we should start seeing interference effects at much lower energies, as we will see shortly.

It has been proposed that we study interference between the electromagnetic and neutral-current interactions by examining possible deviations from the quantum electrodynamics predictions for muon pair production (8.8) in the following three places [91,93]:

i) the magnitude and s-dependence of the total cross-section,

ii) forward-backward asymmetry,

iii) longitudinal muon polarization.

They are sensitive to h_{VV}, h_{AA} and h_{VA}, respectively.

The magnitude of the muon-pair cross-section is expected to deviate from the quantum electrodynamics prediction because the weak neutral current can propagate in the s-channel just like the time-like photon. If s is much smaller than m_Z^2, the deviation is sensitive only to h_{VV}:

$$\frac{\Delta\sigma}{\sigma_{QED}} = (G/\sqrt{2}\,\pi\alpha)\, s\, h_{VV} \simeq 3.5 \cdot 10^{-4}\, h_{VV}\,(s/\mathrm{GeV}^2)\ . \tag{8.14}$$

If the charged-lepton part of the neutral currents has a reasonably large amount of vector, then a colliding beam apparatus of the kind now under design study (or under construction?) - PETRA, PEP, etc. - is guaranteed to uncover a comfortably measurable deviation from the quantum electrodynamics prediction for the muon pair cross-section. For example, in the fermion-current model with $\lambda^2 \simeq 0.2$, we get $\Delta\sigma/\sigma_{QED}$ of 14 % at s = 900 GeV^2 (15 GeV beam energy). In the Weinberg-Salam model, with the currently accepted range of $\sin^2\Theta_W$, the situation is not so favourable, but we still expect $\Delta\sigma/\sigma_{QED}$ of 3 % with $\sin^2\Theta_W = 0.35$ at s = 900 GeV^2. Once a deviation is seen, it is very important to study its s-dependence to see whether the deviation is really due to the weak neutral-current interactions.

Forward-backward asymmetry in the angular distribution for muon pair production arises from the axial vector piece of the weak amplitude interfering with the quantum electrodynamics amplitude. So this is sensitive to h_{AA}. The relevant formula is

$$A(\Theta) \equiv \frac{\sigma(\Theta) - \sigma(\pi-\Theta)}{\sigma(\Theta) + \sigma(\pi+\Theta)} = (G/\sqrt{2}\,\pi\alpha)\, s\, h_{AA}\left(\frac{2\cos\Theta}{1+\cos^2\Theta}\right) \tag{8.15}$$

A two-photon exchange effect (even C in the s-channel) also gives a term in the angular distribution that goes like $\cos\Theta$ but with an s-dependence varying as ln s; this means that a study of the s-dependence is again essential [94]. In the Weinberg-Salam model the quantity $(G/\sqrt{2}\,\pi\alpha)s\, h_{AA}$ is about 0.08 at s = 900 GeV^2 (independent of Θ_W); so the effect should be detectable at PETRA (or PEP) energies as long as

h_{AA} is as large as in the Weinberg-Salam model.

Finally we consider the longitudinal polarization of the muon. This is a genuinely parity-violating effect due to the h_{VA} term. The formula in this case is

$$P_{\mu^+} = -P_{\mu^-} = (G/\sqrt{2}\pi\alpha)\, h_{VA}\, s \left[1 + \frac{2\cos\Theta}{1+\cos^2\Theta - P_{in}^2 \sin^2\Theta \cos^2\varphi} \right] \quad (8.16)$$

where P_{in} is the transverse polarization of the initial electron or positron beam. Alternatively, instead of measuring muon polarization, parity non-conservation can be detected using longitudinally polarized beams. In practice this involves an extra apparatus that rotates the transversely polarized electron or positron beam in the storage ring into a longitudinal one.

Neutral-current effects can also be looked for in electron-positron annihilation into hadrons - both exclusive [95] (e.g. $\Lambda\bar{\Lambda}$, $\pi^+\pi^-$) and inclusive [41,96] (e.g. $\pi^{\pm}$ + any). Theoretical analyses here are much more model-dependent since we do not have reliable theories of electron-positron annihilation into hadrons. But any positive effect found would be of enormous value.

A number of authors [97] have discussed the possibility for observing neutral-current effects in inelastic electron (muon)-nucleon scattering

$$e^{\mp}(\mu^{\mp}) + N \rightarrow e^{\mp}(\mu^{\mp}) + \text{any} \quad (8.17)$$

The simplest experiment along this line is to study the dependence of the cross-section on the helicity and/or the charge of the incident charged lepton. If the neutral currents have reasonable amounts of both vector and axial vector, the expected asymmetry effect is again of order $10^{-4}\, q^2$ (q^2 in GeV^2) where q^2 is now space-like. Another interesting effect, discussed by Stodolsky [98], is the spin rotation of a polarized neutron beam due to a parity non-conserving neutron-electron interaction.

Let me now discuss possible neutral-current effects at extreme low energies - at the atomic physics level. If the neutral-current interactions violate parity, atomic states, in general, contains some parity impurity with an opposite-parity amplitude going like $\langle b|H_{NC}|a\rangle/\Delta E$. Bouchiat and Bouchiat [99] proposed to irradiate Cs atoms to circularly polarized laser beams to induce transitions from $6s_{1/2}$ to $7s_{1/2}$. If the atomic levels are parity-impure, the probability for subsequent emission of fluorescent photons is expected to depend on the direction of the circular polarization of laser photons.

Parity-violating effects may also be studied in muonic atoms [100]. Consider, for definiteness, $2s_{1/2} \rightarrow 1s_{1/2}$, which is an M1 transition. Because of parity-impurity the $2s_{1/2}$ state has a slight admixture of $2p_{1/2}$; the smaller the energy difference between $2s_{1/2}$ and $2p_{1/2}$, the larger the parity-mixing amplitude. As a result, this M1 transition acquires a slight amount of E1. In muonic atoms the vacuum polarization effect lowers $2s_{1/2}$ while the finite size effect raises $2s_{1/2}$, and for $Z \simeq 3$-4 the $2s_{1/2}$ and $2p_{1/2}$ levels lie very close to each other. It has been proposed to look for a parity-violating correlation $\vec{k} \cdot \langle \vec{\sigma}_\mu \rangle$ in radiative transitions between such a pair of closely lying levels. Those who are interested in this subject may consult a recent review by Jarlskog and Simons [101].

In estimating the magnitude of neutral-current effects in these processes not involving neutrinos, one usually assumes that $\bar{e}e$ and $\bar{\mu}\mu$ enter in the neutral-current Hamiltonian with more or less the same strength as $\bar{q}q$ and $\bar{\nu}\nu$. Even though there now exists evidence, as emphasized in Section 7, that $\bar{q}q$ and $\bar{e}e$ enter in the neutral-current Hamiltonian with similar strength, we can still conceive of a model in which the coefficient in front of $\bar{\nu}\nu$ is depressed by some factor r, compared to normal models based on universality while the coefficients of $\bar{q}q$ and $\bar{e}e$ $(\bar{\mu}\mu)$ are both enhanced by the same factor r. If r is sufficiently large, even experiments are relatively low energies might uncover neutral-current effects [102]. The fact that no deviation from quantum electrodynamics has been found in muon pair production and Bhabha scattering at SPEAR I [103] ($s \simeq 20$ GeV^2) may be used to set a limit on r^2, in this particular case on h_{VV}:

$$|h_{VV}| < 6 \quad (2 \text{ standard deviation level}) . \tag{8.18}$$

The possibility of studying neutral-current effects in pure hadronic reactions appears even more difficult. Parity violation in nuclear forces, for example, can arise also from the charged-current interactions, and detailed models are needed to isolate the part due to the neutral-current interactions.

9. Conclusions

It was a little more than two years ago that the dramatic discovery of neutral-current phenomena was announced at the Bonn and Aix-en-Provence Conferences. What have we learned since that time? We can characterize the progress we made in the past two years by saying that, apart from confirming the existence of neutral-current effects, very little has been learned. The situation contrasts sharply with the spectacular progress we made in the field of new particle (ψ/J, etc.) spectroscopy since that fateful Sunday, 10 November, 1974.

Nevertheless it is worth mentioning that some extreme models have been ruled out. As far as the space-time properties of semileptonic neutral-current interactions are concerned, we know from the observed E_{had} distribution that pure S and/or P are ruled out. From the fact that the neutral-to-charged-current ratio is not strongly dependent on the incident neutrino energy, we infer that models with violent scaling violations are unlikely. Models with very different coupling strengths between ($\bar{e}$e) ($\bar{\nu}\nu$) and ($\bar{q}$q) ($\bar{\nu}\nu$) are also ruled out because, when expressed in units of $G^2 s/\pi$, the cross-section for $\bar{\nu}_\mu e$ scattering is similar to that for semileptonic deep-inelastic reactions.

To see whether other possibilities such as pure V-A are really ruled out, we need better data on the y-distribution, which may become available when the experimentalists succeed in studying not only the energy but also the momentum direction of the hadronic debris. It is likely that the isospin properties of the neutral current will be determined by careful studies of specific exclusive channels; we would like to know whether Δ(1236) is excited, whether elastic νp scattering takes place at a rate predicted by one of the various models discussed, whether we start seeing at high energies clean diffractive signals (ρ, A_1, ω, ϕ, etc.) via Pomeron exchange, etc. As for the more distant future, the prospects are extremely bright for detecting neutral-current effects not involving neutrinos at colliding-beam machines like PETRA and PEP.

For those impatient theorists who think that the progress in this field is extremely slow, I have just one thing left to say. It took twenty-five years to establish the vector-axial-vector structure of the charged-current beta decay interactions.

References and Footnotes

1) F.J. Hasert et al., Phys. Letters 46B, 138 (1973).
F.J. Hasert et al., Nuclear Phys. B73,1 (1974).

2) J.H. Christenson et al., Phys. Rev. Letters 13, 138 (1964).

3) J.J. Aubert et al., Phys. Rev. Letters 33, 1404 (1974).
J.-E. Augustin et al., Phys. Rev. Letters 33, 1406 (1974).

4) E. Fermi, Z. Phys. 88, 161 (1934).

5) S. Weinberg, Phys. Rev. Letters 19, 1264 (1967).
A. Salam, in Elementary particle theory; relativistic groups and analyticity: Proc. 8th Nobel Symposium, Aspenäsgården, 1968 (ed. N. Svartholm) (Almquist and Wiksells, Stockholm, 1968), p. 367.
See also A. Salam and J.C. Ward, Phys. Letters 13, 168 (1964).

6) A. Benvenuti et al., Phys. Rev. Letters 32, 800 (1974).
B. Aubert et al., Phys. Rev. Letters 32, 1454 and 1457 (1974).

7) S.J. Barish et al., Phys. Rev. Letters 33, 448 (1974).
L.G. Hyman, Proc. Colloquium La physique du neutrinos à haute énergie, Ecole Polytechnique Paris, 1975 (CNRS, Paris, 1975), p. 183.

8) B.C. Barish et al., Phys. Rev. Letters 34, 538 (1975), Proc. Colloquium La physique du neutrino à haute énergie, Ecole Polytechnique Paris, 1975 (CNRS, Paris, 1975), p. 291.

9) W. Lee et al., Proc. Colloquium La physique du neutrino à haute énergie, Ecole Polytechnique Paris, 1975 (CNRS, Paris, 1975), p. 205.

10) E.G. Cazzoli et al., Proc. Colloquium La physique du neutrino à haute énergie, Ecole Polytechnique Paris, 1975 (CNRS, Paris, 1975), p. 239.

11) Private communication from the ANL Bubble Chamber Group.

12) F.J. Hasert et al.,Phys. Letters 46B, 121 (1973)
Gargamelle Neutrino-Freon Collaboration, Proc. Colloquium La physique du neutrino à haute énergie, Ecole Polytechnique, Paris, 1975 (CNRS, Paris, 1975), p. 257.

13) B. Aubert et al., Phys. Rev. Letters 33, 984 (1974).

14) A. Benvenuti et al., Phys. Rev. Letters 34, 419 (1975).

15) A. Pais and S.B. Treiman, Brookhaven National Laboratory preprint, BNL-20348 (1975).

16) L.M. Sehgal, Phys. Letters 55B, 205 (1975).
G.V. Dass, CERN preprint TH-2020 (1975).

17) H.S. Gurr, F. Reines and H.W. Sobel, Phys. Rev. Letters 28, 1406 (1972).

18) L. Wolfenstein, Nuclear Phys. B91, 95 (1975).
Ya.B. Zeldovitch and A.M. Perelomov, Soviet Phys.-JETP 12, 777 (1961).

19) J. Bernstein and T.D. Lee, Phys. Rev. Letters 11, 512 (1963).
J. Bernstein, M. Ruderman and G. Feinberg, Phys. Rev. 132, 1227 (1963).

20) J.E. Kim, V.S. Mathur and S. Okubo, Phys. Rev. D9, 3050 (1974).
B. Jouvet, Orsay preprint, LPTHE 74/32 (1974).
B.A. Arbuzov, Serpukhov preprint, IHEP 74-98 (1974).

21) Ph. Meyer and D. Schiff, Phys. Letters 8,217 (1964).
C. Bouchiat, J. Iliopoulos and Ph. Meyer, Phys. Letters 42B, 91 (1972).

22) A gauge model of this kind was proposed by M.A.B. Bég and A. Zee, Phys. Rev. Letters 30, 675 (1973).
M.A.B. Bég, Phys. Letters 49B, 361 (1974).

23) For an excellent introduction to charge-current phenomenology, see C. Llewellyn Smith, Physics Reports 3C, No. 5, 261 (1972).

24) J.D. Bjorken, Phys. Rev. 179, 1547 (1969).

25) T. Eichten et al., Phys. Letters 46B, 274 (1973).
H. Deden et al., Nuclear Phys. B85, 269 (1975).

26) C.G. Callan and D.J. Gross, Phys. Rev. Letters 22, 156 (1969).

27) D.C. Cundy, Proc. 17th Internat. Conf. on High-Energy Physics, London, 1974 (Rutherford Lab., Chilton, Didcot, 1974), p. IV-131.

28) J.G. Morfîn, Proc. Internat. Symposium on Lepton and Photon Interactions at High Energies, Stanford, 1975, to be published.

29) Private communication from the HPW Collaboration.

30) E.A. Paschos and L. Wolfenstein, Phys. Rev. D 7, 91 (1973).

31) G. Rajasekaran and K.V.L. Sarma, Phys. Letters 56B, 201 (1975).
M. Gourdin, University of Paris preprint, PAR/LPTHE 75.3 (1975).
E.A. Paschos, Lectures given at the Xth Rencontre de Moriond, Méribel-les-Allues (1975), to be published.

32) G. Rajasekaran and K.V.L. Sarma, Pramana 2, 62 (1974).

33) A. Pais and S.B. Treiman, Phys. Rev. D 6, 2700 (1972).

34) L.M. Sehgal, Nuclear Phys. B65, 141 (1973).
R. Palmer, Phys. Letters 46B, 240 (1973).

35) M. Gronau, Phys. Letters 53B, 260 (1974).

36) G. Rajasekaran and P. Roy, Pramana 4, 222 (1975).

37) Yu.V. Gapanov and I.V. Tyutin, Soviet Phys.-JETP 20, 1231 (1965).
A. Ali and C.A. Dominguez, Neutrino (antineutrino) disintegration of the deuteron, and the structure of the neutral weak current, IPN Mexico preprint (1975).

38) H.S. Gurr, F. Reines and H.W. Sobel, Phys. Rev. Letters 33, 179 (1974).

39) T.W. Donnelly et al., Phys. Letters 49B, 8 (1974).

40) S.L. Adler, Phys. Rev. 135 B, 963 (1964).

41) A. Pais and S.B. Treiman, Phys. Rev. D 9, 1459 (1974).

42) S.P. Rosen, Neutrinos - 1974, Proc. Fourth Internat. Conf. on Neutrino Physics and Astrophysics, Downingtown, Pa., 1974 (ed. C. Baltay) (AIP, New York, 1974), p. 5.
B. Kayser et al., Phys. Letters 52B, 385 (1974).

43) R.L. Kingsley, F. Wilczek and A. Zee, Phys. Rev. D 10, 2216 (1974).
T.C. Yang, Phys. Rev. D 10, 3744 (1974).

44) S. Pakvasa and G. Rajasekaran, Phys. Rev. D 12, 113 (1975).

45) J.S. Bell and G.V. Dass, Rutherford preprint RL-75-151/T.136 (1975).

46) M. Gronau, Nuovo Cimento Letters 14, 204 (1975).

47) See contributions by D. Cline and by J.J. Sakurai in Neutrinos - 1974, Proc. Fourth Internat. Conf. on Neutrino Physics and Astrophysics, Downington, Pa., 1974 (ed. C. Baltay) (AIP, New York, 1974), pp. 201 and 57, respectively.

48) S.L. Adler, Phys. Rev. D 11, 1155 (1975).
S.L. Adler et al., Institute for Advanced Study preprint COO 2220-48 (1975).

49) This is reminiscent of an analogous theorem for Yukawa-type couplings:
G. Feinberg, Phys. Rev. 108, 878 (1957).
S.N. Gupta, Canad. J. Phys. 35, 1309 (1957).
V.G. Soloviev, Nuclear Phys. 6,618 (1958).

50) R.L. Kingsley et al., Phys. Rev. D 11, 1043 (1975).

51) For the status of strangeness-changing neutral currents, see for example K. Kleinknecht, Proc. 17th Internat. Conf. on High-Energy Physics, London, 1974 (Rutherford Lab., Chilton, Didcot, 1974), p. III-23.

52) For the status of isotensor currents in electromagnetic interactions, see for example H.M. Fisher, Proc. 6th Internat. Symposium on Electron and Photon Interactions at High Energies, Bonn, 1973 (eds. H. Rollnik and W. Pfeil) (North-Holland, Amsterdam, 1974), p. 77.

53) A.F. Garfinkel, Proc. Colloquium La physique du neutrino à haute énergie, Ecole Polytechnique Paris, 1975 (CNRS, Paris, 1975), p. 311.

54) L.M. Sehgal, Argonne preprint ANL/HEP/PR 75-45 (1975).

55) J.J. Sakurai, Phys. Rev. D 9,250 (1974).
See also S. Pakvasa and S.F. Tuan, Phys. Rev. D 9, 2698 (1974).

56) A. Rousset, Proc. 17th Internat. Conf. on High-Energy Physics, London, 1974, (Rutherford Lab., Chilton, Didcot, 1974), p. IV-128.

57) F.J. Hasert et al., CERN report TCL/Int. 75-6 (1975).

58) C. Albright et al., Phys. Rev. D 7, 2220 (1973).

59) G. Zweig, in Symmetries in elementary particle physics (ed. A. Zichichi) (Academic Press, New York, 1965), p. 192.
See also S. Okubo, Phys. Letters 5, 165 (1963).

60) J.J. Sakurai, Phys. Rev. Letters 35, 1037 (1975).

61) N. Cabibbo, Phys. Rev. Letters 10, 531 (1963).

62) Y.S. Mathur, S. Okubo and J.E. Kim, Phys. Rev. D 10, 3648 (1974); D 11, 1059 (1975).

63) S.L. Glashow, J. Iliopoulos and L. Maiani, Phys. Rev. D 2, 1285 (1970).

64) S.L. Adler et al., Phys. Rev. D 11, 3309 (1975).

65) J.J. Sakurai and L.F. Urrutia, Phys. Rev. D 11, 159 (1975).

66) D.C. Cundy et al., Phys. Letters 31 B, 478 (1970).

67) S.L. Adler, Phys. Rev. Letters 33, 1511 (1974); Institute for Advanced Study preprint COO 2220-44 (1975).

68) S.L. Adler et al., Institute for Advanced Study preprint COO 2220-50 (1975).

69) B.W. Lee, Phys. Letters 40B, 420 (1972).

70) D.H. Perkins, Proc. 16th Internat. Conf. on High-Energy Physics, Chicago-Batavia, 1972 (NAL, Chicago, 1973), Vol. 4, p. 189.

71) S.L. Adler, Phys. Rev. D 9, 229 (1974).
S.L. Adler, S. Nussinov and E.A. Paschos, Phys. Rev. D 9, 2125 (1974).

72) E.D. Bloom and F.J. Gilman, Phys. Rev. Letters 25, 1140 (1970).

73) F. Ravndal, Nuovo Cimento 18A, 385 (1973).

74) S.L. Adler et al., Fermilab preprint, FERMI-LAB-PUB-75/52-THY (1975).

75) C.A. Piketty and L. Stodolsky, Nuclear Phys. B15, 571 (1970).

76) M.K. Gaillard, S.A. Jackson and D.V. Nanopoulos, CERN preprint TH-2049 (1975).

77) D.Z. Freedman, Phys. Rev. D 9, 1389 (1974).
See also J. Bernabéu, Nuovo Cimento Letters 10, 329 (1974).

78) J.N. Bahcall and S.C. Frautschi, Phys. Rev. 136B, 1547 (1964).

79) J.N. Bahcall, Phys. Rev. 136B, 1164 (1964).

80) W.A. Fowler and F. Hoyle, Astrophys. J. Suppl. No. 91, 9, 201 (1964).

81) S.A. Colgate and R.H. White, Astrophys. J. 143, 626 (1966).
W.D. Arnett, Canad. J. Phys. 44, 2553 (1966).
J.R. Wilson, Astrophys. J. 163, 209 (1971).

82) J.R. Wilson, Phys. Rev. Letters 32, 849 (1974).
D.N. Schramm and W.D. Arnett, Phys. Rev. Letters 34, 113 (1975).
J. Bernabéu, CERN preprint TH-2073 (1975).

83) J.R. Wilson, Lawrence Livermore Laboratory preprint UCRL-76947 (1975).

84) G. t'Hooft, Phys. Letters 37B, 195 (1971).

85) L.M. Sehgal, Nuclear Phys. B70, 61 (1974); Phys. Letters 48B, 60 (1974).
See also H. Terazawa, Phys. Rev. D 8, 1817 (1973).

86) In comparing (7.10) with (3.18) note that for reactions initiated by left-handed neutrinos (right-handed antineutrinos) we can freely make the replacement $\bar{\nu}\,\gamma_\lambda\,\nu \leftrightarrow \frac{1}{2}\,\bar{\nu}\,\gamma_\lambda(1+\gamma_5)\nu$.

87) H.H. Chen and B.W. Lee, Phys. Rev. D 5, 1874 (1972).
A. De Rújula et al., Rev. Mod. Phys. 46, 391 (1974).

88) J.J. Sakurai, Phys. Rev. Letters 34, 56 (1975).
G. Altarelli et al., Nuovo Cimento Letters 11, 609 (1974).

89) R. Schwitters, Proc. Internat. Symposium on Lepton and Photon Interactions at High Energies, Stanford, 1975, to be published.

90) E.A. Paschos, Phys. Rev. Letters 34, 358 (1975).
M.-S. Chen and Y.-P. Yao, Phys. Rev. Letters 34, 628 (1975).

91) N. Cabibbo and R. Gatto, Phys. Rev. 124, 1577 (1961).

92) The spirit of my approach to this subject closely parallels that of L. Wolfenstein AIP Proceedings No. 23, Proc. of the Meeting of The American Physical Society, Division of Particles and Fields, Williamsburg, Sept. 5-7, 1974 (ed. C.E. Carlson), p. 84 who discussed the reaction (8.8) without reference to particular models.

93) A. Love, Nuovo Cimento Letters 5, 113 (1972).
V.K. Cung, A.K. Mann and E.A. Paschos, Phys. Letters 41B, 355 (1972).
J. Godine and A. Hankey, Phys. Rev. D 6, 3301 (1972).

94) The importance of higher-order electromagnetic effects in the reaction (8.8) has been emphasized in:
R.W. Brown et al., Phys. Letters 43B, 403 (1973).
D.A. Dicus, Phys. Rev. D 8, 890 (1973).

95) R. Budny, Phys. Letters 45B, 340 (1973).
M. Kuroda and Y. Yamaguchi, Progr. Theor. Phys. 49, 2160 (1973).
E. Lendvai and G. Pócsik, Phys. Letters 56B, 462 (1975).
H.S. Mani and P. Roy, Pramana 4, 264 (1975).

96) R. Gatto and G. Preparata, Nuovo Cimento Letters 7, 89 (1973).
R. Budny and H. McDonald, Phys. Letters 48B, 423 (1974).
H. McDonald, Nuclear Phys. B75, 343 (1974).
G. Kajon and R. Petronzio, Nuovo Cimento Letters 10, 369 (1974).

97) Ya. B. Zeldovich, Soviet Phys.-JETP 36, 964 (1959).
A. Love, G.C. Ross and D.V. Nanopoulos, Nuclear Phys. B49, 513 (1972).
E. Derman, Phys. Rev. D 7, 2755 (1973).
M. Suzuki, Nuclear Phys. B70, 154 (1974).
S. Berman and J. Primack, Phys. Rev. D 9, 2171 (1974).

98) L. Stodolsky, Phys. Letters 50B, 352 (1974).

99) M.A. Bouchiat and C.C. Bouchiat, Phys. Letters 48B, 111 (1974).

100) J. Bernabéu, T.E.O. Ericson and C. Jarlskog, Phys. Letters B50, 467 (1974).
G. Feinberg and M.Y. Chen, Phys. Rev. D 10, 190 (1974).

101) C. Jarlskog and L.M. Simons, Proc. Fifth Internat. Conf. on Neutrino Science, Balatonfüred, Hungary, June 1975, to be published.

102) C.H. Llewellyn Smith and D.V. Nanopoulos, Nuclear Phys. B78, 205 (1974).

103) J.E. Augustin et al., Phys. Rev. Letters 34, 233 (1975).

Figure Captions

Fig. 1: Neutrino identity in $\bar{\nu}_e e^-$ scattering; Fig. 1(a) and Fig. 1(b) cannot interfere if $\bar{\nu}' \neq \bar{\nu}_e$.

Fig. 2: Diagrams responsible for the electromagnetic charge radius of the neutrino.

Fig. 3: Coupling constant determination in pure isovector (pure isoscalar) models. The curves are computed using the central values of the Gargamelle data.

Fig. 4: The $\pi^0 p$ mass distributions in the reactions (5.9a) and (5.9b) obtained by the Columbia-Illinois-Rockefeller Collaboration.

Fig. 5: Predictions for elastic νp scattering taken from Ref. 65.

Fig. 6: The experimental constraints and theoretical predictions for the purely leptonic reactions (7.1a)-(7.1d).

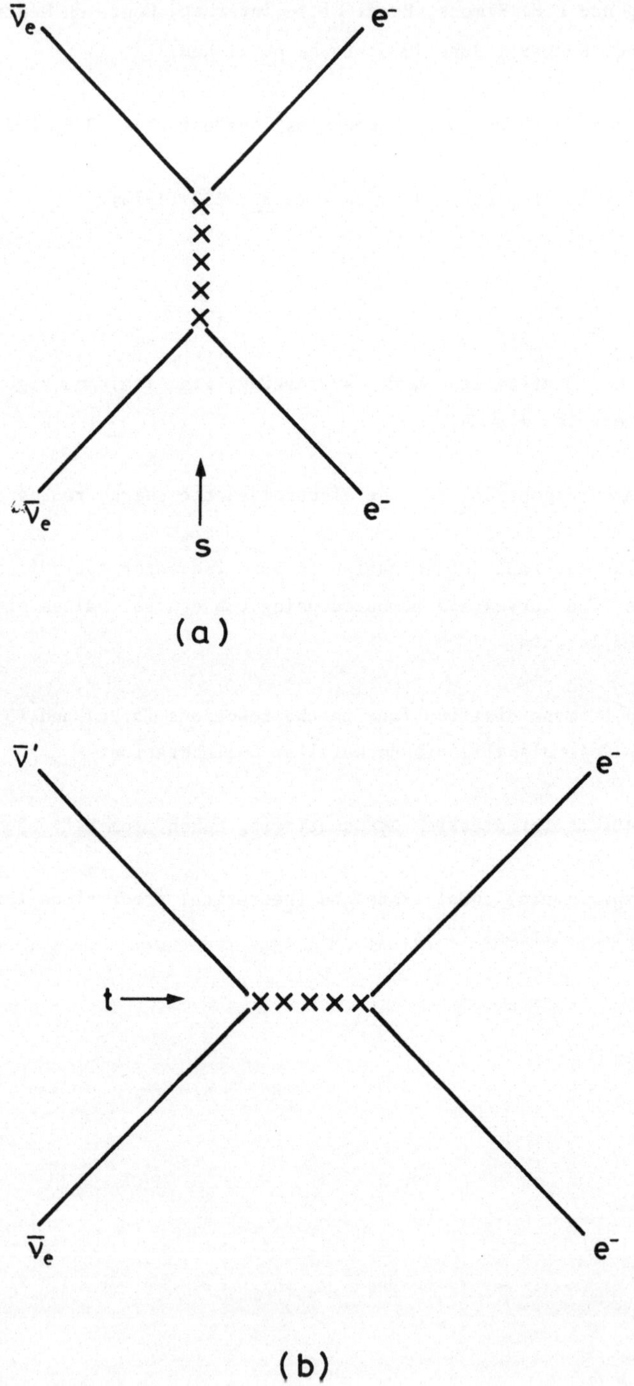

FIG.1

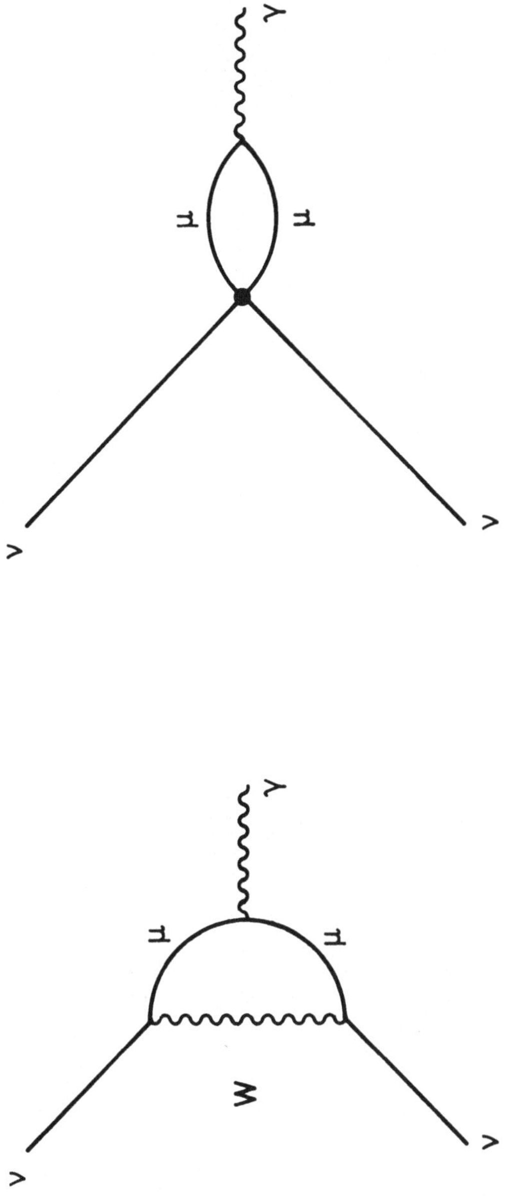

FIG. 2

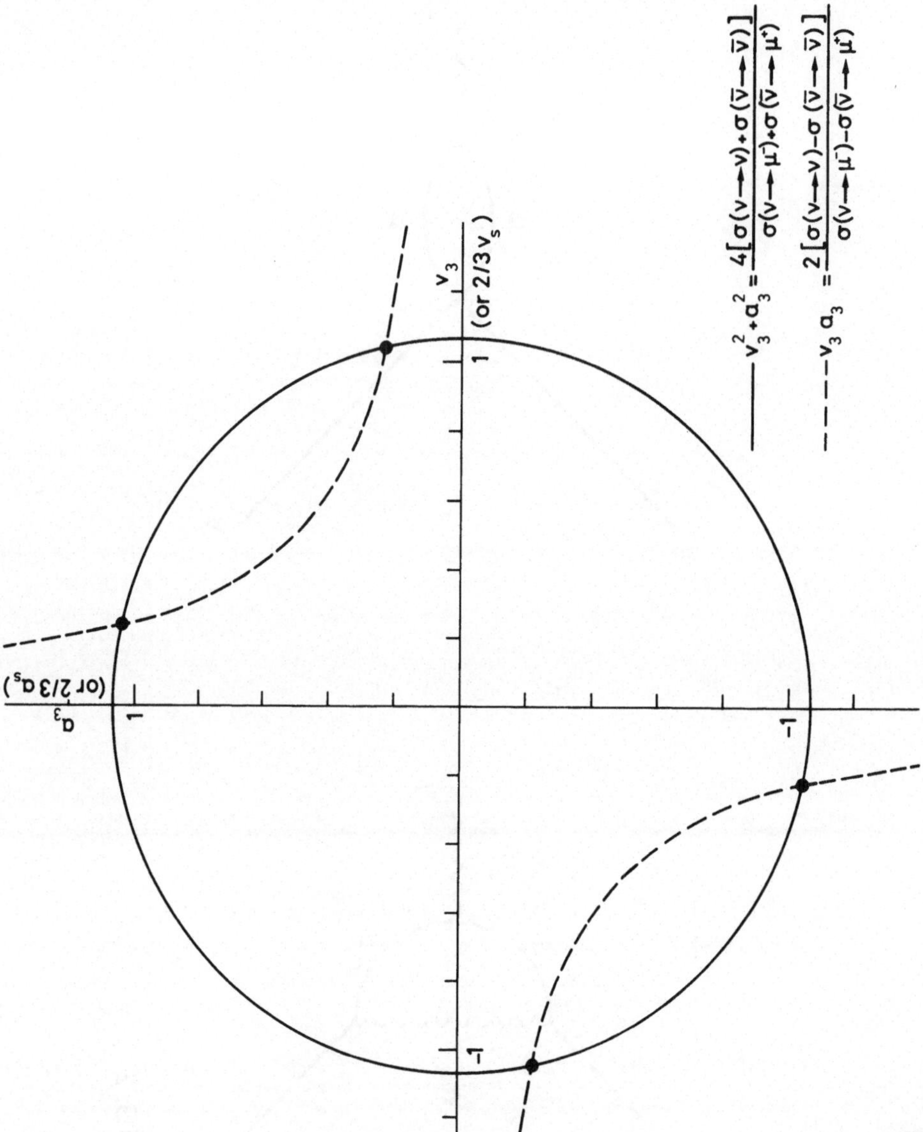

a_3 (or 2/3 a_s)
v_3 (or 2/3 v_s)
1
-1
1
-1
—— $v_3^2 + a_3^2 = \frac{4[\sigma(\nu\to\nu)+\sigma(\bar\nu\to\bar\nu)]}{\sigma(\nu\to\mu^-)+\sigma(\bar\nu\to\mu^+)}$
- - - $v_3\, a_3 = \frac{2[\sigma(\nu\to\nu)-\sigma(\bar\nu\to\bar\nu)]}{\sigma(\nu\to\mu^-)-\sigma(\bar\nu\to\mu^+)}$

FIG.3

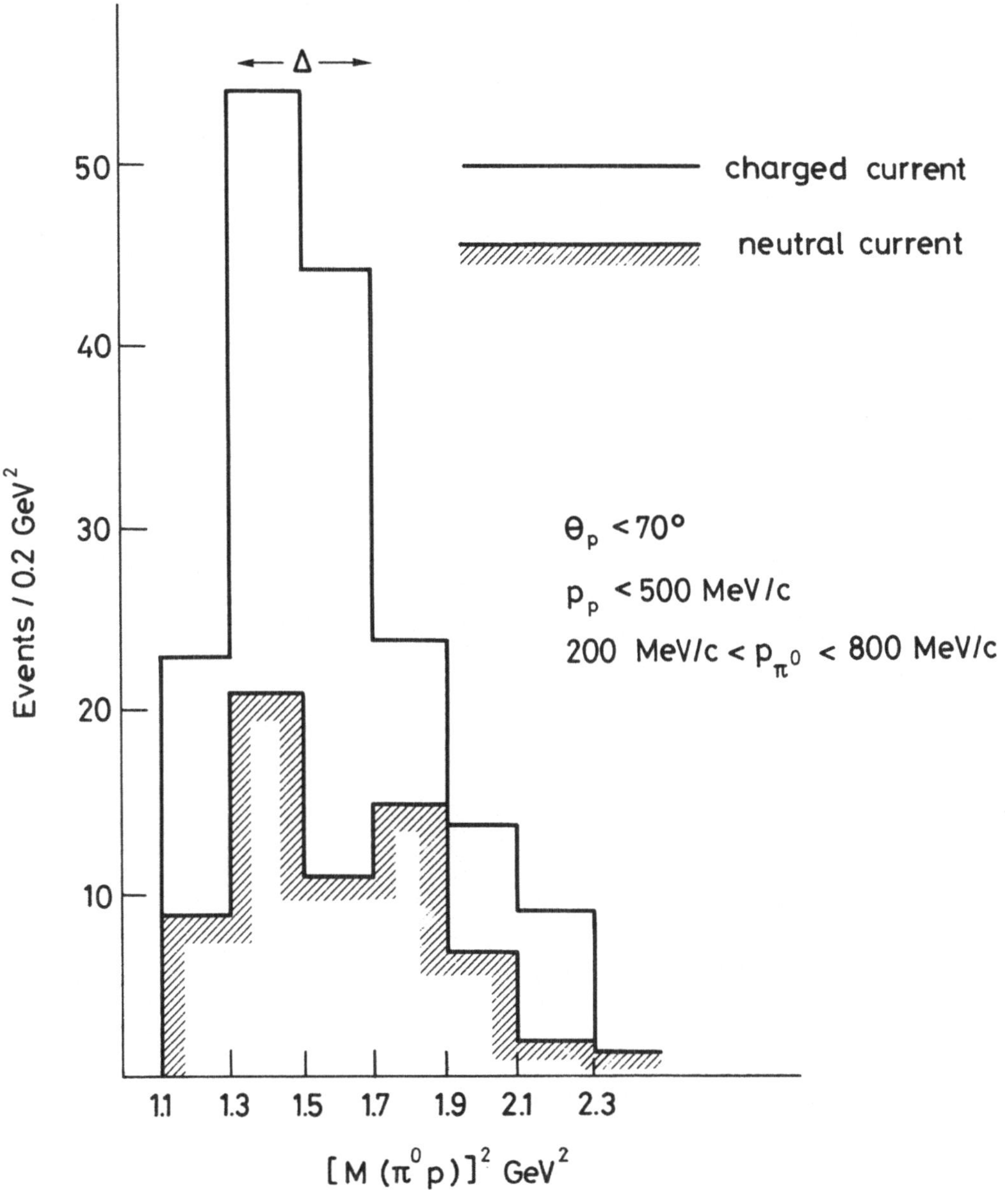
Δ
charged current
neutral current
50
40
30
20
10
Events / 0.2 GeV²
θ_p < 70°
p_p < 500 MeV/c
200 MeV/c < p_π0 < 800 MeV/c
1.1
1.3
1.5
1.7
1.9
2.1
2.3
[M (π⁰p)]² GeV²

FIG. 4

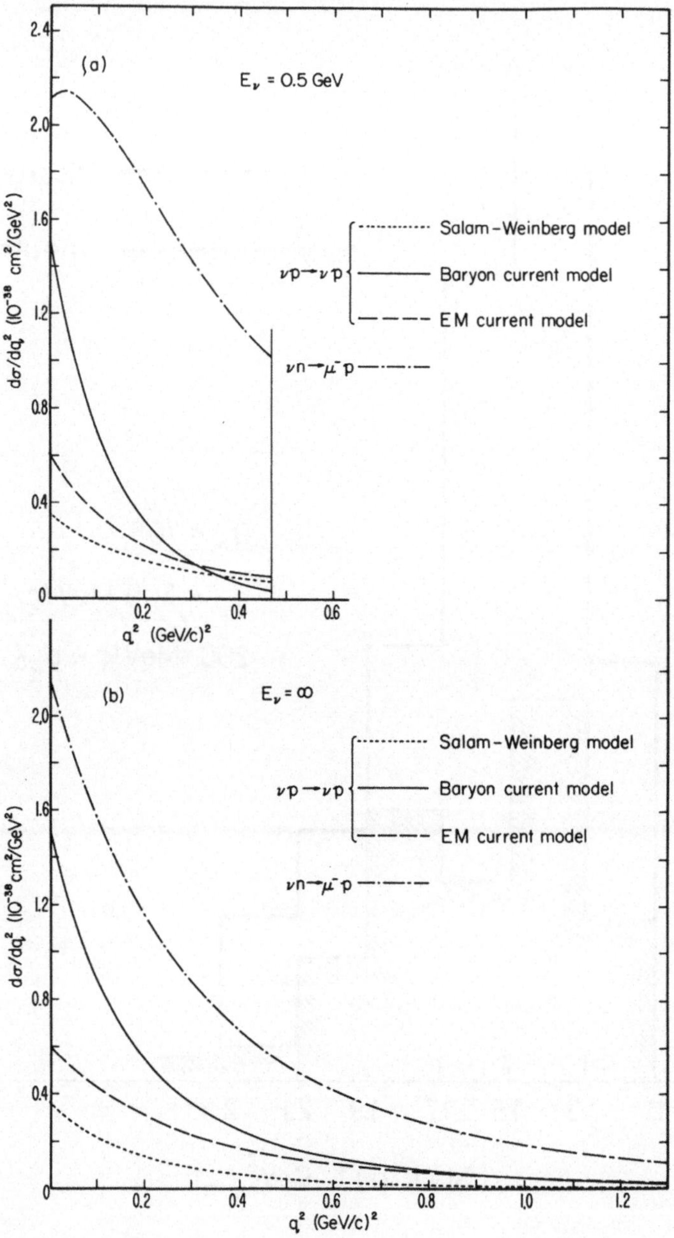
(a)
E$_\nu$ = 0.5 GeV
Salam-Weinberg model
νp→νp
Baryon current model
EM current model
νn→μ⁻p
dσ/dq² (10⁻³⁸ cm²/GeV²)
q² (GeV/c)²
(b)
E$_\nu$ = ∞
Salam-Weinberg model
νp→νp
Baryon current model
EM current model
νn→μ⁻p
dσ/dq² (10⁻³⁸ cm²/GeV²)
q² (GeV/c)²

FIG.5

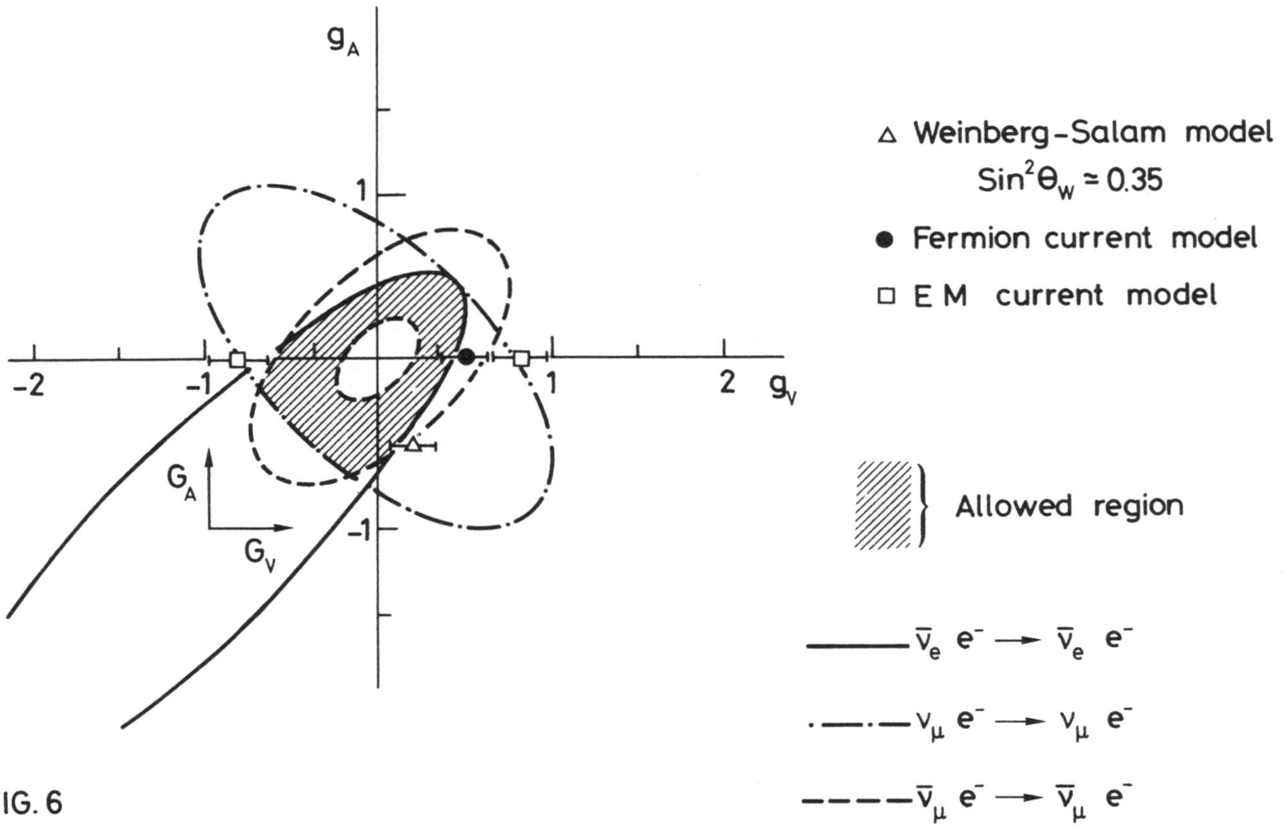

g_A
1
-1
-2
-1
1
2
g_V
G_A
G_V
△ Weinberg-Salam model
$\sin^2\theta_W \simeq 0.35$
● Fermion current model
□ E M current model
Allowed region
$\bar{\nu}_e\, e^- \to \bar{\nu}_e\, e^-$
$\nu_\mu\, e^- \to \nu_\mu\, e^-$
$\bar{\nu}_\mu\, e^- \to \bar{\nu}_\mu\, e^-$

FIG. 6

STATUS OF BROKEN COLOR SYMMETRY

Berthold Stech
Institut für theoretische Physik
der Universität Heidelberg

1. INTRODUCTION

The new particles found at Brookhaven[1)] and Stanford[2)] are usually described as bound states of heavy quarks which carry a new additive quantum number called charm. The recently discovered additional members of the ψ/J family at 3.5, 3.4 and 2.8 GeV found at DESY[3)] and SLAC[4)] apparently support this picture. However, definitive evidence for charm is still lacking and the charm picture is not free of severe problems. It is worthwhile, therefore, to look at other descriptions, in order to see what they predict, what their problems are, and how they could be confirmed or ruled out by experiment.

In the present paper I shall describe the status of a phenomenological model in which it is assumed that the new family of particles are explicit manifestations of color symmetry[5)]. In particular, I will concentrate on the broken color symmetry model for hadrons[6,7)] where color is broken at the level of strong interaction[8)].

Parton models which allow for color breaking and color excitations differ from models in which all observable states are color singlets in a number of important points:

1) color, needed in all parton models, is here an observable degree of freedom. The constituents of hadrons can have integer charge and their confinement is not necessarily complete and permanent.
2) The number of new hadron states is dramatically high. For example, there should be 72 new heavy vector mesons, a few of them stable with respect to the strong and even electromagnetic interactions.

3) The particles of low mass as well as those of high mass are formed by the same constituents. This suggests that the effective masses of partons are dynamical quantities which depend on the state.
4) The weak hadronic current must be color sensitive to allow for the weak decay of hadrons with color isospin different from zero. Lepton-hadron symmetry requires the existence of new leptons[6].

It is evident from these points that broken color symmetry is a highly speculative proposal. It is not developed very far and no field theoretical description has been attempted. This is quite understandable, since a few experimental searches may rule out the model even before the corresponding paper is published. On the other hand, the model gives very interesting predictions for the phenomenology of the new particles. In fact, several predictions have recently found experimental support and thereby drawn attention to the model.

2. COLOR SYMMETRY AND COLOR BREAKING

In this section I give a short description of broken color symmetry[6,7]. The assumptions of the model are:

A. The internal symmetry of hadrons is governed by the Han-Nambu group SU3'xSU3" (color). There are 9 constituent spin 1/2 parton fields which transform according to the $(3, 3^*)$ representation of the group. Charge and hypercharge are defined by the diagonal subgroup of SU3'xSU3" (color).

B. Medium strong symmetry breaking terms break SU3' and SU3" but leave U1 and SU2 subgroups of SU3' and SU3" unbroken: the operators Y', $\vec{I}'$, Y'', $\vec{I}''$ are conserved quantities in strong interactions[6,7,8]. The electromagnetic interactions break SU2' and SU2". At this level the operators Y', I_3', Y'', I_3'' are still constants of motion. The corresponding symmetry is violated by weak interaction only. The "electromagnetic" interactions include "tadpole" terms of unknown origin which are stronger than the usual electromagnetic selfinteraction of order e^2 and which account for the main part of the $|\Delta\vec{I}| = 1$ mass differences. The medium strong symmetry breaking terms as well as the tadpole terms in the phenomenological Hamiltonian transform as the group generators. The (8,8) terms are small.

C. The intrinsic wave functions of low and high mass mesons have a bad overlap such that the usual mesons are color singlets to a good approximation. The vector meson states show nearly ideal ω-ϕ mixing with respect to SU3'[9]. (An alternative possibility is discussed in section 5).

The motivation for point A is well known: It allows the parton fields to obey Fermi statistics and to carry integer charges. One also obtains the correct $\pi^o \to 2\gamma$ rate in a PCAC calculation. The group has the minimal number of constituents which seems possible.

Point B is motivated by the observation that non abelian symmetries are broken in nature. The breaking pattern which has been taken is the simplest generalization of the SU3 and SU2 breaking observed for ordinary hadrons. Indeed, according to B and apart from the electromagnetic self-interaction the effective Hamiltonian has the simple form

$$H = H(1,1) + H(Y,1) + H(1,Y) + H(I_3,1) + H(1,I_3) \quad . \tag{1}$$

The tadpole term $H(I_3,1)$ is well known and small compared to the SU3 breaking interaction but somewhat stronger than the electromagnetic current-current-interaction. The same properties should now hold for the color octet tadpole $H(1,I_3)$. The effective Hamiltonian (1) or part of it could be due to a spontaneous breakdown of the full SU3'xSU3" symmetry.

The basic assumptions contained in A and B are supplemented by two more technical assumptions given in C. They appear plausible. For example, it would seem fortuitous if ideal ω-ϕ mixing in SU3' would hold only for the low mass vector mesons.

The mesons are parton antiparton bound states and thus occur in the (1,1), (8,1), (1,8) and (8,8) representation of the group. The usual vector mesons can be written

$$\omega = (\omega,1) \qquad \phi = (\phi,1) \qquad \rho = (\rho,1) \qquad K^* = (K^*,1) \tag{2}$$

where ω and ϕ describe ideally mixed SU3 wave functions.

The usual pseudoscalar mesons are

$$\eta' = (\eta',1) \qquad \eta = (\eta,1) \qquad \pi = (\pi,1) \qquad K = (K,1) \tag{3}$$

A similar notation is given to the f_2, A_2, B mesons etc. The full set of mesons can now be described by

$$M = (a,b) \tag{4}$$

where the label a runs through ω, ϕ, ρ, K^* for vector mesons, through η', η, π, K for pseudoscalar mesons etc. The label b deonotes the SU3" (color) quantum number of the state. For b we will use universally 1, η_8, π, K irrespective of the spin or parity of the meson under consideration[10]. For example, (ω, η_8) describes a vector meson state which transforms as an ideally mixed ω meson with respect to SU3' and as the $I'' = 0$ member of the SU3" octet. Similarly, (A_2, π^0) is a tensor meson transforming as an $I' = 1$ member of the SU3' octet and at the same time as the $I'' = 1$, $I''_3 = 0$ member of the SU3" octet. One should perhaps note at this place that as long as one is only concerned with the strong interactions the diagonal subgroup of SU3'xSU3" is of little importance and the group SU3' could also be called SU3.

An immediate consequence of the effective Hamiltonian is the stability of the new particles with respect to decays to usual mesons. States with $I'' = 1$, $I''_3 = 0$ can only decay via the tadpole term (1, I_3), via the usual electromagnetic selfinteraction or by radiative transitions. The lowest states with $I''_3 \neq 0$ can only decay by weak interaction. States with $I'' = 0$, on the other hand, can decay strongly to ususal hadrons by means of the (1, Y) term in the effective Hamiltonian. In general, this (1, Y) term will induce a mixing of (a,η_8) states with the color singlet "analog state" $(a,1)_{analog}$ obtained from $H(1,Y) \;|(a,\eta_8)>$. This analog state is a wavepacket formed out of the continuum states from a reasonable large energy region around the resonance mass. According to C it should not contain a sizeable contribution from the low mass meson (a,1), which has the same transformation property but which is far away in energy. The mixed state shall be denoted by

$$(a,\eta) = \cos\Theta''\,(a,\eta_8) + \sin\Theta''\,(a,1)_{analog} \quad , \qquad (5)$$

where θ'' is the singlet-octet color mixing angle.

3. MESON STATES IN THE e^+e^- - CHANNEL

To see which of the color excited states couple directly to the photon we consider the electromagnetic current. This current is a member of the diagonal SU3 group and has the structure

$$j^{e.m.} = (I_3 + \tfrac{1}{2}Y, 1) + (1, I_3 + \tfrac{1}{2}Y) \quad . \qquad (6)$$

Decomposing this current into components which transform as the meson states (a,b) one obtains

$$j^{e.m.} = \frac{1}{\sqrt{6}}(\omega,1) - \frac{1}{\sqrt{3}}(\phi,1) + \sqrt{\frac{3}{2}}(\rho^{\circ},1) - (\omega,\pi^{\circ}) - \frac{1}{\sqrt{2}}(\phi,\pi^{\circ}) + \frac{1}{\sqrt{3}}(\omega,\eta_8) + \frac{1}{\sqrt{6}}(\phi,\eta_8) \; . \tag{7}$$

Thus, four color octet states are directly coupled to the photon. A fifth meson, the (ρ^{o},η) is also excited by the photon because of its mixing with the color singlet analog state. The mesons (ω,π^{o}) and (ϕ,π^{o}) have color isospin I" = 1. Thus, both are stable with respect to strong decays to ordinary hadrons. The remaining three states have I" = 0 and are expected to be non narrow resonances. Ideal mixing in SU3' requires the meson (ρ^{o},η) to be almost degenerate with the (ω,η). Furthermore, all (ϕ,b) states should be somewhat heavier than the (ω,b) states. These particle properties are very welcome for the following reasons:

i) In accordance with the model there is experimental evidence for the existence of just two narrow resonances ψ and ψ'. Also, the decay $\psi' \to \psi + 2\pi$ is the dominante decay mode of the ψ'.
ii) There exists a broad structure ψ'' at 4.1 GeV which looks like two overlapping resonances of rather different heights[11].
iii) A further non narrow resonance has been found[11] at 4.4 GeV.

These results support the identification

$$\psi(3.1) = (\omega,\pi^{\circ}) \qquad \psi'(3.7) = (\phi,\pi^{\circ}) \tag{8}$$
$$\psi(3.95?) = (\rho^{\circ},\eta) \qquad \psi''(4.1) = (\omega,\eta) \qquad \psi'''(4.4) = (\phi,\eta) \; .$$

The state (ϕ,η) was originally expected to lie somewhat higher in energy by assuming the same ω-ϕ splitting for I" = 0 and I" = 1 color octet states. This assumption is not well justified because of the different interactions of these states and the possible existence of a small H(Y,Y) perturbation term in the effective Hamiltonian.

Isospin and G parity of the narrow states ψ and ψ' support the assignments given in (8) as will be discussed in the next section. A verification of the isospin properties of the three non narrow resonances would be very important for the model. In particular, an eventual discovery of the charged partners of $\psi(3.95) = (\rho^{o},\eta)$ would seem decisive. To find them one could look for the process

$$\psi''(4.1) \longrightarrow (\rho^{\pm},\eta) + \pi^{\mp} \tag{9}$$

by using the high mass tail of the 4.1 resonance. One could also look for charged resonances in $\bar{p}n$ collisions. In the meantime the leptonic widths of the five states also provide some check of the model and the assignments (8). From Eq. (7) one obtains[6,7]:

$$
\begin{aligned}
\Gamma^{e^+e^-}(\phi,\pi^0) &= \tfrac{1}{2}\,\Gamma^{e^+e^-}(\omega,\pi^0)\\
\Gamma^{e^+e^-}(\rho^0,\eta) &= \tfrac{3}{2}\sin^2\Theta''\,\Gamma^{e^+e^-}(\omega,\pi^0)\\
\Gamma^{e^+e^-}(\omega,\eta) &= \left(\tfrac{1}{\sqrt{3}}\cos\Theta'' + \tfrac{1}{\sqrt{6}}\sin\Theta''\right)^2 \Gamma^{e^+e^-}(\omega,\pi^0)\\
\Gamma^{e^+e^-}(\phi,\eta) &= \left(\tfrac{1}{\sqrt{6}}\cos\Theta'' - \tfrac{1}{\sqrt{3}}\sin\Theta''\right)^2 \Gamma^{e^+e^-}(\omega,\pi^0)\ .
\end{aligned}
\tag{10}
$$

These equations contain the mixing angle θ'' which has been defined in (5) and which describes to what extent the $I'' = 0$ mesons mix with color singlet wave packets of similar internal structure. Unfortunately, its value is unknown. An educated guess performed in Ref. (6,7) gave a small positive value $\tan\Theta'' \approx 0.2$. In (10) any mass dependence of the decay widths has been neglected. To justify this neglection one can test the formula

$$\Gamma^{e^+e^-}(\omega,\pi^0) = 6\,\Gamma^{e^+e^-}(\omega) \tag{11}$$

which also follows from Eq. (7) and mass independence over a large intervall. It predicts $\Gamma^{e^+e^-}(\psi) = 4.6 \pm .5$ keV in perfect agreement with the reported value of $4.8 \pm .6$ keV[12]. The predictions for the leptonic widths with $\tan\theta'' \approx .2$ are given in table 1.

Table 1: Leptonic widths in keV as predicted from Eq. (10) and $\tan\Theta'' \approx .2$

	ψ(3.1)	ψ'(3.7)	ψ(3.95?)	ψ"(4.1)	ψ'''(4.4)
experiment	4.8 ± .6	2.1 ± .3	≈ .5	≈ 2	≈ .5
model	4.8 ± .6 input	2.4 ± .3	≈ .3*	≈2.0	≈ .4*

* sensitive to mixing angle θ"

The experimental values are taken from a preliminary analysis of the SLAC results performed by G. Wolf and reported at this school. It is seen that the leptonic width of ψ' relative to ψ is correctly predicted. This was recognised by several authors[9] and gives some confidence to the ideal mixing used for the color states. Interestingly, the new data for the leptonic widths of the non narrow resonances also fit to the pattern predicted by Eq. (10). We may, therefore, add a fourth point in favour of the model:

iv) The leptonic widths of the narrow and non narrow resonances in the e^+e^- channel support the assignments made for these states.

The hadron and photon decay widths of the sharp resonances will be discussed in the next section. The total widths for the much broader resonances are partly due to their strong decays to instable meson states of different spin and parity if this is energetically possible. An important part of the width will also be due to the direct decays to ordinary hadrons via the (1,Y) term in the effective Hamiltonian. The strength of the term H(1,Y) is known from the mass difference between $\psi'' = (\omega,\eta)$ and $\psi(\omega,\pi^0)$ or $\psi''' = (\phi,\eta)$ and $\psi' = (\phi,\pi^0)$. The corresponding widths are given - in perturbation theory - by the formula

$$\Gamma(\psi''') = \frac{1}{2m(\psi''')}\int d^4x \langle(\phi,\eta_8)|\mathcal{H}_{1,Y}(x)\mathcal{H}_{1,Y}(0)|(\phi,\eta_8)\rangle \tag{12}$$

and by similar expressions for the other two non narrow resonances. To obtain a rough estimate one can replace the time integration in (12) by an effective time interval $1/\Delta M$ where ΔM is typical for the level distance between old and new mesons with I" = 0. One arrives at an equal time matrix element which can be estimated in a naive parton model calculation[13]:

$$\Gamma(\psi''' \to \text{usual hadrons}) \simeq \frac{1}{\Delta M}\,\frac{1}{2}\,(m(\phi,\eta_8) - m(\phi,\pi^0))^2 \quad . \tag{13}$$

With $\Delta M \simeq 3.3$ GeV one obtains for the decay width to usual hadrons

$$\Gamma(\psi''') \approx 80 \text{ MeV} \quad . \tag{14}$$

A somewhat larger width is obtained if the mass difference between (ω,η_8) and (ω,π_0) is used. In any case one obtains the correct order of magnitude. The result (14) is only a factor of about ≃2 or 3 too large if the widths of the 3 non narrow resonances due to the term H(1,Y) is taken to be ≃ 30 MeV[14]. Accordingly, one may state:

v) The model predicts direct decays of the non narrow resonances to usual hadrons with a width of the correct order of magnitude.

Thus, if the model is correct, the non narrow resonances ψ(3.95?), ψ''(4.1) and ψ'''(4.4) should lead to usual hadron states with the SU3 properties of ρ, ω and ϕ mesons, respectively.

I conclude this section with a remark on possible cascade transitions of the states above 4 GeV into ψ(3.1) and ψ'(3.7). Although no published branching ra-

tios exist only few transitions from ψ'', ψ''' to ψ, ψ' would be compatible with the data. Fortunately, in the model, such transitions are not allowed by color isospin conservation. On the other hand, the continuum states with color isospin 1 have the possibility to decay to ψ and ψ'. However, at energies where this could be seen many different channels are open since the color model predicts a large variety of states besides ψ and ψ'.
A preliminary result is, therefore:

vi) Also the cascade processes seem to follow the rules set forth by the model.

4. DECAYS OF THE $\psi(3.1)$ AND $\psi'(3.7)$

The ψ particle is stable with respect to strong interaction. Its decay to ordinary hadrons can proceed by means of the $(1,I_3)$ tadpole term and by the usual second order electromagnetic interaction. The latter interaction should be dominated by the graph

Fig. 1

which contributes $R\Gamma^{\mu^+\mu^-}(\psi) \simeq 12$ keV[15] to the decay width, $\simeq 9$ keV to G-parity $G = +1$ final states and $\simeq 3$ keV to $G = -1$ states. R is the ratio of hadron production to μ-pair production in e^+e^- collisions outside the resonance. Additional second order electromagnetic transitions like the one described by the graph

Fig. 2

contain form factors and overlap integrals and are presumably much smaller. In the model, the graph of Fig. 2 leads again mainly to final hadron states of G-parity +1.

Direct decays of the $\psi(3.1)$, i.e. decays without an intermediate photon, proceed via the $(1,I_3)$ tadpole term. It is evident from the assignment $\psi = (\omega,\pi^0)$ that these transitions give final hadron states with G-parity -1 and isospin zero[16]. This agrees with the experimental findings[15]. Which widths do we expect from the tadpole interaction? The decay width due to the tadpole is

$$\Gamma_{\text{tadpole}}(\Psi) = \frac{1}{2m(\Psi)} \int d^4x \langle \Psi | \mathcal{H}_{1,I_3}(x)\, \mathcal{H}_{1,I_3}(0) | \Psi \rangle \quad . \tag{15}$$

Replacing again the time integration by an effective time interval $1/\Delta M$ and performing a parton model calculation one obtains:

$$\Gamma_{tadpole}(\Psi) \simeq \frac{1}{\Delta M} s^2 \tag{16}$$

Here, s denotes the tadpole contribution to the $|\Delta \vec{I}''| = 1$ mass difference. By taking $s \approx 10$ MeV and $\Delta M \approx 2.3$ GeV one finds $\Gamma_{tadpole}(\psi) \approx 40$ keV. Of course, this number is not to be taken literally. If we use our previous experience with total widths in section 2 formula (16) overestimates the width by a factor 2 or 3. On the other hand, the $|\Delta \vec{I}''| = 1$ mass differences could also be somewhat larger than 10 MeV. One can also argue that the tadpole width should be a factor of about 3 or 4 bigger than the width due to the virtual photon intermediate state since also the usual tadpole interaction is known to be at least so much stronger. According to these reasonable expectations the probability for transition to $G = -1$ final states will be an order of magnitude stronger than the few transitions to $G = -1$ states which are due to processes with an intermediate photon (Figs. 1 and 2). Since this corresponds to the experimental situation[15)] it is possible to state:

vii) Direct transitions of the $\psi(3.1)$ lead to $G = -1$ $I = 0$ final hadron states. The experimental rate for these transitions is roughly the one expected from the model.

It should be noted at this place that an (I_3, I_3) tadpole of similar strength would spoil the agreement with experiment.

The direct hadronic decays test the properties of the product of the tadpole $H(1,I_3)$ operator with the particle state. Hence, the final hadron states in direct decays of the ψ should be the same as if resulting from the decay of a fictitious heavy ω-meson. With regard to SU3 one expects, therefore, a mixture between singlet and octet corresponding to the ω-meson. Using naive parton model relations one finds for direct decays

$$\begin{aligned}
&\Gamma(\psi \to \phi\eta) \simeq 0 \\
&\Gamma(\psi \to \omega\eta)/\Gamma(\psi \to K^*K) \simeq .5 \\
&\Gamma(\psi \to \omega\eta')/\Gamma(\psi \to K^*K) \simeq .5 \qquad (\eta\text{-}\eta' \text{ mixing angle} \simeq 10°) \\
&\Gamma(\psi \to \rho\pi)/\Gamma(\psi \to K^*K) = 3
\end{aligned} \tag{17}$$

The decay $\psi \to \phi\eta$ is suppressed by a weak form of the Okubo-Zweig-Iizuka rule[17]) only. Unfortunately, there is so far no published information on the $\omega\eta$ and $\phi\eta$ decay mode. The $\rho\pi$ and K^*K decay modes have however, been studied. It is remarkable that the data support the relation (17) rather than the charm model prediction

$$\Gamma(\psi \to \rho\pi)/\Gamma(\psi \to K^*K) = 3/4 \quad .$$

One has experimentally[15])

$$\Gamma(\psi \to \rho\pi)/\Gamma(\psi \to K^*K) = 2.4 \pm 1.0 \tag{18}$$

Phase space differences between the K^*K and the $\rho\pi$ decay modes are not severe and may be neglected. Presumably, it would even be wrong to apply phase space corrections without taking the decrease of the matrix element with increasing particle momenta into account. For energetic transitions where the wave lengths of the outgoind particles are small compared with the interaction region such an effect can over-compensate the phase space factors[6,7]). Thus I can state:

viii) The $\rho\pi$ to K^*K decay ratio of the ψ support the "ω" property of the ψ in direct hadronic decays.

A word of caution is appropriate here. The experimentally found smallness or absence of the direct decay modes of the ψ into $K_S K_L, K^+K^-, K^*\bar{K}^*$ etc. which should be mentioned here remains unexplained. If ψ would transform as a pure SU3 singlet these decays are forbidden[18]). In this case, however, the ratio (18) would be mysterious.

The cascade decays of the $\psi'(3.7)$ in particular the process $\psi' \to \psi + \eta$ shows very clearly that ψ and ψ' have the same isospin and G-parity. About the non cascade decay processes of the ψ' only very little is known. They have a total width of about 90 keV[19]). From the model one expects that a sizeable fraction of these decays ($\simeq$ 40 keV) goes via the $(1,I_3)$ tadpole directly to ordinary hadrons. The rest could be hadron transitions to other states with color isospin 1 and photon transitions to colored and uncolored states. The tadpole transitions should be different from the tadpole transitions of the ψ: the ψ' particle should behave as a fictitious heavy ϕ-meson. Hence, one obtains the predictions for direct decays

$$\begin{aligned}
&\Gamma(\psi' \to \rho\pi) = 0 \\
&\Gamma(\psi' \to \omega\eta) \simeq \Gamma(\psi' \to \omega\eta') \simeq 0 \\
&\Gamma(\psi' \to \phi\eta)/\Gamma(\psi' \to K^*K) \simeq .5 \qquad (\eta\text{-}\eta' \text{ mixing angle} \simeq 10^\circ) \\
&\Gamma(\psi' \to \phi\eta')/\Gamma(\psi' \to K^*K) \simeq .5
\end{aligned} \tag{19}$$

No precise test exists so far. Nevertheless, one can state:

ix) The smallness of the $\rho^0\pi^0$ decay mode of the $\psi'(3.7)$ ($\lesssim .2$ keV) as well as the fact that the Dalitz plot of ψ' decays (after taking out the cascade decays) looks different from the Dalitz plot of the ψ may be taken as a first hint that the direct decays of the ψ' are indeed different from the direct decays of the ψ.

RADIATIVE DECAYS

In color models the photon can carry away color. One then expects that the first order radiative transitions should dominate over purely hadronic transitions which are of order e^2. The experimental fact that this is not so gave rise to much scepticism with regard to the color explanation for the ψ[20]. However, already the decay modes of the η-meson provide us with an example in which the first order γ-transitions have a smaller width than the two photon decay and the tadpole transition $\eta \to 3\pi$. For a specific channel the γ-transition probability depends very sensitively on the size of the radiating object compared to the photon wave length, on the overlap between the internal wave functions of initial and final states and on the effective parton mass[21]. Therefore, even an order of magnitude estimate is unreliable. For the total radiation width, on the other hand, a better formula can be derived since here only the global properties of the decaying particles are relevant. The total width for first order radiative decays may be obtained from the expression

$$\Gamma^{\gamma}_{total}(\psi) = \frac{e^2}{2m(\psi)} \int \frac{d^3k}{2k} \int d^4x\, e^{ikx} \langle \psi | j_\perp^{e.m.}(x)\, j_\perp^{e.m.}(0) | \psi \rangle \quad . \tag{20}$$

For an estimate of the total radiative width for the decay to usual hadrons one can again replace the time integration by $1/\Delta M$. This leads to a manageable equal time matrix element of the two currents. Denoting by R_c the equal time correlation radius one finds in a naive parton model calculation

$$\Gamma^{\gamma\ \text{to}\ I''=0\ \text{hadrons}}_{(\psi)} \simeq \frac{4}{3\pi} \frac{e^2}{4\pi} \frac{1}{\Delta M} \frac{3}{2R_c^2} \quad . \tag{21}$$

Taking $R_c^2 \simeq 3\, R_o^2$, which is valid for nucleons[22], one finds with $R_o \simeq 1$ F and $\Delta M \simeq 2.3$ GeV

$$\Gamma^{\gamma\ \text{to}\ I''=0\ \text{hadrons}}_{(\psi)} \simeq 26\ \text{keV} \tag{22}$$

which is a tolerable result. It may even be an overestimate by a factor 2 or so as was discussed before. Eq. (21) depends only quadratically on the particle radius. Thus it is much less sensitive to the radius than a specific high energy transition. Nevertheless, a small radius of the ψ particle could bring the color model in difficulties. On the other hand, Eq. (21) does not include the effect of the slow motion of heavy mass partons. This leads, perhaps, to a suppression factor $(m(\omega)/m(\psi))^2$ which could even compensate the effect of a smaller radius[23]. A possible resumé of this discussion is:

x) The absence of pronounced radiative decays of the ψ - although not favourable to the color model - is still compatible with it.

In radiative decays of the ψ and ψ' to usual hadrons the photon is an SU3' singlet. Hence, the final state must have isospin zero. The decays $\psi,\psi' \to \pi^0\gamma$ are forbidden. The experimental upper limit on the corresponding decay width of the ψ is $\lesssim .3$ keV[15]. The decay $\psi \to \eta\gamma$ has been seen with a width of $\approx$ 100 eV[24]. One also knows the ratio[25]

$$\Gamma(\psi \to \eta'\gamma)/\Gamma(\psi \to \eta\gamma) = 4 \pm 2.5 \tag{23}$$

and the limit[24] $\Gamma(\psi' \to \eta\gamma) \lesssim 35$ eV.
The ratio (23) should be equal to 2 if η and η' are pure SU3 octet and singlet states. A singlet octet mixing with $\theta \simeq 10^o$ reduces this ratio to $\simeq 1$. The γ-transition rates must be corrected for unknown mass and momentum dependent effects. To see that such effects are drastic let me consider the decay $\psi \to \eta\gamma$ which should be dominated by the graphs

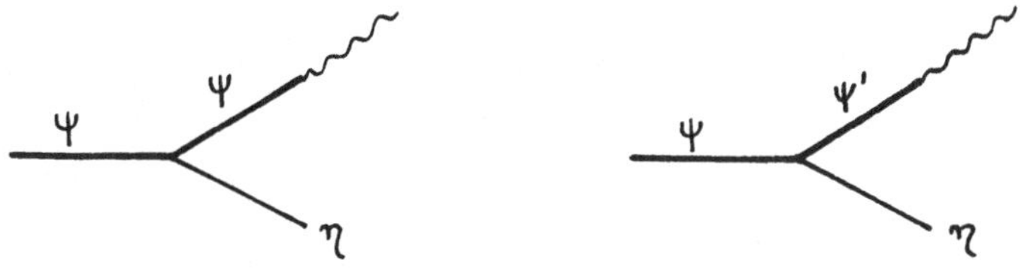

Fig. 3

From known on mass shell couplings one can only calculate the second graph which should give a much smaller contribution than the first graph. Nevertheless, the resulting width comes out too large by a factor 30! One must conclude that the coupling constants occuring in Fig. 3 are strongly q^2 dependent. Naive vector meson dominance is not applicable. A q^2 dependence of the photon couplings of the ψ particles was suggested by Bars and Peccei[26].

In radiative decays of ψ particles to colored states the isospin I' may or may not change. These decays should be more pronounced than radiative decays to usual hadrons since the effective level distance in a formula analog to Eq. (21) is much smaller. A suppression of radiative transitions due to the slow motion of heavy mass partons is, of course, possible.

5. DIFFICULTIES AND A POSSIBLE MODIFICATION OF THE MODEL

As shown in the previous sections several properties of the new particles and their decays agree well with the predictions of the model. In the form presented the model also predicts the existence of the states $(\rho^{\pm}, \pi^{o})$. According to the assumption of singlet octet degeneracy and ideal mixing in SU3', this state should be almost degenerate in mass with the $\psi = (\omega, \pi^{o})$. It has been looked for in the reaction

$$\psi'(3.7) \rightarrow (\rho^{-}, \pi^{o}) + \pi^{+} \tag{24}$$

but not seen[24]. The upper limit of the corresponding branching ratio is 5 %. A second difficulty is the non observation of hadronic transitions to the pseudoscalar meson (π, π^{o}). Although the mass of this state is not known it is reasonable to expect it near or somewhat below 3 GeV. One expects to see this state in the reaction

$$\psi'(3.7) \rightarrow (\pi, \pi^{o}) + 2\pi \quad . \tag{25}$$

Depending on the masses, the (ρ, π^{o}) could be instable and decay by π emsission to the (π, π^{o}). The newly discovered[25] X-meson at 2.8 GeV could be the (π, π^{o}) or the (η, π^{o}) pseudoscalar meson. In these cases the hadronic decay of this meson would proceed via the tadpole interaction to final states with G = -1 and I = 1 or G = +1, I = 0 respectively, with a decay width of the order of 100 keV. With these assignments for the X-meson a new difficulty arises: Why does the P_c meson at 3.4 GeV which has natural spin parity not decay to the X? It is obvious from these remarks and questions that within the color model there is not yet an understanding of the severe suppression of certain hadronic decays. I believe that this is the most serious objection which can be raised against the model.

The transitions (24) and (25) are forbidden by the Okubo-Zweig-Iizuka rule in its strong form while the observed decays $\psi' \rightarrow \psi + \varepsilon$, $\psi' \rightarrow \psi + \eta$ are only weakly Zweig forbidden. (ε and η contain both strange and non strange partons). A strong Zweig suppression could be the reason why these decays have not been seen; but, presumably, an additional mechanism is required, for example, mass dependent couplings which suppress single π emission between states which are not related by generators of SU6 etc.

From the point of view of the color model the most urgent question refers to the isospin of the particles at 3.95? GeV and 2.8 GeV. If the enhancement at 3.95? has definitely $I' = 1$, the model would be in good shape even if one does not understand the suppression of certain hadronic transitions. On the other hand, if there is no particle with $I' = 1$ around, the model has to be modified or abandoned. One would certainly have to give up the assumption of ideal mixing (point C of section 2) and to assume instead that the (8,8) states lie higher in energy than the (1,8) states and are unbound. Let me shortly discuss this modified form of the model. In this case the observed new mesons are all SU3' singlet states with some SU3' octet admixture due to the (Y,1) term in the Hamiltonian. The $\psi(3.1)$ and $\psi'(3.7)$ mesons would still have color isospin 1. However, the $\psi'(3.7)$ would have to be identified with the first radial excited state of the ψ as in the charm model. Similarly, $\psi''(4.1)$ and $\psi'''(4.4)$ would still be color octet states with color isospin zero, but the $\psi'''(4.4)$ would have to be the lowest radial excitation of the $\psi''(4.1)$ resonance. The enhancement at 3.95? GeV would have to go away or be the second radial excitation of the ψ. Some of the prediction made for ψ and ψ' decays would change, in particular the ones given in Eq.(17) and (19).

A further objection to color models is connected with deep inelastic scattering. A sizeable rise of the scaling functions is expected when the threshold for the excitation of colored particles is surpassed[27]. No strong effects have been seen. One excuse is that a new Bloom-Gilman type variable appropriate for the color threshold of ≈ 4 GeV will shift the effect to small values of x where the scaling function is smaller[28]. However, this excuse might not be sufficient. The absence of large threshold effects seems to require a better explanation or else may be held against the color case. On the other hand, nothing is known about possible color excitations of baryons and their classification under the color group. A very preliminary discussion of baryons is given in the appendix.

6. WEAK INTERACTION AND BROKEN COLOR SYMMETRY

Already before the discovery of the new particles the need for new hadronic degrees of freedom was evident from weak interactions. A very appealing solution to the problem of the absence of $\Delta S=1$ neutral currents was achieved by adding to the Cabibbo current a new piece with carries charm[29]. The success of this approach gave support to the charm picture. Only recently a difficulty seems to arise: The experimentally found rate for dilepton production in neutrino reactions[30] is larger than expected. The questions to be asked are: can color replace charm also in weak interactions? Can it be as successful as charm or even better?

I will again stay at the phenomenological level and disregard problems connected with the renormalizability of the theory. From the nine parton fields many currents can be formed. To reduce the possibilities I am using 3 assumptions[6]:

a) The weak currents are generated by left handed SU2 (weak) and obey the corresponding commutation rules (principle of universality and V-A currents). The neutral current component may mix with the electromagnetic current.

b) The weak hadronic currents are linear combinations of the currents of the strong interaction symmetry apart from a unitary transformation (generalized conserved vector current hypothesis).

c) The charged current has $\Delta Y = 0$ and $\Delta Y = 1$ pieces, the neutral current has only $\Delta Y = 0$ parts. (The hypercharge operator Y is - like the charge operator - taken from the diagonal subgroup of SU3'xSU3").

Assumption b) is very restrictive. It implies that the symmetries of weak and strong interactions are closely related.

From the requirements a) to c) and without the unitary transformation the weak hadronic current must have the structure[31]

$$\vec{j}^{\,\text{hadronic}} = (\vec{I},1) + (1,\vec{V}) \quad . \tag{26}$$

This solution is unique apart from phases and the interchange of I and V (I and V denote Isospin and V-spin SU3 operators). Notably, the two pieces in Eq. (26) have to have equal weights in analogy to the $\bar{\nu}_e e^-$ and $\bar{\nu}_\mu \mu^-$ pieces in the leptonic current

$$j_+^{\text{leptonic}} = (\bar{\nu}_e e^-) + (\bar{\nu}_\mu \mu^-) \quad . \tag{27}$$

The form (26) can, however, not be entirely correct since it cannot describe the strangeness changing decays of the usual color singlet mesons. One must allow for a unitary transformation of the weak hadronic current. In fact, it appears plausible that this current adjusts itself by a unitary transformation of the nine parton fields such that all mesons can decay weakly. Unlike the Cabibbo rotation, however, this transformation must respect hypercharge as well as charge. By writing the nine partons in a matrix

$$\underset{SU3'}{\updownarrow}\quad \overset{\longleftarrow\ SU3''\ \longrightarrow}{\begin{pmatrix} p_1 & p_2 & p_3 \\ n_1 & n_2 & n_3 \\ \lambda_1 & \lambda_2 & \lambda_3 \end{pmatrix}} \begin{matrix} \}\ (I^{\pm},1) \\ \\ \end{matrix} \qquad \underset{(1,V^{\pm})}{} \tag{28}$$

the charge and hypercharge matrices are

$$Q = \begin{pmatrix} 0 & 1 & 1 \\ -1 & 0 & 0 \\ -1 & 0 & 0 \end{pmatrix} \qquad Y = \begin{pmatrix} 0 & 0 & 1 \\ 0 & 0 & 1 \\ -1 & -1 & 0 \end{pmatrix} \quad . \tag{29}$$

Thus, the linear transformation of the parton fields which is allowed involves the 3 fields p_1, n_2, λ_3 only. The transformation I am using conserves the isospin of the diagonal subgroup of SU3'xSU3" and contains the reflexion $\lambda_3 \to -\lambda_3$[31]:

$$\begin{aligned} (p_1 - n_2)' &= p_1 - n_2 \\ \left(\frac{p_1+n_2}{\sqrt{2}}\right)' &= \cos\vartheta\ \frac{p_1+n_2}{\sqrt{2}} + \sin\vartheta\ \lambda_3 \\ \lambda_3' &= \sin\vartheta\ \frac{p_1+n_2}{\sqrt{2}} - \cos\vartheta\ \lambda_3 \end{aligned} \tag{30}$$

After this transformation the charged weak current takes the form ($U^{\pm}$ denotes strangeness changing U-spin operators)

$$\begin{aligned} j_+^{\text{hadronic}} \simeq\ & \left(1 - \tfrac{1}{3}(1-\cos\vartheta)\right)(I^+,1) - \frac{\sin\vartheta}{\sqrt{2}}\left((U^-,V^+) + (V^+,U^-)\right) \\ & + \tfrac{1}{3}(1,V^+) + 2\,(Y,V^+) + \frac{\sin\vartheta}{\sqrt{2}}\left\{ \tfrac{2}{3}(V^+,1) + (V^+,I_3) \right. \\ & \left. - \tfrac{1}{2}(V^+,Y) + (U^+,I^+) \right\} \quad . \end{aligned} \tag{31}$$

For simplicity, terms proportional to $1-\cos\vartheta$ have been neglected except in the first term. A comparison with the Cabibbo expression gives

$$\begin{aligned} \frac{2}{3}\,\frac{\sin\vartheta}{\sqrt{2}} &= \sin\Theta_{\text{Cabibbo}} \\ \text{i.e.}\qquad \sin\vartheta &\simeq \frac{1}{2} \quad . \end{aligned} \tag{32}$$

With this value $1 - \frac{1}{3}(1-\cos\vartheta)$ differs from $\cos\theta_{\text{Cabibbo}}$ only by 2 %. Thus, the color singlet part of Eq. (31) describes the known leptonic and semileptonic decays correctly. It is also seen that all new heavy mesons can decay to color singlet states by weak interactions. Interestingly, there is no Cabibbo suppression in the neutrino production of particles with a strange color quantum number, for example of heavy mesons of the type $(\omega, K^{\pm})$ and $(\phi, K^{\pm})$.

These mesons are expected to have masses around 3.8 and 4.2 GeV respectively. In the color model one expects, therefore, a relatively large production rate for $\mu^+\mu^-$ pairs due to the creation of heavy particles with a strange color quantum number $K^{\pm}$ and their subsequent decays. This mechanism could be responsible for the results recently reported[30].

Qualitatively, also the production of $\mu^-\mu^-$ pairs in neutrino interactions can be understood in the model: Particles of the type $(a,\bar{K}^o)$ are also produced by neutrinos (with a suppression factor $\frac{1}{2}\sin^2\vartheta \approx \frac{1}{8}$). Depending on their SU3' quantum number and the corresponding mass degeneracy such particles will rapidly convert to states (a,K_1^o) or (a,K_2^o) by second order weak interactions and can then decay by μ^- emission as well as μ^+ emission. Altogether one has again a point in favour of broken color symmetry:

xi) The experimentally observed rates for the production of dilepton pairs in neutrino reactions are in qualitative agreement with the expectations from the broken color symmetry model.

Because the symmetries of strong interaction seem to manifest themselves in the weak hadronic current the symmetries of weak and strong interactions are likely to be related. In particular, the similarity of the currents (26) and (27) suggests a lepton-hadron symmetry. To have a complete correspondence one has to assume nine lepton fields[6] and to arrange them in the following way (see the parton matrix (28)):

$$\begin{pmatrix} \nu_e & L^+ & M^+ \\ e^- & L^o & M^o \\ \mu^- & N^o & \nu_\mu \end{pmatrix} \Big\rangle (I^{\pm},1) \quad . \qquad \underbrace{\quad\quad}_{(1,V^{\pm})} \tag{33}$$

Hence, the lepton matrix contains 2 new positively charged leptons L^+ and M^+ and 3 new neutral leptons. The charged leptonic V-A current has now the form

$$J_+^{leptonic} = (\bar{\nu}_e e^-) + (\bar{L}^+ L^o) + (\bar{M}^+ M^o) + (\bar{\nu}_\mu \mu^-) + (\bar{M}^o e^-) + (\bar{M}^+ \nu_e) \quad . \tag{34}$$

The neutral leptonic current is

$$2 j_3^{\text{leptonic}} = (\bar{\nu}_\mu \nu_\mu) - (\bar{\mu}^- \mu^-) - 2(\bar{e}^- e^-) + 2(\bar{M}^+ M^+) + (\bar{L}^+ L^+) - (\bar{L}_0 L_0) \tag{35}$$

In both formulas we left out the $\gamma^\mu(1-\gamma_5)$ matrices. Of course, the neutral current (35) may in general mix with the electromagnetic current and thereby loose its pure V-A form for the charged lepton fields. According to (34, 35) an essential prediction of broken color symmetry together with lepton hadron symmetry is the existence of two positively charged new leptons, one of them coupled to the electron neutrino and a new neutral lepton, the other to another new neutral lepton. The neutral current has the pecularity that electron neutrinos do not appear in it. Irrespective of these details[32)] one may state:

xii) The experimental indication[33)] for the existence of a new charged lepton gives some support to a color model with lepton hadron symmetry.

Deep inelastic electron and neutrino scattering below color threshold essentially involves the (Q,1) and $(\vec{I},1)$ parts of the currents only. Hence, the results of the model are practically identical to the results of the fractionally charged quark parton model. Above color threshold, however, the remaining currents should manifest themselves. In particular, if the model holds, the ratio R should approach the value R = 6 in the asymptotic region where color is fully excited and the two positively charged leptons are produced. The deep inelastic electron and neutrino scattering processes involve baryon matrix elements of current products. Since the baryons are least understood in color models and we are already on very speculative grounds we defer a preliminary discussion of baryons to the appendix.

7. CONCLUSION

Broken color symmetry is an attempt of a fresh and unconventional look at particle physics. Numerous interesting predictions are obtained, the tested ones in surprising agreement with experiment. Several details of this very speculative model will surely require severe modifications and additions as soon as more experimental information is available. On the other hand it would be surprising if the successful predictions made by broken color symmetry turned out to be purely accidental. Several of them have been stated before the actual experiments were performed. The model should therefore be kept in mind in the discussion of the wealth of new phenomena we are confronted with today.

Acknowledgement

The author takes pleasure in thanking Otto Nachtmann und Nicolae Marinescu for interesting discussions.

NOTE ADDED IN PROOF:

According to recent reports from SLAC the enhancement refered in the paper as $\psi(3.95?)$ is now better established. Furthermore, the structure refered in the paper as $\psi''(4.1)$ appears to contain a peak at $\simeq$ 4.03 GeV and another maximum near 4.11 GeV. One can identify as before

$$\psi(3.95) = (\rho^o,\eta) \quad , \quad \psi''(4.03) = (\omega,\eta)$$

and try to explain the maximum near 4.11 GeV as a π-threshold effect $\psi(3.95) + \pi$ in the tail of the state $\psi''(4.03)$ (see footnote 14). The experiment suggested by Eq. (9) appears now even more important and informative (Of course, one of the three peaks near 4.0 GeV could also be a radial excited state. In this case the model would predict $\psi(3.1)$ and $\psi'(3.7)$ particles in the decay products of the corresponding resonance).

8. APPENDIX

If color symmetry is indeed broken in strong interaction, the color classification of baryons will be important for the baryon spectrum and for the structure functions of baryons. The simplest possibility is to describe the usual baryons as color singlets and allow for color excited states such as color octet and color decuplet particles. In this case all baryons have triality zero in SU3' and SU3" color). They are composed of integer charged Han-Nambu quarks with (average) baryon number 1/3. In the color threshold region a change of the nucleon structure functions occurs. It is due to a color singlet contribution which arises from the product of two color octet (1,8) currents. The expected change is rather drastic. In spite of the V-part in the weak current (Eq.(26)) no charge symmetry violations occur in deep inelastic neutrino interactions. The sum rule of Gross and Llewellyn-Smith becomes

$$-\frac{1}{2}\int_0^1 \left(F_3^{\nu}(x) + F_3^{\bar{\nu}}(x) \right) dx = 3 + \frac{1}{2}(1 + 3\cos^2\vartheta) \qquad (36)$$
$$\sin\vartheta \simeq \frac{1}{2}$$

If at very high energies the confining forces between the quarks could be overcome - in e^+e^- annihilation for example - the produced quarks would be stable. They could only decay to hadrons if in addition to the interactions introduced so far also triality violating interactions exist.

A completely different and rather peculiar situation arises if one wants to stick to the point of view[34] that besides color also the parton field is "observable" in strong interactions, in other words, that the parton field has sizeable hadron matrix elements and baryon number 1. Then, the baryon composed of $(3,3^*)$ partons must have SU3' and SU3" (color) quantum numbers with triality different from zero. This appears to be ruled out experimentally because of the known octet and decuplet properties of baryons. Curiosity and the need for unconventional steps in attacking the quark puzzle may suffice, however, to try to go on and see to what type of model one is led. Usual baryons would consist of 3 partons with integer baryon number, the gnomes[34,31], which are coupled like the quarks to form a "56" in SU6 and a singlet in SU3" (color). In the potential of these 3 valence gnomes there are - besides a sea - two antignomes. Each of those should be, presumably, in a state of color isospin zero, SU3' isospin zero, and charge zero. This configuration of the antipartons is a special $(6^*,6)$ state, namely $\overline{\lambda_3}\overline{\lambda_3}$, a truly strange and, if taken literally, unrealistic structure. It would certainly be a mystery why this parton configuration is preferred and if so why states of similar mass and different SU3' configurations do not seem to exist. Very peculiar forces, for which baryon number must play an essential role, are needed. I have nothing to say about this but I will simply state some consequences of such a baryon model with non-zero triality quantum numbers.

i) The non-narrow resonances at 3.95?, 4.1 and 4.4 GeV are expected to decay preferentially to baryon antibaryon states or resonances. The narrow states ψ (3.1) and ψ'(3.7) remain stable in strong interaction because of color isospin conservation.

ii) A jet structure is expected at high e^+e^- energies. Each jet of low sphericity should contain a baryon and have a net total charge +1 or -1 [6].

iii) In deep inelastic electron scattering below and above color isospin 1 threshold the two antipartons do not contribute but carry the fraction 2/5 of the total momentum.

iv) The results for deep inelastic neutrino and antineutrino scattering below the color isospin threshold are practically identical to the fractionally charged quark parton model. Above the color isospin 1/2 threshold a charge symmetry violation will occur due to the V-spin term in the charged weak hadronic current (Eq. (26)). By writing the cross-section for the charged current reactions in the form

$$\begin{aligned} \frac{2\pi}{sG^2}\frac{d\sigma^{\nu}}{dx\,dy} &= g_{\nu}(x) + \bar{g}_{\nu}(x)(1-y)^2 \\ \frac{2\pi}{sG^2}\frac{d\sigma^{\bar{\nu}}}{dx\,dy} &= g_{\bar{\nu}}(x)(1-y)^2 + \bar{g}_{\bar{\nu}}(x) \end{aligned} \tag{37}$$

one obtains

$$\begin{aligned} g_{\nu}(x) &= g_{\bar{\nu}}(x) \\ \bar{g}_{\nu}(x) &= 4\,\frac{1+\cos^2\vartheta}{7+3\cos^2\vartheta}\,g_{\nu}(x) \\ \bar{g}_{\bar{\nu}}(x) &= 8\,\frac{1-\cos^2\vartheta}{7+3\cos^2\vartheta}\,g_{\nu}(x) \end{aligned} \tag{38}$$

In these formulas (which should hold asymptotically) x and y are the usual scaling variables and $\sin\vartheta \simeq 1/2$.
The effect of an additional parton antiparton sea has been ignored. For the sum rule of Gross and Llewellyn-Smith one finds

$$-\frac{1}{2}\int_0^1 \left(F_3^{\nu}(x) + F_3^{\bar{\nu}}(x)\right)dx = \frac{1+5\cos\vartheta}{2} \tag{39}$$

The Adler sum rule takes the form

$$\frac{1}{2}\int_0^1 \left(F_2^{\nu}(x) - F_2^{\bar{\nu}}(x)\right)\frac{dx}{x} = -2\cos\vartheta\, I_3 + 3\cos^2\vartheta - 1 \tag{40}$$

The present experimental situation relevant to the point iv) has been reviewed at the Stanford meeting and by J. von Krogh at this school. It is interesting to note that the new neutrino data reported by the HPW group indicate deviations from the usual sum rules which are qualitatively similar to the ones suggested by (39) and (40).

The color threshold effects remain severe in all color models. The expected increase of $g_\nu(x) = g_{\bar{\nu}}(x)$ is 67 %. In deep inelastic electron scattering, on the other hand, the (1,Y) term in the Hamiltonian may cause the effective parton charges to be different from the usual fractional quark charges already somewhat below the color isospin 1 threshold. One has in the corresponding energy region $Q = (Q,1) + \frac{1}{2}(1,Y)$, i.e.

$$Q = \begin{pmatrix} 1/2 & 1/2 & 1 \\ -1/2 & -1/2 & 0 \\ -1/2 & -1/2 & 0 \end{pmatrix}$$

Hence, $R = \Sigma_i Q_i^2$ can approach the value 2.5 already before the more significant rise of R sets in which is due to the color isospin 1 threshold. Also the change of the nucleon structure function will appear somewhat less drastic: the color isospin 1 threshold causes an increase of 50 % (for an isoscalar target). This number may be compared with the total effect of color which is 80 %.

REFERENCES AND FOOTNOTES

1) J.J. Aubert et al., Phys. Rev. Lett. 33, 14o4 (1974)

2) J.E. Augustin et al., Phys. Rev. Lett. 33, 14o6 (1974)
G.S. Abrams et al., Phys. Rev. Lett 33, 1453 (1974)

3) B.H. Wiik, DORIS results presented at the Lepton Photon Symposim, Stanford, 1975
W. Braunschweig et al., Phys. Lett. 57B (1975) 4o7
J. Heintze, results from the DESY-Heidelberg-Collaboration, Desy-report 75/34

4) G.J. Feldman: SLAC report, presented at the Lepton Photon Symposium, Stanford, 1975

5) Color symmetry as an observable degree of freedom in strong interactions has been used under different names by several authors, notably
M.Y. Han and Y. Nambu, Phys. Rev. 139, B 1oo6 (1965)
Y. Nambu and M.Y. Han, Phys. Rev. D10, 674 (1974)
A.N. Tavkhelidze et al., IAEA, ICTP Trieste, p. 763 (1965)

After the discovery of the new particles several color models have been proposed. Review papers are:
O.W. Greenberg, Proceedings of the Orbis Scientiae at the University of Miami, Coral Gables, Florida, January 1976
D. Schildknecht, DESY 75/13; Invited talk presented at the Xth Rencontre de Moriond, March, 1975
F.E. Close, Ref. Th 2041 - CERN, July, 1975
Lectures at the XVth Cracow School of Physics, Zakopane, Poland, June 1975
R.D. Peccei, Stanford ITP-5o1, 7/75. Invited talk given at the Argonne Summer Symposium on New Directions in Hadron Spectroscopy, July 1975

6) B. Stech, Heidelberg preprints HD-THEP 74,75-3,75-6; Invited talk, Proceedings of the XII. International Winterschool, Karpacz, Poland, February 1975

7) N. Marinescu and B. Stech, HD-THEP-75-4, 75-7, Phys. Rev. D12, 1356 (1975)

8) Symmetry breaking terms and particle assignments similar to the ones used in Refs. 6,7) have been proposed independently by
W. Alles, Lett. al Nuovo Cim. 12, N. 8, p.280 (1975)
G. Feldman and P.T. Matthews, Imperial College Report 74/75
P.T. Matthews, report given at the Palermo Conference 1975
B.G. Kenny, D.C. Peaslee and L.J. Tassie, Phys. Rev. Letters 34, 1482 (1975)

9) W. Alles, Lett. Nuovo Cimento 12, 185 (1975)
M. Krammer, D. Schildknecht, and F. Steiner, Phys. Rev. D12, 139 (1975)
S. Kitakado and T.F. Walsh, DESY 74/64
I. Bars and R.D. Peccei, Phys. Rev. Lett. 34, 985 (1975)
S.B. Gerasimov and A.B. Govorkov, Dubna E2-8656 (1975)

10) This notation is slightly different from the one used in Refs. 6,7)

11) R. Schwitters , SLAC report, presented at the Lepton Photon Symposium, Stanford, 1975

12) A.M. Boyarski et al., Phys. Rev. Lett. 34, 1357 (1975)

13) Eq. (13) differs by a factor 2/3 from the formula given in Refs. 6,7) due to the fact that only the width to usual hadrons is estimated.

14) It is not clear why the 4.1 GeV resonance is wider than the others. Perhaps, the high energy tail of this resonance decays preferentially to the $\psi(3.95?)$ by π-emission.

15) V. Lüth, Invited lecture given at the Palermo Conference, June 1975, and SLAC-PUB-1599 (1975)

16) The ψ itself has G-parity +1 and I = 1 if the isospin operators are taken from the diagonal subgroup of SU3'xSU3".

17) S. Okubo, Phys. Lett. 5, 165 (1963)
G. Zweig, CERN-TH 4o1 and TH-412 (1964)
J. Iizuka, Prog. Theor. Phys. Suppl. 37-38. 21 (1966)

18) F. Gilman, SLAC-PUB-1600 (1975)

19) G. Abrams, SLAC report presented at the Lepton Photon Symposium, Stanford 1975

20) CERN ψ-notes, Ref.TH 2964-CERN
H. Harari, ψ -chology notes, SLAC-PUB-1514

21) O.W. Greenberg, loc. cit. ref. 5)
M. Kuroda and Y. Yamagouchi, report UT-248 (1975)
See also Refs. 6,7)

22) M.G. Schmidt and B. Stech, Nucl. Phys. B52, 445 (1973)

23) A mass dependence of particle radii is suggested by the mass independence of the leptonic widths.

24) B. Wiik, loc. cit. ref. 3).

25) J. Heintze, loc. cit. ref. 3).

26) I. Bars and R.D. Peccei, loc. cit. ref. 9).

27) J. Ellis; Invited talk presented at the Weekend Meeting on Deep Inelastic Phenomena, Rutherford Lab., 1974
Ref. TH.1880-CERN

28) H.C. Llewellyn-Smith; Invited talk, Lepton Photon Symposium, Stanford, 1975

29) S.L. Glashow, J. Iliopoulos and L. Maiani, Phys. Rev. D2, 1285 (1970)

30) B. Aubert et al., Proceedings of the London Conference on High Energy Physics, 1974,
A. Benvenuti et al., Phys. Rev. Lett. 34, 419 (1975)
See also the reports given by B.C. Barish and C. Rubbia in Stanford (1975)

31) B. Stech, Phys. Lett. 49B, 471 (1975)

32) A somewhat different leptonic current is used by T.C. Yang, Maryland report 76-o25, July, 1975

33) G.J. Feldman and M.L. Perl, loc. cit. ref. 4).

34) B. Stech, Nucl. Phys. B64, 194 (1973)

PRODUCTION OF NEW PARTICLES BY p-N INTERACTIONS

Ulrich Becker

presented by

Samuel C.C. Ting

Massachusetts Institute of Technology
Laboratory for Nuclear Science
Cambridge, Massachusetts 02139

PART I: DISCOVERY OF THE J PARTICLE

The discovery of the J particle[1,2,3] in proton-proton collisions by the MIT-BNL group at Brookhaven National Laboratory follows a decade of experiments associated with e^+e^- pair productions from photon or hadron interactions at high energies. From the reaction:

$$\begin{pmatrix} \gamma \\ \text{or} \\ p \end{pmatrix} + p \rightarrow e^+e^- + X .$$

three different areas of physics can be studied:

a) Using a 7.5 GeV bremsstrahlung photon beam one can verify the e^+e^- yield predictions of QED at large momentum transfer or small distances, $< 10^{-14}$ cm.[4)]

b) One can study the e^+e^- decay of vector mesons. With spin 1 and negative parity and charge conjugation, such as the ρ,[5)] ϕ,[6)] ω,[7)] and ρ' [8)] the coupling strengths between these particles[9)] and photons can be measured. One can also study the production mechanisms of these vector mesons produced by photons.

c) Search for particles which decay to e^+e^- by investigating $pp \rightarrow e^+e^- + X$ or $pp \rightarrow \mu^+\mu^- + X$.

Whereas a) and b) has been studied at DESY extensively, let's concentrate on c) originally intended as search for more vector mesons.

An early experiment at the AGS[10)] studied the continuum of $\mu^+\mu^-$ from

from p+Uranium → $\mu^+\mu^-$ + X, giving approximately the size of $\mu^+\mu^-$ yield. In the last ten years there have been many experiments[11] at Brookhaven, at CERN I.S.R., at Fermi Lab., etc., to study the inclusive e(μ) production p + p → e(μ) + X. Again, these experiments gave no indication of a long lived particle.

Using the reaction pp → e^+e^- + X , detecting narrow width particles, in a high sensitivity experiment, over a wide mass region, requires precautions to overcome the high background. We note:

1) Since the e^+e^- come from photon decay, the yield of e^+e^- is lower than hadron pairs ($\pi^+\pi^-$, K^+K^-, $p\bar{p}$, $K^+\bar{p}$, etc.,) roughly by a factor

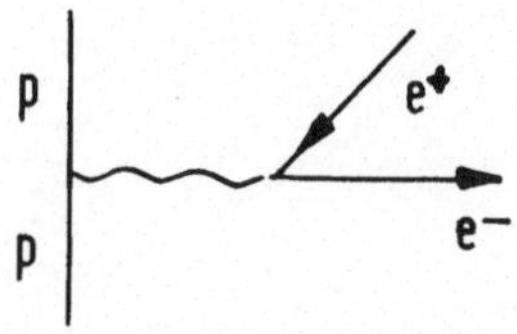

$$\frac{\alpha^2}{m^4} F^2 (m^2) \simeq 10^{-6}$$

The factor α^2 comes from the virtual photon decay, m^{-4} is the photon propagator and $F(m^2)$ the form factor of the target proton, where m is the invariant mass of the e^+e^- pair.

2) This means a low production cross section; hence the detector must survive a high flux of hadrons to obtain a sufficient yield of e^+e^- pairs.

3) It also implies that the detector must be able to reject hadron pairs by a factor of 10^{-6} - 10^{-8}.

4) In choosing the best kinematic region to detect the decay of new particles, one notes that at high energies, inclusive production of ρ, π and ω from p-p interactions can all be described in the c.m. system by a dependence of the form

$$\frac{d^3\sigma}{dp_\parallel^* \, dp_\perp^2} \propto \frac{a\, e^{-b p_\perp^*}}{E^*}$$ independent of $p_\parallel^*$,

where $p_\parallel^*$, $p_\perp^*$ and E^* are CMS momenta and energy.

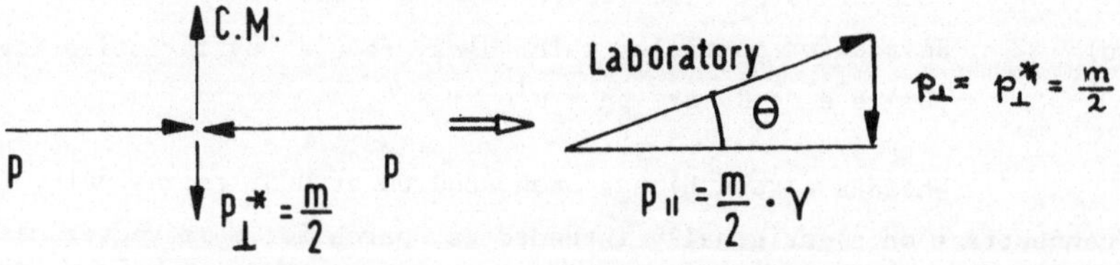

Thus the maximum yield will occur when the particle is produced at rest in the center of mass system. If we look at the 90° decay of the $e^{+}e^{-}$ pair, we note that they emerge at an angle θ = arc tan $(1/\gamma)$ = 14.6^{o} in the lab system for an incident proton energy of 28.5 GeV, independent of the mass of the decaying particle.

Experimental Set-up. Figure 1 shows the plan and side views of the spectrometer and detectors. The intense proton beam of up to $2 \cdot 10^{12}$ proton/sec was focused by quadrupoles to a 4 x 3 mm image spot size on a 10% interaction length target. This consisted of several beryllium pieces, equidistantly spaced over about 60 cm. Bending is done vertically to decouple angle (θ) and momentum (p). The field of the magnets in their final location was measured with a three-dimensional Hall probe at a total of 10^{5} points. C_B, C_o and C_e are gas threshold Cerenkov counters. C_B is filled with isobutane at 1 atm., C_o with hydrogen at 1 atm. and C_e with hydrogen at 0.8 atm. A_o, A, B and C are proportional wire chambers with 2 mm spacing and a total of 4,000 wires on each arm. Situated behind chambers A and B are two planes of hodoscopes, 8 x 8 pieces to improve the time resolution.

Behind the last chamber C, there are two orthogonal banks of lead glass counters of three radiation lengths each, the first containing twelve elements, the second thirteen, followed by one horizontal bank of lead lucite shower counters, seven in number, each ten radiation lengths thick, to further reject hadrons from electrons and improve track identification.

Special details incorporated to match the outlined problems are:

a) To obtain a rejection against hadrons of 10^{8} or better, the two gas Cerenkov counters in each arm, C_o and C_e (Fig.2a, b) are filled with hydrogen and made with thin mylar windows to reduce knockon electrons and scintillation effects. The counters are painted black inside and are decoupled by the strong magnetic fields of M_1 and M_2, so that knock-on electrons produced in C_o do not enter C_e. Furthermore, only electrons along the beam trajectory will emit Cerenkov light which is focussed onto the photomultiplier tube. Special high gain, high efficiency phototubes of the type RCA C31000M were used so that the counters C_o and C_e can be operated at 100% efficiency with very low voltage. The counter C_o collects an average of 9 photoelectrons. To ensure the voltage was set on a single electron and not on $e^{+}e^{-}$ pairs from π^{o}'s, which would give 18 photoelectrons, the counter C_o was filled with He and the location of the single photoelectron peak was found (Fig.2c).

b) To be able to handle a high intensity of 2×10^{12} protons per pulse with consequent single arm rates of 20 MHz, there are eleven planes of proportional wires ($2 \times A_o$, $3 \times A$, $3 \times B$, and $3 \times C$) rotated 22.5^o with respect to each other, as shown in Fig.3a, to reduce multitrack ambiguities. The chambers have a 100% uniform efficiency at low voltage (Fig.3b) and a long live time in the highly radioactive environment, because specially a Argon-Methylal mixture at 2^oC was used. With 2mm spatial resolution they enable for a mass reconstruction as precise as 5 MeV.

c) To reduce multiple scattering and photon conversion, the material along the path of the particles is kept small. The front and rear windows of C_o are 125 μm thick respectively, both mirrors of C_o and C_e are made of 3mm aluminized lucite and the hodoscopes are only 1.6 mm thick.

The thickness of one piece of Beryllium target is 1.8 mm and the nine pieces are each separated by 7.5 cm so that the particles produced in one piece and accepted by the spectrometer do not pass through the next piece. This is important for γ conversion. None of the counters sees the target directly. This reduces the γ and neutron contamination.

d) To improve the rejection against Dalitz decays $\pi^o \to \gamma e^+e^-$, a very directional Cerenkov counter C_B was placed close to the target and below a specially constructed magnet M_o (Fig.4a). If the high energetic e^+ is accepted by the spectrometer, the low energetic e^- will be deflected by M_o down into C_B, sensitive to electrons of 10 MeV/c and pions only above 2.7 GeV/c. The coincidence between C_B and C_o, C_e together with shower counters and the hodoscopes indicated the detection of an e^+e^- pair from the process $\pi^o \to \gamma e^+e^-$, and such events are rejected. A typical plot of the relative timing of this coincidence is shown in Fig.4b. Conversely, one can trigger on C_B and provide a pure electron beam to calibrate C_o , C_e and the shower counters.

e) The spectrometer has a very large mass acceptance of 2 GeV/c^2 and enables us to study the entire mass region from 1.5 to 5.5 GeV/c^2 in three overlapping settings. For a point target the acceptance in θ is $\pm 1^o$, in ϕ it is $\pm 2^o$ and in momentum it varies from $0.6 \times p_o$ to $1.8 \times p_o$ (where p_o is the principal axis momentum), all in the lab system.

The following table summarises the unique features of the MIT-BNL experimental apparatus:

1. It can operate at the maximal incident flux of 2×10^{12} proton/pulse, or 20 MC single rates, such that the sensitivity is $\sigma B \to 10^{-36}$ cm^2/week.

2. Sort out 8 tracks per arm, with good $\Delta p/p$.
3. Rejection of hadrons $<< 10^{-8}$
4. Mass resolution $\pm$ 5 MeV/c^2.
5. Mass acceptance 2 GeV/c^2.

First Results. The first data from August 1974 are shown in Fig.5a as shaded areas. There is a clear, sharp enhancement at a mass of 3.1 GeV/c^2.

This was a total surprise and to ensure that the observed peak is a real particle J, from August to November experimental checks were made on the data. I list six examples:

1) The magnet currents were decreased by 10%, the peak remained fixed at 3.1 GeV/c^2. See Fig.5a, clear area.
2) To check second order effects on the target, the target thickness was increased by a factor of two. The yield increased by the expected factor of two, not by four.
3) To check the calibration of the lead glass and shower counters, runs were made with different voltage settings on the counters. No effect was observed on the yield of J's.
4) To ensure that the peak is not due to scattering from the sides of the magnets, cuts were made in the data to reduce the effective aperture. No significant reduction in the J yield was found.
5) To check the read-out system of the chambers and the triggering system of the hodoscopes, runs were taken with a few planes of chambers deleted and with sections of the hodoscopes omitted from the trigger. No effect was observed on the J yield.
6) Runs with different beam intensity were made and the yield did not change.

These and many other checks convinced us that we have observed a real massive particle $J \to e^+e^-$.

Partial analysis of the width of the J particle shown in Fig.5b indicates it has a width smaller than 5 MeV/c^2.

If we assume a production mechanism for J to be

$$\frac{d^3\sigma}{dp_\perp^{*2}\, dp_\parallel^*} = \frac{e^{-6p_\perp^*}}{E^*}, \quad \text{independent of } p_\parallel^*$$

and an isotropic decay in the rest system of the J , we obtain a $J \to e^+e^-$ cross section of 10^{-34} cm^2/nucleon at 28.5. GeV incident proton energy.

Fig.6 shows the yield of e^+e^- in the region 3.2 to 4.0 GeV/c^2 with about equal protons on the targets. The acceptance in this region is a smooth function and varies at most by a factor of two. The observed events are consistent with purely random coincidences. To a level of 1% of the J yield, with a confidence level of 95%, no heavier J particles were found. We note that this upper limit is independent of any production mechanism of the J . With the J production mechanism above, we obtain an upper limit of 10^{-36} cm^2/ nucleon for the production of heavier J's with a 95% confidence level for 28.5 GeV proton energy.

PART II: HADRONIC PRODUCTION OF J PARTICLES

The data on hadron production of J also shows an increase in yield as function of energy. I list three experiments.

At BNL the MIT-BNL group measured the yield of J with a 20 GeV incident proton beam on a Be target and found the J yield down by almost a factor of 10.

The preliminary $P_\perp^2$ dependence at 28.5 GeV is shown in Fig.7. The data are consistent with $e^{-1.6\ P_\perp^2}$.

J production at I.S.R.:[12] The CERN - Columbia - Rockefeller - Saclay group performed an experiment at I.S.R. with a two-arm spectrometer shown in Fig.8a).

Each spectrometer consisted of a system of wire spark chambers with magnetostrictive read-out and a magnet with a bending power of 3.4 kΓm providing a momentum measurement with a standard deviation of $\Delta p/p = \sqrt{(0.025)^2 + (0.02\ )^2}$ (p in GeV/c).

Electron identification was achieved by means of threshold gas Cerenkov counters and electromagnetic shower detectors. Counter C was filled with isobutane (C_4H_{10}) at atmospheric pressure and C in the other arm simply with air. The corresponding pion momenta at threshold were 2.7 and 5.6 GeV/c respectively.

The electromagnetic shower detector in Arm 1 has an energy resolution of $\frac{\Delta E}{E} = 30\%$. The r.m.s. energy resolution of Arm 2 measured to be $\Delta E/E = 0.017 + 0.064/\sqrt{E}$ (E in GeV). The detection efficiency for Arm 1 is 84 ± 4% and the hadron rejection factor is $>3 \times 10^3$, the corresponding numbers in

Arm 2 being 45 ± 5% and $> 2 \times 10^4$ respectively.

Each arm also contained three scintillation counter hodoscopes H_1, H_2, H_3, in Arm 1, and H_1', H_2', H_3', in Arm 2.

Hodoscopes H_1 and H_1' were equipped with pulse-height measuring electronic electronics.

To compute the acceptance they assume $\langle P_\perp \rangle = 0.67$ which is consistent with their data and an $e^{-3P_\perp}$ dependence. The decay angular distribution was taken to be isotropic in the rest system of the pair.

Fig.8b) shows the observed events. In accordance with the resolution a signal at 3.1 GeV is observed. The acceptances from Arm 1 and 2 are shown, too. They exhibit a reassuringly flat behavior in the region of interest.

The tables below summarize all the data taken at various energies. Y is as usual the rapdity variable.

$\sqrt{s}$ (GeV)	$M(e^+e^-)$ (GeV/c^2)	P_T^* (GeV/c)	Y
44.8	2.93	0.88	-0.165
52.7	2.99	0.78	-0.033
52.7	3.00	0.20	0.239
44.8	3.02	0.33	0.191
44.8	3.07	0.72	-0.007
52.7	3.07	0.30	0.157
30.5	3.12	0.32	0.318
44.8	3.18	0.33	0.095
52.7	3.29	1.43	-0.001
52.7	3.46	1.18	0.193
52.7	3.79	0.36	0.121

Differential cross section for the reaction

$$p + p \longrightarrow J(3.1) + \text{anything}$$
$$\quad \hookrightarrow e^+e^-$$

for various values of assumed $\langle p_T \rangle_J$

$\langle p_T \rangle_J$ (GeV/c)	$B_{ee} \frac{d}{dy}$ (p + p → J + anything) (cm^2)
0.67	$(7.5 \pm 2.5) \times 10^{-33}$
1.0	$(1.2 \pm 0.4) \times 10^{-32}$
1.5	$(2.1 \pm 0.7) \times 10^{-32}$
2.0	$(3.2 \pm 1.1) \times 10^{-32}$

The cross section, for example, increases by 60%, if one calculates the acceptance with $\langle P_\perp \rangle = 1.0$ GeV.

The value of the cross section for J production at I.S.R. energies is two orders of magnitude higher than the corresponding value quoted at BNL energies. A comparison with Fermi Laboratory measurement is more difficult since the experiments are done in disjoint regions of rapidity, but the cross sections appear to be of the same order of magnitude.

A very interesting experiment has just been carried out by the CERN-Hamburg-Orsay-Wien collaboration[13)]. In designing a detector they directed their interest towards the fragmentation region ($x \simeq 0.3$). They installed two roads of absorbers in one arm of the Split-Field-Magnet (SFM) facility, at angles of $+23^{o}$ with respect to the outgoing proton beam.

The SFM detector is shown schematically in Fig.9a; it consists of two forward telescopes equipped with 24 multiwire proportional chambers of 2mm wire spacing, 1 m high and 2 m wide; a central vertex detector around the interaction region consists of two densely packed chamber units and four additional chambers. Each chamber has a vertical and a horizontal wire plane. The magnetic field varies from 5 to 10 k Γ. In the central region the field has a quadrupole configuration. Each absorber road starts with a section of 0.7 interaction length (Λ_π), between the central vertex detector and the first chamber of the forward telescope, at a distance of 1.10 m from the interaction region. The following sections of $1.4\Lambda_\pi$, $1.4\Lambda_\pi$ and $4\Lambda_\pi$ length, respectively, are inserted in front of the second, third and fourth chamber.

The probability that a pion, travelling inside a road, does not strongly interact is 0.55×10^{-3}; the probability for decay in flight is $2.33\ 10^{-2} \times \frac{1\ \text{GeV/c}}{p}$. The solid angle of each road is 0.074 sr. The mass resolution of the spectrometer is dominated by multiple scattering caused by the absorber road. A Monte Carlo simulation predicts $\Delta M/M = \pm\ 10\%$ at $M = 3\ \text{GeV/c}^2$.

The mass spectrum of 153 opposite-charge muon parts is shown in Fig.9b, together with the expected background, calculated as described before. The background at low masses is attributed to the tail of the ρ-meson. The acceptance is steeply rising in this region, thereby producing a peaked distribution. For $1.6 < M < 2.7\ \text{GeV/c}^2$ there are 11±5 events above background. In the mass region $2.7 < M < 3.4\ \text{GeV/c}^2$ they observe 14 events and expect 2.6 background events. They conclude that 11 $\mu^+\mu^-$ pairs in this mass interval are genuine $J \to \mu^+\mu^-$. Two events are observed in the mass interval $3.6 \leq M \leq 4\ \text{GeV/c}^2$, where one expects 0.26 background events, and none above.

In this experiment the acceptance is a slow variation function of m and P in the range, $2 < M < 8\ \text{GeV/c}^2$, $0 < P_T < 3$ GeV/c. The acceptance in x is strongly correlated with M and P_T.

Weighting each event in the region $2.8 < M < 3.4\ \text{GeV/c}^2$ with its acceptance probability they obtain:

for $P^+P \to J + X$, $J \to \mu^-\mu^+$ $\qquad \langle P_T \rangle = (1.0\pm0.2)$ GeV

$\qquad \langle P_T^2 \rangle = (1.2\pm0.3)$ GeV

and the mean transverse kinetic energy

$$\langle T \rangle = \langle \sqrt{P_T^2 + m_J^2} \rangle - m_J = (0.18\pm0.05)\ \text{GeV}.$$

The authors noted that the J(3.1) particle is produced with the same mean transverse kinetic energy as pions, kaons and protons, suggesting that it is produced like a hadron.

They have determined the cross section for inclusive J(3.1) production at $\sqrt{s}$ = 52 GeV and a mean rapidity y = 1.6 with the result

$$B_{\mu\mu} \cdot \frac{d\sigma}{dy} \; [pp \rightarrow J(3.1) + \text{anything}] = (7.2 \pm 2.4) \; 10^{-33} \; cm^2 ,$$

where $B_{\mu\mu}$ is the branching ratio for J(3.1) → μμ decay. We note close agreement with a previous result obtained at the ISR if $\langle P_T \rangle$ = 1.0±0.2 GeV/c is used

$$B_{ee} \cdot \frac{d\sigma}{dy} \; (y = 0) = (1.2 \pm 0.4) \; 10^{-32} \; cm^2 .$$

After numerical integration and inserting the branching ratio $B_{\mu\mu}=B_{ee}$=0.069±0.009 we find

$$\sigma[pp \rightarrow J(3.1) + \text{anything}] = (0.61 \pm 0.26) \; \mu\text{barn, hence}$$

a factor 500 higher than at BNL energy, $\sqrt{s}$ = 7.5 GeV.

A comparison of the invariant cross section for J production at y = 1.6 and p_T = 0.5 GeV/c

$$E \frac{d^3\sigma}{dp^3} = (3 \pm 2) \; 10^{-32} \; cm^2/GeV^2$$

with pion, kaon and proton data at the same value of the transverse energy, $E_T = \sqrt{P_T^2 + m^2}$ in Fig.10, shows close agreement, again supporting the conclusion that the J(3.1) is produced hadronically. Tom Ferber[14)] has also made the interesting observation that the increase of average multiplicity <n> as function of the available Q values of J particle follows the same trend as K^-, $\bar{p}$, etc., (Fig.11).

These properties of J would indicate that it is a strong interaction particle.

PART III: SEARCH FOR MORE NEW PARTICLES

There have been many theoretical papers appearing in journals explaining the new particles.

The model of Glashow[15)] views the J particle as a bound Ortho state of Charm (C) and Anti-Charm. Then, necessarily, a para state J' should exist, too, with a mass of about 3 GeV, coupled more strongly to hadron, such that J' → $p\bar{p}$ is expected. Furthermore "charmed" baryons and mesons, D, are predicted with D → Kπ or semileptonic decays. D is expected in the 1.5 - 2.5 GeV mass region.

The model of Yang and Wu[16)], which views the J particle production from pp reaction to be via $p + p \to J\theta + X$ to conserve the possible additional quantum numbers of the J particle, would predict a long lived $\theta \to \pi + p$ particle near the mass of 2 GeV/c^2.

Independent of theories, one can ask two important experimental questions.
How many other narrow resonances exist?
Why is the ratio $e/\pi = 10^{-4}$ for inclusive production at high $p_\perp$
To answer these questions, the MIT group has just completed a systematic search of

$$
\begin{array}{llll}
p + Be \to & \pi^- p & + X & A \\
& \pi^+\pi^- & + X & \\
& \bar{p}\, p & + X & B \\
& K^- p & + X & \\
& K^+\pi^- & + X & \\
& K^+K^- & + X & \\
& K^-\pi^+ & + X & C \\
& K^+\bar{p} & + X & \\
& \pi^+\bar{p} & + X &
\end{array}
$$

in the pair mass region 1.2 - 5.0 GeV with a mass resolution of ±5 MeV.

The experimental set up for this experiment is very similar to the original e^+e^- experiment. However, since there are much more hadrons than electrons the random accidentals are more serious. To reduce the accidentals a new target system was put in, consisting of five pieces of 4 mm x 4 mm x 4 mm Be target spaced by six inches. Limiting the vertex point to ±2 mm from the beam axis matches the wire chamber resolution and yields the best possible mass resolution. The targets are supported by thin piono wires. This arrangement enables one to locate the point of intersection between the two trajectories. Genuine events will originate from the same target piece, accidentals usually not (Fig.12a). To further reduce the accidentals additional scintillation counters were installed to tighten the two arm coincidence to 0.9 ns (Fig.12b).

Two high pressure (300 psi of ethylene) Cerenkov counters were installed in front of the shower counters to identify K's. The counters C_e (Fig.1) were filled with 1 atm isobutane to identify π's. The Cerenkov counters need momentum cuts and thereby limit the mass acceptance to ≈ 1.5GeV. In this way all nine combinations are measured simultaneously. To avoid systematic errors, six overlapping magnet settings were made.

Figures 13-21 show the results of the nine reactions. The data are displayed directly to give the relevant statistical evidence. The shape of these mass spectra originates from an increase in acceptance at low masses and then decreases due to the decrease in production cross sections. As seen there are no sharp narrow resonances in any of the nine reactions.

There may be, of course, wide "ordinary" resonances with widths of 200 MeV or more. A search for these depends on exact calculations of acceptance and has not yet been made. To obtain a feeling of the sensitivities of the measurements, we take the production of the nine reactions to be the same as the assumed J production mechanism. From this we obtain the following table:

TABLE II SENSITIVITY (cm^2) FOR NARROW RESONANCES

h^+X^- \ m	2.25 GeV	3.1 GeV	3.7 GeV
$\pi^+\pi^-$	8×10^{-33}	5×10^{-34}	3×10^{-35}
π^+K^-	1×10^{-33}	4×10^{-35}	1×10^{-35}
$\pi^+\bar{p}$	2×10^{-33}	4×10^{-35}	7×10^{-36}
$K^+\pi^-$	4×10^{-33}	8×10^{-35}	4×10^{-35}
K^+K^-	1×10^{-33}	5×10^{-35}	1×10^{-35}
$K^+\bar{p}$	2×10^{-33}	4×10^{-35}	8×10^{-36}
$p\pi^-$	4×10^{-32}	4×10^{-33}	5×10^{-34}
pK^-	7×10^{-33}	4×10^{-34}	3×10^{-35}
$p\bar{p}$	--	4×10^{-34}	2×10^{-35}

The sensitivity is the product of branding ratio and cross section yielding a five standard deviation effect in one of the 12.5 MeV mass bins. These upper limits are a smooth function of mass; of particular interest, for example, may be $p\pi^-$ at 2.25 GeV, π^+K^- at 2.25 GeV and $p\bar{p}$ at 3.1 GeV. Whereas the spectra do not show any narrow resonances, the cross sections $d^2\sigma/dmdx$ $(x = 0)$ plotted vs the pair mass as shown in Fig.22 display remarkable features:

1) They drop exponentially.

2) All decrease in about the same way: $\propto e^{-5m}$.

3) There is group structure like:

$$p + Be \to \pi^- + p + X \ ,$$

$$p + Be \to \begin{cases} \pi^+ + \pi^- + X, \\ \bar{p} + p + X, \\ K^- + p + X; \end{cases}$$

$$p + Be \to \begin{cases} K^- + \pi^+ + X, \\ \pi^+ + \bar{p} + X. \end{cases}$$

Over the limited range of transverse momenta of the pair $0 < p_\perp < 1$ GeV/c measured no simple exponential behavior was observed. However, the ansatz

$$E\, d^3\sigma/dp^3 \propto \exp(-5E) = \exp(-5m)\, \exp(-5T) \ ,$$

where $T = (m^2 + 3/2p_\perp^2)^{1/2} - m$ described the data well. For $m \to 0$ this yields the well known $\exp(-5T) = \exp(-6p)$ behavior, whereas at $m = 3.1$ GeV the slope is compatible with the data in Fig.7.

To search for multibody final states additional reactions:

$$P + Be \to \begin{matrix} \pi^-\pi^- \\ K^-K^- \\ \bar{p}\,\bar{p} \\ K^-\bar{p} \\ \pi^-\bar{p} \end{matrix} + X$$

were measured, too. One measures the two particle spectrum to search for multi-body decays like $D \to K^-\pi^+\pi^-\pi^+$, ect. one would expect a discontinuity in the $K^-\pi$ spectrum near the mass of C:

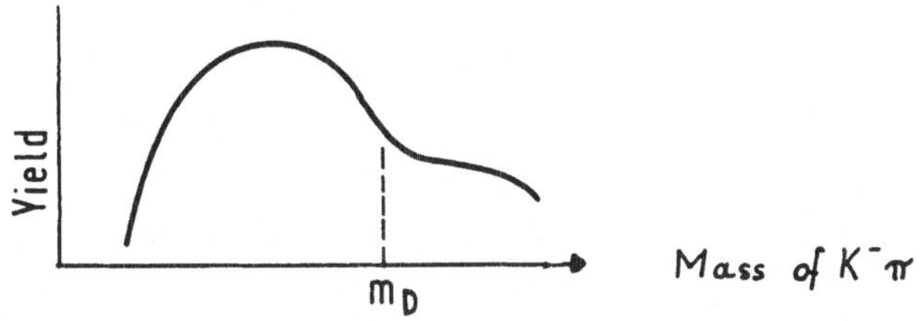

No such discontinuities were found.

There are many other experiments at I.S.R., Fermilab. etc. designed to find charmed particles. So far no definite positive results have been reported.

A more sensitive experiment is presently underway from MIT-BNL looking for e^+e^- coincidences in search for $D\to e^+h^-$ due to low background the sensitivity level will be better by orders of magnitude compared to Table II.

Furthermore, since under 14.6^o emitted electrons in the lab correspond to 90° in CMS, one may expect to get insight into the ratio $e^-/\pi^- \simeq 10^{-4}$ by detecting hadrons on the opposite side.

REFERENCES

(1) J.J. Aubert et al., Phys. Rev. Lett. 33, 1404 (1974)

(2) J.E. Augustin et al., Phys. Rev. Lett. 33, 1406 (1974)

(3) C. Bacci et al., Phys. Rev. Lett. 33, 1408 (1974)

(4) J. G. Asbury et al., Phys. Rev. Lett. 18, 65 (1967)
H. Alvensleben et al., Phys. Rev. Lett. 21, 1501 (1968)

(5) J.G. Asbury et al., Phys. Rev. Lett. 19, 869 (1967)

(6) U. Becker et al., Phys. Rev. Lett. 21, 1504 (1968)

(7) H. Alvensleben et al., Phys. Rev. Lett. 25, 1373 (1970)
P. Biggs et al., Phys. Rev. Lett. 24, 1197 (1970)

(8) G. Barbarino et al., Lett. al Nuovo Cimento 3, 693 (1972)

(9) U. Becker, Thesis Hamburg (1966)
J. G. Asbury et al., Phys. Rev. Lett. 19, 865 (1967)
H. Alvensleben et al., Phys. Rev. Lett. 24, 786 (1970)
H. Alvensleben et al., Phys. Rev. Lett. 28, 66 (1972)
K. Gottfried, 1971 International Symposium on Electron/Photon Interactions at High Energies, Cornell (1971); see reference listed therein

(10) J.H. Christenson et al., Phys. Rev. Lett. 25, 1523 (1970)

(11) R. Burns et al., Phys. Rev. Lett. 15, 830 (1965)
J.P. Boymond et al., Phys. Rev. Lett. 33, 112 (1974)
J.A. Appel et al., Phys. Rev. Lett. 33, 722 (1974)
F.W. Büsser et al., Phys. Lett. 53B, 212 (1974)
L.B. Leipuner et al., Phys. Rev. Lett. 34, 103 (1975)

(12) F.W. Büsser et al., Contribution to this conference

(13) E. Nagy, et al., Contribution to this conference. I thank Dr. K. Winter and M. Vivargent for interesting discussions on their experiment.

(14) T. Ferber: private communication

(15) A. de Rujula and S.L. Glashow, Phys. Rev. Lett. 34, 46 (1975)

(16) H.T. Nieh, T.T. Wu, and C.N. Yang, Phys. Rev. Lett. 34, 49 (1975)
C.G. Callam, R.L. Kingsley, S.B. Treiman, F. Wilczek and A. Zee, Phys. Rev. Lett. 34, 52 (1975)
J.J. Sakurai, Phys. Rev. Lett. 34, 56 (1975)

(17) J.J. Aubert et al., Phys. Rev. Lett. 35, 639 (1975).

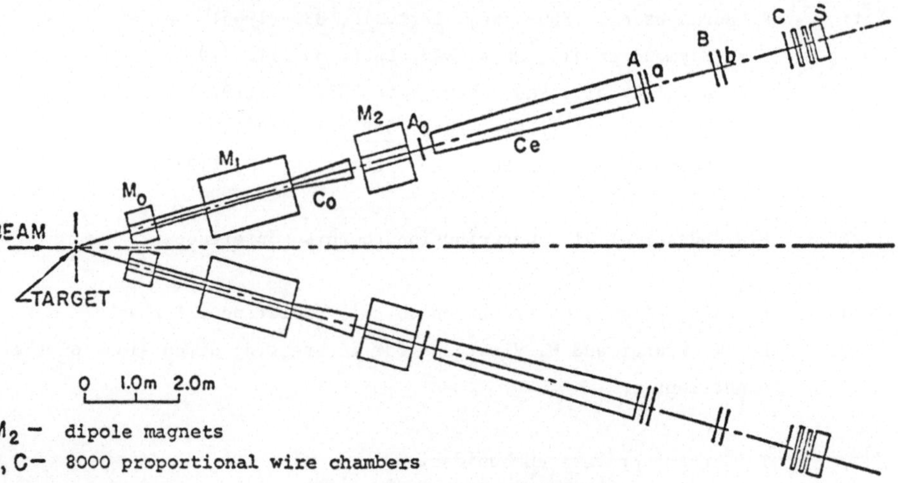

M_0, M_1, M_2 — dipole magnets

A_0, A, B, C— 8000 proportional wire chambers

a, b — 8x8 hodoscopes

S:— 3 banks of pb-glass shower counters

C_B, C_0 C_e— gas cerenkov counters

Plan view of the detector

Fig. 1a

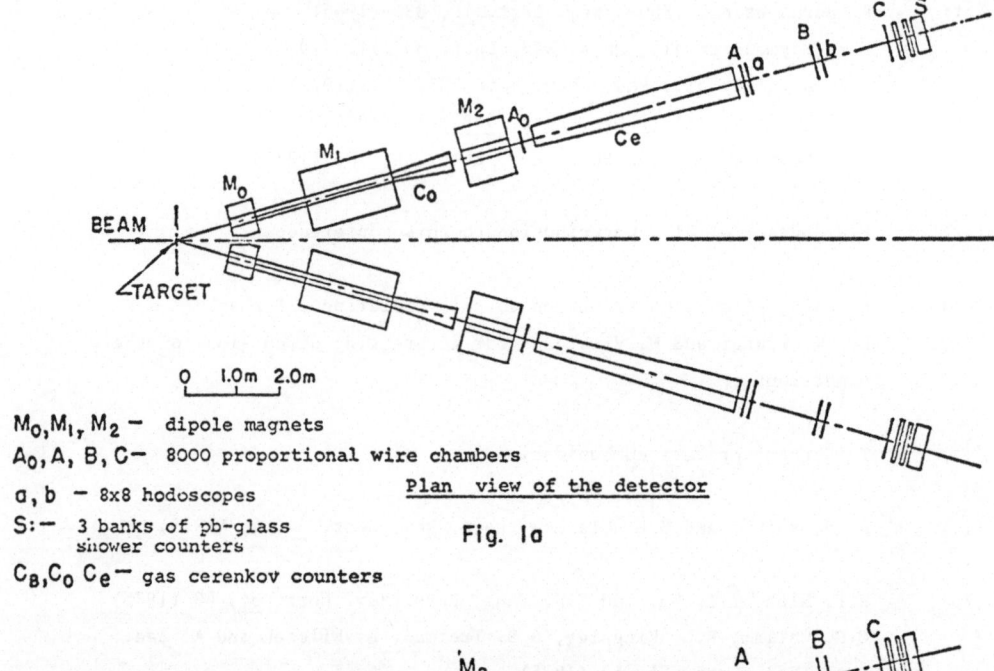

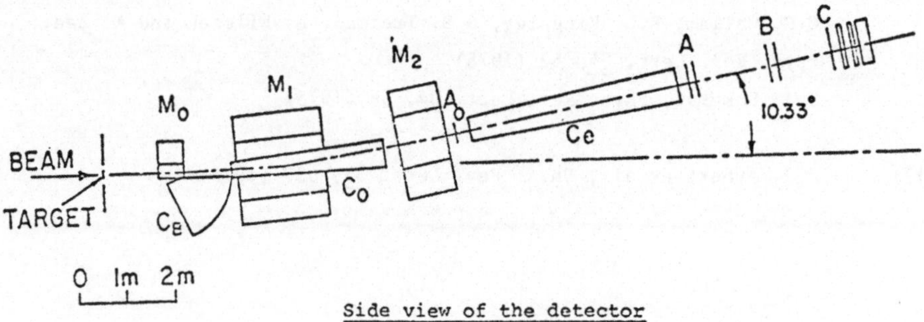

Side view of the detector

Fig. 1b

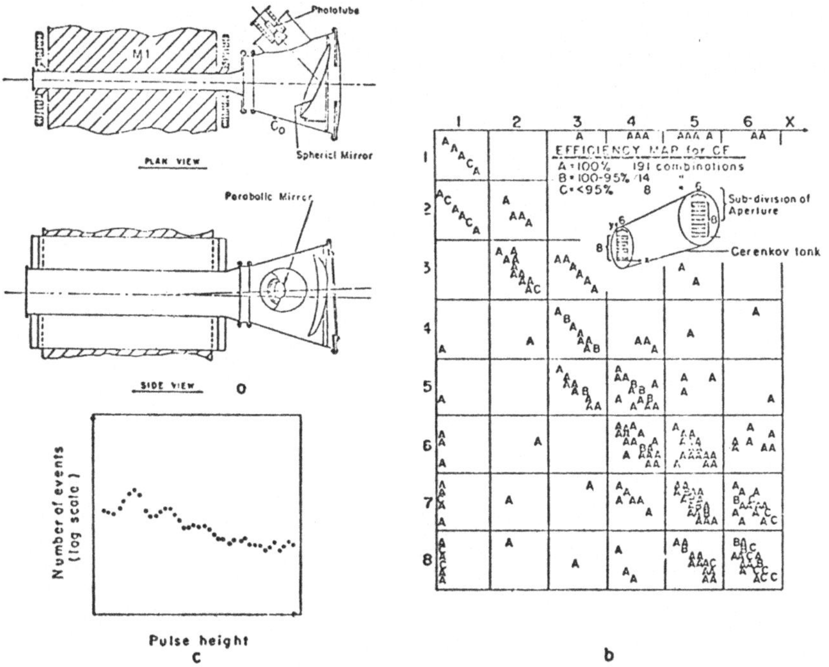

Fig. 2

a. Plan and side view of the C_o counter shown in its location in the experiment.
b. Mapping of the efficiency of the C_e counter over its whole phase space. The letters on the plot refer to efficiencies measured for trajectories between the corresponding points marked on the grid at each end of the counter.
c. Pulse height spectrom from the photo tube (RCA C31000M) of the C_o Cerenkov counter. Clearly visible are the one, two or three photoelectron peeks.

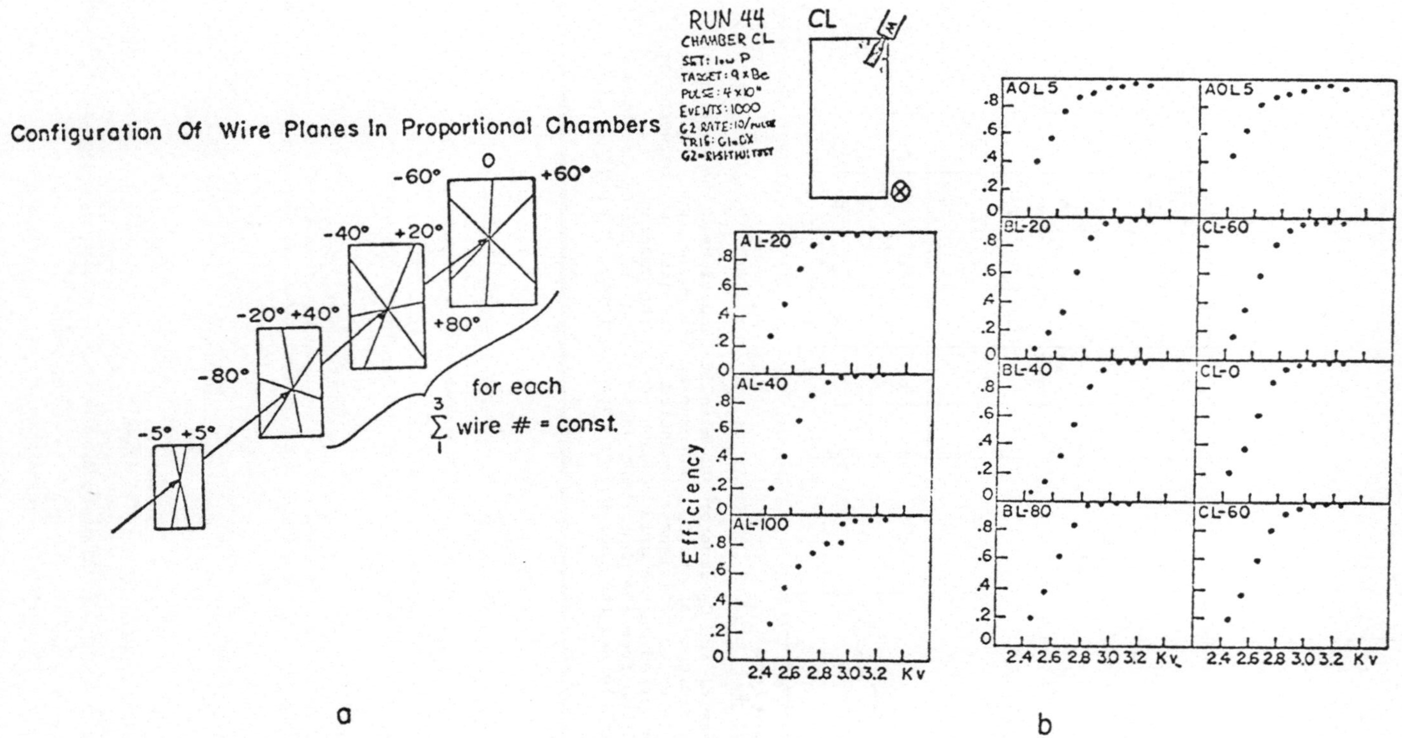

Fig 3

(a) Relative orientation of the planes of wires in the proportional chambers.

(b) Efficiency of all the wire planes as a function of the applied voltage. The position where this measurement was made is indicated by the sketch in the top left hand corner.

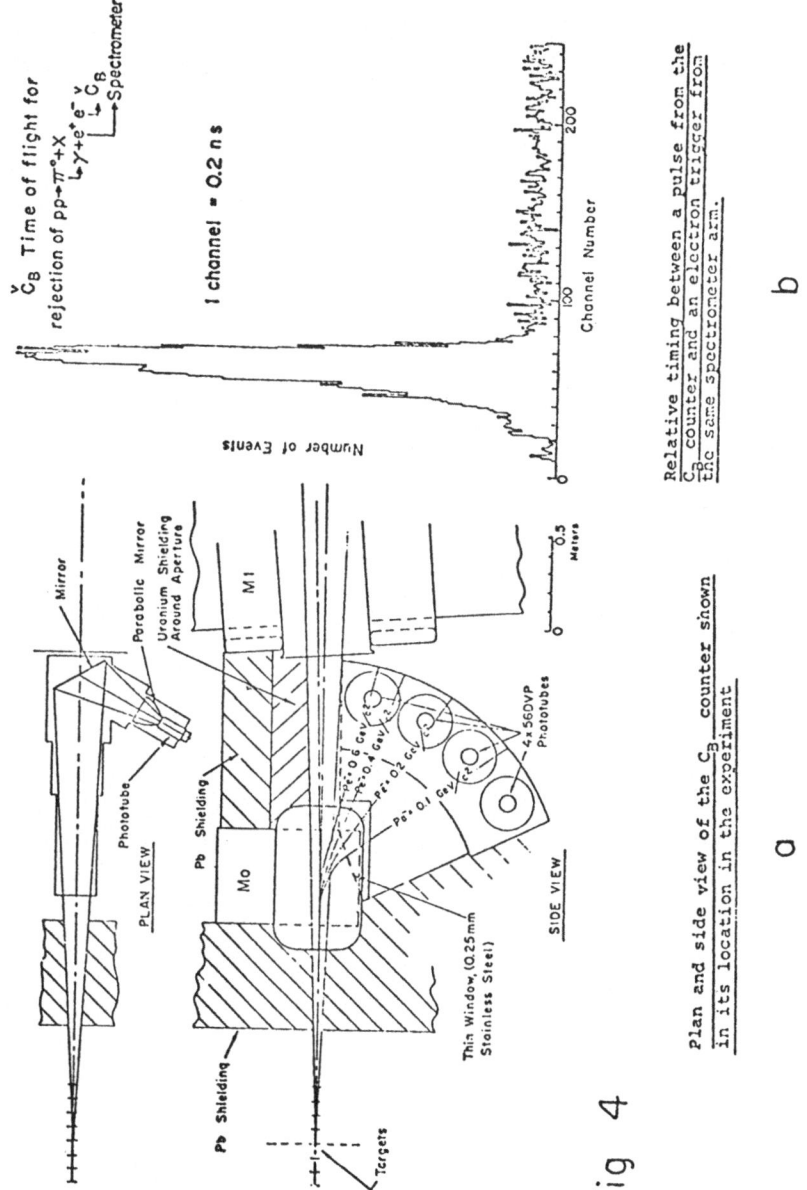

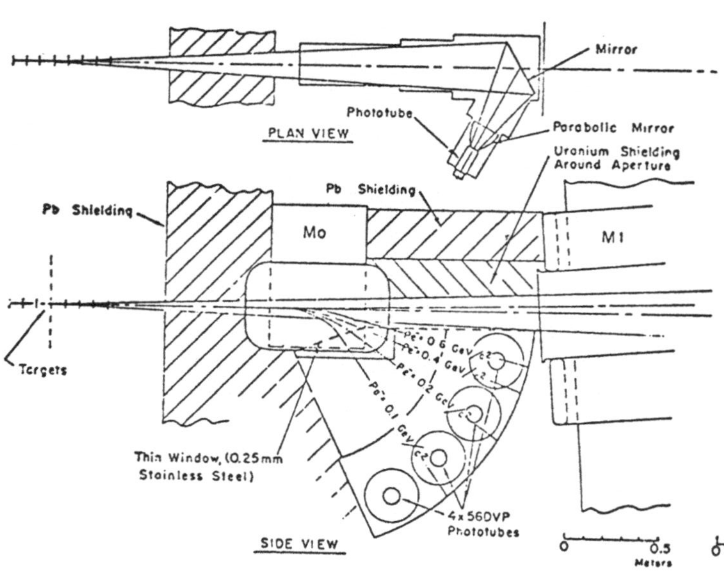

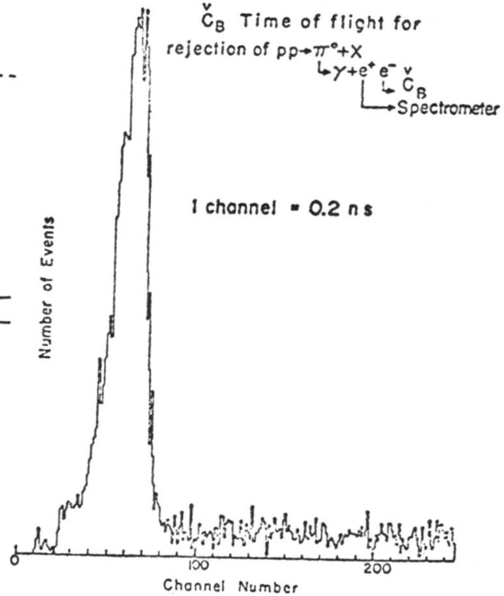

Fig 4

Plan and side view of the C_B counter shown in its location in the experiment

a

Relative timing between a pulse from the C_B counter and an electron trigger from the same spectrometer arm.

b

Fig 5

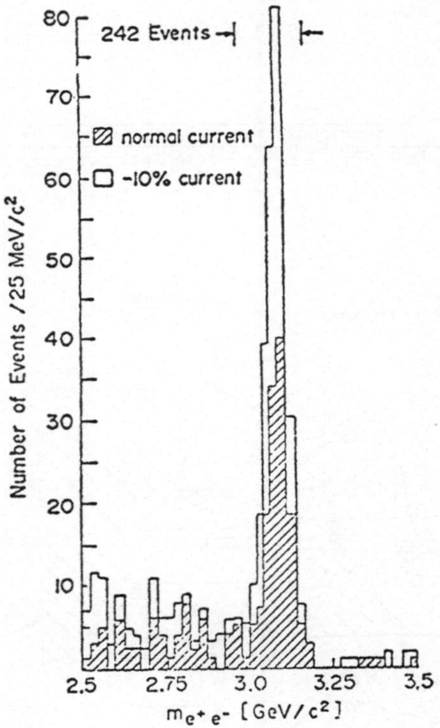

Mass spectrum for events in the mass range $2.5<m_{ee}<3.5$ GeV/c^2. The shaded events correspond to those taken at the normal momentum setting, while the unshaded ones correspond to a momentum setting 10% below normal. The acceptance is a smooth function of m.

a

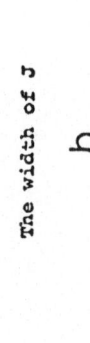

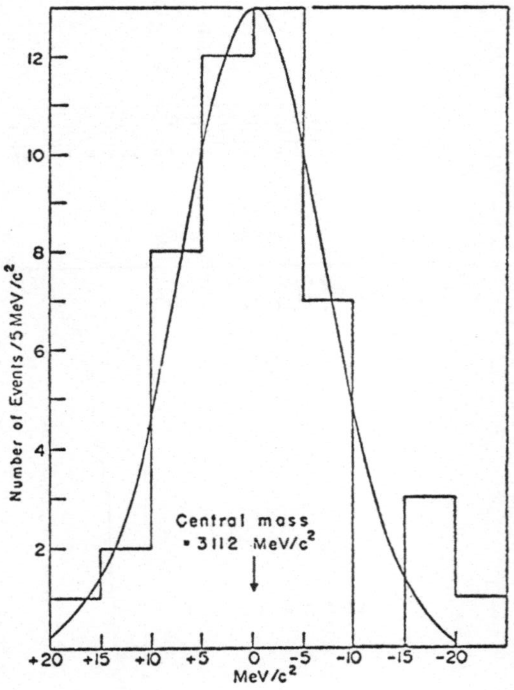

The width of J

b

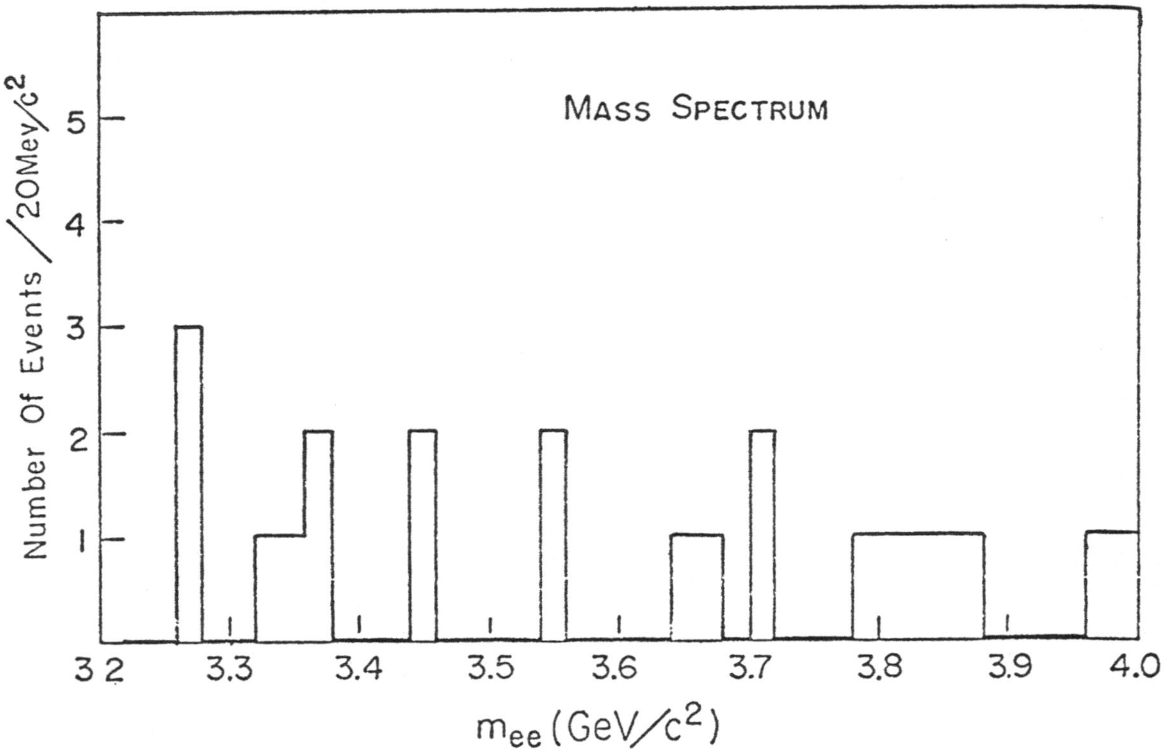

Fig. 6

Mass spectrum for events in the mass range $3.2<m_{ee}<4.0$ GeV/c^2, normalized to fig. 5a.

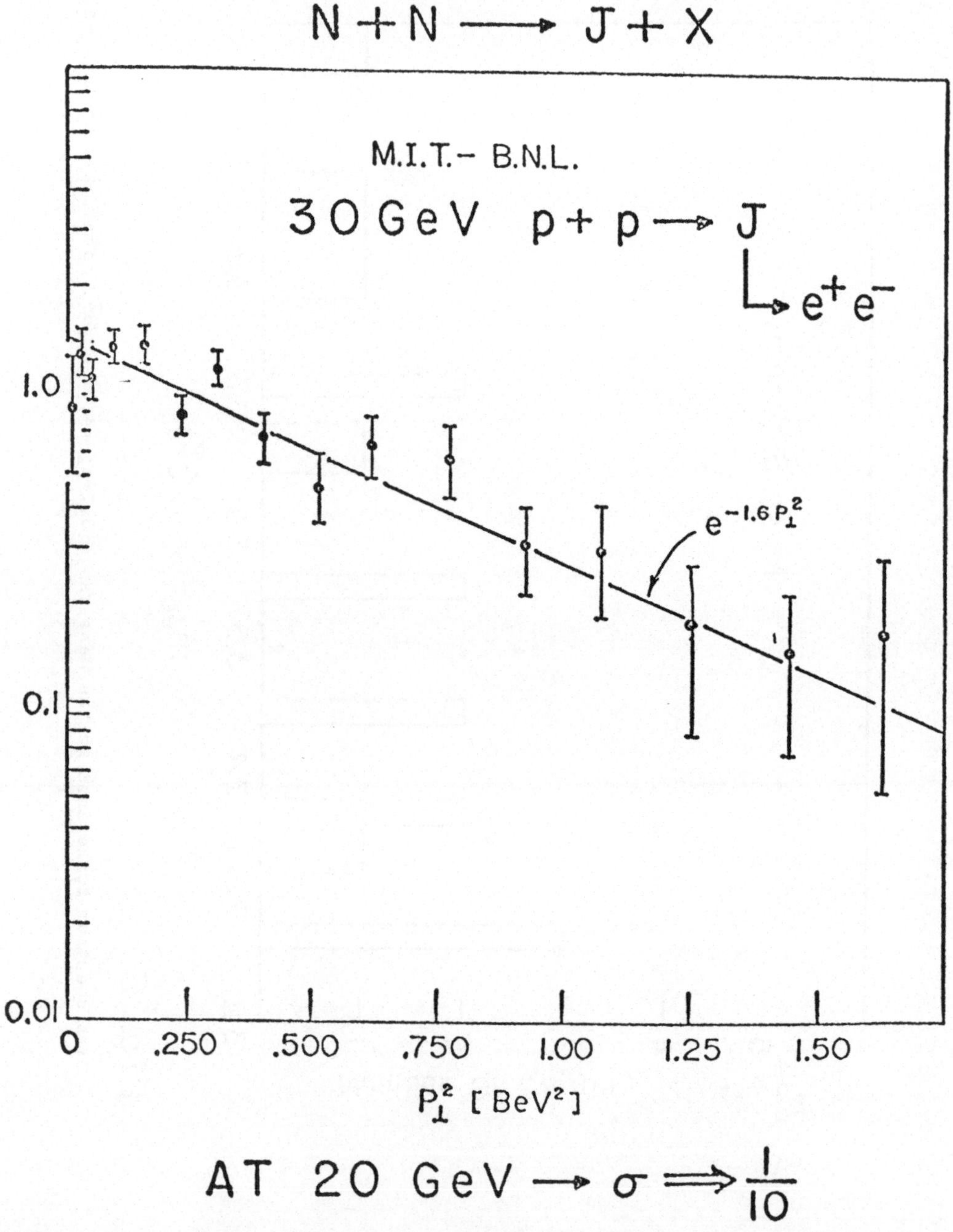

Fig. 7

Preliminary Analysis of $P_\perp^{\ 2}$ dependence of the MIT-BNL data.

I.S.R. (Di Lella Group)

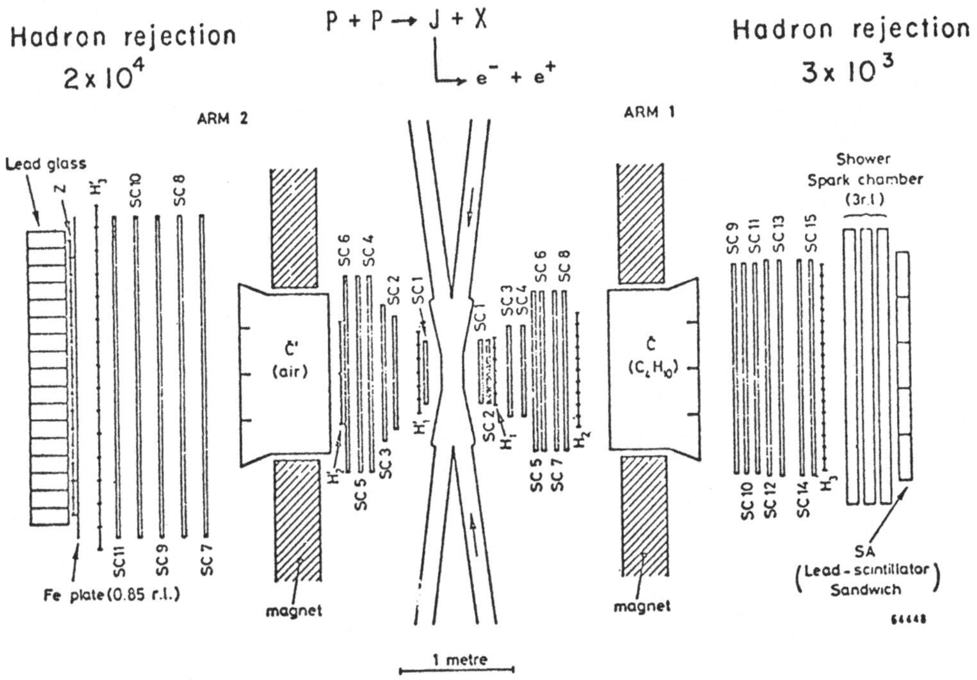

Fig. 8a) Apparatus

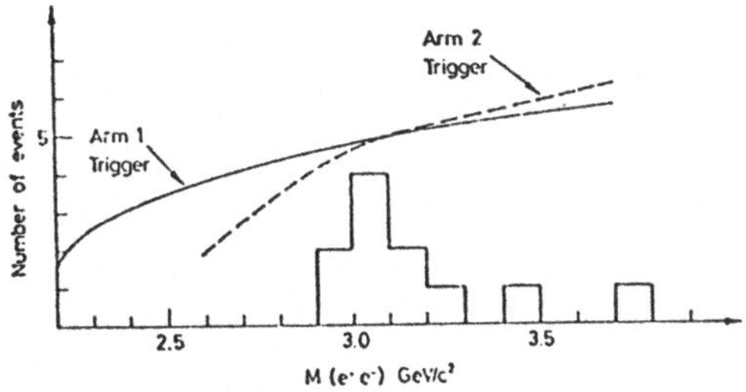

Fig. 8b) Mass spectrum

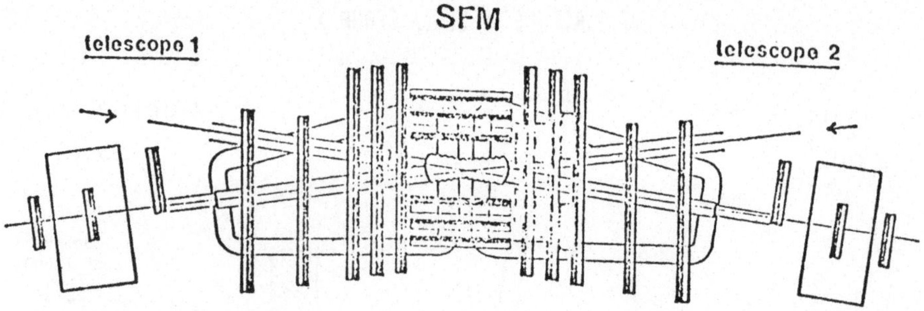

The Split Field Magnet detector of the CERN ISR.

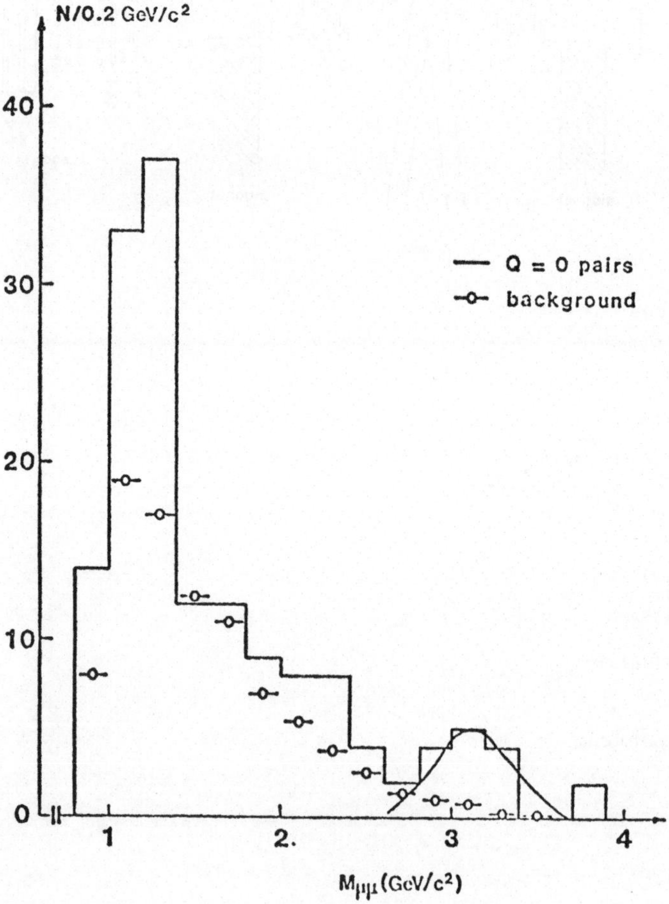

Fig.9b

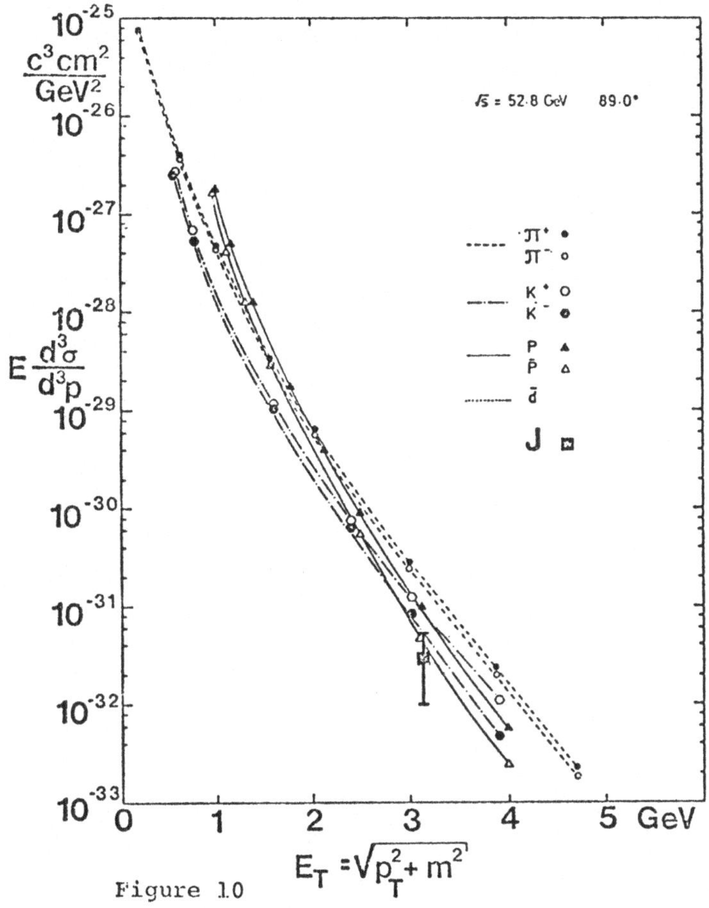

Figure 10

Comparison of the J production cross section at $y = 1.6$ and $P_\perp = 0.5$ GeV/c with pion, kaon and proton data.

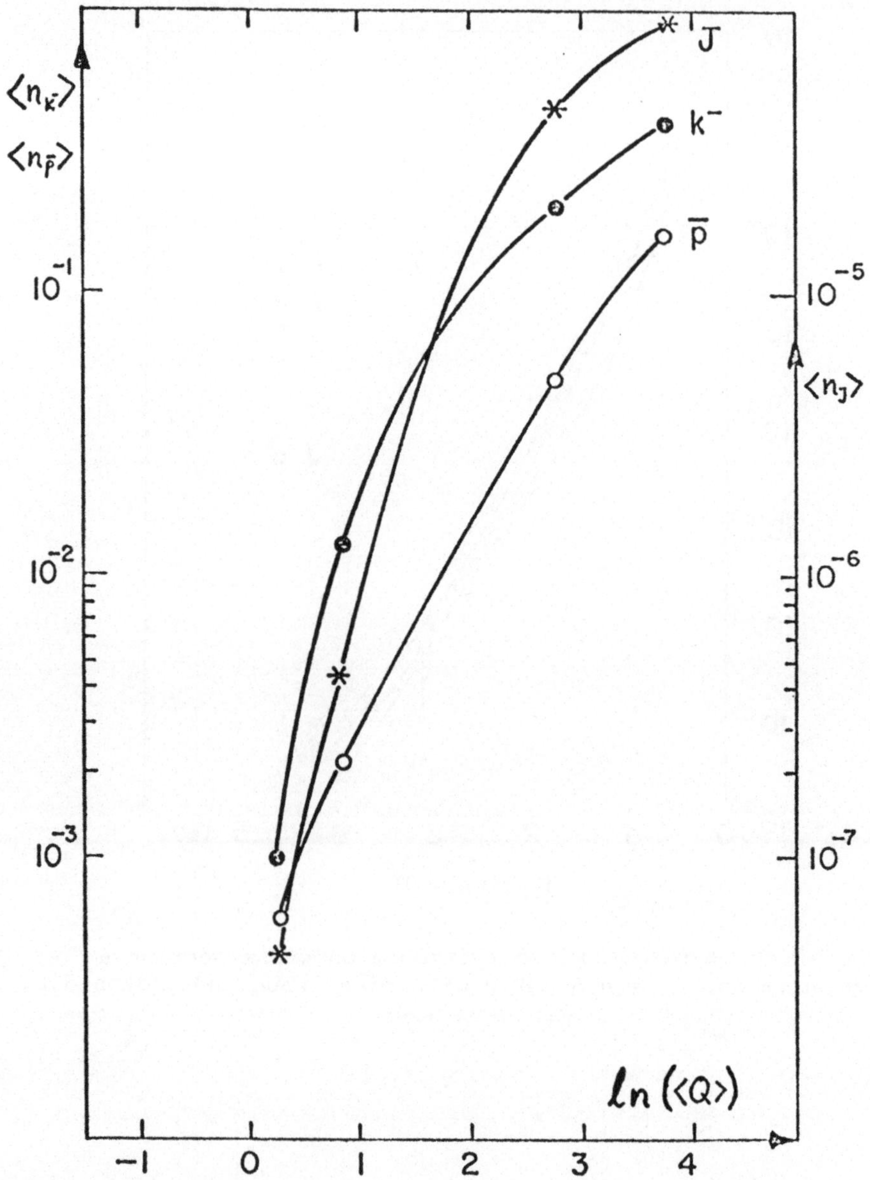

Fig. 11

Behavior of average multipicity vs $\langle Q \rangle$ value

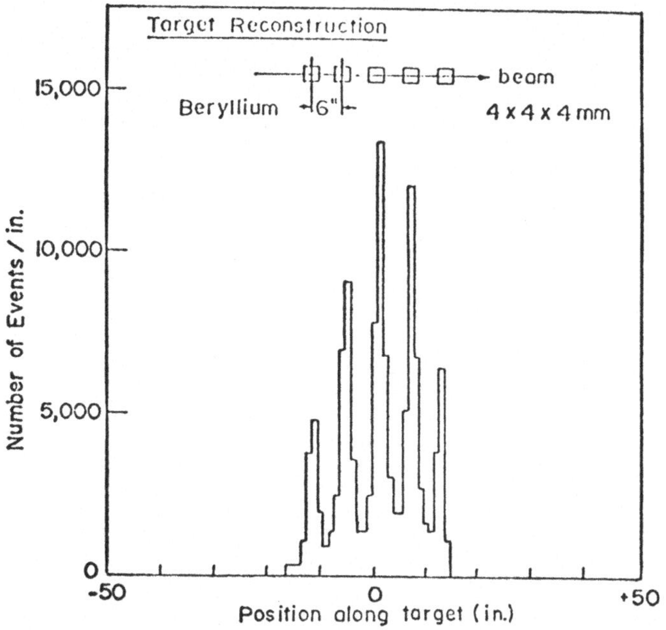

a. Reconstruction of the pair vertex at the target; using information from the proportional chambers. The five pieces of beryllium are seen clearly.

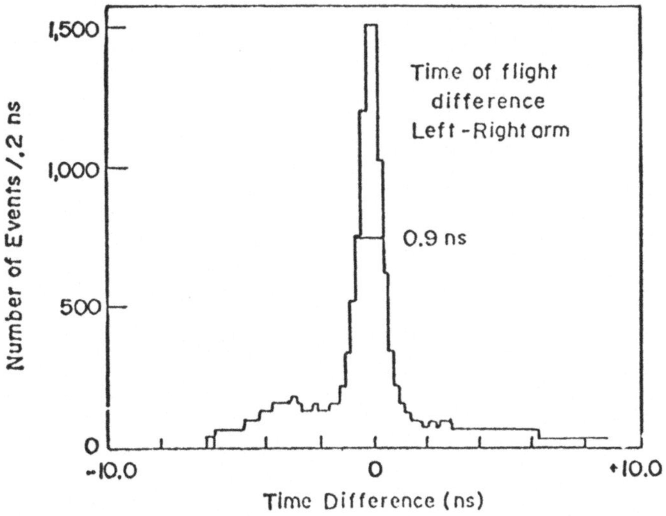

b. Time difference between additional scintillation counters in the left and right arms. The resolution obtained is 0.9ns and little background is present.

Fig. 12

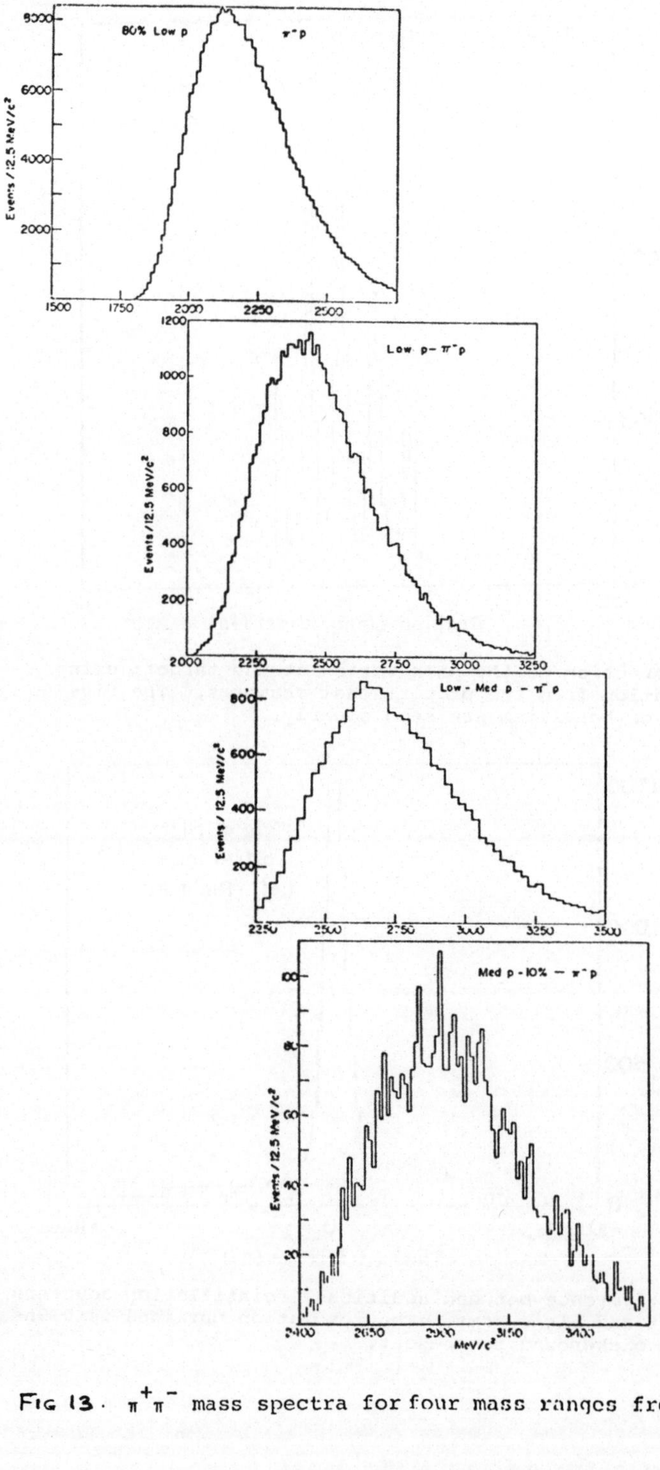

FIG 13. $\pi^+\pi^-$ mass spectra for four mass ranges from MIT-BNL

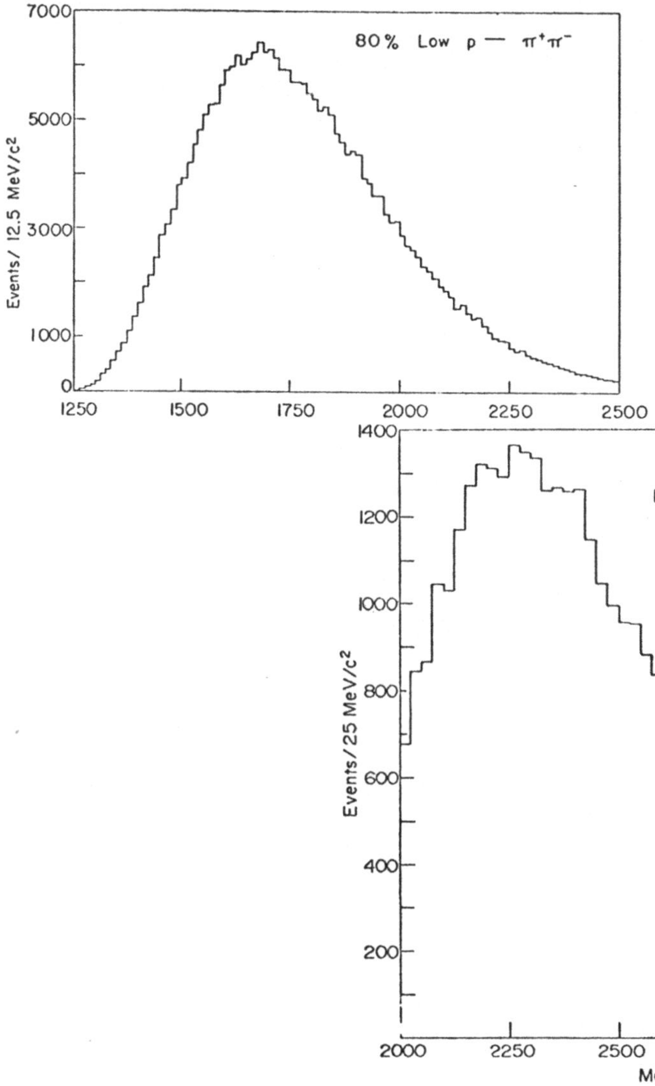

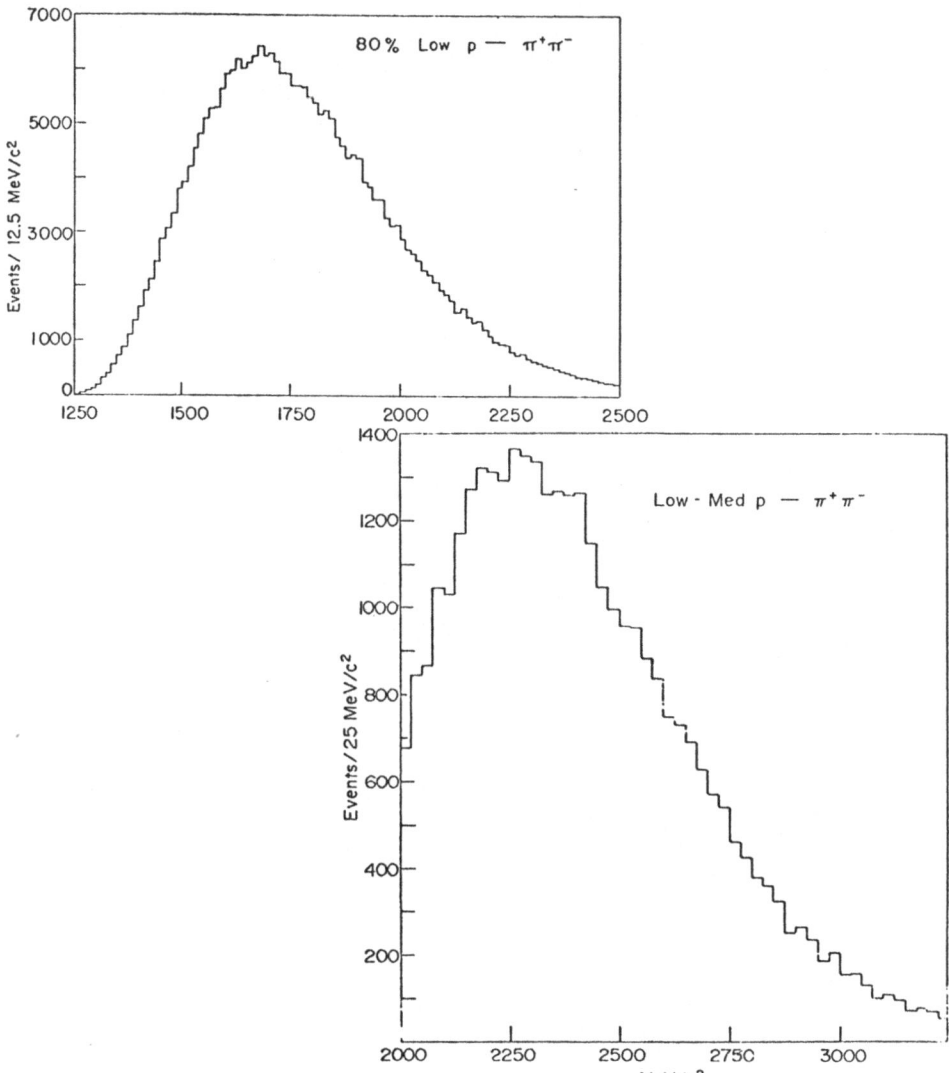

Fig. 14 $\pi^+\pi^-$ mass spectra for two mass ranges from MIT-BNL

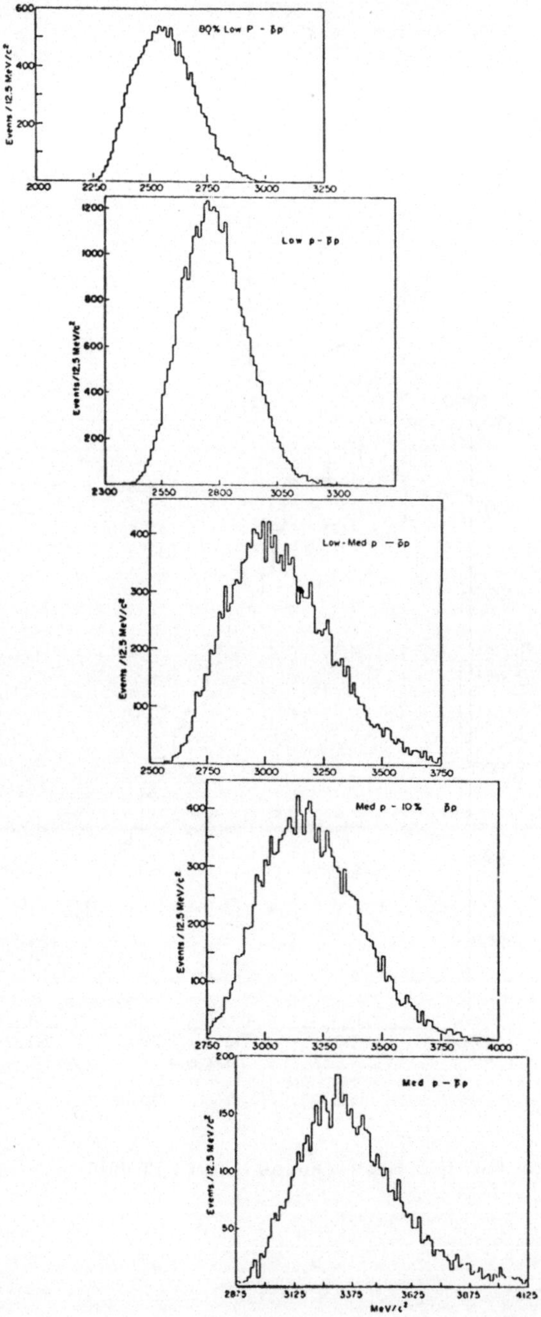

Fig. 15 $\bar{p}$ p mass spectra for five mass ranges from MIT-BNL

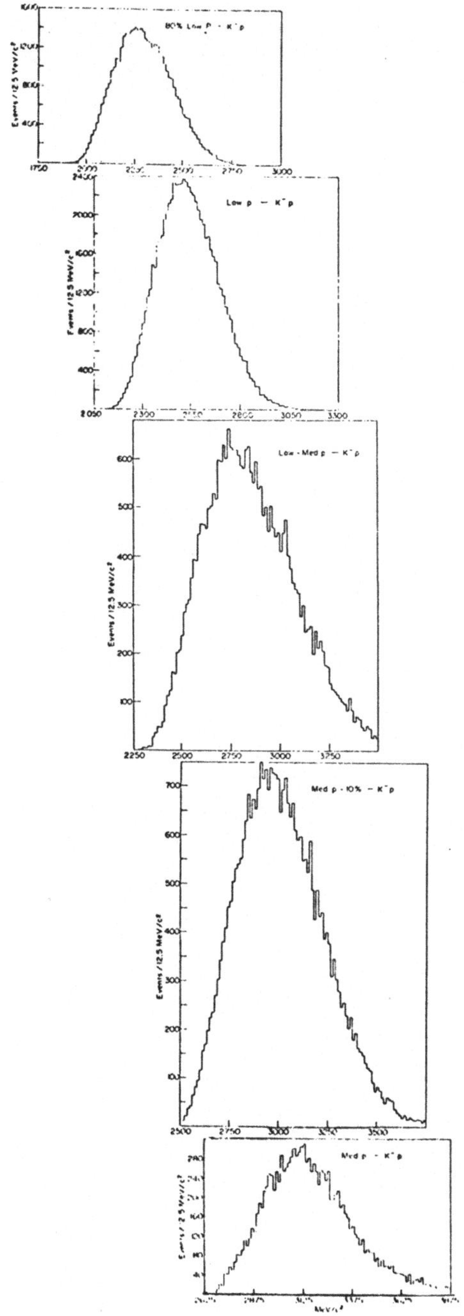

Fig. 16 K^-p mass spectra for five mass ranges from MIT-BNL

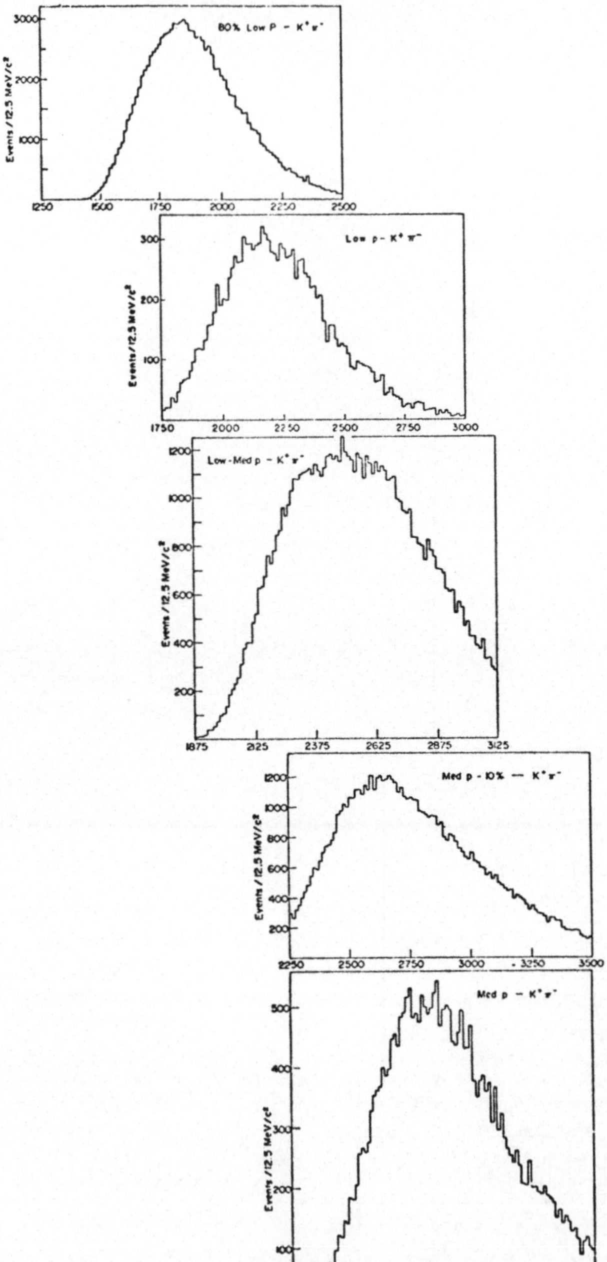

Fig. 17 $K^+\pi^-$ mass spectra for five mass ranges from MIT-BNL

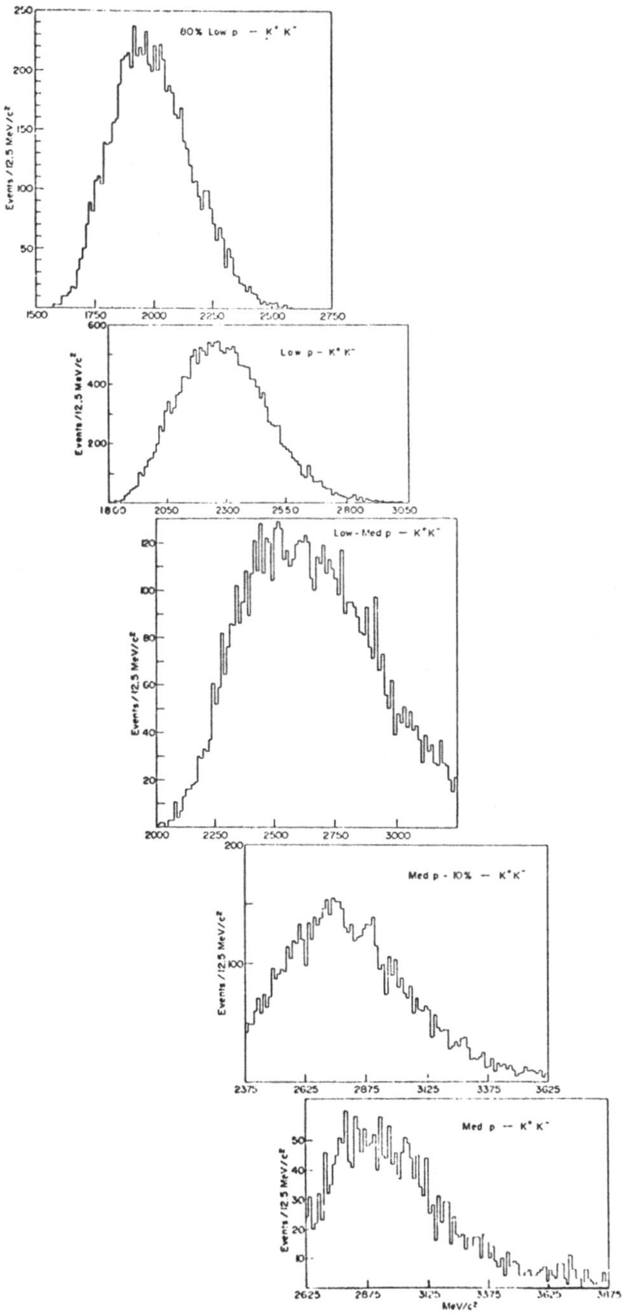

Fig. 18 K^+K^- mass spectra for five mass ranges from MIT-BNL

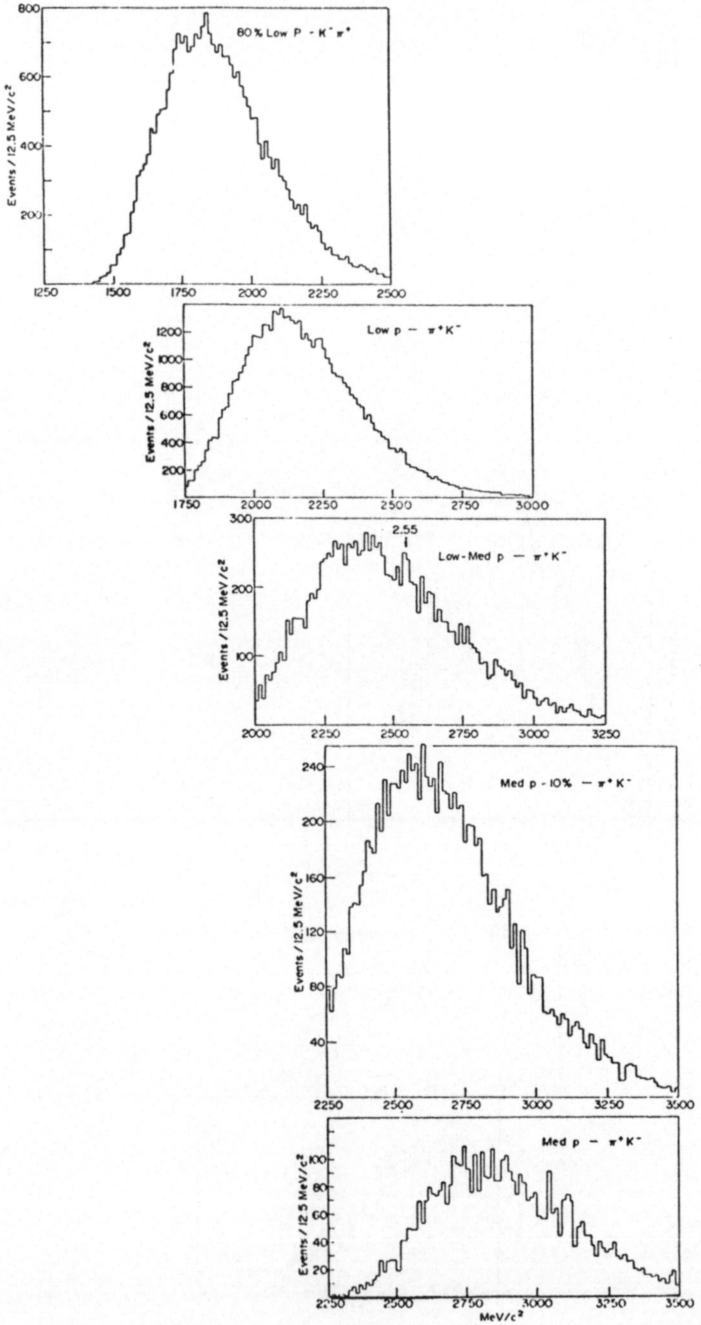

Fig. 19. $K^-\pi^+$ mass spectra for five mass ranges from MIT-BNL

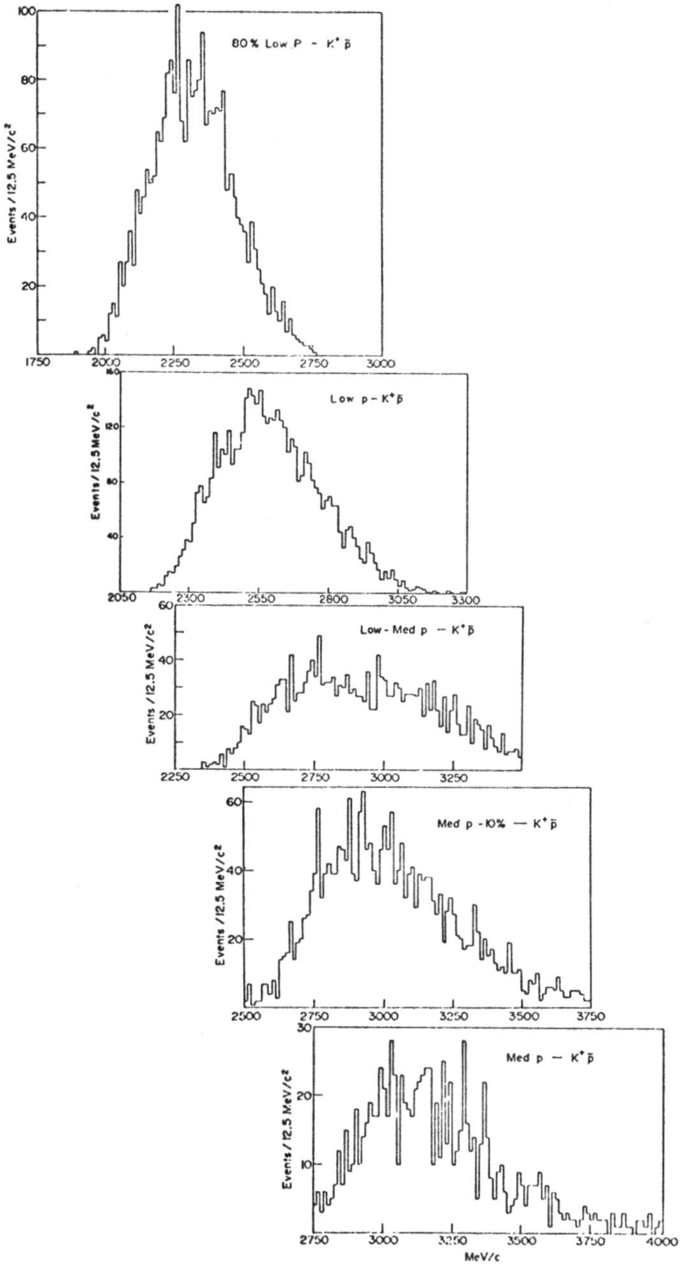

Fig. 20 $K^+\bar{p}$ mass spectra for five mass ranges from MIT-BNL

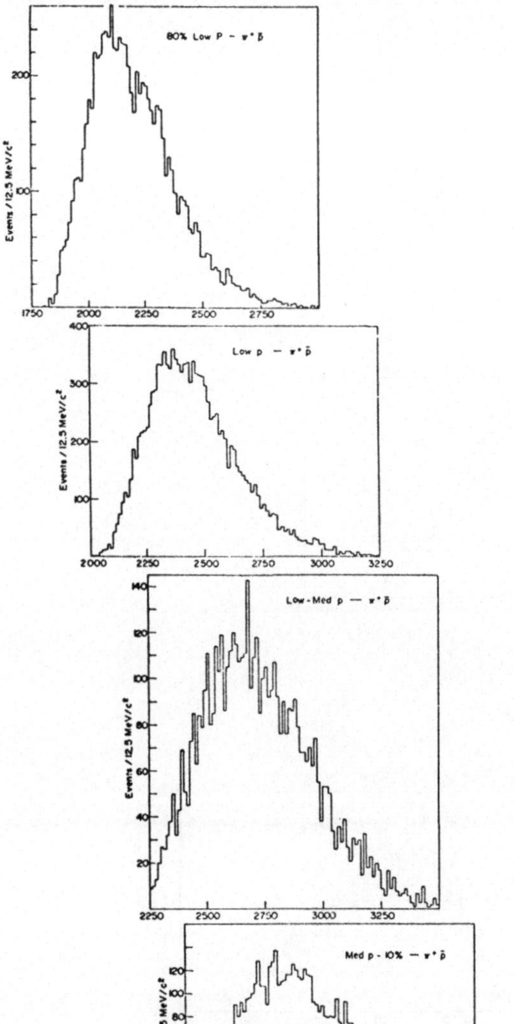

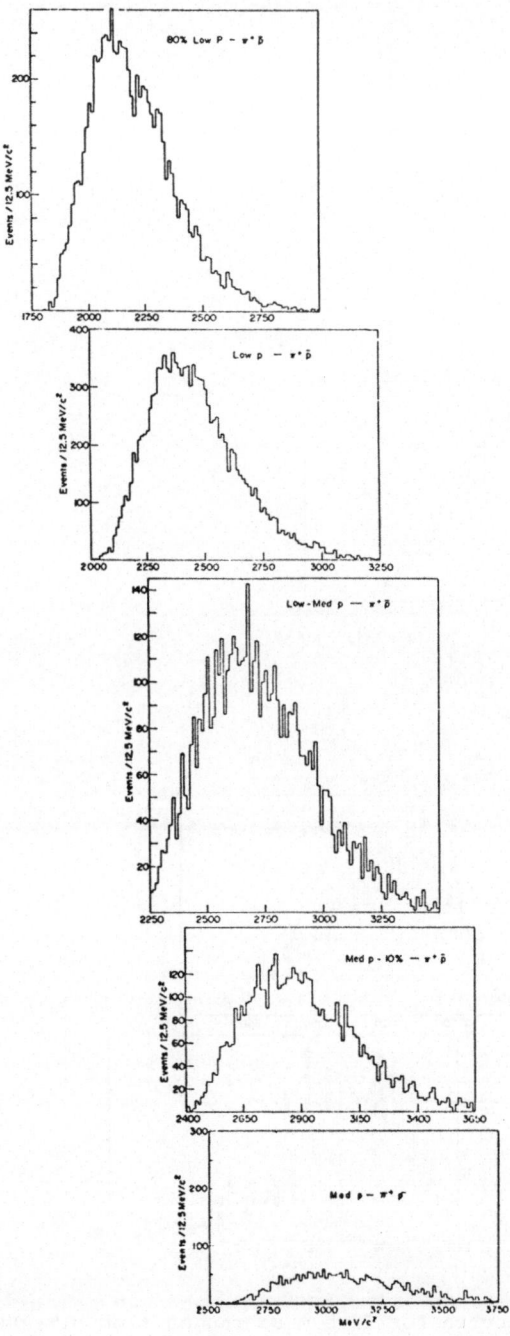

Fig. 21 $\pi^+\bar{p}$ mass spectra for five mass ranges from MIT-BNL

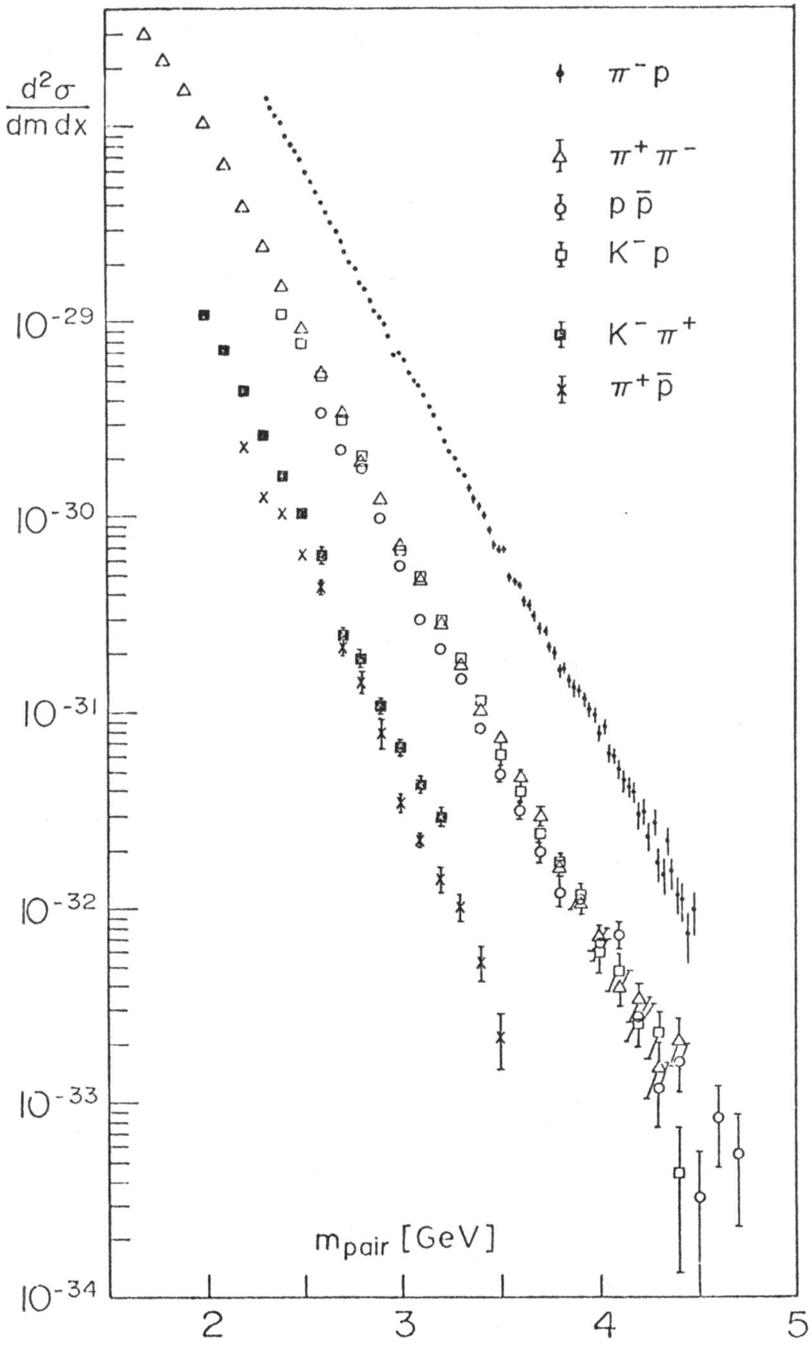

Fig. 22 Production cross section as a function of mass from MIT-BNL

OBSERVABLE QUARKS, GLUONS AND THE LEPTONS

Jogesh C. Pati
Center for Theoretical Physics
Department of Physics and Astronomy
University of Maryland
College Park, Maryland 20742

1. INTRODUCTION

Within a unified theory[1,2] of quarks and leptons the composition of the massless photon in terms of the canonical gauge fields and therefore the charges of the fermions depend in general upon the nature of spontaneous symmetry-breaking. If the symmetry is nonabelian and if in addition we demand that "color" be preserved (at least) as a good[3] global symmetry, then there are only two alternatives[4]: (i) quarks are fractionally charged: the octet of color gluons are neutral and must remain massless; (ii) quarks are integer-charged; the octet of color gluons possess charged members and acquire mass through spontaneous symmetry breaking. In case of the former, to account for the fact that no massless gluons or stable fractionally charged objects have been observed, one must assume[5] that quarks and all color suffer from "infra-red slavery" and are confined. If, on the other hand, quarks are integer-charged, there is the intriguing possibility that quarks have already been produced in pairs in hadronic and leptonic collisions, but have not yet been detected, since they decay[1] rapidly into leptons (violating baryon and lepton numbers). There is thus no need to assume confinement in this case. Color is physical and observable like flavor.

The two alternatives lead to identical physics at short distances, since both are at least temporarily asymptotically free[6]. In particular both theories enjoy the bonus of "scaling". The differences between them arise only in the infrared or long distance regime. For the former the attractive quark-quark potential should grow indefinitely, if indeed confinement has to materialize. For the latter, one would expect a turning point at large distances (e.g. $r \gg m_{\mathrm{gluon}}^{-1}$) for such a rising potential, beyond which the potential should begin to die. The recent calculations[7] of charmonium spectroscopy based on the confinement hypothesis would apply to this nonconfining potential as well (at least qualitatively) since the potentials for the two cases are not too different from each other in the region where the bulk of the wavefunction resides.

In order to distinguish between these two alternatives experimentally it is important, therefore, to search explicitly for unconfined physical color - in particular the unstable quarks and color gluons. With this in view I devote a major part of this talk to a discussion of the second alternative spelling out the expected production and decay signatures of integer-charge quarks and gluons as they arise under the restrictions of a unified theory, which Salam and I proposed some time ago.

I confine my remarks solely to the unified gauge theory approach to physical color (rather than adopting a more general phenomenological approach) for several reasons:

(i) First one has an aesthetic bias in this regard.

(ii) Second, the consequences of a unified theory (in contrast to a phenomenological theory), are fairly restrictive as we will see, and therefore its basic premises can be rejected or supported experimentally without much difficulty.

(iii) Third, the hypothesis of physical color (integer-charge quarks) together with the assumption that color-threshold is relatively low ($\lesssim$ few GeV), appears to be incompatible with existing electro- and muonproduction data <u>unless</u> we adopt the gauge theory approach to color. Within a phenomenological theory based on Han-Nambu quarks, electro- and muonproduction of color would proceed through one-photon exchange similar to flavor production; one would therefore expect in this picture color to be produced on par with flavor above color-threshold, which should result in a large rise (~ 100 %) in structure functions, as epxressed by several authors[8]; no such rise is, however, observed experimentally. Within the gauge theory approach[1], on the other hand, it has recently been shown by Salam and myself[9] and by Rajasekharan and Roy[10] that lepto-production of color must proceed through exchange of two gauge particles - the photon $\tilde{A}_\mu$ as well as its orthogonal color gauge partner $\tilde{U}_\mu$; the contributions from photon and U-exchange <u>cancel</u> each other except for the difference between their propagator functions. This cancellation-effect, which may ultimately be traced in this theory[1] to the spontaneously broken gauge-symmetry-origin of the "colored" photon, explains why color-production in electro- and muon-production experiments is not as bright as flavor-production even though photon is "colored" and even if color-threshold is relatively low. Incorporating this special feature of cancellation between photon and U-contributions, we predict that color should be visible at 10 - 20 %-level for its production-ratio in electro- and muon-production experiments in the presently available kinematic region. Analogous remarks are found to apply to neutrino-production of color. I disucss in Sec. III these results as well as their implications on

(i) dimuon-production, (ii) low x high y-anomaly observed in $(\nu,\bar{\nu})$-scatterings, (iii) nonvanishing of (σ_L/σ_T) and (iv) scaling violations observed in lepto-production experiments.

The second and main remark[11] I wish to make in this talk is that if quarks (at least the p,n,λ-quarks) are as light as about 2 GeV and if we make one further assumption that colored vector mesons lighter than quarks exist, then it appears that quark-lepton-decays[1] could provide an interesting explanation of the (μe)-events[12] as well as of the jet-structure[13] recently observed at SPEAR. Within this hypothesis one may also understand why charmed particles are not being observed at SPEAR, even if we assume that charm-excitation is responsible for a significant fraction of the observed value of R. There are several experimental distinctions between the quark versus heavy lepton-origin for the (μe)-events, which should help distinguish between these two alternative hypotheses.

II. UNIFICATION WITH LEPTON NUMBERS AS THE FOURTH COLOR: A REVIEW

1. The Basic Model

Given that quarks possess three colors (red, yellow and blue) and at least three flavors (p,n,λ), the smallest multiplet containing quarks and leptons as its members is given by a sixteen-fold[1] of four-component fermions $F_{L,R}$, which transforms as a $(4,\bar{4})$ representation of a global non-chiral SU(4) x SU(4)'-group; SU(4) acts on the four flavor-indices (p,n,λ,χ = charm), while SU(4)' acts on the four color-indices (red, yellow, blue and lilac); the fourth color "lilac" being lepton-number. Automatically the multiplet contains four leptons, which one may identify with $(\nu_e, e^-, \mu^-, \nu_\mu)$.

$$F_{L,R} = \begin{bmatrix} p_r & p_y & p_b & p_\ell = \nu_e \\ n_r & n_y & n_b & n_\ell = e^- \\ \lambda_r & \lambda_y & \lambda_b & \lambda_\ell = \mu^- \\ \chi_r & \chi_y & \chi_b & \chi_\ell = \nu_\mu \\ \text{red} & \text{yellow} & \text{blue} & \text{lilac} \end{bmatrix}_{L,R} \qquad (1)$$

All "low energy" phenomena may be described by gauging a minimal non-abelian symmetry of this multiplet:

$$G = SU(2)_L \times SU(2)_R \times SU(4)'_{L+R} \tag{2}$$

where $SU(2)_{L,R}$ gauges flavor indices $\{(p,n) + (\lambda,\chi)\}_{L,R}$, while SU(4)' gauges the four colors (red, yellow, blue and lilac) = (r,y,b,ℓ). This is the minimal symmetry capable of uniting baryons and leptons and providing a unified description of weak, electromagnetic and strong interactions. It satisfies the left $\leftrightarrow$ right discrete symmetry and thus incorporates a desirable milliweak-theory[14] of CP-violation (to be discussed later). Being non-abelian, it also provides a rationale for the quantization of electric

$$W_{L,R} = \frac{1}{2}\begin{bmatrix} \vec{\tau}\cdot\vec{W} & 0 \\ 0 & \tau_1(\vec{\tau}\cdot\vec{W})\tau_1 \end{bmatrix}_{L,R}, \quad V = \begin{bmatrix} V_{11} & V_\rho^- & V_{K^*}^- & \bar{X}^o \\ V_\rho^+ & V_{22} & \bar{V}_{K^*}^o & X^+ \\ V_{K^*}^+ & V_{K^*}^o & V_{33} & X'^+ \\ X^o & X^- & X'^- & \sqrt{3/4}\,S^o \end{bmatrix} \tag{3}$$

where $V_{11} = 1/\sqrt{2}\,(V_3 + V_8/\sqrt{3} - S^o/\sqrt{6})$; $V_{22} = 1/\sqrt{2}\,(-V_3 + V_8/\sqrt{3} - S^o/\sqrt{6})$, $V_{33} = 1/\sqrt{2}\,(-2V_8/\sqrt{3} - S^o/\sqrt{6})$ and $S^o \equiv V_{15}$. The color-octet of gauge-particles (gluons) V(8) given by ($V_\rho^\pm$, $V_{K^*}^\pm$, $\bar{V}_{K^*}^o$, $V_{K^*}^o$, V_3 and V_8) appearing in the topleft 3 x 3 block of V are coupled to (red, yellow and blue) - colors only (i.e. only to quarks). {The charges of V-particles shown above correspond to the integer-charge quark-model (see below). For the fractionally-charged quark model, V(8)-gluons would be electrically neutral and the X's would carry charges $\pm$ 2/3 e}. The effective low energy coupling parameters $g_{L,R}$ and f associated with the gauge groups $SU(2)_{L,R}$ and SU(4)' are:

$$g_L^2/4\pi = g_R^2/4\pi \approx 2\alpha \quad ; \quad f^2/4\pi \sim 1 \tag{4}$$

Subject to the nature of spontaneous symmetry-braking, the model described above leads to two distinct possibilities for quark-charges with a unique prediction for the lepton charges (0,-1,-1,0):

$$[Q]_F = \begin{bmatrix} 0 & +1 & +1 & 0 \\ -1 & 0 & 0 & -1 \\ -1 & 0 & 0 & -1 \\ 0 & +1 & +1 & 0 \\ \text{red} & \text{yellow} & \text{blue} & \text{lilac} \end{bmatrix}, \text{or} \begin{bmatrix} 2/3 & 2/3 & 2/3 & 0 \\ -1/3 & -1/3 & -1/3 & -1 \\ -1/3 & -1/3 & -1/3 & -1 \\ 2/3 & 2/3 & 2/3 & 0 \\ \text{red} & \text{yellow} & \text{blue} & \text{lilac} \end{bmatrix} \tag{5}$$

In addition, it is anomaly-free. The gauge fields of the theory $W_{L,R}$ and V generated by $SU(2)_{L,R}$ and SU(4)' respectively are:

$$Q = (F_3 + \tfrac{1}{\sqrt{3}} F_8 - \sqrt{2/3}\, F_{15}) + (F_3' + \tfrac{1}{\sqrt{3}} F_8' - \sqrt{2/3}\, F_{15}') \quad \text{(Integer charge quarks)} \tag{6}$$
$$= (F_3 + \tfrac{1}{\sqrt{3}} F_8 - \sqrt{2/3}\, F_{15}) - \sqrt{2/3}\, F_{15}' \quad \text{(Fractionally charged quarks)}$$

Note the symmetrical pattern of charges and the symmetrical combination of flavor and color-generators for the integer charge case. Defining for convenience:

$$Q_{flav.} \equiv (F_3 + \tfrac{1}{\sqrt{3}} F_8 - \sqrt{2/3}\, F_{15}) - \sqrt{2/3}\, F_{15}'$$
$$Q_{col.} \equiv F_3' + \tfrac{1}{\sqrt{3}} F_8' \tag{7}$$

the charge formula for the two cases are:

$$Q = Q_{flav} + Q_{col} \quad \text{(Integer-charge-quarks)}$$
$$= Q_{flav} \quad \text{(Fractionally charged quarks)} \tag{8}$$

From now on we discuss the masses of gauge-particles of the basic model and mass-eigen states for the case of integer-charge quarks only. The analogous discussions for the case of fractionally charged quarks (and therefore massless octet of color gluons) may be found in the third paper of Ref. 1.

2. Gauge-Masses and Eigenstates for the Case of Integer Charge Quarks

The masses of gauge particles for the basic model and the relevant physical processes, which restrict these masses, are listed below (taking $f^2/4\pi = 1$):

Mass	Process
$m(V(8)) \approx (1 \sim \text{few})$ GeV	(q-q strong force)
$m(W_L) \approx (50 \sim 100)$ GeV	V-A Int.
$m(W_R) \gtrsim 300$ GeV	V+A Int.
$m(S^0) \gtrsim 1000$ GeV	$\nu + p \to \nu +$ Hadrons
	$\nu_\mu + e \to \nu_\mu + e$
$m(X) \gtrsim 3 \times 10^4$ GeV	$K_L \not\to \bar{\mu} + e$

The masses shown above can be obtained in the integer-charge quark-model with simple representations of Higgs-Kibble-multiplets and an allowed pattern of spontaneous symmetry breaking[2]. (For example, the multiplets A = (2+2, 2+2, 1), B = (1, 2+2, $\bar{4}$), C = (2+2, 1, $\bar{4}$), D = (1, 1, 15) and E = (1, 3, 1) are adequate to generate the desired mass-pattern for the gauge particles.)

Spontaneous symmetry breaking, while giving masses to the gauge bosons induces mixing between several of them, so that the physical gauge-particles are in general mixtures of the canonical gauge particles: The nature of such mixing is determined to a large extent by the composition of the photon and the values of the effective gauge coupling constants g and f in the W and V-sectors. I list below the five neutral and six charged eigenstates for the integer-charge-model for the simplest case of spontaneous symmetry breaking (ignoring correction terms of order (g^2/f^2) or smaller and setting $g_L = g_R = g$ (for convenience of writing)):

$$\begin{aligned} A &= (e/fg)\ [f\, W_{flavor} + (2/\sqrt{3})g\, U^o]; & m_A &= 0 \\ \tilde{U} &= (3f^2 + 2g^2)^{-1/2}\ [\sqrt{3}\, f\, U^o - g\, W_{flavor} \\ &\quad + O(\varepsilon^2)]; & m_U &\sim 1 \text{ to few GeV} \\ V^o &= 1/2[\sqrt{3}\, V_8 - V_3]; & m_{Vo} &\sim 1 \text{ to few GeV} \\ Z^o &= 1/\sqrt{2}\ [(W_R^3 - W_L^3) - \sqrt{2/3}\ (g/f)S^o]; & m_{Zo} &\approx 100 \text{ GeV} \\ S &= [S^o + \sqrt{2/3}\, gW_R^3]; & m_S &\gtrsim 1000 \text{ GeV} \end{aligned} \tag{9}$$

where,

$$\begin{aligned} U^o &= (1/2)\ (\sqrt{3}\, V_3 + V_8) \\ W &= W_{3L} + W_{3R} - \sqrt{2/3}\ (g/f)S^o \\ 2e^2 &= g^2f^2/(g^2 + f^2) \simeq g^2 \\ O(\varepsilon^2) &= O[(m_U/m_{W_L})^2] \ll 1 \end{aligned} \tag{10}$$

The charged particle eigenstates (ignoring W-X mixing for this purpose and corrections of $O(\delta^2)$; $\delta \approx 10^{-4}$) are given by:

$$\begin{aligned} \tilde{V}_\rho^{\pm} &\approx \cos\beta\, V_\rho^{\pm} + \sin\beta\, W_L^{\pm} \\ \tilde{V}_K^{\pm} &\approx \cos\alpha\, V_{K*}^{\pm} + \sin\alpha\, W_L^{\pm} \\ \tilde{W}^{\pm} &\approx W_L^{\pm} - V_\rho^{\pm} \sin\beta - V_{K*}^{\pm} \sin\alpha \end{aligned} \tag{11}$$

where

$$\begin{aligned} \sin\alpha &= -\sin\ (\theta_L + \phi_L)\ (m_V/\ m_{W_L})^2\ (g/f) \\ \sin\beta &= -\cos\ (\theta_L + \phi_L)\ (m_V/m_{W_L})^2\ (g/f) \end{aligned} \tag{12}$$

Here θ_L and ϕ_L are the Cabibbo rotations in $(n,\lambda)_L$ and $(p,\chi)_L$-spaces respectively, the observed Cabibbo-angle being $\theta_c = \theta_L - \phi_L$. Note the mixings of the color-gluons with the weak valency gauge mesons W's both in the neutral and in the charged sectors. These mixings are inevitable, if the photon is to contain a mixture of valency and color-gauge meson (eq. (9)), i.e. if quarks are integer-charged. Such mixings have profound physical implications on the one hand on lepto-production of color (i.e. $e + p \to e + X_{col}$; $\nu_\mu + p \to \mu^- + X_{col}$) ; on the other hand on the decays of lowest lying color-octet states (which, without such mixings, would contain stable members by conservation of color I_3' and Y'-quantum numbers). It should be stressed at the same time that the above mixings being small do not lead to any undesirable effect either as regards parity or strangeness violation in hadronic processes or as regards q.e.d.-processes such as (g-2) for the electron and the muon[15]. We return to a detailed investigation of the consequences of such mixing in the next section.

3. Extended Local Symmetry-Vector-Like Gauges - The Mirror Fermions:

The minimal symmetry $G = SU(2)_L \times SU(2)_R \times SU(4)'_{L+R}$ requires two basic coupling constants g and f. If the unification is to be carried to a stage, where all basic forces (weak, electromagnetic and strong) are governed by one generating coupling constant, one must imbed the minimal symmetry G within a higher unifying symmetry G and interpret the former to be the "low-energy" manifestation of the latter. A natural candidate for such an extended symmetry (and one which preserves the interpretation[1] of lepton number as the fourth color) is the symmetry[16] $G = SU(4)_1 \times SU(4)_2 \times SU(4)'_1 \times SU(4)'_2$; this is the chiral extension of the symmetry $SU(4) \times SU(4)'$; $SU(4)_1$ and $SU(4)_2$ act respectively on the left and right flavor-indices of F, while $SU(4)'_1$ and $SU(4)'_2$ act respectively on the left and right color-indices of F. The symmetry G permits two discrete symmetries:

$$\begin{aligned} \text{left} &\leftrightarrow \text{right} \\ \text{flavor} &\leftrightarrow \text{color.} \end{aligned} \qquad (13)$$

Imposing these, we may obtain a unified theory with one coupling constant[17]; in addition the basic lagrangian is symmetric between left and right as well as between flavor and color. If indeed leptons define just a new color but have the same flavors as the quarks (this would imply that quarks (with three colors) are three times as many as leptons), then it appears[18] that the symmetry $G = [SU(4)]^4$ is the unique minimal choice for uniting known particle forces. Personally, I regard this symmetry to be the most elegant choice unless Nature reveals new kinds of fermionic matter (with perhaps new class of low-energy effective interactions)[19], which may not be classified either as quarks (light or heavy) or as leptons (light or heavy). We have, of course, no reasons yet to believe that such new kind of matter exists.

If we extend the minimal symmetry $G = SU(2)_L \times SU(2)_R \times SU(4)'_{L+R}$ to $G = [SU(4)]^4$, there would however have to be one important change; to secure freedom from anomalies one must supplement the basic set of sixteen fermions F by a mirror set F', the two sets F and F' being coupled with opposite chiral projections to the same set of gauge bosons; thus

$$\begin{aligned} F_L &= (4,1,\overline{4},1);\ F_R = (1,4,1,\overline{4}) \\ F'_L &= (1,4,1,\overline{4});\ F'_R = (4,1,\overline{4},1) \end{aligned} \qquad (14)$$

The resulting gauge theory is invariant under the mirror transormation

$$F_L \leftrightarrow F'_R \text{ and } F_R \leftrightarrow F'_L \qquad (15)$$

This suggestion of extension of the minimal symmetry (for the sake of complete unification) and of the introduction of the mirror fermions F' was made in Ref. 16, where it was assumed that F' is heavy and normal matter is composed of F only. Note that F' introduces four new flavors (p',n',λ',c') and thus $(4x3) = 12$ heavy quarks and four new leptons (E^o,E^-,M^-,M^o); at least the charged members of the new lepton-set must be heavier than the electron and the muon. Recently certain consequences of this mirror-hypothesis as regards the interpretation of the J/ψ-particles and the new complexions, which arise due to the presence of the two sets of fermions F and F', have been considered by Salam and myself[2o] and by Branco, Hagiwara and Mohapatra[21]. I should like to make a few brief remarks regarding the similarities and the differences between the mirror-hypothesis on the one hand and the recent work of several authors[22] based on heavy quarks and vector-like theories on the other:

(i) At a stage where the fermions are massless, the mirror-theory is automatically vector-like from the start, as is evidnet from the mirror-symmetry (eq. (15)) of the gauge-Lagrangian. To see how the "weak"-gauge Lagrangian looks like before fermions acquire mass, it is useful to write down the coupling of the gauge-mesons belonging to the $SU(2)_1 \times SU(2)_2$ subgroup of $SU(4)_1 \times SU(4)_2$:

$$\begin{aligned} & g_{4\overset{\Sigma}{colors}}\, g\, \vec{W}_1 \cdot [\binom{P}{n}_L + \binom{\chi}{\lambda}_L + \binom{P'}{n'}_R + \binom{\chi'}{\lambda'}_R] \\ + & g_{4\overset{\Sigma}{colors}}\, g\, \vec{W}_2 \cdot [\binom{P}{n}_R + \binom{\chi}{\lambda}_R + \binom{P'}{n'}_L + \binom{\chi'}{\lambda'}_L] \end{aligned} \qquad (16)$$

Note that p' denotes the mirror of p, while χ denotes charm. For every left-current, there is a right-current coupled to the same gauge boson; thus the theory is vector-like. Yet there are important differences between this theory and the standard vector-like theories[22] based on $SU(2) \times U(1)$-symmetry and six quark flavors:

(a) The SU(2) x U(1)-symmetry permits a vector-like-theory consistent with known data provided there are 2n flavors with $n \geq 3$. The mirror-theory on the other hand can not be sustained with six flavors; it requires the existence of eight and only eight flavors (independently of the development of the new data).

(b) Standard vector-like theories predict vectorial neutral current-neutrino interactions; the mirror-theory by contrast predicts[2o] in general a mixture of vector and axial vector neutral current interactions (mediated by the fmailar Z-particle). Recent experimental data[23] seem to disfavor pure vectorial neutral current interactions.

(ii) The second remark concerns the nature of physical weak-interactions. These depend upon the Fermi-mass matrix, which in general can induce Cabibbo-like ($n \leftrightarrow \lambda$, $p \leftrightarrow \chi$, $n' \leftrightarrow \lambda'$, $p' \leftrightarrow \chi'$) as well as $F \leftrightarrow F'$-mixings (e.g. $p \rightarrow p'$, $n \leftrightarrow n'$, $\lambda \leftrightarrow \lambda'$, $\chi \leftrightarrow \chi'$). I wish to emphasize however that within a unified theory (in constrast to a SU(2) x U(1)-gauge theory), the Fermi mass-matrix is quite restrictive and it does not seem to permit the sort of skew-physical couplings, which have been actively considered in the literature. (e.g. $\vec{W}_1$ being coupled to the doublets $|\binom{p}{n_\theta}_L + \binom{\chi}{\lambda_\theta}_L + \binom{\chi}{\lambda_\phi}_R + \binom{p}{n'}_R + \ldots|$ where $\lambda_\phi = \lambda\cos\phi + n\sin\phi$ with $\phi \approx 0^o$ or 90°, and (p,λ,n,c) now denoting "physical" fields.) Although these skew-orientations would have important implication (as pointed out by several authors) as regards the origins of (i) $\Delta I = 1/2$-rule, (ii) the dimuon-events and (iii) the low-x high y-anomaly; believing in a unified theory, I tend to feel, that the origins of these phenomena lie outside of such skew-interactions. (Note this is not a remark against vector-like interactions and the mirror-hypothesis as such, but just possibly against the skew-orientations of the physical interactions.) I will argue in the course of this talk that the various features of the (nonleptonic weak decays ($\Delta I = 1/2$-rule and all that)) can be understood within a more conventional framework without the introduction of the skew-interactions. Analogously I will argue that the arguments in favor of skew-right handed currents on the basis of dimuon events and y-anomaly are not compelling if we allow for the excitation of one new flavor (for example charm) as well as physical color through conventional currents.

4. Left $\leftrightarrow$ Right Symmetry, CP-Violation and Cabibbo-Rotations:

It is of interest to review the nature of CP-violation arising within the minimal (or the extended) symmetry in view of the recent improvement[24] in the measurement of the electric dipole moment of the neutron. The gauge interactions generated by the minimal symmetry $G = SU(2)_L \times SU(2)_R \times SU(4)'_{L+R}$ as well as the extended symmetry $G = [SU(4)]^4$ are nonabelian and symmetric under the interchange left $\leftrightarrow$ right. This has the consequence[25] that parity as well as CP-violation can arise in this theory entirely through spontaneous symmetry breaking. Parity Parity violation arises because W_R's (or W_2's) become heavier[1] than W_L's (or W_1's). CP violation arises[14] because Fermi-mass matrix, which is generated through spontaneous symmetry breaking and which gives rise to Cabibbo-rotations, acquires a complex form. The CP violation arising in this manner has the following desirable features:

(i) We obtain (for $m_{W_R} >> m_{W_L}$)

$$|\eta_{+-}| \approx (m_{W_L}/m_{W_R})^2 \quad |(\frac{\sin 2\theta_R}{\sin 2\theta_L}) \sin (\delta_R - \delta_L)| \tag{17}$$

where θ_L and θ_R denote Cabibbo-rotations in the left and right-handed fermion-sectors, while δ_L and δ_R are the associated phase angles needed for the diagonalization of the complex Fermi-mass matrix. Note that the observed smallness of CP-violation can be linked in this theory to the observed smallness of V+A-interactions compared to V-A interactions rather than to an arbitrarily small parameter (like the difference of phase angles). Note also that CP violation, thus arising, is intimately linked to Cabibbo-rotations in the left and right sectors; in particular the exact expression (for η_{+-}) yields that CP violation vanishes as Θ_L and/or $\Theta_R \to 0$.

(ii) Independently of the values of the phase angles, we obtain

$$\begin{aligned} \eta_{+-} &= \eta_{oo} \\ \phi_{+-} &\approx \tan^{-1} (2\Delta m/\Gamma s) \\ \eta_{+-0} &= \eta_{ooo} \\ \eta_{+-0} &\neq \eta_{+-} \end{aligned} \tag{18}$$

Note that these relations (except the last inequality) nearly coincide with the prediction of the superweak-theory; the top two relations are experimentally verified. The basic theory is however not a super-weak theory, since it predicts for example milliweak CP violation in $\Lambda \to N + \pi$-decay.

(iii) Finally the theory predicts the electric dipole moment of the neutron to lie in the region $10^{-24} - 10^{-29}$ ecm, the range corresponds to relevant Higgs masses lying in the region 10 to 10^4 GeV.

Thus the theory may clearly be distinguished from the superweak theory if the e.d.m. of the neutron is found to lie in the region 10^{-24} to 10^{-26} ecm (which, we learn, may be probed in the near future).

I wish to stress that there are three essential ingredients for the theory to permit CP violation with the desirable features as noted above. They are (a) the nonabelian nature, (b) the left-right-symmetric character of the gauge structure and (c) a minimum of four flavors. Had we chosen a $SU(2)_L \times U(1)$ gauge symmetry with four quark flavors (rather than $SU(2)_L \times SU(2)_R$ or $SU(4)_1 \times SU(4)_2$), it is easy to see that one could have rotated away the complex phase from the Fermi mass matrix and not realize this scheme of CP violation. Thus only within a limited gauge structure (like SU(2) x U(1)), one might feel the need for 6 or more flavors to generate a (desirable) theory of CP violation, as advocated recently by a number of authors. Insofar as one regards SU(2) x U(1) as only part of a super structure, however, such a view for profileration of quarks (on the basis of CP violation) appears to be unnecessary.

5. Non-leptonic Weak Decays: $\Delta I = 1/2$-Rule

There are two brief remarks, which I wish to make in connection with the problem of nonleptonic weak decays, since they have implications on model-building. First, consider the old problem that within the standard Cabibbo-theory based on (V-A) currents (with or without the GIM-modification), $K_s \to 2\pi$-amplitude is forbidden in the SU(3)-limit. The question has often been raised as to whether this calls for a departure from the familiar (V-A) interaction. Here I wish to remark[27] that $K_s \to 2\pi$-decay is especially sensitive to SU(3)-breaking; simple dynamical analysis of this decay yields a suppression factor like $(m_K^2-m_\pi^2)/(m_K^2+m_\pi^2)$, which vanishes in the SU(3)-limit, but is of order unity for physical masses. Therefore, the SU(3)-forbiddeness of this decay must not be taken too seriously.

Second, consider the problem of $\Delta I = 1/2$-rule for nonleptonic weak decays. The relevant question here is: how does one understand the relative (and "absolute") enhancement of $\Delta I = 1/2$ compared to $\Delta I = 3/2$ or octet compared to 27-plet-amplitude? Within the standard Cabibbo-theory (with or without the GIM-current), the said enhancement must be attributed one way or another to dynamical mechanisms. The fact that there exist very good reasons to expect qualitatively such an enhancement of the octet amplitude has been amply disucssed in the literatur[28] over the last 15 years. The question now is one of quantitative nature. In the light of recent developments, there are two distinct sources for the enhancement of the octet relative to the 27-plet-amplitude. (i) First, making use of asymptotic freedom for strong interactions, the short distance character of W-boson exchange is found to lead to an effective lagrangian for which the coefficient associated with the octet operator is enhanced[29] relative to that associated with the 27-plet operator. Typically this enhancement factor[29,3o] is found to be nearly 4-5 for four quark-flavors with familiar (V-A)-currents and is nearly 7 to

8 for the mirror-theory with 8 quarks. (ii) Second, above and beyond this short distance enhancement, the matrix elements of the octet operators between low-lying color-singlet states have been shown[31] to be enhanced relative to that of the 27-plet operator; the arguments here are based on current Algebra, PCAC and color-quark-model or duality. These two sources of enhancement within the minimal theory (and even better within the mirror-theory) appear to be adequate to explain the observed enhancement (of a factor $\approx$ 20 for K-decays and nearly 50 for hyperon-decays) of octet transitions relative to the 27-plet-transitions. In summary, it seems that there is no compelling need for skew right-handed current interactions on the basis of the observed $\Delta I = 1/2$ rule and the SU(3) forbiddeness of the $K_s \to 2\pi$ amplitude of the minimal theory. (See also remarks in Sec. II.3.)

III. QUARK-CHARGES; COLOR BRIGHTENING

1. Unphysical Versus Physical Color

The unification hypothesis outlined in the previous section uniquely fixes the charges of the leptons; but it does not a priori fix the quark-charges. It allows, in general (as also other alternative approaches) two very distinct possibilities:

(i) Quarks are fractionally charged; the octet of color-gluons are massless. In this case, from experimental considerations, one must assume that quarks and all color are confined. In short by hypothesis color is unphysical (although there exists no theoretical proof for such a hypothesis at present).

(ii) Alternatively; quarks are integer-charged. The octet of color-gluons (possessing charged members) acquire mass ($\approx$ 1 to few GeV). Local color-symmetry is broken, yet a global color-symmetry is preserved up to order α corrections. As mentioned before, there is no need to assume confinement in this case, since integer charge quarks, following the quark-lepton-unification hypothesis[1], may decay rapidly ($\tau_q \lesssim 10^{-10}$ sec.) into leptons ($q \to \ell + \ell + \bar{\ell}$, $q \to \ell +$ pions) and such short lived quarks are likely to have been missed by standard particle and quark-searches, even if their production cross section was not insignificant ($\gtrsim 10^{-32}$ cm^2 say). In short, with integer-charge quarks color is as physical as valency (or flavor); there lies a rich world of color to be discovered.

It is quite possible in fact that color has already appeared. We point out that the anomalous $(\bar{\mu}e)$-events[12] seen at SPEAR may owe their origin to pair production of quarks in e^-e^+-annihilation of quark-like objectS (see later) -

(rather than heavy leptons), followed by quark and antiquark decays into leptons. In addition, of course, one or several of the multitude of J/ψ-particles may represent color. Familiar arguments[33] against the color interpretations for the J/ψ-particles do not apply to the cases[20] where some of them represent color, others representing new flavor quantum numbers.

2. Familiar Arguments Against Physical Color

The hypothesis of decaying integer-charge-quarks (physical color) is an attractive alternative to the more commonly held hypothesis of fractionally charged quarks (unphysical color), since it does not need to hide the basic particles of the theory and yet it provides a reason for the "missing quark". Despite these possible advantages, there have been a set of standard objections[33] to the twin hypothesis that (i) quarks carry integer-charges (color is physical) and (ii) color-threshold is not very high (i.e. it is within reach of lepto-production experiments at least at Fermilab). We first list below these familiar objections and then point out a flaw in them; which invalidate these objections:

(i) If quarks carry integer-charges, sufficiently above color-threshold, the color-part of quark-charges (Q_{col}) would be expected to contribute to electro (or muon)-production structure functions on par with the valency part of quark-charges (Q_{flav}). If color-threshold was as low as 2 to 5 GeV, then color ought to have brightened in present electro- or muon-production experiments with a large rise (~ 100 %) in the structure functions, contrary to observations.

By the same token, sufficiently above color-threshold, familiar parton model based sum rules should have exhibited integer-character for quark-charges rather than fractional. For example, (ignoring gluon-contributions) the famous ratio

$$r \equiv \frac{4G_F^2 M_N E_\nu}{2\pi(\sigma_\nu + \sigma_{\bar{\nu}})} \int F_2^{\gamma N} \, dx \tag{19}$$

of electro and neutrino-production cross-sections should have acquired a value of 0.5, if quarks are integer-charged, rather than[33,34] a value ≈ 5/18 appropriate for fractional charges.

(ii) It is known that the gluon-content of the nucleon in the deep inelastic regime is non-neglible since nearly 50 % of the nucleon-momentum resides within gluons. If quarks are integer-charged, then sufficiently above color-threshold, the Yang-Mills coupling of the charged members ($V_\rho^\pm$, $V_{K^\times}^\pm$) of the spin-1 gluons to the photon would be expected to lead to a non-vanishing (large contribution) to σ_L/σ_T, which should be growing[35] with $|q^2|$. This too is qualitatively not supported by the data.

Accepting the above arguments, one would be inclined to conclude that either quarks are fractionally charged, or that color-threshold is sufficiently above 10 GeV beyond the reach of present experiments. This has been the standard point of view over the years. If this were true, none of the observed new phenomena (such as J/ψ-particles, or dimuons, or $(\bar{\mu}e)$-events could have been attributed to excitations of color. We show[9,10] below that the above arguments, while they might have been sound within a phenomenological approach to color, do not hold within the gauge theory approach.

3. A Flaw in the Familiar Arguments

First, we argue that even though photon carries color, color cannot brighten on par with flavor for all (spontaneously broken) gauge theories, satisfying the following two criteria: (1) flavor and color are gauged independently with weak interactions associated with valency gauging, strong with color-gauging and (2) leptons are treated as singlets of SU(3)'-color.

The argument is simple: express the gauging pattern above symolically in the form:

$$L_{int} = g\, W_\mu \,(J_\mu^{flavor} + J_\mu^{leptons}) + f\, V_\mu\, J_\mu^{color}$$

where W stands for the weak and V for the strong gauges. Notice now that before spontaneous symmetry breaking (when all gauge gields are massless), leptons interact with the quark-valency-current J_μ^{flavor} through the intermediacy of W_μ's: but there is no interaction between $J^{leptons}$ and J^{color}. Due to spontaneous symmetry breaking valency and color gauge mesons mix. This generates (through diagonalisation of fields) the massless photon A_μ as a mixture of W_3 and the color-gluon $U^o = 1/2\,(\sqrt{3}\,V_8 + V_8)$, but inevitably also the orthogonal color-gauge partner $\tilde{U}_\mu$ (with mass m_U), both of which contribute to lepton-color-interaction. The two contributions would exactly cancel each other, except for the difference between the photon and $\tilde{U}$-propagators. This has the consequence that while lepton-flavor interaction has the matrix element $J^{lepton}\,(1/q^2)\,J^{valency}$; the corresponding expression[36] for lepton-color-interaction equals $J^{lepton}\,(1/q^2 - 1/(q^2-m_U^2))\,J^{color}$, where the local symmetry is spontaneously broken. It is the crucial negative sign between the two propagators in the second case, resulting from a diagonalization of fields, which suppresses color-brightening effects. The resulting matrx element for lepto-production of color thus acquires a new kinematic factor within the gauge-theory approach (compared to the matrix-element for lepton-flavor-interaction):

$$\Delta(q^2) \equiv q^2 \left(\frac{1}{q^2} - \frac{1}{q^2 - m_U^2}\right) = - \frac{m_U^2}{q^2 - m_U^2} \tag{20}$$

It is this kinematic Δ^2-factor (in the cross-section) that provides the sharp distinction between lepto-production of color and flavor. Note that asymptotically $\Delta^2(q^2) \to 0$. In other words, leptons having started life as color-singlets, are not efficient to probe the color-octet part of quark-charges.

The familiar arguments outlined in subsection 2 were based upon the presumption that lepto-production of color (like that of valency) proceeds through one-photon-exchange only. They do not, therefore, take into account the cancellation effect between the photon and the U-gluon contributions. In the next section, we discuss in detail the consequences of this cancellation-effect on the color-contribution to structure functions.

4. Lepto-Production of Color Through Quark and Gluon Partons

A. Though the results stated in the previous subsection are general, they may explicitly be verified within the minimal symmetry model[2] $G = SU(2)_L \times SU(2)_R \times SU(4)'_{L+R}$, for which the mass matrix and the eigenstates are available in detail. We write below the coupling of the two low-mass eigenstates[37] - the photon A_μ and the color-gluon $\tilde{U}_\mu$ - which are coupled both to leptons and to the color-octet current (see eq. (9)); we have dropped $O(\epsilon^2)$-corrections):

$$L_I = e A_\mu [J_\mu^{flav} + J_\mu^{col}] + \tilde{U}_\mu [(\sqrt{3/2})\, f\, J_\mu^{col} - (2/\sqrt{3})\,(e^2/f)\, J_\mu^{val}] \tag{21}$$

where the fermionic contents of the currents are:

$$\begin{aligned} J_\mu^{flav} &= \sum_{\alpha=r,y,b} \{2/3\, \bar{p}_\alpha p_\alpha - 1/3\, \bar{n}_\alpha n_\alpha - 1/3\, \bar{\lambda}_\alpha \lambda_\alpha + 2/3\, \bar{\chi}_\alpha \chi_\alpha\}_{L+R} \\ &\quad - (\bar{e}e + \bar{\mu}\mu)_{L+R} \\ J_\mu^{col} &= \sum_{q=p,n,\lambda,\chi} \{-2/3\, \bar{q}_r q_r + 1/3\, \bar{q}_y q_y + 1/3\, \bar{q}_b q_b\}_{L+R} \end{aligned} \tag{22}$$

Note α runs over three colors (red, yellow, blue) = (r,y,b) and q runs over four flavor indices (p,n,λ,χ). The contributions to J_μ^{flav} and J_μ^{col} arising from the Yang-Mills coupling of the gauge fields $\vec{W}$ and the color-gluons V(8) respectively are not exhibited, but should be understood.

Note that on account of the W_{flavor}-component, the physical gauge particle $\tilde{U}_\mu$ becomes directly coupled to the electron. Such a coupling must be present within the gauge theory approach, if the photon must couple to J^{col}. The strength and relative sign of such a coupling are determined by the composition of the massless photon and the renormalized effective gauge coupling parameters.

Note the crucial feature of eq. (21). The strength factor for the current correlation $(J_\mu^{lep}(x)\, J_\nu^{col}(x'))$ arising due to photon-interaction in second order is $-e^2$, while that arising due to $\tilde{U}$-interaction is $(2/\sqrt{3})\,(e^2/f)\,(\sqrt{3/2})f = +e^2$. The two contributions, as mentioned before, thus exactly cancel each other except for the difference between the photon and the $\tilde{U}$-propagators.

Treating the leptons (but not hadrons) perturbatively and including the contribution of only one photon-exchange[38] for color-singlet (flavor)-production, but photon as well as U-exchanges[39] for color-production, we obtain

$$M(e + N \to e + X_{flav}) = -e^2\,(\bar{e}\gamma_\mu\, e)\, D_\gamma(q^2) < X_{flav}\,|J^{flav}|\, N>$$

$$M(e + N \to e + X_{col}) = -e^2\,(\bar{e}\gamma_\mu\, e) D_\gamma\,(q^2)\,\Delta(q^2) < X_{col}\,|J^{col}|\, N > \qquad (23)$$

where

$$\Delta(q^2) = (D_\gamma(q^2) - D_U(q^2))\, /\, D_\gamma(q^2) = \frac{m_U^2}{q^2 - m_U^2}$$

Here $D_{\tilde{\gamma}}(q^2)$ and $D_U(q^2)$ are the renormalized propagator functions for the photon and $\tilde{U}$ respectively. Thus there is the additional kinematic factor $\Delta(q^2)$ for the matrix element for lepto-production of color relative to that of flavor production. (Quite clearly an identical factor would arise for time-like processes ($q^2 > 0$) such as e^-e^+-annihilation.)

We thus deduce that for integer-charge quarks the structure-functions $F_{1,2}^{eN}(q^2,\nu)$ (representing cross sections) are sums of two pieces:

$$F_i^{eN}(q^2,\nu) = F_i^{flav}\,(q^2,\nu) + \bar{F}_i^{col}\,(q^2,\nu) \qquad (24)$$

where

$$\bar{F}_i^{col}\,(q^2,\nu) = \Delta^2\,(q^2)\,F_i^{col}\,(q^2,\nu) \qquad (25)$$

$F_i^{flav}(q^2,\nu)$ and $F_i^{col}(q^2,\nu)$ are defined in the usual manner by the Fourier transforms of the current correlation matrix elements.

Above threhold for color-production (i.e. $M_x^2 \equiv M_N^2 + 2M_N\nu - |q^2| > M_{col}^2$ F_i^{col} should be non-vanishing; however we do not expect parton-model (light-cone or asympotic-freedom) considerations to apply to F_i^{col} unless $|q^2|$ and $M_N\nu$ are sufficiently above characteristic color-octet masses. Noting that characteristic masses for flavor and color transitions (in the q^2-variable) are of order m_ρ^2 and m_U^2 respectively, and using the empirical fact that $F_i^{flav}(q^2,\nu)$ are damped[36] for small $|q^2|$ and that they acquire their "full weight" (scaling value) for $|q^2| > 2m_\rho^2$, we may thus expect parton-model considerations for $F_i^{col}(q^2,\nu)$ to apply only when $|q^2|$ and $M_N\nu \gtrsim 2m_u^2$. (For lower values of $|q^2|$, $F_i^{col}(q^2,\nu)$ should be damped, the damping becoming progressively more severe as $|q^2| \to 0$.)

Including the Δ^2-factor, the net asymptotic contributions of quark as well as charged spin-1 gluon-partons[40] to electro-production structure functions and to the R-parameter (for e^-e^+-annihilation) are given by:

$$F_1^{eN} = \{F_1^{flav}\} + (1+\xi)^{-2}[F_1^{col}]$$
$$= \{\tfrac{1}{2}\sum_{q_i=p,n,\lambda,\chi} Q^2_{flav}(q_i)[q_i(x)+\bar{q}_i(x)]\} \tag{26}$$
$$+ (1+\xi)^{-2}\left[\tfrac{1}{3}\sum_{q_i}\{q_i(x)+\bar{q}_i(x)\} + \tfrac{16}{3}(1+\tfrac{\xi}{4})v(x)\right]$$

$$F_2^{eN} = \{F_2^{flav}\} + (1+\xi)^{-2}[F_2^{col}]$$
$$= \{\sum_{q_i} x\,Q^2_{val}(q_i)[q_i(x)+\bar{q}_i(x)]\} \tag{27}$$
$$+ (1+\xi)^{-2}\left[\tfrac{2x}{3}\sum_{q_i}\{q_i(x)+\bar{q}_i(x)\} + x\,V(x)(4+\tfrac{4}{3}\xi+\tfrac{1}{3}\xi^2)\right]$$

$$R \equiv \sigma(e^+e^- \to \text{hadrons})/\sigma(e^+e^- \to \mu^+\mu^-)$$
$$= \{R_{flav} + (1+\xi)^{-2}[R_{col}] \tag{28}$$
$$= \{\sum_{q_i} Q^2_{flav}(q_i)\} + (1+\xi)^{-2}\left[\tfrac{2}{3}\cdot(\text{No. of quark valencies}) + \tfrac{1}{8}(1-\tfrac{4}{\xi})^{3/2}(12+20\xi+\xi^2)\right]$$

where $\xi = |q^2|/m_U^2$; $x = 1/\omega = |q^2|/2M_N\nu$

$q_i(x)$ = momentum-distribution function for the ith type quark (within nucleon)

$v(x)$ = momentum-distribution function for any one of the octet of color-gluons (within nucleon)

$$Q^2_{val}(q_i) = \tfrac{4}{9}, \tfrac{1}{9}, \tfrac{1}{9}, \tfrac{4}{9} \text{ for } q_i = (p,n,\lambda,\chi). \tag{29}$$

Note the curly bracket denotes color-singlet (flavor production), while square-brackets denote color-octet production. The first and second terms inside the square bracket of eq. (24) denote respectively the contributions to the R-parameter from the color-octet part of quark-charges and the gluon-charges (If there are quark-flavors in addition to (p,n,λ,χ), the sum over q_i should include these flavors as well).

Asymptotically ($\xi >>1$, i.e. $|q^2| >> m_U^2$), the color-structure functions $\bar{F}_i^{col}(q^2,\nu)$ and $\bar{R}_{col}$ reduce to:

$$\bar{F}_1^{col}\ (eN) \rightarrow 0$$

$$\bar{F}_2^{col}\ (eN) \rightarrow (1/3)\ x\ v(x) \qquad (30)$$

$$\bar{R}_{col} \rightarrow \bar{R}(\text{Gluons}) = 1/8$$

Thus asymptotically (i.e. for $|q^2| >> m_U^2$), the contributions of the color-octed part of quark-charges to electro-production structure functions as well as to R-parameter die out because of the Δ^2-factor. The contributions of the charged spin-1-color-gluons ($V_\rho^\pm$, $V_{K*}^\pm$) to F_1, F_2 and R (instead of growing like $|q^2|$, q^4 and q^4 respectively) scale (again because of the Δ^2-factor). Thus color-production survives in a scale-invariant manner only due to the gluon-parton-contributions.

It should be stressed that the damping due to the kinematic factor Δ^2 is not quite as rapid in the time-like process ($q^2 > 0$, i.e. for e^-e^+-annihilation) as it is in the space-like-processes ($q^2 < 0$) especially for intermediate or semi-asymptotic $|q^2|$ (i.e. $2m_U^2 < |q^2| < 4m_U^2$); and in fact for $|q^2| < 2m_U^2$ the Δ^2-factor acts as an enhancement[41] for time-like processes; where as it provides a damping for space-like processes for all q^2. Thus, in the intermediate region ($m_U^2 < q^2 < 3m_U^2$), the color-contribution to R-parameter (just from the quark-charges) would still be significant. Exactly what is the contribution to R from color-part of quark-charges in this semi-asymptotic region and how it varies depends upon the thresholds of color-resonant states (which should be opening in this range) and their dynamics. Note, however, that the color-gluon $\tilde{U}$ is produced through its direct coupling to the leptons ($e^-e^+ \rightarrow \tilde{U}$) without the intermediacy of the photon, and thus without the Δ^2-factor.

It should also be noted that because of the relatively large coefficients associated with the non-leading terms in ξ compared to the leading term (see eqs. (26, 27) and especially eq. (28)), the asymptotic values of the color-gluon contributions to F_1, F_2 and R (eq. (30)) are not attained until ξ is much greater than unity. This is especially true for their contribution to the R-parameter, as it may be judged from the following numerical values of R_{col} for $\xi > 4$ (taken from eq. (28)):

$$\begin{aligned}\bar{R}_{col} &= \bar{R}_{col}\,(\text{quarks}) + \bar{R}_{col}\,(\text{gluons}) \\ &\approx \tfrac{1}{6} + \tfrac{0.8}{8} \approx .27 \quad (\xi = |q^2|/m_u^2 = 5) \\ &\approx .03 + 1.95/8 \approx .27 \quad (\xi = 10) \\ &\approx .00 + 1.2/8 \approx .15 \quad (\xi = 100)\end{aligned} \tag{31}$$

Thus the net contribution[42] from color-gluon-pairs ($V_\rho^+ V_\rho^-$ and $V_{K*}^+ V_{K*}^-$) to R-parameter is at most about 1/4 (for $\xi \approx 10$), which drops to 1/8 for sufficiently large ξ. We later point out distinct signatures for charged color gluons (or, in general, lowest lying charged color-octet states).

B. Rise in Structure Functions Due to Color-Production

The contributions from quark and gluon-partons to the color-part of the structure functions, as listed in sec. A., should lead to a rise in these functions at energies (and $|q^2|$ and $M_N\nu$) sufficiently above color-threshold. Numerical values of such a rise depend upon the values of the kinematic variables $\xi = |q^2|/m_U^2$ and $x = |q^2|/(2M_N\nu)$ on the one hand, and the quark and gluon momentum-distribution functions $q_i(x)$ and $v(x)$ on the other. While the quark-distribution functions are somewhat known[43]; there is at present no information on the detailed properties of $v(x)$. It is, however, known[44] (using momentum-conservation) that the gluons carry about 50 % of the nucleon's momentum. Thus,

$$\int_0^1 8x\, v(x)\, dx \approx 0.5 \tag{32}$$

Subject to this condition, we made two models for $v(x)$.

Model I: $v(x)$ has a shape similar to that of the neutron-quark distribution function within the proton. This yields:

$$\begin{aligned} x v(x) &\approx 0.04 \quad (x \approx .5) \\ &\approx 0.1 \quad (x \approx .2) \\ &\approx 0.11 \quad (x \approx 0)\end{aligned} \tag{33}$$

Model II: v(x) has a shape similar to that of the sea (being very small for x > .1 and rising steeply as $x \to 0$). This gives:

$$\begin{aligned} x\,v(x) &\approx 0.01 \qquad (x \approx .5) \\ &\approx 0.05 \qquad (x \approx .2) \\ &\approx 0.65 \qquad (x \approx 0) \end{aligned} \tag{34}$$

Below, we present numerical values of $(\bar{F}_2^{col}/F_2^{flav})$ for ep-scattering for models I and II for two values of x and two values of ξ. The quark-distribution functions are taken from Ref.[43]

Table I

		Model I		Model II	
		ξ = 3	ξ = 5	ξ = 3	ξ = 5
$F^{ep}(col)/(F^{ep})_{flav}$	x = .5	.25	.18	.14	.10
	x = .2	.30	.20	.20	.14

We notice qualitatively the following features:

(i) In either model (I or II), the rise in F_2 due to color-production is expected to decrease as $|q^2|$ (or ξ) increases until fairly large values of ξ >> 1, which would manifest as a scale-violating contribution. These arise due to non-leading terms (in ξ) in the structure functions.

(ii) For large x >.1 (i.e. ω < 10), the rise in F_2 due to color-production is expected to be more prominent for model I (for which the rise is of order 25 to 30 % for $\xi = |q^2|/m_U^2 = 3$) than it is for model II (for which the rise lies between 10 to 20 %).

(iii) For low values of x <.1 (i.e. ω >10) on the other hand, we would expect model II to exhibit a significant rise (much more so than model I) reflecting the relatively high concentration of gluons at low x for this model.

As we explained before, we, of course, do not expect color-contributions to structure-functions to acquire their scaling "weight" in any case until $|q^2|$ and $M_N\nu \gtrsim 2$ to $3m_U^2$ (i.e. ξ ≳2 to 3). Thus if $m_U^2 \approx 9$ to 17 $(GeV)^2$ (corresponding to J/ψ-particle masses), the above considerations for color-contributions to structure-functions are expected to apply only for $q^2 \gtrsim 20$ to 30 $(GeV)^2$. Thus in order to verify the expected large rise (≳ 50 %)of F_2 (for model II) at small x ≲ .05 (say) (i.e. ω ≳ 20), one would need energies such that $2M_N\nu \gtrsim 400$ to 600 $(GeV)^2$. For presently explored regions of q^2 and ω, the expected rise of F_2 for model II (and probably even for model I) is consistent with the data.

To summarize; the color-gluon contributions to structure functions, though not uncomfortably large for presently explored regions of q^2 and ω at MIT, SLAC and Fermilab, are observable at present energies, assuming that color-threshold and color-gluon-masses are $\lesssim$ 5 GeV. It is conceivable that the observed[45] decrease in structure functions with increasing q^2 for $\omega < 6$ could be partly or entirely attributed to the color-gluon-contributions. More refined data would be needed to separate color-gluon contributions from the logarithmic scale-violating terms, expected on the basis of asymptotically free theory[46]. However a large increase (~ 50 %) in structure function at very large ω(> 20) with $|q^2| \gtrsim 20$ (GeV)^2, if observed, is likely to signal color-gluon contribution favouring their "sea-like"-momentum distribution within the nucleon (model II) over "quark-like"-momentum dsitribution (model I).

C. The Ratio σ_L/σ_T

As is well known (σ_L/σ_T) is expected to receive finite contributions from charged spin-1 partons. With a Yang-Mills type coupling, the contributions of these spin-1 partons to (σ_L/σ_T) should in fact have grown with $|q^2|$. Including the Δ^2-factor, the ratio (σ_L/σ_T) becomes a function of the scaling variable x only. Using eqs. (22) and (27), we obtain,

$$\frac{\sigma_L}{\sigma_T} = \frac{F_2^{ep} - 2x F_1^{ep}}{2x F_1^{ep}}$$

$$= \frac{x V(x)}{(1+\xi)^2}\left[(4 + \tfrac{4}{3}\xi + \tfrac{1}{3}\xi^2 - \tfrac{32}{3}(1+\tfrac{1}{4}\xi)\right]\Big/$$

$$\Big[\sum_{q_i} x\, Q^2_{val}(q_i)\{q_i(x) + \bar{q}_i(x)\} \tag{35}$$

$$+ \frac{1}{(1+\xi)^2}\,\frac{2}{3}\sum_{q_i} x(q_i(x) + \bar{q}_i(x)) + \frac{32}{3}(1+\frac{\xi}{4})\, x V(x)\Big]$$

$$\xrightarrow[(\xi \gg 1)]{} \frac{\frac{1}{3} x V(x)}{\sum_{q_i} x\, Q^2_{val}(q_i)\{q_i(x) + \bar{q}_i(x)\}} \tag{36}$$

Substituting typical values of x = .2, .5 and $\xi = |q^2|/m_U^2$ = 2, 3, 5, we obtain (σ_L/σ_T) lying between 0.1 and 0.2 (for either model I and II) consistent with the data. In other words, the cancellation-factor $\Delta^2 = (1/1+\xi)^2$ softens the color-gluon contribution thereby allowing lepto-production data to be compatible with the twin notion that color-threshold is not very high ($\lesssim$ 5 GeV) and that photon carries color (quarks are integer-charged). There is, of course, the strong experimental prediction of this notion that asymptotically (σ_L/σ_T) can not vanish (especially at low x) and that it is a function of x only.

D. Neutrino-Production of Color

Within the gauge theory approach, as stressed in Sec. II, if the gauge-meson-mass matrix induces a mixing between the neutral members $(W_3)_{L,R}$ and U^0 to make the photon, it must inevitably induce either $W_L^\pm$ or $W_R^\pm$ (or both) to mix with the charged members of the color-gluon-octet $V_\rho^\pm$(or $V_{K*}^\pm$or both).

For the case, where V_ρ^+ mixes with W_R^+ (rather than W_L^+), there will be no neutrino-production of color in charged current interactions (barring the tiny $W_L^\pm - W_R^\pm$-mixing), since ν_{eL} and $\nu_{\mu L}$ do not couple to W_R.

The case, where $V_\rho^\pm$ and $V_{K*}^\pm$ mix with W_L, is given in Sec. II (see eq. (11)). For this case, exchanging V_ρ^+ and neglecting W_L-exchange insofar as color-production is concerned (since $m_{W_L}^2 \gg m_{V_\rho}^2$) , we obtain (using the mixing angles given in eq. (12) and $G_F/\sqrt{2} = g^2/2m_W^2$):

$$M(\nu_\mu + N \to \mu^- + X_{col}) = \frac{G_F}{\sqrt{2}} \cos(\theta_L + \phi_L)(\bar{\mu}\gamma_\mu(1+\gamma_5)\nu_\mu) \cdot \left(\frac{m_{V_\rho}^2}{q^2 - m_{V_\rho}^2}\right) \langle X_{col} | J_{col}^+ | N \rangle \tag{37}$$

Similarly V_{K*}^+-exchange contributes an amplitude proportional to $\sin(\theta_L + \phi_L)$. To compare, with the above, the amplitude for color-singlet (valency)-production is given by[47]:

$$M(\nu_\mu + N \to \mu^- + X_{val}) = \frac{G_F}{\sqrt{2}} (\bar{\mu}\gamma_\mu(1+\gamma_5)\nu_\mu) \langle X_{val} | J_{val}^+ | N \rangle \tag{38}$$

Adding the two separate contributions (due to V_ρ^+ and V_{K*}^+ exchanges) to cross-sections, we note the emergence of the same $\Delta^2 = (m_V^2/q^2 - m_V^2)^2$-factor for neutrino-production of color in charged-current interactions, as it appears for electro-production of color. Hence the discussion of sec. C (see Table) regarding the relative importance of color versus valency-production in the various kinematic regions ($|q^2|$ and x) wou would apply to neutrino-production of color, as it does to electro or muon-production of color (assuming, of course, that it is W_L (rather than $W_R^\pm$), which mix with $V^{\pm}_{\rho}$ and $V^{\pm}_{K^*}$). Thus, once again, the rise in structure functions due to color-production in neutrino-reactions are also limited to (10 to 30) %-level for the presently explored regions of q^2 and x.

Taking an average value $\langle\epsilon_{col}\rangle \lesssim (1/5)$ for an estimate of the suppression of color relative to valency-production, we may now evaluate the ratio r (eq. (19)):

$$r = \frac{4G_F^2 M_N E_\nu}{2\pi(\sigma_\nu + \sigma_{\bar\nu})} \int F_2^{\gamma N} dx \tag{39}$$

$$= \frac{\langle Q^2_{flav}\rangle_{eN} [1 + \langle\epsilon_{col}\rangle_{eN}]}{\langle Q^2_{flav}\rangle_{\nu N} [1 + \langle\epsilon_{col}\rangle_{\nu N}]}$$

where $\langle Q^2_{flav}\rangle_{eN} = 5/18$; $\langle Q^2_{flav}\rangle_{\nu N} = 1$. Even though $\langle\epsilon_{col}\rangle_{eN}$ need not be exactly equal to $\langle\epsilon_{col}\rangle_{\nu N}$ (since the data is folded for the eN and νN scatterings for different kinematic regions), since both are small ($\lesssim$ 1/5), it is clear that r will not deviate from 5/18 by more than 10 %. Thus, within the gauge theory approach to color, we expect the value of r (as extracted from the eN and νN-data) to be $\approx$ .28 ± .03 for the case of integer charge-quarks and 5/18 $\approx$.28 for the case fractionally charged quarks. A value of r anywhere in the range .25 to .31 is certainly consistent with the data[34]. The moral is that, contrary to familiar expectations (based on phenomenological approach to color), this sum rule (eq. (39)) involving ratio of eN to νN-data is potentially not sensitive to quark-charges within the gauge theory approach[1] to color.

The novel qualitative features, which emerge due to the cancellation effect between the photon and $\tilde{U}$-contributions may now be summarized below:

(a) For lepto-production experiments, integer-charge quarks behave asymptotically just as if they carried their fractional flavor-charges.

(b) Color-gluon-contributions to structure functions, instead of growing like q^2 or q^4, <u>scale.</u> The ratio (σ_L/σ_T), though non-vanishing is small and becomes a function of x only.

(c) Color does not brighten on par with flavor. The rise in structure functions due to color-production is limited to the (10 ~20) % level at least for presently explored regions of q^2 and ω. Correspondingly, the parton-model based sum rules, do not acquire modifications any larger than (10-20) % to take account of color production.

These new results take away the only objection that existed on experimental and theoretical grounds to the hypotheses that color is physical (quarks are integer charged) and that color-threshold is "low" ($\lesssim$ 5 GeV). There are <u>two</u>

striking predictions of the integer-charge-hypothesis (within the gauge theory approach):

(i) Asymptotically (σ_L/σ_T) should be nonvanishing and a function of x only;

(ii) there should be scaling violations at 10~20 % level due to color-production in the semi-asymptotic region $\xi \lesssim 8$, which should decrease with $|q^2|$ for fixed $\omega \lesssim 10$, above and beyond the logarithmic violations expected from asymptotic freedom-considerations. By contrast, for fractionally charged quarks $(\sigma_L/\sigma_T) \to 0$ (Callan-Gross sum rule) and scaling violations should be limited to the said logarithmic terms only. Here lies an experimental method of distinguishing between the hypotheses of unphysical versus physical color. I now turn to a discussion of the distinctive signatures for color, which should be helpful in its search.

E. Distinctive Signatures for Color: Di-muons

One expects to produce charged members of the color-gluons (or similar color-octet states) carrying color-quantum numbers (I_3' and Y') in pairs either in hadronic collisions through strong interactions (the crosssection in this case may lie in the range of $10^{-31} \sim 10^{-32} cm^2$ at Fermilab energies for $M_V \simeq 2$-3 GeV), or in e^-e^+-annihilation:

$$p + p \to V_g^+ + V_g^- + \text{Hadrons} \tag{40}$$

$$e^+e^- \to V_g^+ + V_g^- + (\text{Known Hadrons}) \tag{41}$$

In addition for the case of $W_L^\pm$ mixing with $V_\rho^\pm$ and $V_{K^\times}^\pm$, the may also be produced singly in charged-current neutrino interactions with an amplitude of order G_{Fermi} (see eqs. (37),(26) and (28)), the net effect of the total color-production (using the Δ^2-factor) being limited to the 10~20 % level relative to the total valency (color-singlet)-production in accordance with Table I. Thus, for example

$$\nu_\mu + N \to \mu^- + V_g^+ + (\text{Hadrons}) \tag{42}$$

For e^-e^+-annihilation, the net contribution[48] to R from pair-production of charged color-gluon-partons $(V_\rho^+ V_\rho^-, V_{K^\times}^+ V_{K^\times}^-)$ at $q^2 > 4m_{V_\rho}^2$ is at most 1/4 (for $q^2 \simeq 10\, m_{V_\rho}^2$), which drops to (1/8) for sufficiently large q^2 (see eqs. (28) and (31)). Even though this is a small percentage of R_{total} varying from 2 to 5.5 for q^2 varying from 9 to 60 $(GeV)^2$, the charged color-gluons possess distinct decay modes, which can help identify their production especially through a detailed study of non-collinear lepton-pair production in e^-e^+-annihilation (and, of course, they may also be identified by a study of their production in pairs in p+p-collicions).

Note that in contrast to heavy leptons, the charged color gluons may be produced (in pairs) in e^-e^+-annihilation <u>together</u> with other known hadrons (such as pions) above threshold.

The charged color-gluons ($V^\pm_\rho$, $V^\pm_{K^\times}$), if they are the <u>lightest</u> color-octet states (transforming like (1,8) under $SU(3) \times SU(3)'_{color}$ in the absence of their small components), would decay into leptons and known color-singlet hadrons only through their small components $W^\pm_L$ (or $W^\pm_R$) as given by eq. (11). Thus, their allowed decay modes are:

$$\begin{aligned}(V^+_\rho, V^+_{K^*}) &\to \mu^+\nu_\mu, e^+\nu_e \\ &\to \pi\pi e\nu \\ &\to K\bar{K} e\nu \\ &\to \eta\eta e\nu \\ &\to \pi\pi, 3\pi, 4\pi, K\bar{K} \text{ etc.}\end{aligned} \qquad (43)$$

(the $\pi\pi e\nu$, $K\bar{K}e\nu$, $\eta\eta e\nu$ modes: with hadrons in I=0 state and also in SU(3)-singlet state to the extent SU(3) is a good symmetry.)

The leptonic and semi-leptonic decay modes are expected to be a significant fraction of all decay modes ($\approx$ 20 to 40 % for the electron-modes and similarly for the muon-modes). The life time of the charged gluons is roughly proportional to the inverse fifth power[48] of the mass of the gluon. We obtain[49] (for the strong gauge coupling parameter $f^2/4\pi \simeq 1$):

$$\begin{aligned}\tau(V^+_\rho) &\approx 2 \times 10^{-15} \text{ sec.} && (m_V \approx 1.5 \text{ GeV}) \\ &\approx 5 \times 10^{-17} \text{ sec.} && (m_V \approx 3 \text{ GeV}) \\ &\approx 10^{-17} \text{ sec.} && (m_V \approx 4.1 \text{ GeV})\end{aligned} \qquad (44)$$

Note the important selection-rule that <u>the semi-leptonic decay modes of $V^\pm_\rho$ and $V^\pm_{K^\times}$ can not involve a single π, a single K or a single η</u> (i.e. $V^+_\rho \to \pi^0\mu^+\nu_\mu$, $K^0\mu^+\nu_\mu$ and $\eta\mu^+\eta_\mu$ are forbidden to $O(\alpha)$ in the matrix element relative to the allowed decay modes exhibited in eq. (43)). By contrast such decay modes are allowed for color-singlet mesons carrying new flavor quantum numbers such as charmed D and F-mesons. Thus, if one discovers new charged short lived objects, which on the one hand have large (leptonic + semileptonic) decay-branching ratios, and on the other hand exhibit semileptonic decays involving only two pions, or two kaons, or two η's etc. in the final state (but no single π, or single K or single η), <u>then beyond doubt they must either represent color-gluons (or analogous color-octet states lighter than the color-gluons).</u> Of course, other characteristic features for color-gluons (and analogous color-octet states) are that there must be eight[52] of them (nearly degenerate with each other within few to 50 MeV (say)) with charges 0,0,0,0,+,+,-,-). They may all be produced in <u>pairs</u> in hadronic collisions; the $\tilde{U}$-member may be produced singly in e^-e^+-annihilation through its direct coupling to leptons.

IV. THE NEW PHENOMENA AND PHYSICAL COLOR

(1) Dimuons: The dimuons[19] of opposite charge ($\mu^-\mu^+$) recently produced in neutrino and anti-neutrino reactions can be attributed to production of charged color-gluons (or analogous color-octet particle) as in reaction (42), followed by their (leptonic + semileptonic)-decays as in (43). Given that the total color-production is of order 10 to 20 % compared to non-color-production (see above) and that the net (leptonic + semileptonic)-decay-branching-ratio of the lightest charged color-octet members to be of order (20 to 40 %), which is divided between electron and muon modes, we expect a dimuon event rate of order (1 to 4) % compatible with the observed 1 % rate compared to the single muon-events.

(2) $\bar{\mu}e$-Events: Production of Integer-Charge-Quarks and Color-Gluons by e^-e^+-Annihilation: We expect integer-charge quarks and color-gluons to be produced in pairs in general in association with other known hadrons (sufficiently above threshold)

$$
\begin{aligned}
e^+e^- &\to p^{0,+,+}_{r,y,b} + \bar{p}^{0,-,-}_{r,y,b} \\
&\to n^{-,0,0}_{r,y,b} + \bar{n}^{+,0,0}_{r,y,b} \quad \text{etc.} \\
&\to V^+_{\rho} + V^-_{\rho} \\
&\to V^+_{K^*} + V^-_{K^*}
\end{aligned}
\tag{45}
$$

A large or fair fraction of the quark-pairs treated as partons would recombine to form known hadrons, while the color-gluon-pairs (see Sec. III.E) would recombine to form color-octet hadrons together with known hadrons. The total contribution to R from the various quark-parton-pair production (which gets divided between many available channels) may be obtained by noting that their charges receive ccontributions from two sources Q_{flav} = (2/3, -1/2, -1/3, 2/3) for (p, n, λ,χ) and Q_{col} = (-2/3, 1/3, 1/3) for (red, yellow, blue) (see eq. (8)). The charges from the second source do not asymptotically contribute to R (because of the Δ^2-factor, (Sec. III). Thus, interestingly enough, asympotitcally all three proton-quark-pairs (including the neutral pair $p^0_r\,\bar{p}^0_r$) contribute the same amount to R:

$$R(p^0_r\bar{p}^0_r) = R(p^+_y p^-_y) = R(p^+_b p^-_b) = 4/9$$

Similarly,

$$
\begin{aligned}
R(n^-_r n^+_r) &= R(n^0_y \bar{n}^0_y) = R(n^0_b \bar{n}^0_b) = 1/9 \\
R(\lambda^-_r \lambda^+_r) &= R(\lambda^0_y \bar{\lambda}^0_y) = R(\lambda^0_b \bar{\lambda}^0_b) = 1/9 \\
R(\chi^0_r \bar{\chi}^0_r) &= R(\chi^+_y \chi^-_y) = R(\chi^+_y \chi^-_b) = 4/9
\end{aligned}
\tag{46}
$$

Some of the allowed decay modes[54,55] for the yellow and blue quarks arising due to W-X mixing are listed below:

$$\left.\begin{aligned} p^{+}_{y,b} &\rightarrow \nu_e + \text{pions} \\ &\rightarrow \nu_\mu + K^0 + \text{pions} \end{aligned}\right\} \text{(dominant)} \tag{47}$$

$$\rightarrow e^- + \pi^+ + \pi^+ \quad \text{(likely to be suppressed}^{55}\text{)}$$

$$\left.\begin{aligned} &\rightarrow \nu_e + (e^+ + \nu_e) \\ &\rightarrow \nu_e + (\mu^+ + \nu_\mu) \end{aligned}\right\} \text{(suppressed}^{57}\text{)} \tag{48}$$

$$\not\rightarrow \mu^- + \pi^+ + \pi^+ \quad \text{(forbidden}^{55}\text{)}$$

$$\left.\begin{aligned} n^{0}_{y,b} &\rightarrow \nu_e + \text{pions} \\ &\rightarrow \nu_\mu + K^0 + \text{pions} \end{aligned}\right\} \text{(dominant)} \tag{49}$$

$$\rightarrow e^- + \pi^+ \quad \text{(likely to be suppressed}^{55}\text{)}$$

$$\left.\begin{aligned} \lambda^{0}_{y,b} &\rightarrow \nu_\mu + \eta \\ &\rightarrow \nu_e + \bar{K}^0 + \text{pions} \end{aligned}\right\} \text{(dominant)}$$

$$\rightarrow p^{+}_{y,b} + \pi^- \quad \text{(ordinary weak decay)} \tag{50}$$

$$\chi^{+}_{y,b} \rightarrow \nu_e + (\text{charmed } D^+)$$

$$\rightarrow \nu_\mu + (\text{charmed } F^+)$$

$$\left.\begin{aligned} &\rightarrow p^{+}_{y,b} + \pi^0 \\ &\rightarrow \lambda^{0}_{y,b} + \pi^+ \end{aligned}\right\} \text{(ordinary weak decay)} \tag{51}$$

While for the red quarks (whose decays may need to utilise both $V_\rho^+ - W^+$ and $W^+ - X^+$-mixings), the allowed decay modes are:

$$p_r^0 \to n_y^0 + \gamma \quad (\text{If } m_{p_r} > m_{n_y}) \quad \Big\} \text{(dominant modes)}$$

$$\to V_\xi^- + \pi^+ + \nu_e \quad (\text{If } m_{p_r} > m_{V_\xi} + m_{\pi^+})$$

$$\to \nu_e + \pi^0 \, , \; e^- + \pi^+ \quad \text{etc.} \tag{52}$$

$$n_r^- \to V_\xi^- + \nu_e \quad (\text{If } m_{n_r} > m_{V_\xi}) \quad \Big\} \text{(dominant mode)}$$

$$\quad \hookrightarrow (\mu^- \bar{\nu}_\mu) \text{ or } (e^- \bar{\nu}_e) \tag{53}$$

$$\to \nu_e + \pi^- \, , \; e^- + \pi^0 \tag{54}$$

$$\lambda_r^- \to V_\xi^- + \nu_\mu \quad (\text{If } m_{\lambda_a} > m_{V_\xi}) \quad \text{(dominant mode)}$$

$$\to \nu_e + K^- \, , \; \mu^- + \eta \tag{55}$$

$$\chi_r^0 \to \lambda_y^0 + \gamma \tag{56}$$

In above, we have exhibited few quark-number-conserving-decay modes arising via familiar weak interactions (e.g. $\lambda \to p + \pi$, $\chi \to \lambda + \pi^+$ and $\chi \to p + \pi^0$ etc.), which may have rates in the range of 10^{10} to 10^{11} sec^{-1} depending upon quark-mass differences. Such decays may be the dominant decay modes for the charm-quarks, but they are not important for the lighter quarks, especially the p and n-quarks.

In the basic model, quarks of <u>all three colors</u> (red, yellow and blue) with mass as low as 2 to 3 GeV can be relatively shortlived ($\tau \lesssim 10^{-11}$ sec.) without conflicting with the proton lifetime provided[56] that the quarks are heavier than some color-octet states like the color-gluons. For this model, the semileptonic decay modes of the yellow and blue quarks (e.g. $q \to \ell$ + pions, see (47) and (49)) strongly dominate[57] over their pure leptonic decay modes ($q \to \ell + \ell + \ell$, see ( )). Thus at least within the basic model, the pair production of yellow and blue quarks with their predominant semileptonic decays cannot account for the $(\bar{\mu}e)$-events seen at SPEAR. On the other hand, if quarks are heavier than the color-gluons, the charged red-quarks (n_r^- and λ_r^-) would predominantly decay into (color gluon + lepton) (see (53)), which followed by rapid decay of the color-gluon into a lepton-pair (see eqs. (43) and (44)) would appear like a three body leptonic-decay[63] of the quarks. Thus, pair-production of $(n_r^- n_r^+)$ and $(\lambda_r^- \lambda_r^+)$-quarks, followed by their sequential decays as above, can provide within our basic model a consistent explanation of the anomalous $(\bar{\mu}e)$-events seen at SPEAR:

$$\begin{array}{llll} e^+ + e^- \rightarrow & n_r^- & + \; n_r^+ & \\ & \downarrow & \;\; \hookrightarrow V_g^+ + \bar{\nu}_e & \\ & V_g^- + \nu_e & \qquad \hookrightarrow (e^+ + \nu_e) & \\ & \hookrightarrow \mu^- + \bar{\nu}_\mu & & \end{array} \qquad (57)$$

An important question arises: What are the sources of these anomalous events? Are they hadronic constituents such as[59] quarks, or are they heavy leptons?

To distinguish between these possibilities, among other means, one must search for pair-production of these objects in hadronic collisions. While neutral as well as charged quark-pairs (with quark-mass $\approx$ 2 to 3 GeV) are expected to be produced via strong interactions with cross sections $\approx 10^{-32}$ cm^2 at Fermilab energies; the neutral and charged heavy lepton-pairs are expected to be produced via weak and electromagnetic interactions respectively (with cross sections $\lesssim 10^{-38}$ cm^2 for the neutral pair and $\approx 10^{-34}$ cm^2 for the charged pair at Fermilab energies for heavy-lepton-mass $\approx$ 2 GeV). We, therefore, urge a search for anomalous di-lepton-production (involving $e^+e^-\mu^\pm e$, $\mu^+\mu^-$-pairs) in hadronic collisions. Such a search may help decide whether the $(\bar{\mu}e)$-events seen at SPEAR have hadronic or leptonic parents.

Furthermore, above threshold, quark-pairs can be produced by e^-e^+-annihilation in association with other known hadrons, something not possible for heavy leptons. Secondly, if the semileptonic decay modes involving charged lepton emission (i.e. $p^+_{y,b} \rightarrow e^- + \pi^+ + \pi^+$; $n^o_{y,b} \rightarrow e^- + \pi^+$ etc.) are not suppressed[55], (which would be the case if pion emission from quark-lines are not associated with form-factors (see Ref. 60)), an important consequence of the quark-hypothesis for the $(\bar{\mu}e)$-events would be that there would be semileptonic signals of the form $e^- + e^+ \rightarrow e^- + e^+ + (\pi^+\pi^-\pi^+\pi^-)$ with energetic (e^-e^+)-pairs[60] in the final state. Allowing for separate conservations of electron and muon numbers such semileptonic signals would not be produced via decays of heavy leptons. Finally a third necessary consequence of the quark-hypothesis for the $(\bar{\mu}e)$-events is that a significant fraction of the total hadronic cross section must involve real quark-antiquark pair production. Noting that the yellow and blue quarks (as well as χ^o_r and p^o_r) disappear into neutrinos plus mesons, this may on the one hand help understand the energy-crisis and on the other hand the jet-structure[13] observed at SPEAR.

3. Direct Leptons: We expect quarks as well as color-gluons to be produced in pairs in hadronic collissions

$$p+p \rightarrow q+\bar{q} + \text{Hadrons}$$
$$\rightarrow V_g^+ + V_g^- + \text{Hadrons} \qquad (58)$$
$$\rightarrow \tilde{U} + \tilde{U} + \text{Hadrons}$$

Pair production of quarks (mass $\simeq$2 GeV) and/or color-gluons (which may even be lighter [51] than the quarks), followed by their predominant semi-leptonic or leptonic decays (eqs. (48) - (56) and (43)) thus provide an obvious source of excess direct leptons (electrons as well as muons), which may account for the observed direct leptons[61] especially at higher incident energies ($\sqrt{s} \gtrsim$ 10 GeV). For reasons mentioned in the previous subsection, it is important to search for anomalous varying invariant mass deleptons ($\mu^{\pm}e^{\mp}$, $\mu^+\mu^-$ and e^+e^-) in (p+p)-collisions.

4. The J/ψ-Particles

At present the multitude of the J/ψ-particles (J/ψ_1(3.1), ψ_2(3.7), ψ_3(4.1), ψ_4(4.4) and possibly others) together with the C-even states (3.5, 3.4, 2.8) recently discovered at DESY and SPEAR allow alternative interpretations under the new quantum numbers: color, charm, and mirror (or heavy-quark flavors). As we have stressed elsewhere[2o], it appears, however, that none of these quantum numbers by themselves (especially neither color nor charm alone) can account for the multitude of phenomena, which include not only the J/ψ-particles and their C-even analogs, but also dimuons (of both unlike and like charges) and the anomalous ($\bar{\mu}$e)-events.

Here we note one distinct possibility: J/ψ_1(3.1) is the ground-state $3S_1$ mirror-antimirror quark composite with mirror-isospin[64] I_m = 1; ψ_2(3.7) is its radial excitation; while ψ_3(4.1) (which may represent a superposition of several resonances) and ψ_4(4.4) represent color (such as the color-gluon $\tilde{U}$) and/or charm-anticharm-composites ϕ_c.

The C-even ^1S_o and $^3P_{o,1,2}$-composites of mirror (heavy) quarks $\bar{p}'p'$ and $\bar{n}'n'$ are expected to accompany the C-odd $^3S_1$-states analogous to the charmonium-picture, except that for the present case additional states are expected, their mirror-isospin I_m being either 0 or 1.

With the above assignment, the mirror analogs of the charmed D and F should lie around 2 GeV, while the D and F themselves should lie around 3 GeV or higher. The decays of mirror-D and F-particles will not preferentially involve K-mesons (unlike D and F).

In this picture color and charm threshold start above 4 GeV. Alternatively an intriguing possibility is that color-threshold and color-gluon $\tilde{U}$ lies below[51] 3 GeV. Depending upon where lies $\tilde{U}$, there should of course exist its seven partners around the mass of $\bar{U}$. If the $(\bar{\mu}e)$-events seen at SPEAR are to be attributed to quark-decays, it is preferable that color-gluon mass lies below 2 GeV (see Sec. IV). We urge an exhaustive search for narrow resonant states[63] in (e^-e^+)-system in the 1 to 3 GeV region.

We stress that a clear choice between the allowed interpretations of the new particles (including the one mentioned above) can be made only after a proper search for particles carrying the new quantum numbers (color, charm and/or mirror) is carried out. To this end, we have noted in Sec. III that there are distinct decay modes of color-carrying particles (such as $V^{\pm}_{\rho}$, $V^{\pm}_{K^+}$) which sould enable one to distinguish experimentally between particles carrying color on the one hand and those carrying new valency-quantum numbers (such as charm and mirror) on the other.

V. SUMMARY AND CONCLUDING REMARKS

The main theoretical remark of this talk is that the commonly held objections to the concept of integer-charge quarks with relatively low mass phycisal color do not hold within the gauge-theory approach to color[1].

On the experimental side, we point out:

(i) The dimuon-events $(\mu^-\mu^+)$ with their observed rate receive a simple explanation in terms of charged-color-gluon-production (or production of similar color-octet states).

(ii) Production of quark-pairs followed by their leptonit decay provides a viable explanation of the anomalous $(\bar{\mu}e)$-events observed at SPEAR. In order to distinguish between quark and heavy lepton-hypothesis for these events, one needs to study the rate of production of analogous events in hadronic collisions on the one hand and the production of hadron-associated $(\bar{\mu}e)$-events in e^-e^+-annihilation at $E_{CM} \geq 5$ GeV on the other.

(iii) Direct lepton production in hadronic collisions especially at higher energies finds an explanation in terms of quark and/or gluon-pair production. Such a hypothesis may be tested by studying production of anomalous like and unlike lepton-pairs in pp-collisions.

(iv) Asymptotically, for lepto-production experiments, integer charge quarks (treated as partons) would behave as though they carried their fractional valency-charges. A major prediction of the integer-charge-quark hypothesis (within the gauge theory framework) is that (σ_L/σ_T) should be nonvanishing and asymptotically a function of x only; in the semi-asymptotic region ($|q^2| \lesssim 8m_U^2$), there should be scaling violations due to color-production at the 10 to 20 % level, which should decrease with q^2 for fixed $\omega \lesssim 10$, above and beyond logarithmic scaling violations expected from asymptotic freedom considerations.

(v) Charged color-gluons ($V_\rho^\pm$, $V_{K^+}^\pm$), which may be produced in pairs both by pp-collisions and by e^-e^+-annihilation, possess distinct decay modes - their semileptonic decays (eq. (39)) would involve two pions, $\bar{K}K$ and $\eta\eta$, but no single pion, or single kaon, or single η in sharp contrast to the decay modes of particles carrying new valency quantum numbers (such as charm).

(vi) Last but not the least, within the gauge-theory approach, one intriguing feature is that asymptotically the neutral red-quark-pair ($p_r^o\bar{p}_r^o$) will be produced by e^-e^+-annihilation on par with the charged quark pairs ($p_y^+p_y^-$ and $p_b^+p_b^-$), all three pairs possessing charged-particle decay modes.

I am grateful for several helpful discussions to O.W. Greenberg, G.A. Snow, J. Sucher and C.H. Woo.

REFERENCES AND FOOTNOTES

1. J.C. Pati and Abdus Salam; Phys. Rev. D8, 1240(1973); Phys. Rev. Lett. 31, 661(1973); Phys. Rev. D10, 275(1974) and Physics Letters (1975)

2. H. Georgi and S.L. Glashow, Phys. Rev. Lett. 32, 438 (1974), H. Fritzsch and P. Minkowski, Annals of Physics 93, 222 (1975); F. Gürsey and P. Sikivie (Preprint, 1976) and P. Ramond (Preprint, 1976)

3. By "Good" we mean that the symmetry breaking terms are of order α or few times α. Operationally we require that "color" as a global symmetry be at least as good as SU(3).

4. R.N. Mohapatra, J.C. Pati and Abdus Salam, Univ. of Md. Tech. Rep. No. 1975; Phys. Rev. D (to appear)

5. See for example the comprehensive summary of this point by R. Dashen, SLAC Conference Proceedings (August, 1975)

6. In both cases asymptotic freedom is lost only due to the quartic terms of the Higgs-potential (note at least the weak gauge bosons have to be massive in both cases). Since the renormalized values of these quartic couplings may typically be chosen to be less than e (at low energies), such loss of asymptotic freedom would not manifest itself until much higher energies. In this sence both theories are "temporarily" asymptotically free in the present energy regime (see D. Politzer, Physics Reports, 1974). Alternatively if Higgs-Kibble fields arise dynamically as composite fields, both theories would be truly asymptotically free.

7. See, for example, A. de Rujula's talk at this conference. Other references may be found in here.

8. See, for example, C.H. Llewellyn Smith, Rapporteur's talk at SLAC-Lepton-Photon-Symposium, August, 1975, and A. De Rujula, H. Georgi and S.L. Glashow, Phys. Rev. 12, 147(1975).

9. J.C. Pati and Abdus Salam, Phys. Rev. Lett. 36,11 (1976); J.C. Pati, Report of a Talk presented at the Conference on Gauge Theories and Modern Field Theory at North Eastern University (Sept., 1975), Univ. of Md. Tech. Rep. No. 76-071 (to appear in the proceedings).

10. G. Rajasekharan and P. Roy; TIFR Preprints TH75-38 and TH/75-42.

11. J.C. Pati, Abdus Salam and S. Sakakibara, U. of Md. Tech. Rep. (1975); J.C. Pati (North Eastern Talk, Sept. 1975, see ref. 9).

12. M.L. Perl et al., Phys. Rev. Letters 35, 1489(1975)

13. G. Hanson et al., Phys. Rev. Lett. 35, 1609(1975)

14. R.N. Mohapatra and J.C. Pati, Phys. Rev. D11, 566(1975).

15. For example, the contribution to (g-2) of muon due to $\tilde{U}$-exchange is $(1/8\pi^2)$ $(2e^2/\sqrt{3f})^2$ $(m_\mu/m_U)^2 \approx 10^{-8}$ for $m_U \approx 2$ GeV and $f^2/4\pi \approx 2$. The present value of (expt.-(q.e.d.) theory) for $(g-2)_\mu$ is $\lesssim 3 \cdot 10^{-8}$.

16. J.C. Pati and Abdus Salam, Phys. Rev. D11, 1137(1975); Physics Letters, R.N. Mohapatra and J.C. Pati, Phys. Rev. D11, 2558; H. Fritzsch and P. Minkowski, Annals of Physics 93, 222(1975).

17. The effective low-energy coupling constants (g and f), which are operative in the flavor and SU(3)'-color sectors can still differ due to finite renormalization effects which are in general different in different sectors due to mass differences between the gauge particles. A practical realization of this low-energy disparity between coupling constants in different sectors within a unified theory remains to be shown.

18. If the octet of color gluons are massive and we demand that SU(3)-color be preserved as a good global symmetry, then it appears that $[SU(4)]^4$ is essentially the unique unifying symmetry barring proliferation of fermions for example through new colors. (This will be discussed in a forthcoming paper by Mohapatra and myself).

19. This may arise if there exist fermions with new colors, the unifying symmetry in this case might be for example $[SU(5)]^4$ or $[SU(6)]^4$ with the basic fermionic multiplet being a 25-plet or a 36-plet (there would still have to be in addition the mirror set F' to cancel anomalies). All these cases still preserve the interpretation of lepton number as the fourth color (Ref. 1) and put quarks and leptons in the same multiplet. The fermions with new colors (fifth or sixth) do not douple to the SU(3)'-octet of color gauge mesons; thus they max exhibit a new class of interactions (in addition to weak and electromagnetic interactions), which would be characterized by the masses of gauge mesons belonging to SU(6)' but outside of SU(4)'-subgroup. There is an alternative possibility: quarks and leptons may belong to distinct but parallel multiplets and yet provide a unified theory, this would again need a proliferation of the basic fermions.

20. J.C. Pati and Abdus Salam, Physics Letters 58B, 333(1975)

21. G.Branco, T. Hagiwara and R.N. Mohapatra (preprint, 1975); Phys. Rev. (to be published).

22. See for example A. DeRujula, H. Georgi and S.L. Glashow, Phys. Rev. D12, 3589 (1975); H. Fritzsch, M. Gell-Mann and P. Minkowski, Phys. Letters 59B, 256 (1975); F. Wilczek, A. Zee, R.L. Kingsley and S.B. Treiman, Phys. Rev. D (to be published) and S. Pakvasa, W.A. Simmons and S.F. Tuan, Phys. Rev. Letters

22. 35, 702(1975)

23. F. Sciulli, Talk at Coral Gables Conference, Jan.1976, to appear in the proceedings.

24. The most recent experimental value of $d_{neutron}$ is $(.4 \pm 1.1) \cdot 10^{-24}$ ecm. (N.F. Ramsey; Invited Talk at New York APS meeting, Febr., 1976).

25. Ref. 1; R.N. Mohapatra and J.C. Pati, Phys. Rev. D11, 2558(1975); G. Senjanovic and R.N. Mohapatra, Phys. Rev. D12, (1975)

26. See for example S. Pakvasa and H. Sugawara, Hawaii Preprint (1975); L. Maiani, Rome Preprint (1975).

27. J.C. Pati and S. Oneda, Phys. Rev. 140, 1351(1965).

28. See for example S. Oneda, J.C. Pati and B. Sakita, Phys. Rev. 119, 482 (1960); S. Coleman and S.L. Glashow, Phys. Rev. 134, B681 (1964) and A. Salam and J.C. Wan, Phys. Letters 8, 217(1964).

29. M.K. Gaillard and B.W. Lee, Phys. Rev. Lett. 33, 1o8(1974); G. Altarelli and L. Maiani, Phys. Lett. 52B, 351(1974).

3o. F. Wilczek, A. Zee, R.L. Kingsley and S.B. Treiman, Phys. Rev. (to be published).

31. J.C. Pati and C.H. Woo, Phys. Rev. D3, 292o(1971); S. Nussinov and J. Rosner, Phys. Rev. Letter, 23, 1264(1969); C.A. Nelson and K.J. Sebastian, Phys. Rev. D8, 3144(1973).

32. See for example H. Harari, Rapporteur's Talk at SLAC-Lepton-Photon-Symposium, August, 1975 (to appear in the proceedings). Harari's arguments against color-interpretation are confined to the case where the only new quantum number excited is color.

33. See for example F. Gilman, proceedings of the 17th International Conference, London (1974); C.H. Llewellyn Smith, Rapporteur's talk at SLAC-Lepton-Photon-Symposium, August 1975; A. DeRujula, H. Georgi and S.L. Glashow, Phys. Rev. 12, 147(1975); S.L. Glashow, after dinner talk at Northeastern University Conference on Gauge Theories and Modern Field Theory (Sept., 1975).

34. See for example, B.C. Barish, Invited Talk at the American Physical Society (Div. of Particles and Fields), Sept. 1974.

35. See for example, J.D. Bjorken, Proceedings of 1973 Bonn Conference, P.25 (1974).

36. We neglect the logarithmic corrections to the propagators, which are not important at present energies (see Ref. 17).

37. We do not exhibit V^o (see eq.(9)), since it is not couplet to leptons. Even if $\tilde{U}$ and V^o mix, our conclusions in this section are not affected.

38. $\tilde{U}$-exchange-contribution to color-singlet-production is smaller than the one-photon-contribution by a factor = $(2/\sqrt{3})\ (e^2/f^2) \ll 1$.

39. Note that multiple $\tilde{U}$-exchanges for color-production may be neglected to the same extent as multiple-photon-exchange.

40. The results for the present case may be obtained straightforwardly (by invoking the Δ^2-factor) from those of N. Babibbo and R. Gatto, Phys. Rev. 124, 1577(1961) and M.A. Furman and G.J. Komen, Nucl. Phys. B84, 323(1975).

41. If one of the J/ψ-particles is identified with the $\tilde{U}$-gluon, it is possible to verify that no undue enhancement takes place even though $\Delta^2 \gg 1$ as $q^2 \to m_U^2$, provided $\tilde{U}$ is the lowest mass color-octet state with $J^{PC} = 1^{--}$. (See Ref. 9 for details.)

42. This net contribution would exhibit as a sum over several possible color-octet states.

43. See for example, J.D. Bjorken, Proceedings of the second International Conference on Elementary Particles, Aix-en-Provence, 1973. Aposteriori, since color-contributions (with the Δ^2-factor) turns out to be small $\lesssim$ 10 to 20 % compared to flavor-contribution (see later), the determination of the quark-distributions functions does not alter significantly with the inclusion of color-production.

44. C.H. Llewellyn-Smith, Phys. Rev. D4, 2392(1971).

45. C. Chang et al., Phys. Rev. Lett. 35, 9o1(1975); R. Taylor, Report of MIT and SLAC data at the SLAC-Lepton-Photon-Symposium(August, 1975).

46. Asymptotic freedom applies if spontaneous symmetry breaking is dynamical, or else "temporarily" if quartic scalar couplings are small as mentioned before.

47. For simplicity of writing, we do not exhibit the Cabibbo-angle factors ($\sin\theta_c$ and $\cos\theta_c$) in eqs. (37) and (38), which are immaterial for total cross sections.

48. Strictly speaking this is divided between different color-octet final states, not all of which need contain a pair of charged gluons.

49. These selection rules and decay modes in fact apply to the lightes color-octet states with quantum numbers of $V_\rho^\pm$ and $V_{K^+}^\pm$.

5o. This is because mixing angle is proportional to m_V^2 (eq. (11)) and phase space αm_V.We take $\sin(\theta_L + \theta_L) \simeq \cos(\sigma_L + \sigma_L) \simeq 1/\sqrt{2}$ for simplicity.

51. With the new results on color-brightening (sec. III), there is the intriguing possibility that color-gluons may in fact be relatively light ($m_V \simeq 1$ to 2 GeV) (J.C. Pati, J. Sucher and C.H. Woo (forthcoming preprint).

52. For decays of ($\tilde{U}$, V^o, $\tilde{V}^o_{K^+}$) see Refs. 1 and 9; W.R. Franklin, Nucl. Phys. B91, 16o (1975).

53. A. Benvenuti et al., Phys. Rev. Lett. 34, 419, 597(1975).

54. See Ref. 9 and W.R. Franklin, Ref. 52.

55. In particular, see.J.C. Pati, S. Sakakibara and A. Salam (Trieste Preprint IC/75/93, to appear). The semi-leptonic decay modes such as $p^+_{y,b} \to e^- + \pi^+ + \pi^+$ involving emission of a charged lepton require that one of the pions be emitted from a quark-line inside the loop, which would be suppressed by two large masses (m_X^2 and m_W^2), if pion-emission is associated with a form factor. (Note pions are composites in the theory.) In this case neutral lepton-emission (i.e. $p^+_{y,b} \to \nu_e$+pions, etc.) would be the dominant modes. A second point worth noting is the intricate selection rules, which arise for quark-decays. For example, transitions such as $p^+_{y,b} \to \mu^- + \pi^+ + \pi^+$, $n^o_{y,b} \to \mu^- + \pi^+$, $\lambda^o_{y,b} \to e^- + \pi^+$ etc. are forbidden (neglecting corrections of order G_{Fermi}).

56. Otherwise, at least the red-neutron quark ($\bar{n}_r$) would be longer lived ($\tau(\bar{n}_r) \simeq 10^{-6}$ to 10^{-7} sec. for $m(\bar{n}_r) \simeq 2$ GeV); even though the yellow and blue-quarks would still be shortlived ($\tau \lesssim 10^{-11}$ sec.).

57. The rates of semileptonic-decay modes ($q \to \ell$ + Mesons), when allowed (see Ref. 55), exceed those of leptonic decay modes ($q \to \ell + \ell + \bar{\ell}$) by a factor $\sim O(m_W^4/m_q^4) > 10^5$ within the basic model. This is because the former receive contribution from (convergent) loop-diagrams, while the latter receive contributions from tree-diagrams only.

58. Although with sufficient data, the sequential decay (Ref. 57) might be distringuishable from the genuine three-body leptonic decays of the parent particles.

59. In addition to the production of $q\bar{q}$-pairs, by $e^- e^+$-annihilation, production of charged color-gluon pairs ($V^+_\rho V^-_\rho$ and $V^+_{K^+} V^-_{K^+}$), which is limited by the net contribution from color-gluons to R = 1/8 (see Sec. III), followed by their two-body leptonic decays would also contribute to the leptonic ($\bar{\mu}e$)-events. The available SPEAR data is not inconsistent with three and two body-decays of parent particles (see M.L. Perl, SLAC-PUB-1644, Nov. 1975).

60. It is important to note that due to selection rules involved in quark-decays (see Ref. 55), yellow and blue-quark-pair production does not give rise to semi leptonic signals with either ($\bar{\mu}e$) or ($\mu^+\mu^-$) in the final state. The red-quark pair production ($e^-e^+ \to n_a^- + n_a^+$) can give rise to such signals depending upon the semi-leptonic decay-branching ratios of the charged color-gluons (see (43) and (57).

61. See L. Lederman, Rapporteur's talk at SLAC Conference, August, 1975.

62. The heavy-quark-(mirror)-interpretation with $I_m = 1$ for the lowest lying states has the advantage (R.M. Barnett, Phys. Rev. Lett. 34, 41(1975) that their decays into hadrons would be suppressed by mirror-isospin selection rule as well as by the Zweig-rule-factor. This provides a natrual explanation of their extreme narrowness without invoking an unusual Zweig-suppression-factor (as is needed for the charm-anticharm interpretation of 3.1).

63. The $\tilde{U}$-color-gluon may be searched for both in e^-e^+-annihilation and in photoproduction experiments allowing for good resolution.

MORE THAN FOUR QUARKS?

by

T.F. Walsh
Deutsches Elektronen-Synchrotron DESY, Hamburg

The discovery of narrow resonances in pN and e^+e^- has stimulated a lot of theoretical activity. There are two poles to this activity: explain the world or explain the data. So far, the theoretically popular marriage of charm and asymptotic freedom dogma has been most successful at the former. Many of the difficulties this scheme faces may be only transitory, but to list them:

(i) The $J/\psi - X(2.8)$ splitting and the splitting of the P_c/χ states are about ten times what was originally predicted (1). This is probably a consequence of hubris and does not reflect on the 4 quark charm scheme itself. There may be trouble with the rate for $J/\psi \to \gamma + X(2.8)$; this may be serious in any model (it is a supposedly well-understood magnetic transition). Also, the decay rates for $\psi' \to \gamma + P_c/\chi$ are about a tenth of what was predicted (2). This may also be serious.

ii) There are a number of curiosities about the decays $\psi' \to J/\psi + \pi\pi$ and $\psi' \to J/\psi + \eta$ (3). They involve only small momentum transfers and the former (seemingly) ought to be big. One of the explanations of the J/ψ width was that the strong Zweig-rule violating interactions are supposed to be small at large mass or momentum transfer and large at small mass or momentum transfer. If so, we would expect $\psi' \to J/\psi + \pi\pi$ at the MeV level. One out here is simply to disregard explanation and problem both. The decay $\psi' \to J/\psi + \eta$ may be too large relative to $J/\psi + \pi\pi$, considering that ψ' and J/ψ are supposedly SU_3 singlets and η mostly octet. But maybe SU_3 is badly broken; it does not yet seem the proper time to worry about this.

iii) The ratio $R = \sigma_{had} / \sigma_{\mu^+\mu^-}$ in e^+e^- shows a lot of unexpected structure for 3.9 GeV $\lesssim \sqrt{s} \lesssim$ 4.6 GeV. To explain this with resonances we need to assign $c\bar{c}$ states roughly as in the table:

$J/\psi(3.1)$	1^3S_1
$\psi'(3.7)$	$\approx 2^3S_1$
$\psi''(3.95)$	$\approx 1^3D_1$ (or 2^3D_1 ?)
$\psi'''(4.15)$	$\approx 3^3S_1$
$\psi^{(iv)}(4.45)$	$\approx 2^3D_1$ (or 3^3D_1)

The D-wave states only couple to e^+e^- via a (small) S-wave admixture. Maybe they couple to hadrons this way too. Then they give a spike in e^+e^- with ΔR roughly that for a mostly broad S-wave $c\bar{c}$ state. Maybe something even more involved occurs (interferences for example). At any rate, the structure is much more complicated than expected. Besides this complexity of the data, R goes to rather large values at $\sqrt{s}$ = 5-7 GeV, R $\sim$ 5-5.5. This wasn't expected either. There is also as yet no sign of the large K fraction the model predicts for e^+e^- above 4 GeV.
(iv) There are events of the form $e^+e^- \to \mu^\pm e^\mp$ + nothing visible. If these are not due to charm (e.g. missing K_L), they may be due to a heavy lepton, $e^+e^- \to L^+L^- \to \mu^\pm e^\mp$ + neutrinos. This may solve the problem of a large R, but creates the problem of an uninvited new lepton.
(v) The dimuons $\nu_\mu N \to \mu^-\mu^+ + \cdots$; if the presumed new hadrons from which the second muon comes do not have large ($\gtrsim$10 %) semileptonic branching ratios to muons, there is a problem explaining the apparent fact that the dimuons come from events where ν_μ strikes a valence quark (u or d), and that the rate is large ($\gtrsim$ 20 % of $\sin^2\theta_c$).

These may not be completely compelling grounds to change the model yet, but it is instructive to try, and it is always a good idea to explore options when something new like the discovery of the psions has occurred. We want to try a minimal extension of charm [(4)]. The reason is to preserve its original and still best feature: the absence of $|\Delta S| = 1$ neutral currents.

The original model added a singlet c quark to the original u,d,s (in analogy to the strange quark). Let us add a doublet (in analogy to isospin). This gives us two models (c - c' and c - f) Q(c)=2/3, Q(c')=2/3 or Q(c)=2/3Q(f) = - 1/3. We avoid models with $|Q|$ = 4/3, and use the c quark as in the charm scheme.

There are two possibilities for the spectroscopy. Either ψ' and J/ψ involve both new quarks or only one new quark. What happens if we use both? The leptonic widths allow us to establish the mixing of the $^3S_1$ states [4,5]

$$J/\psi = .981\, c\bar{c} + .189\, c'\bar{c}'$$
$$\text{or } c \rightleftarrows c' \qquad (cc')$$
$$\psi' = .981\, c'\bar{c}' - .189\, c\bar{c} \tag{1}$$

$$J/\psi = .990\, c\bar{c} + .139\, f\bar{f}$$
$$(cf)$$
$$\psi' = .990\, f\bar{f} - .139\, c\bar{c}$$

where we choose the desirable case with small mixing away from pure quark content (small violation of Zweig's rule). Of course there are $^1S_0$ states X near J/ψ and ψ'. Now the large bump at 4.2 GeV should be the $\approx c\bar{c}$ (or $\bar{c}'c'$) radial excitation. Then for a roughly harmonic potential we expect ^3P_J + $^1P_1$ states with centrum nearly halfway from J/ψ to 4.2 GeV - i.e. near ψ'. Spin-orbit splittings as in the non-relativistic quark model (or the low energy data) lead us to expect a 0^{++} $^3P_0$ and a 1^{++} $^3P_1$ $c\bar{c}$ (or $c'\bar{c}'$) below ψ' and a $^3P_2$ above ψ'. This model now has the nice feature that radiative decays $\psi'(\approx f\bar{f}) \rightarrow \gamma + {}^3P_J(\approx c\bar{c})$ or $\gamma + {}^1S_0(\approx c\bar{c})$ are all suppressed because of the small mixing. Empirically this is a factor ~ 10, consistent with (1) when one remembers that both the ^3P_J and ψ' wave functions enter. We have also eliminated the $\psi' \rightarrow J/\psi + \pi\pi$ problem this way too. The price paid is that a number of small unknown angles appear. We do however have a prediction of sorts: 4.2 $\rightarrow$ 3.1 + hadrons is larger than in the charm model by the same factor $\sim$10 mentioned earlier. In the cf model we have $c\bar{q}$ + $\bar{c}q$ mesons produced in e^+e^- above the $c\bar{c}$ threshold and $f\bar{q}$ + $\bar{f}q$ mesons above the $f\bar{f}$ threshold (q denotes u,d,s). If the 4.45 GeV resonance is interpreted as a radial excitation of ψ' it is narrow because the $f\bar{f}$ threshold is near or above this mass. In the latter case this state decays through its $c\bar{c}$ admixture. As to the value of R in the region explored so far, it is crudely consistent with the value expected in the cc' model ($R = 4\frac{2}{3}$) or in the c-f model if a heavy lepton is also produced ($R = 3\frac{2}{3} + 1 = 4\frac{2}{3}$). All this can account for most of the troubles alluded to except (so far) for the dimuons and the complexity of R around 4.1 GeV.

What about weak interactions? So as to leave V-A current algebra untouched, we introduce V+A currents ([V+A, V-A] = o). Of course, we might try V-A currents too, but extra care is needed. Then the charged currents can involve the terms

$(\bar{u}f)_R$ or $(\bar{c}f)_R$ (c-f model)

$(\bar{c}'d)_R$ or $(\bar{c}'s)_R$ (c-c' model)

where the subscript means that the bracket is to be understood as $\bar{q}\gamma_\mu(1-\gamma_5)q$. It is obviously possible to choose a current so that new hadrons with a c' or f quark do not yield a strange particle final state unless the initial new particle has S = o. There is one more condition (if one likes to have a renormalizable theory of weak and electromagnetic interactions): for each new quark doublet we introduce we need a lepton doublet [6]. In the above cases we need $(\nu', L^-)_R$ ($\nu' = \nu_e$ or ν_μ is possible or ν' could be a new neutrino). This condition gives a home to a heavy lepton if that is responsible for the SLAC-LBL μe events. We have the prediction that L^- decays via a V+A interaction.

An important test of any model is whether it delivers an acceptable weak neutral current. The c-f and c-c' models involving quarks not present in the nucleon except at small $x < .1$ (namely $(c's)_R$ and $(c,f)_R$) are automatically acceptable. The other options give terms $(\bar{u}u)_R$ or $(\bar{d}d)_R$ in the neutral current. It is not so obvious that these are acceptable, but in fact they are [7].

Notice that if the f quark is heavier than the c quark and the charged current is of the form $(\bar{c}f)_R$ we would expect long lived "fancy" hadrons decaying to charmed hadrons. This could produce multilepton final states in e^+e^-, but not in $\nu_\mu/\bar{\nu}_\mu N$.

The charged currents involving valence quarks all lead to large effects in either $\nu_\mu N$ or $\bar{\nu}_\mu N$ reactions at high energy. If such effects are not found then these models would be excluded. In the cf model with $(\bar{u}f)_R$, $\sigma(\bar{\nu}_\mu N)$ goes up by a factor 4 far above threshold compared to the value below the fancy threshold. For the cc' model with $(c'd)_R$, $\sigma(\nu_\mu N)$ goes up by a factor 4/3. In both cases we can accomodate dimuons produced off valence quarks; the rate depends on the (unknown) semileptonic branching ratios for the new hadrons containing c' or f.

What if the new fifth quark is much more massive than the c-quark ($> m_\psi/2$)? Then we expect new narrow resonance + threshold structure in e^+e^-. The predicted $\bar{\nu}_\mu N$ or $\nu_\mu N$ behavior will then appear only at high energy. Maybe the discovery of a heavy lepton will herald the discovery of more new hadronic degrees of freedom in e^+e^- and $\nu_\mu N$, $\bar{\nu}_\mu N$.

This is the minimal extension of charm. It can accomodate a new heavy lepton and is not yet excluded by data. Bigger (and maybe better) models can be built with more new quarks (three in six quark models of charges 2/3, 2/3, - 1/3 [8]) or 2/3, -1/3, -1/3 [(9)] if $|Q| \geq 4/3$ is excluded). These bigger models have the nice feature that they can incorporate vectorlike weak currents [(1)]. In the five quark sector they reduce to one of the models we've discussed. An alternative is to use these models but stick to V-A weak currents.

It is useful to study such models (e.g., so as to see what kind of new phenomena might be discovered at higher energies than now available). But one has to admit that they have their unhappy side: the original motivation of the quark model was to reduce the confusion of hadronic spectroscopy by postulating small number of fundamental constituents. Now it appears that a spectroscopy of the constituents may develop.

Acknowledgements

I want to thank Y. Achiman and K. Koller for their collaboration on the ideas expressed here.

References

(1) A. de Rujula, these proceedings
(2) Stanford Lepton-Photon Symposium proceedings
(3) Crude charm estimates give $\psi' \to \omega\, x(2.8)$ = .1-1 MeV; this decay has not yet been seen
(4) Y. Achiman, K. Koller and T.F. Walsh, Phys. Lett. 59B (1975) 261
(5) H. Fritzsch, Phys. Lett. 59B (1975)
(6) C. Bouchiat, J. Illiopolous and P.H. Meyer, Phys. Lett. 38B (1972) 519
(7) Constraints on 5 and 6 quark models arising from neutral current data are being studied with Y. Achiman
(8) H. Harari, SLAC-PUB 1568 (1975)
see also Ref. (1)
(9) P. Fayet, Nuclear Physics B78 (1974) 14
R.M. Barnett, Phys. Rev. Lett. 34 (1975) 41

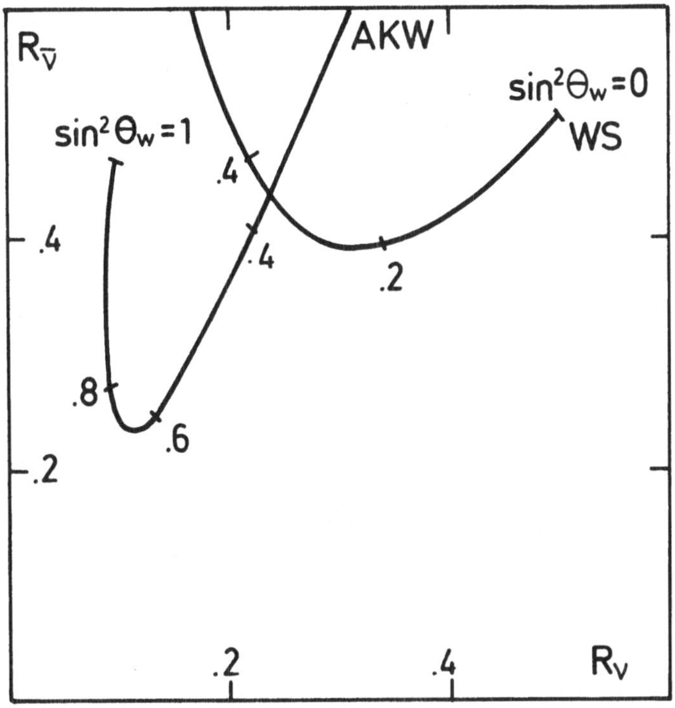

A plot of

$$R_{\bar{\nu}} = \frac{\sigma_{TOT}(\bar{\nu}_\mu N \to \bar{\nu}_\mu X)}{\sigma_{TOT}(\bar{\nu}_\mu N \to \mu^+ X)} \qquad \text{versus} \qquad R_\nu = \frac{\sigma_{TOT}(\nu_\mu N \to \nu_\mu X)}{\sigma_{TOT}(\nu_\mu N \to \mu^- X)}$$

for an isoscalar target in the Weinberg-Salam model and in the AKW cf model with a new $(\bar{u}f)_R$ piece in the charged current. Assumptions

(i) The nucleon is three quarks (no parton "sea")

(ii) We are below new thresholds for the charged current

(iii) We take the SU(2) x U(1) model with $M_Z = M_W / \cos\theta_W$

QUARK CONFINEMENT

by

Hans Joos
Deutsches Elektronen-Synchrotron DESY, Hamburg

1. Introduction

1.1 The power of the quark model in explaining the hadron spectrum, the general features of deep inelastic electron and neutrino scattering together with the other problems of elementary particle physics is rated so highly by many high energy physicists, that they considered the old Gell-Mann-Zweig quarks as "observed" long before the recent "discovery" of the "charmed quarks" in the new particles. This way of speaking about quarks is of course only a semantic solution to the main puzzle of the quark hypothesis: In spite of a big experimental effort no free particles with fractional charges have been found [1]. It is the aim of this review to be a guide to the theoretical attempts to reconcile this experimental fact with the successful applications of the elementary quark model.

1.2 For a long time there was the general feeling that a fundamentally puzzling discovery was necessary in order to induce essential progress in our understanding of elementary particle physics. "Essential" in this context denotes a step forward comparable to the transition from classical to quantum mechanics. The puzzling quarks, which sometimes behave like free particles but are never observed as isolated particles, the "confined" quarks might represent such a discovery. Therefore, our understanding of elementary particle physics, being close to a phase transition, shows wild fluctuations of ideas for the explanation of the quark puzzle. For example, I mention the tentative introduction of quark operators with a completely unorthodox product definition [2]. In this general review, we are somewhat more conservative. We assume that quantum field theory based on quark fields is a good candidate for hadron dynamics. Therefore we restrict ourselves to confinement models which are developed from phenomenological considerations, or which were made possible by the progress of quantum field theory.

1.3 From a phenomenologist's point of view, quark confinement is more a question of principle and not one of practical relevance. Since the main topic of this summer school is the "new physics" coming from e^+e^- storage rings, one should ask about the relation of our subject to these general themes. I mention two examples for which I adopt the hidden charm scheme [3]:

(a) The small width of the new resonances is explained by Zweig's rule [4]. Since this rule states that quarks - visualized as quark lines - can leave a hadron only combined with an anti-quark (-line), this rule has to be explained by any confinement scheme. Hence, the understanding of Zweig's rule means the understanding of important aspects of confinement and new particles [5].

(b) In the simplest way, confinement of quarks is described with the help of a strong infinite range potential, a "confining potential". The discussion of the J/ψ -states of the charmonium spectrum [6], derived from such potentials sheds some light on the confinement dynamics.

However, applications are not the main goal of these lectures. I rather try to inform you on some recent theoretical developments in order to provide you with a broader view on possible dynamical explanations of the new phenomena.

1.4 Quark confinement is a challenging question for quantum field theory [7]. Of course, this phenomenon cannot be described in the framework of renormalized perturbation theory. But progress in understanding the structure of quantum field theory, for instance with the help of the renormalization group techniques [8], lead to interesting suggestions for the solution of our problem. Most important, we learned from the close relations between statistical mechanics and quantum field theory [9], that quantum field theory might describe quite different phases of matter than that related to the usual particle interpretation of relativistic field theory. The following examples may illustrate this point: The "dual strings" conjectured from the interpretation of the Veneziano formula might be considered as something like magnetic vortexes in superconductors of the second kind [10]. The "quark bags" can be related to stable classical solutions of non-linear field equations, like solitons first discussed in hydrodynamics [11]. There are models considering hadrons as superfluid droplets in equilibrium with a gaseous phase [12] etc. etc.

1.5 Certainly, the solutions suggested for the problem of quark confinement are still in a very speculative state. I think, in this situation it is not yet appropriate to try a critical assessment of the different models. Hence, in these lectures I shall simply give an introduction to the new concepts related to quark confinement, and I shall illustrate these with help of simple examples. However, I shall try to arrange the different ideas in such a way, that, hopefully, a vague, but coherent picture of quark confinement becomes visible. In this spirit I treat the following topics in the next Sections:

2. The phenomenological approach to quark confinement
3. The standard Lagrangian of hadrondynamics
4. Lagrangian field theory and quark confinement
5. Classical soliton solutions in a simple model. - Physical examples
6. Quantization of extended systems
7. Colour charge screening and quantization on a lattice
8. Short remarks on applications and summary

Even with the modern tools of information [13)], it is nowadays nearly impossible to quote the literature completely. But I hope that the references added to these lectures make these a useful guide to this new field of problems.

2. The phenomenological approach to quark confinement

2.1 The first and most simple idea to explain the absence of free quarks is to assume quarks having a large mass which are bound strongly in hadrons. This was suggested by G. Morpurgo [14)] in the framework of the non-relativistic quark model. In this model it is easy to consider the limiting case of infinitely strong binding by potentials singular at infinite distances, like the linear potential or the harmonic oscillator potential. These interactions do not allow approximately free particle states at large distances at any energy; even the remainder of the free particles, the one particle poles in the bound state wave function disappear. Such potentials are nowadays called "confining potentials". Many phenomenological questions are still discussed in this nonrelativistic framework where sometimes the form of the potential is guessed from field theoretical models, like the linear potential [15)] from the string model with colour charge screening (cf. 7.1e).

How can one formulate these simple ideas relativistically? The most consistent framework of relativistic quantum mechanics is general field theory, it is well suited to treat phenomenological problems. It has been shown [16)] that a relativistic model with heavy quarks and strong binding, formulated in the language of general field theory, can give the known results of the non-relativistic quark model for the spectrum of the mesons and their properties. The mesons are described here by Bethe-Salpeter amplitudes of the quark fields, and the strong binding is described by a phenomenological B.S.-kernel which compensates the large quark mass. Such a phenomenological discussion in a model with relativistic strong binding rises questions which are typical for the dynamics of quark confinement. For this we shall discuss some of the detailed aspects of the relativistic heavy quark model later. But before that I would like to mention the first attempts [17] of discussing the limit of strict confinement: infinite strong binding within general field theory. There are conjectures [18] of phenomenological interactions singular at large distances which make disappear the one quark pole in the hadron scattering amplitudes. In such models quark production by hadron collision is forbidden. The smoothness of Green's functions following from the absence of quark poles might be related to many phenomena of high energy processes [19]. However, in all these early models, the confining interaction is mostly guessed phenomenologically and there were only vague ideas [20] about their origin. In this respect, and not yet by the reliability or variety of their applications, the Lagrangian models in non-perturbative treatment brought some important progress.

2.2 Now I shall describe briefly the heavy quark model with strong binding [16], formulated in the language of general field theory (GFT). We consider the meson resonances as approximate single-particle states: $\left| {M \atop P} \; {j^{\pi c} \atop j_3} \; {I \atop I_3} \; Y \right\rangle$.

In the quark model they are described as $q\bar{q}$ bound states. The following matrix elements of the quark fields $\psi(x)$, $\bar{\psi}(x)$ between the vacuum $|0\rangle$ and the meson states, the B.S. amplitudes [21]

$$\chi(q,P) = (2\pi)^{3/2} \int dx\, e^{iqx} \langle 0 | T\, \psi(\tfrac{x}{2}) \bar{\psi}(-\tfrac{x}{2}) | {M \atop P} \, j^{\pi c} \, q\bar{q} \rangle \qquad (2.1)$$

(χ : 4 x 4 matrix in Dirac indices, when the quark label p, n, λ is suppressed) play a role similar to the Schrödinger wave functions in non-relativistic quantum mechanics. However, their meaning is completely different (no "probability interpretation") and given by the interpretation rules of GFT [22]. Structure analysis [23] of GFT tells that there is a Bethe-Salpeter equation for these amplitudes

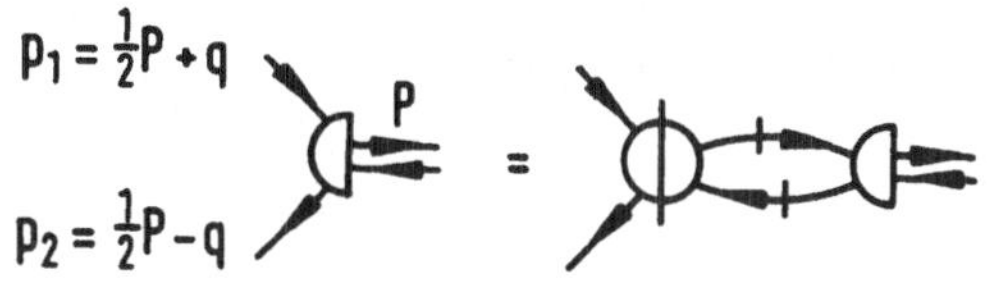

Analytically this bound-state B.S. equation reads

$$S^{-1}(p_1)\, \chi(q,P)\, \bar{S}^{-1}(p_2) = i \int d^4q'\, \mathfrak{K}(q,q';P)\, \chi(q',P) \qquad (2.2)$$

There is a normalization condition [24] for χ following from GFT. In a specific Lagrangian field theory, the B.S. kernel $\mathfrak{K}(q,q';P)$ describing the $q\bar{q}$-interaction, and the propagator S(p) are determined by the field equations. Since we use GFT only as a general framework for a model, we have to determine (like in non-relativistic quantum mechanics) $\mathfrak{K}$ ($\sim$ potential) and S by considerations guided by phenomenology and simplicity.

We give now, along this line, the arguments which lead to a tractable B.S. equation and to a model with many applications:

a) We assume a large quark mass m for practical quark confinement.

b) The propagator we set as $S^{-1}(p)=\gamma p-m$ for reasons of simplicity. This might have the meaning of an approximation for $\langle p^2\rangle \ll m^2$.

c) We assume convolution type, energy independent kernels for simplicity:

$$\mathfrak{K}(q,q';P) = \mathfrak{K}(q-q')$$

d) For the spin dependence of the kernel we start with the general ansatz:

$$\mathfrak{K} = K^S\, 1\times 1 + K^V \gamma_\mu\times\gamma_\mu + K^T \sigma_{\mu\nu}\times\sigma_{\mu\nu} + K^A \gamma_5\gamma_\mu\times\gamma_5\gamma_\mu + K^P\, \gamma_5\times\gamma_5 \tag{2.3}$$

$$(\Gamma_i\times\Gamma_i)\chi = \Gamma_i\chi\Gamma_i$$

The following phenomenological considerations lead to a restriction of this generality: i) The spectrum derived from the B.S. equation should show the singlet-triplet structure of the non-relativistic quark model. ii) "spin saturation" of the super-strong $q\bar{q}$-interaction implies a kernel which is dominantly of $(-\gamma_5\times\gamma_5)$-type [25]. (See 2.3b).

e) Wick rotation [26]: Analytic continuation $q_0 \to iq_4$ is allowed for B.S. equations with kernels which are superpositions of one particle exchanges. The "Wick rotated" equation can be more easily approximated.

f) Bound state mass zero approximation: Since $M \ll m$, it is natural to begin with an equation for M = o, and then use an expansion in $(M/m)^2$. The advantage is that the zero-order Wick-rotated equation has the higher symmetry of $O^{\pi C}(4)$.

g) Smooth potentials: The Fourier transform of a Wick-rotated convolution type B.S.-kernel might be considered as a 4-dimensional potential. In order for the level spacing to be small compared to the quark mass, this potential much be smooth.

h) Oscillator approximation: For low-lying states, smooth potentials in Wick-rotated B.S. equations can be approximated by harmonic forces [27]. The resulting wave functions are good approximations for space-like relative momenta; their analytic continuation into the time-like region is not justified. This understanding of the relativistic oscillator differs from that of other models not based on field theory [28].

i) The approximate dynamical equation: When we take all these considerations together, we get

$$(\gamma q - im)\chi(\gamma q - im) = -K\gamma_5\chi\gamma_5$$
$$K = \alpha - \beta\Box_q \quad , \quad M = 0$$

as a first approximation to the M ≠ 0 B.S. equation

$$(\tfrac{i}{2}M\gamma_4 + \gamma q - im)\chi(-\tfrac{i}{2}M\gamma_4 + \gamma q - im) = -(\alpha - \beta\Box_q)\gamma_5\chi\gamma_5 + K_i\Gamma_i\chi\Gamma_i$$
$$K_i\Gamma_i\chi\Gamma_i = O(m^o) \qquad (2.4)$$

k) Spectrum and solutions: We can solve Equ. (2.4) in $(M/m)^2$- approximation, and we get for the meson-mass spectrum:

$$M^2 = 4(\alpha^i + m^2 + 2(n+2r+2)\sqrt{\beta^i}) \qquad (2.5)$$

The quantum numbers j("spin"), Π ("parity"), C("charge parity") of the states and their multiplicities are related by the following scheme to the O(4)-orbital momentum n = 0, 1, 2, 3, .., and the radial excitation number r = 0, 1, 2, ... :

"spin-singlets": $\Pi = (-1)^{j+1}$, $C = (-1)^j$, $j = n, n-2, \ldots$

"spin-triplets":

$$\Pi = (-1)^{j+1}, \quad C = (-1)^{j+1}, \quad j = n+1, n-1, \ldots$$
$$\Pi = (-1)^j, \quad C = (-1)^j, \quad j = n, n-2, \ldots \quad n \neq 0$$
$$\Pi = (-1)^{j-1}, \quad C = (-1)^{j-1}, \quad j = n-1, n-3, \ldots \quad n \neq 0 \qquad (2.6)$$

The mesons of the "leading trajectory" j = n, or j = n ± 1, n respectively, have the quantum numbers of the non-relativistic quark model. There is singlet-triplet splitting, if the potential parameters α^i, β^i, (i = t(riplet), singlet) are modified in order $O(m^o)$. Because of the underlying four-dimensional oscillator, there are more states than in the non-relativistic quark model, namely those with j = n-2, ... etc. Seen intuitively, they correspond to excitations of the "sea of quark pairs and gluons". An example is the case of the vector mesons with M ≈ 1600 MeV. In the relativistic quark model solving Equ. (2.4) provides us with a list of approximate B.S. amplitudes for all the meson states of the spectrum. As an example we give the amplitude for the radially excited vector mesons

($j^{\pi c} = 1^{--}$, $n = o$, $r = o,1,..$; e_μ polarization vector)

$$\chi^{V}_{r}(q,P) = \frac{4\pi}{\sqrt{\beta(r+1)}}\left[\not{e}\left(1-\frac{\not{P}}{2m}\right) - \frac{ieq}{m}\right] e^{-q^2/2\sqrt{\beta}}\, L^{1}_{r}\left(\frac{q^2}{\sqrt{\beta}}\right) |q\bar{q}\rangle_V \tag{2.7}$$

1) Phenomenological applications: This model allows the discussion of the meson spectrum, the strong and leptonic decays of mesons and of their e.m. form factors. It relates the Regge slope to the range parameters of the hadronic decays and the leptonic decay constants, and thus gives an example for the solution of the dynamics of the meson spectrum. For a comprehensive comparison of these results with the experiment we refer to the literature [29].

2.3 The phenomenological relativistic quark model with strongly bound quarks reveals some features as being typical for the confinement dynamics. We want to elaborate a little bit more on this part of the model.

a) In the mass formula Equ. (2.5) the depth of the potential α has to compensate the large quark mass: $\alpha \approx - m^2$. Hence in this very deep potential, we expect many bound states - resonances -, as expressed by the mass formula (2.5) describing linear Regge trajectories, which are infinitely rising if $m^2 \to \infty$. We have to consider this as a typical feature of confinement dynamics. In addition, strong binding in field theory implies apparently more degrees of freedom in resonating states. This is a common feature of confinement models, too.

b) The formulation of our relativistic quark model in the notion of general field theory enables us to make a consistent ansatz for the calculation of mesonic coupling constants, in which all mesons are treated symmetrically [30]. We consider the approximation in which these constants are calculated from the triangle graph with the help of our B.S. amplitudes [31]:

M = 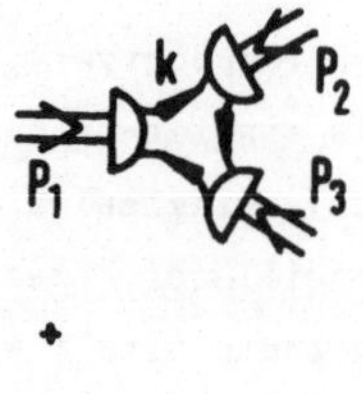

$$= Tr_{SU_3} \cdot (2\pi)^{-9/2}\, i \int d^4k\, Tr\Big[\chi^{(1)}(k,P_1) \cdot \left(\not{k} - \frac{\not{P}_1}{2} - m\right)\chi^{(2)}\left(k+\frac{P_3}{2}, P_2\right)\left(\not{k} - \frac{\not{P}_2 - \not{P}_3}{2} - m\right) \cdot \chi^{(3)}\left(k - \frac{P_2}{2}, P_3\right)\left(\not{k} + \frac{P_1}{2} - m\right)\Big] \tag{2.8}$$

$P_1 + P_2 + P_3 = 0$

The forces between the mesons are then derived from the interaction between quarks. They will increase in general with the quark mass - Equ. (2.7) suggests an increase with m^3. But for the spinor structure of the χ imposed by an $-\gamma_5\times\gamma_5$ -interaction, all the terms of order m^3, m^2, m in M (Equ. (2.7)) vanish. Thus, in spite of the superstrong forces between quarks, the mesonic forces are of moderate strength. We call this phenomenon the spin saturation of the superstrong binding forces.

Because of saturation, the $O(m^o)$ terms in Equ. (2.5) are now important for the calculation of $\mathcal{M}$. Therefore we cannot determine the mesonic couplings from the Regge slope alone, but they depend also on some other potential parameters. Thus we can use only the relative coupling strength of particles along a Regge slope as a test of our model. For example, we get $\Gamma(\rho\to 2\pi)$ = 146 MeV (input), $\Gamma(f(1270)\to 2\pi)$ = 140 MeV (Exp. 130 MeV), $\Gamma(g(1620)\to 2\pi)$ = 76 MeV (Exp. 64 MeV).

This discussion should give some impression of the problems posed by confinement dynamics for the meson decays. "Spin saturation of quark forces" may a possible way to solve these. There are similar ideas in an attempt to derive Zweig's rule in the framework of Wilson's confinement scheme [5].

c) Large series of resonances is a characteristic feature of a relativistic model of strongly bound quarks. On the other hand, at small distances we expect partonlike behaviour of the quarks. It is one of the main problems of confinement dynamics to reconcile these two aspects of the quark model. Our phenomenological model can also give a hint for the solution of this problem. For this we consider the reaction $e^+e^-\to$ hadrons. The cross-section of this process is related to the Green's function of the e.m. current $\langle 0|T j_\mu(x) j_\nu(y)|0\rangle$; $j_\mu(x) = Z\bar{\Psi}(x)\gamma_\mu Q\Psi(x)$. In our model of strongly bound heavy quarks this Green's function should be approximated by the intermediate bound states with the correct quantum numbers, i.e. the vector mesons states [32].

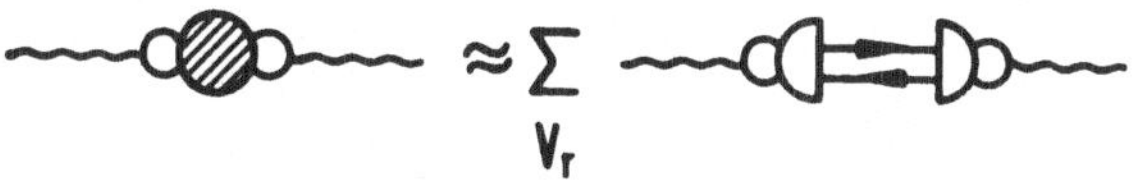

(2.9)

On the assumption that the individual vector mesons dominate locally, the cross-section is given by

$$\sigma_{tot}(e^+e^-\to \text{hadr.}) = \frac{16\pi^2\alpha^2}{s^2}\sum_{V_r} g_{V_r}^2 \frac{M_V\Gamma_V}{(s-M_V^2)^2 + M_V^2\Gamma_V^2} \qquad (2.10)$$

With the mass spectrum Equ. (2.5) and the photon-vector-meson couplings derived from the B.S. amplitudes Equ. (2.7):

$$\langle 0 | j_\mu(0) | V \rangle = g_V e_\mu(V) = (2\pi)^{-11/2} \, Z \; \mathrm{Trace}\, Q \gamma_\mu \int d^4q \, \chi(q,P)$$

$$g_V = (-1)^r \, Z \, \frac{4}{\pi} \sqrt{r+1} \sqrt{\beta} \, \langle Q_V \rangle \qquad \text{for radial excitations}$$

$$g_V = 0 \qquad \text{otherwise} \tag{2.11}$$

We calculate a smoothed cross-section

$$\sigma_{tot}(e^+e^- \to \text{hadr.}) = \frac{\pi \alpha^2}{s} \, Z^2 \sum_V \langle Q_V \rangle^2 \left(1 + \frac{1.6\ \text{GeV}^2}{s}\right) \tag{2.12}$$

It shows partonlike behaviour for large energies $E = \sqrt{s}$ [33].

3. The "Standard Lagrangean" of Hadron Dynamics

3.1 In the preceding Section 2, we tried to show how a phenomenological quark model, which takes into account the experimental absence of free quarks together with the principle of relativistic quantum mechanics, may bring about a first impression on the problems of the dynamics of quark confinement. In the remaining sections we want to consider this problem embedded in a more fundamental approach to hadron dynamics. Of course, a Lagrangean field theory of hadron dynamics is not yet really established. But the quark field theory with a non-abelian gauge invariant, renormalizable interaction incorporates a large part of the experimental and theoretical experience in elementary particle physics, hence it is an educated guess, that hadron dynamics might be based on a "Standard Lagrangian" [34] similar to that of the successful QED:

$$\mathcal{L} = -\frac{1}{4} F_{\mu\nu} F^{\mu\nu} + i \bar{q} \not{D} q + \bar{q} M q \tag{3.1}$$

Here $F_{\mu\nu}(x)$ denotes the "gluon-field", the analogue of the electromagnetic field; the quark fields $q(x), \bar{q}(x)$ take the place of the electron field in QED; D_μ; $\not{D} = D_\mu \gamma^\mu$ corresponds to the gauge invariant derivative $\partial_\mu + ieA_\mu$; M is the bare mass-matrix.

3.2 The intricacies of $\mathcal{L}$ are hidden in the details suppressed by the notation. I shall sketch these details as a reminder of the ideas and experimental discoveries which are focussed in the simple expression (3.1);

a) The quark field q(x) is coloured and flavoured: $q(x) \equiv q_{fc}(x)$

b) The "Flavour" f = $p, n, \lambda, \ldots c$ (?) denote the quark degrees of freedom, derived from the SU(3) structure of the hadron spectrum and confirmed by the properties of the valence quarks in deep inelastic e-ν -scattering etc. The discovery of the new particles adds new flavours like charm [3] c, Han-Nambu colour [35] (?) etc. to the quark fields. There is a general prejudice that the flavour degrees of freedom don't play an essential rôle in the dynamics of quark confinement.

c) The "Colour" C = red, white and blue is introduced [36] to solve the statistics problem of the baryon spectrum in the quark model. It is used to satisfy the condition that the sum of the electric charges Q of all hadron and lepton fields should vanish, a condition necessary to avoid the socalled "triangle anomalies" [37]. Colour is welcome to improve the calculation [36] of $\Gamma(\pi^0 \to 2\gamma)$ and to give larger values of R = $\sigma(e^+e^- \to \text{hadr.})/\sigma(e^+e^- \to \mu^+\mu^-) = \sum_{f,c} Q^2_{f,c}$.
The postulate that free particle states are colour singlets, guarantees that hadrons have conventional charges. Therefore the coupling of the interaction to the colour degrees of freedom is essential for quark confinement.

d) The "Gluons" [38] were introduced in order to mediate the interaction between quarks within the framework of renormalizable field theory. They form a coloured octet of vector fields $A_\mu^c(x)$:

$$C = \bar{r}w,\ \bar{r}b,\ \bar{w}r,\ \bar{w}b,\ \bar{b}w,\ \bar{b}r,\ (r\bar{r}-\bar{b}b),\ (\bar{r}r+\bar{b}b-2\bar{w}w)$$

which are singlets with respect to flavour.
As coloured objects, gluons are confined. Since gluons carry no electric or Cabbibo charge, their "observation" in deep inelastic electron or neutrino scattering is even more indirect than the "observation" of quarks. The energy sum rule:

$$1-\epsilon = 9\left[\int dx\, F_2^{\gamma N}(x) - \frac{1}{6}\int dx\, F_2^{\nu N}(x)\right] \approx 1/2$$

ϵ = percentage of 4-momentum p_μ carried by gluons. [39]
F(x) Scaling structure functions of deep inelastic scattering is considered as an indication of the reality of gluons.

Clearly more direct experimental evidence of these basic fields is of the utmost importance for the justification of hadron dynamics based on the standard Lagrangian.

The standard Lagrangian is constructed from the quark and gluon fields, taking into account the following symmetry and dynamical principles:

e) The postulate of "renormalizability" of a Lagrangian field theory, reduces possible Lagrangians to a very restricted class. Renormalizable field theories represent the only complete and consistent relativistic dynamical system. Therefore it is theoretically rational to postulate renormalizability. However, the physical meaning of this assumption is unclear.

The renormalization group techniques [8] allows the characterization of the asymptotic behaviour of renormalizable field theories for large momenta (small distances), and small momenta (large distances). In particular, a relation for Green's functions in momentum space: $\tau(q_1,\cdots,q_n)$ of the type

$$\tau(\lambda q_1,\cdots\lambda q_n;g,m) = \lambda^{\dim\tau}\, e^{-n\int_g^{g(\lambda)}\frac{\gamma(\lambda')}{\beta(\lambda')}d\lambda'}\,\tau(q_1,\cdots q_n;g(\lambda),m) \tag{3.2}$$

$$\frac{dg(\lambda)}{d\log\lambda} = \beta(g) \tag{3.3}$$

follows from the solution of the Callan-Symanzik equation. It relates a dilatation of the momenta by a factor λ to a change of the coupling constant from g to g(λ). The function β(g), according to which Equ. (3.3) determines the λ-dependence of the coupling constant, and γ(g) responsible for the dynamical dimensions of the fields, might in principle be calculated from the Lagrangian. From a practical point of view, it is more important that already the qualitative features of β(g) have serious implications on the general structure of the theory. Thus zeros of β(g) imply the existence of limits of the "running" coupling constant $g(\lambda)$: $g(\lambda)\rightarrow g_0$ for $\lambda\rightarrow\infty$, or 0, if $\beta(g_0)=0$ and $\beta'(g_0)<0$, or $\beta'(g_0)>0$ respectively. In these cases the theory is called ultraviolet or infrared asymptotically stable.

Relations of type (3.2) hold only for non-exceptional Euclidean momenta. This has the practical consequence that conclusions from asymptotic stability apply mainly to averages over physically observed quantities [40].

f) Asymptotic Freedom: Field theories, which approach for large momenta the free field theory, i.e. for which $g(\lambda)\rightarrow 0$ for $\lambda\rightarrow\infty$; are called

asymptotically free.
In asymptotic free theories, Bjorken scaling for $\sigma(e^+e^- \to \text{hadrons})$ [41)] and for deep inelastic e, ν -scattering [41)] can be proven - if formulated appropriately. Hence the experimental discovery of scaling (!) and the success of the parton model suggests the postulate of asymptotic freedom [41b)].

g) Non-abelian gauge invariance. The only renormalizable, asymptotic free theories are the non-abelian gauge theories [41)] with Lagrangians like Equ. (2.1), i.e. an interaction between fermions and gauge vector mesons minimally induced by the gauge invariant derivative D. Assuming the colour SU(3) as gauge group, the coupling becomes explicitely

$$F_{\mu\nu}^{c} = \partial_\mu A_\nu^{c} - \partial_\nu A_\mu^{c} - g f^{cc'c''} A_\mu^{c'} A_\nu^{c''}$$

$$D_\mu = \partial_\mu + i g A_\mu^{c} \lambda^{c}/2$$

(λ^c the Gell-Mann matrices, $f^{cc'c''}$ the SU(3)-antisymmetric octet tensor).
We have motivated this assumption under (c).

h) The SU(N)-symmetry in the flavour degrees (SU(3), SU(4)) of freedom might be broken by the mass term $\bar{q}Mq$ in $\mathcal{L}$. In the limit $M \to 0$ the standard Lagrangian is SU(N) symmetric. This implies the conservation laws of charge, hypercharge, isospin etc., as well as the existence of the corresponding conserved or partially conserved currents and axial currents, on which current algebra is based. M is not expected to be related to the mass of free quarks. On the contrary, it is conjectured that the correct standard Lagrangian leads to a theory which is infrared instable: $g(\lambda) \to \infty$ for $\lambda \to 0$, and that the strong interaction at large distances assures quark confinement.

i) This conjecture is sometimes called: "Infrared Slavery". [42)]

It is this aspect of the dynamics based on the standard Lagrangian, on which we have to elaborate further.

4. Lagrangian Field Theory and Quark Confinement

4.1 Assuming now the standard Lagrangian Equ. (3.1) as a basis for hadron dynamics; how should one organize the "solution" of the quantized field equations derived from it? The renormalization group technique mentioned in 3.2.e., suggests to arrange the solution in three parts:

a) On the one hand we have to consider the small distance limit. This should explain the properties of the partons, like their symmetry properties expressed by sum rules for their quantum numbers (f.e. $R = \sum Q_i^2$, etc.) together with their dynamical behaviour observed in the scaling laws of deep inelastic e and ν scattering etc.

b) On the other hand we have to discuss the large distance limit. Here one has to consider the hadronic stable particles and resonances, their exact and approximate symmetries, like isospin, SU(3), spontaneously broken chiral symmetry, etc. Further we have to explain dual dynamics, combining the dynamics of the meson spectrum and the peripheral part of reaction dynamics. Last but not least we have to understand the screening of the charges of colour symmetry, i.e. quark confinement.

c) The remaining problem is the connection of the two limiting cases. In a sense, this is the real problem. Here the Lagrangian approach to hadron dynamics should show its full power in expressing the dynamical parameters of the scaling region as well as those of the hadron spectrum by the renormalized coupling constant g and the mass parameters related to the normalization point. This is also the "general problem of quark confinement", of which the answer to the question: Are there particles with the quantum numbers of the partons? is concerned with a particular aspect of the relation between the large and small distance limit.

4.2 Since the Standard Lagrangian describes an asymptotically free field theory: $g(\lambda) \to 0$ for $\lambda \to \infty$ (cf. 3.2.f), the problems related to the small distance limit can in principle be handled relatively easily. Equation (3.2) relates the Green's function at large momenta, to those of a theory with a small coupling constant. However, for small coupling constants we may apply conventional perturbation theory. Therefore we may get along this line reliable asymptotic expressions of Green's functions describing deep inelastic e, ν -scattering and e^+e^- annihilation [41] into hadrons. However, because our main problem here is quark confinement related to the large distance limit, we have to refer to the literature and to other lectures of this school [43] for further discussion of this topic.

4.3 It is guessed that the field theory or the standard Lagrangian is infrared instable: $g(\lambda) \to \infty$, for $\lambda \to 0$ (cf. 3.2.i). Therefore, the infrared limit is related by Equ. (3.2) to theories with large coupling constants. The standard Lagrangian with large g may describe quite different phases of hadronic matter compared to the approximate free particle states at small g. But we understand strong coupling theories much less than weak coupling perturbation theory. Hence our knowledge of the phenomen appearing in the region of strong coupling is much more speculative. The following may illustrate such a line of speculations:

(a) Strong coupling may induce dynamical spontaneous symmetry breaking [41a], which is related to a shift in the vacuum fluctuations $\langle \bar{\Psi}(x) \Psi(x) \rangle \neq 0$. This process is visualized as an analogue of the formation of Cooper pairs in superconductivity.

(b) In order to make spontaneous symmetry breaking more obvious, the standard Lagrangian is supplemented by a scalar Higg's field $\Phi(x)$, which via the Higgs mechanism, produces spontaneous symmetry breaking (cf. 5.1.d). The modified $\mathcal{L}$ becomes, according to some speculative calculations [44],

$$\begin{aligned} \mathcal{L} = & -\tfrac{1}{4} Z_3(\phi) F_{\mu\nu} F^{\mu\nu} + \tfrac{1}{2} \partial_\mu \phi \, \partial_\mu \phi - V(\phi) \\ & + i \bar{q} \not{D} q + \bar{q} M q \end{aligned} \tag{4.1}$$

This Higgs Yang Mills Lagrangian [45] is renormalizable, but now longer asymptotically free. But as already mentioned, this $\mathcal{L}$ has to be considered only as an approximate model for describing the long distance - strong coupling-phenomena induced by the somewhat obscure mechanism of dynamical symmetry breaking. It should not be considered in the small distance limit.

(c) This Lagrangian (4.1) describes the relativistic version of the Landau-Ginsburg theory of superconductors of the second kind [46] (for $Z(\phi) = 1$, $V(\phi) = -\frac{m^2}{2}\phi^2 + \frac{\lambda}{4}\phi^4$), where $\phi(x)$ is a phenomenological field describing Cooper-pairs. Hence one expects the existence of stable classical solutions of the field equations derived from (4.1) representing the magnetic vertex lines of the super conductor. One suspects [10] that these strings are the objects on which the Veneziano dual dynamics is based (cf. 5.2.c). But quite generally, there is the wide spread hope that the new phenomena appearing for large coupling constants might be described with the help of stable classical solutions, of which the main features survive the quantization procedure. It is expected that the quarks are confined in such strongly bound structures ("SLAC-Bag" [47] "MIT Bag" [48] etc. (cf. 5.3).

(d) There is another approach to the strong coupling limit of the standard Lagrangian which is based on a lattice approximation to the space time continuum [49]. Also this method seems to indicate the existence of stringlike structures [50].

As already emphasized, these conjectures on the large distance limit of the standard Lagrangian field theory are highly speculative and mainly guided by the desire to explain successfully the phenomena of high energy physics. But one should consider them seriously as possibilities and one might study them in more detail in models, which do not have the full complexity of the standard $\mathcal{L}$. In this spirit we shall consider stable classical solutions, their quantization, and quark confinement in stringlike structures in very simplified versions.

4.4 With so little knowledge on the large distance limit, there is little to say about the connection of the both limits of the standard Lagrangian. (For a phenomenological comment cf. 2.3c). However, in order to give a first impression on the problems showing up here, we may have a look on the recently discussed relation [51)]between the 2-dimensional sine Gordon equation [52)]

$$\partial^\mu \partial_\mu \phi + (\mu^2/\beta) : \sin \beta \phi : = 0 \tag{4.2}$$

and the Thirring model: [53)]

$$(-i\gamma^\mu \partial_\mu - m_0)\psi = g : (\gamma^\mu \bar{\psi} \gamma_\mu \psi) : \psi \tag{4.3}$$

It is claimed [54)] that in the quantized form the two models are equivalent, if one relates

$$\frac{\beta^2}{4\pi} = \frac{1}{1 + g/\pi} \tag{4.4}$$

In this equivalence the spinor particle of the Thirring model is described by the stable classical soliton solution of (4.2):

$$\phi_{so}(x) = \frac{4}{g} \operatorname{tang}^{-1} e^{\mu x} \tag{4.5}$$

This non-perturbative solution of (4.2) (leading term in $1/\beta$) allows a good description of the excitations, form factors etc. of this particle in the limit of small coupling constants of the sine Gordon theory, i.e. for strong coupling in the Thirring model. For further study we have to refer to the literature.

5. Classical Models

5.1 In order to give an impression on the new type of bound state structures described by stable solutions of classical field equations (cf. 4.3.c), we discuss briefly a simple $1+1$ -dimensional example [55]. This model is the starting point of the so called "SLAC-BAG" describing strongly bound quarks [47]. In spite of being simple and explicitely solvable, it shows many characteristic features which allow the illustration of new concepts used in the discussion of quark confinement models.

The fields of this model consist of a colourless quark field $\psi(x)$ of only one flavour, a Higg's field $\phi(x)$, but no gluons. The model is described by

a) The field equations

$$(-\partial_t^2+\partial_x^2)\phi + m^2\phi - \lambda\phi^3 = -g\psi^+\beta\psi \tag{5.1}$$

$$i(\gamma^0\partial_t+\gamma^1\partial_x)\psi = g\phi\psi$$

$$\beta = \gamma^0 = \begin{pmatrix}0&1\\1&0\end{pmatrix}; \quad \gamma^1 = \begin{pmatrix}0&1\\-1&0\end{pmatrix} \tag{5.2}$$

b) These are derived from the Lagrange density

$$\begin{aligned}\mathcal{L} = {} & \tfrac{1}{2}(\partial_\mu\phi)^2 + \tfrac{1}{2}m^2\phi^2 - \tfrac{1}{4}\lambda\phi^4 \\ & + i\bar{\psi}\gamma^\mu\partial_\mu\psi + g\phi\bar{\psi}\psi \quad .\end{aligned} \tag{5.3}$$

or the Hamiltonian $\quad \pi(x) = \partial_t\phi(x)$

$$\begin{aligned}H = \int d^1x \, \{ & \tfrac{1}{2}(\pi^2(x) + (\tfrac{d}{dx}\phi)^2 + \tfrac{\lambda}{4}(\phi^2(x) - \tfrac{m^2}{\lambda})^2 \\ & + \psi^+(\alpha\tfrac{\partial}{\partial x} + g\beta\phi)\psi(x) \ \}\end{aligned} \tag{5.4}$$

c) There are the trivial classical solutions of Equs. (5.1,2)

$$\phi^{\pm}_{cl.} = \pm\frac{m}{\sqrt{\lambda}} \quad , \quad \psi = 0 \tag{5.4}$$

corresponding to the minimum of the "potential energy" $V(\phi)$:

$V(\phi) = \frac{\lambda}{4}\int(\phi^2(x) - \frac{m^2}{\lambda})^2 d^1x$, normalized to zero.

d) The energy degeneracy of the classical ground state solutions (5.4) implies spontaneous symmetry breaking: None of the two ground state solutions $\phi_{cl.}^{+}$ and $\phi_{cl.}^{-}$ has the symmetry $\phi(x) \to -\phi(x)$, $\psi \to -\psi$ of the field equations (5.1,2).

e) Expanding around the stable vacuum solutions (5.4):
$\phi(x) = \varphi(x) \pm \frac{m}{\sqrt{\lambda}}$ leads to

$$(-\Box - 2m^2)\varphi \pm 3m\sqrt{\lambda}\,\varphi^2 - \lambda\varphi^3 = -g\psi^+\beta\psi \tag{5.5}$$

$$(-i\not{\partial} \pm g\frac{m}{\sqrt{\lambda}})\psi + g\,\varphi(x)\,\psi = 0 \tag{5.6}$$

thus determining the classical mass parameters of the quark field m_q and of the Higgs field μ :

$$m_q = g\frac{m}{\sqrt{\lambda}} \quad , \quad \mu = \sqrt{2}\, m \; . \tag{5.7}$$

f) "Kink solution": There is another stationary solution, with $\psi = 0$, of

$$(\frac{d^2}{dx^2} + m^2 - \lambda\phi^2)\,\phi(x) = 0 \tag{5.8}$$

namely

$$\phi_{KI}(x) = \frac{m}{\sqrt{\lambda}} \tanh \frac{mx}{\sqrt{2}} \tag{5.9}$$

It has some characteristic features of the classical solutions, of the type we are always talking about:

(i) $\phi_{KI}(x)$ is classically stable (cf. 5.1g,6).

(ii) $\phi_{KI}(x)$ connects the two degenerate vacuum solutions:

$\phi_{KI}(x) \to \pm\frac{m}{\sqrt{\lambda}}$ for $x \to \pm\infty$. This illustrates the relation between spontaneous symmetry breaking and stable classical solutions.

(iii) The total energy of $\phi_{KI}(x)$ according to (5.4) is finite

$$E(\phi_{KI}) = \frac{\sqrt{8}}{3}\frac{m^3}{\lambda} \tag{5.10}$$

the energy density is concentrated in a finite region around x = o.
(iv) The kink solution $\phi_{KI}(x)$ is singular for $\lambda \to 0$, i.e. "non-perturbative".

It has been suggested [56] to generally call solutions of non-linear field equations with such properties "solitons" (cf. Equ. (4.4)).

g) "Topological stability" [57]: The kink solution $\phi_{KI}(x)$ is stable, because it is the lowest energy solution with the conserved topological number $\mathcal{K}=1$:

$$\mathcal{K} = \frac{\sqrt{\lambda}}{2m} \left(\phi(\infty) - \phi(-\infty) \right) \tag{5.11}$$

Note the relation between spontaneous symmetry breaking and the existence of such "topological quantum numbers" (cf. 5.1f (ii)). According to the definition (5.11) $\mathcal{K}$ is conserved, because the spatial asymptotic behaviour of solutions $\phi(x,t)$ of the field equation (5.1) remain fixed. The topological property of the 2-dimensional spacetime allows to relate $\mathcal{K}$ to a conserved current:

$$\begin{aligned} B^{\mu}(x,t) &= \frac{\sqrt{\lambda}}{2m} \epsilon^{\mu\nu} \partial_{\nu} \phi(x,t) \\ \mathcal{K} &= \int B_0(x,t)\, d^1x \\ \epsilon^{\mu\nu} &= -\epsilon^{\nu\mu}, \quad \epsilon^{01} = 1 \end{aligned} \tag{5.12}$$

f) "Kink with trapped quark": There is a simultaneous solution of Equs. (5.1) and (5.2) with   :

$$\begin{aligned} \phi(x) &= \phi_{KI}(x) = \frac{m}{\sqrt{\lambda}} \tanh \frac{mx}{\sqrt{2}} \\ \psi_{tr}(x) &= C \cdot \left[\cosh \frac{xm}{\sqrt{2}} \right]^{-\sqrt{2/\lambda}\, g} \begin{pmatrix} 1 \\ -i \end{pmatrix} \end{aligned} \tag{5.13}$$

and with a classical energy according to Equ. (5.4) of

$$E(\phi_{KI}, \psi_{tr}) = E(\phi_{KI}) = \frac{\sqrt{8}}{3} \frac{m^3}{\lambda} = \frac{\sqrt{8}}{3} \frac{m^2}{g\sqrt{\lambda}} m_q \; . \tag{5.14}$$

The trapping of the quark does not add to the kink energy. Anyhow, for an appropriate choice of the parameters, the trapped quark may have a very small mass compared to the "free" quark (cf. Equ. (5.7)). Thus the state "kink with trapped quark" described by the classical solutions (5.13) represents an example of a "field theoretical bound state" with strong binding. Such type of states may describe hadrons with strongly bound or confined quarks. In contrast to our phenomenological model (cf. 2.2), strong binding is derived from an invariant, local, renormalizable Lagrangian Equ. (5.3).

5.2 The 1 + 1-dimensional models are very helpful for getting acquainted with the properties of soliton type solutions of non-linear field equations. But of course, physical models should be described in 3 + 1 dimensions. There are several attempts to describe hadrons by such type of solutions, called bags and strings, in which quarks are strongly bound or confined. We shall mention now in a cursory manner some of the more popular ones. Several example will also be treated in the lecture by Prof. Y. Nambu.

a) The dynamical structure of the "SLAC-BAG" [47)] is derived from the 2-dimensional model discussed above. This model is considered in 3 + 1 dimensions. It is argued that

$$\begin{aligned} \phi(\vec{x}) &= \phi_{KI}(r-R) \\ \psi(\vec{x}) &= \psi_{tr}(r-R) \end{aligned} \tag{5.15}$$

with $r = |\vec{x}|$ and $\begin{pmatrix} 1 \\ -i\frac{\vec{\sigma}\cdot\vec{x}}{r} \end{pmatrix}$ substituted for $\begin{pmatrix} 1 \\ -i \end{pmatrix}$ in 5.13

represent good approximation of the 3 + 1-dimensional equations (5.1) and (5.2), if $g \gg \sqrt{\lambda} \gg 1$. Since we have seen that the interaction of the quarks with the Higgs-field $\phi(x)$ does not perturb the structure of the kink solution, one may put N quarks in the corresponding 3-dimensional bag without disturbing significantly its structure.

The radius R of the bag is determined by minimalizing the energy, it results:

$$R_N = N^{1/3} R_0 \, , \quad E_N = \frac{2}{2R_0} N^{2/3} \, , \quad R_0 = \frac{\lambda^{1/6}}{2m} \tag{5.16}$$

The stability of the SLAC-Bag is no longer topological, but dynamical. Comparing the energy of N free quarks: $Ngm/\sqrt{\lambda}$ with E_N, we see that for small λ and large g the SLAC-bag is a field theoretical bound state with strong binding: "It is for a quark energetically favourable to dig a hole from one classical Higgs vacuum to the other. The kinetic energy necessary for that guarantees the finite size of the hole" [47]. This mechanism does not explain why the bound states are colour neutral. For this an effective coupling to the gluons is considered. This can be arranged in such a form that according to rough estimates the coloured states have a much higher energy than the colourless ones - a mechanism suggested first by Y. Nambu [58]. The SLAC-Bag describes only approximate confinement of coloured objects.

b) The MIT-Bag [48]: Different types of bag structures can be produced by assuming different types of "potentials" $V(\phi)$ for the Higgs field. Using a potential with a metastable vacuum first suggested by Vinciarelli [59] and applying an appropriate limit procedure, it can be shown [60] that the socalled "MIT-bag" is a limit of a field theoretical bag. In this model the classical fields of quarks and gluons are confined in the inner of the bag. It is described by the Lagrangian

$$L = \int_{\mathrm{Bag}} d\vec{x} \left\{ -\frac{1}{4} F_{\mu\nu} F^{\mu\nu} + i \bar{q} \not{D} q - B \right\} \tag{5.17}$$

where the volume term B is the only remainder from the pressure of the Higgs field from the outside of the bag. The field equations derived from (5.17) are

$$\left.\begin{aligned} D^{\mu}_{cc'} F^{c'}_{\mu\nu}(x) &= -g \bar{q} \lambda^{c} \gamma_{\nu} q(x) \\ i \not{D} q(x) &= 0 \end{aligned}\right\} \text{ in the bag} \tag{5.18}$$

$$q = \bar{q} = F_{\mu\nu} = 0 \qquad \text{outside.}$$

Since it is assumed that the coupling of the gluons inside the bag is very weak, the MIT-bag model realizes the ideas of infrared slavery and asymptotic freedom in an almost exaggerated form. As a limiting case of a field theoretical bag, its surface is determined dynamically, expressed by the following boundary conditions on the surface of the bag:

$$\begin{aligned} n^{\mu} F_{\mu\nu}(x) &= 0 , \qquad i \not{n} q(x) = q(x) \\ -\frac{1}{4} F_{\mu\nu} F^{\mu\nu} &+ \frac{1}{2} n_{\mu} \partial^{\mu} \bar{q} q - B = 0 \end{aligned} \tag{5.19}$$

n_{μ} : normal vector of the surface.

It follows from field equations and boundary conditions, that the colour charges of the bags are zero; therefore we have exact confinement of coloured objects.

c) Dual strings: The similarity of the Higgs Yang-Mills Lagrangian to the Ginzburg-Landau Lagrangian for the phenomenological description of superconductivity (cf. 4.3b,c) lead Nielsen and Oleson [(61)] to suggest, that there might be classical solutions which correspond to the vertex lines in superconductors of the second kind. In order to get finite vertex lines, the "gluon magnetic" flux has to be absorbed by the quarks as magnetic monopoles with respect to gluon coupling. The question of the stability of such solutions is open [62)]. There is the conjecture that non-abelian gauge theories allow topologically stable solutions, in which quarks are trapped. This conjecture is supported by the existence of a solution of this kind for a single monopol [63)]. (But this does not trap quarks [64)]).

Anyhow, stringlike soliton solutions are very much preferred for physical reasons: First, because vertices would provide a physical model for dual strings [65)]. Second, one-dimensional systems may show more easily colour charge screening [66)], (cf. 7.), and hence confinement. Under this aspect several forms of Lagrangians of type Equ. (4.1) are discussed in the literature [67)].

6. Quantization of extended Systems

We pursue the idea that the main phenomen in the large distance, strong coupling limit is the formation of bags or strings to which quarks are confined. These structures might be described by classical soliton solutions of non-linear field equations. It is assumed, of course, that such types of stable classical solutions survive, together with their main features the quantization procedure. However, in view of the long and incomplete story of the quantization of extended relativistic systems, this hypothesis is difficult to check.

6.1 In order to get a feeling for the problems involved here, we consider our simple field equation (cf. 5.1; $\Psi \equiv 0$)

$$(-\partial_t^2 + \partial_x^2 + m^2 - \lambda\phi^2)\,\phi(x,t) = 0 \qquad (6.1)$$

with the classical "kink" solution

$$\phi_{KI}(x) = \frac{m}{\sqrt{\lambda}} \tanh \frac{x m}{\sqrt{2}} \qquad (6.2)$$

a) For a first step towards quantization, we consider the oszillations around this solution. Therefore we expand

$$\phi(x,t) = \phi_{KI}(x) + \sum_n e^{i\omega_n t} \varphi_n(x) \tag{6.3}$$

which leads to the linearized equation

$$\left(-\frac{d^2}{dx^2} - m^2 + 3\lambda\phi_{KI}^2(x)\right)\varphi(x) = \omega_n^2 \varphi_n(x) \tag{6.5}$$

These equations can also be solved explicitely [68], which makes our simple model so illustrative. We get for

$$\begin{aligned}
\omega_0 &= 0 \,, & \varphi_0(x) &= \left[\cosh \frac{mx}{\sqrt{2}}\right]^{-2} \sim \frac{d}{dx}\phi_{KI}(x) \\
\omega_1 &= \frac{\sqrt{3}}{2}\mu \,, & \varphi_1(x) &= \sinh \frac{mx}{\sqrt{2}} \cdot \varphi_0(x) \\
\omega_2 &= \mu \,, & \varphi_2(x) &= \ldots \\
\omega_k &= \sqrt{\mu^2+k^2} \,, & \varphi_k(x) &= e^{ikx}\left[3\tanh \frac{xm}{\sqrt{2}} + \left(\frac{k}{\mu}\right)^2 - 1 - 3i\frac{k}{\mu}\tanh\frac{xm}{\sqrt{2}}\right] \\
& & &\rightarrow e^{\pm i\delta(k)} e^{ikx} \quad \text{for } x \rightarrow \pm\infty
\end{aligned} \tag{6.6}$$

These solutions have the following meaning:

$\omega_0 = 0$	represents the translation mode of the kink.
ω_1, ω_2	are classical excitation frequencies of the kink.
$\omega_k = \sqrt{\mu^2+k^2}$	are the frequencies of the scattering modes. It gives the correct energy momentum relation of the asymptotically free mesons.

b) The zero frequency translation mode is classically unstable. Therefore it must be separated from the other modes with the help of a canonical transformation, as has been extensively discussed by N.H. Christ and T.D. Lee [56]. The discussion of this transformation is rather involved, we sketch here only the main features. We complete the expansion (6.3) around the static kink solution to a canonical transformation between the canonical field coordinates $\phi(x), \dot{\phi}(x)$ at $t = 0$, and the new coordinates Z ("kink position"), P ("total momentum") $\Pi(\bar{k}), q(\bar{k}), \bar{k} = 1,2,k$ (the canonical coordinates of the modes 6.6):

$$\phi(x) = \phi_{KI}(x-z) + \int \varphi_{\bar{k}}(x-z)\, q(\bar{k})\, d\mu(\bar{k})$$

$$\dot{\phi}(x) = -\frac{1}{M_{KI}} P \frac{d\phi_{KI}(x-z)}{dx} + \int \varphi_{\bar{k}}(x-z)\, \Pi(\bar{k})\, d\mu(\bar{k}) + \ldots \tag{6.7}$$

$M_{KI} = \frac{(\sqrt{2}m)^3}{3\lambda}$ is the classical rest energy of the kink.

This transformation has to be considered as an expansion in powers of λ. The Hamiltonian becomes now, up to zero order in λ

$$H = \frac{1}{2}\int dx \left\{ \dot{\phi}^2 + \left(\frac{d}{dx}\phi\right)^2 + \frac{1}{2}\left(\phi^2 - \frac{m^2}{\lambda}\right)^2 \right\}$$

$$= M_{KI} + \frac{1}{2}\int \left(\Pi^2(\bar{k}) + \omega^2(\bar{k})\, q^2(\bar{k})\right) d\mu(\bar{k}) \tag{6.8}$$

which agrees with the classical interpretation of the modes discussed above. Note that in this order of approximation, the recoil of the kink is neglected in the meson-kink-scattering. Therefore it is relatively easy to achieve relativistic covariance by boosting the kink solution and the expansion (6.7) [56].

c) The quantum mechanical kink state $|KI\rangle$ is characterized by the absence of excited modes:

$$\left(\sqrt{\omega_{\bar{k}}}\, q(\bar{k}) + i\,\Pi(\bar{k})/\sqrt{\omega_{\bar{k}}}\right)|KI\rangle = 0 \tag{6.9}$$

Formally implementation of the canonical transformation by a "dressing" operator, leads to an approximate representation of $|KI\rangle$ by a coherent state:

$$|KI\rangle \sim \, :\exp\left\{-i\int dx\, \phi_{KI}(x)\overleftrightarrow{\partial}\phi(x)\right\}: |0\rangle \tag{6.10}$$

It follows from (6.9) that the vertex $\langle KI|\phi(x)|KI\rangle$ is essentially given by the classical kink solution [69]

6.2 We want to add to this sketchy introduction to the quantization problem further references and some general remarks:

a) Because of the topological stability, the transition amplitudes between the vacuum $|0\rangle$ and $|KI\rangle$ of polynomials of $\phi(x)$ vanish: $\langle 0|\phi(x_1)\dots\phi(x_n)|KI\rangle = 0$ This rises the question of the appropriate fields, which describe the transition between the vacuum sector and the kink sector (cf. 7.1.d).

b) Coherent states like (6.10) as trial states in a variational method were used by several groups as a starting point of quantization [70].

c) The characteristic properties of $|KI\rangle$: $\langle KI|\phi(x)|KI\rangle \sim \phi_{KI}(x)$; $\langle 0|\phi(x_1)\dots\phi(x_n)|KI\rangle = 0$ were used by Goldstone and Jackiw [69] to derive from a Green's function formulation the classical kink solution for the vertex.

d) Besides these different attempts of canonical quantization [71] the 2-dimensional ϕ^4 -model, the Sine-Gordon-model, etc. were also studied by the WKB-methods [72]. In the context with quantization, the conjecture on the relation between the massive Thirring model and the Sine-Gordon model should be mentioned again [51] (cf. 4.3).

7. Colour Charge Screening

In hadron dynamics based on the standard Lagrangian $\mathcal{L}$ (cf. 3.1), the idea of quark confinement is generalized to the conjecture of the non-existence of all kinds of free coloured particles. Since $\mathcal{L}$ is invariant under SU(3)-colour gauge transformations, there are conserved currents of colour $j^c_\mu(x)$. However, it is the meaning of the conjectured "colour charge screening", that all physical states have colour charge

$$Q^c = \int j^c_0 \, d\vec{x} = 0$$

The phenomenon of charge screening was discovered by J. Schwinger [73] in his investigation of two-dimensional, massless QED. A. Casher, J. Kogut and L. Susskind [66] proposed it as a model of quark confinement. It was presented very clearly by L. Susskind at the previous "International Summer School in Theoretical Physics" at Bonn. Only for rounding off our introduction to the ideas on quark confinement, we want to add some short remarks on colour charge screening.

7.1 The Dirac-Maxwell equations for a massless Dirac field in two dimensions are

$$i\gamma^\mu \partial_\mu \Psi(x) = -e\gamma^\mu A_\mu(x)\Psi(x)$$

$$= \frac{1}{2} \lim_{\substack{\epsilon \to 0 \\ \epsilon^2 < 0}} \gamma^\mu \left[A_\mu(x+\epsilon)\Psi(x) + \Psi(x) A_\mu(x-\epsilon) \right] \tag{7.1}$$

$$\partial^\nu F_{\mu\nu}(x) = e j_\mu(x)$$

$$= e \lim_{\epsilon \to 0} \Big\{ \bar\Psi(x+\epsilon)\gamma_\mu \Psi(x) - f^{-1}(\epsilon) \langle 0 | \bar\Psi(x+\epsilon)\gamma_\mu\Psi(x)|0\rangle \cdot$$

$$\cdot (1 - e\epsilon^\mu A_\mu(x)) \Big\} \tag{7.2}$$

For the definition of two-dimensional QED one has to complete these equations by the known commutation and anticommutation relations. It is well-known, that a covariant representation of QED by local fields requires the introduction of a Hilbert space with indefinite metric. However, those vectors of this space, which have negative or zero norm, do not have a gauge-invariant, physical meaning. However, there are special gauges of the fields, which allow the restriction to physical states only, whereby the fields loose locality and relativistic covariance. The following operator solutions of Equ. (7.1), (7.2), expressed by a free massive field are of this type [74]:

$$\Psi_a(x) = N_1 : \exp i\sqrt{\pi}\gamma_5 \phi(x) : \chi_a \tag{7.3}$$

$$j_\mu(x) = -\frac{1}{\sqrt{\pi}} \epsilon_{\mu\nu}\partial^\nu \phi(x)$$

$$F_{\mu\nu}(x) = \partial_\mu A_\nu(x) - \partial_\nu A_\mu(x) \tag{7.4}$$

$$A_\mu(x) = -\frac{\sqrt{\pi}}{e} \epsilon_{\mu\nu}\partial^\nu \phi(x) \tag{7.5}$$

The numerical 2-spinors χ_a are phases $|\chi_a| = 1$, the free field satisfies

$$-\partial_\nu \partial^\nu \phi(x) = \mu^2 \phi(x) \quad , \qquad \mu = e/\sqrt{\pi}$$

$$[\phi(x), \phi(x')] = i\,\Delta(x-x') \tag{7.6}$$

It is easy to verify that (7.3) is really a solution of the Dirac-Maxwell equations (7.1), (7.2). On the other hand, it was shown [75] by a detailed mathematical analysis, that all irreducible solutions of these properly quantized equations are essentially equivalent by operator gauge transformations to the special solutions given by Eqs. (7.3) and (7.4). In particular, this means that the gauge invariant observables expressed by the solution (7.3) describe completely and uniquely the physical content of two-dimensional, "massless" QED. As examples of such observables we mention the electric field tensor $F_{\mu\nu}(x)$ and the bilocal operator T(x,y) [74]:

$$F_{\mu\nu}(x) = -\frac{e}{\sqrt{\pi}}\,\epsilon_{\mu\nu}\,\phi(x)$$

$$\begin{aligned} T(x_1,x_2) &= \Psi(x_1)\exp\Big\{ie\int_{x_1}^{x_2} A^{\nu}(t)\,dt_{\nu}\Big\}\overline{\Psi}(x_2) \\ &= N(x_1-x_2) :\exp i\sqrt{\pi}\Big\{-\int_{x_1}^{x_2}\epsilon^{\mu\nu}\partial_{\nu}\phi(t)\,dt_{\mu} + \gamma_5^1\phi(x_1) - \gamma_5^2\phi(x_2)\Big\}: \end{aligned} \tag{7.7}$$

Let us now draw some conclusions from this analysis, which illustrate some of the main features of charge screening:

a) The <u>physical states</u> consist of free massive pseudoscalar particles ("mesons").

b) The <u>charge</u> of these particles is <u>zero:</u> $Q = \int_{-\infty}^{+\infty} j_0(x,t)\,dx = \int_{-\infty}^{+\infty}\partial_x\phi(x,t)\,dx = 0$

There are no charged fermions.

c) The fermion fields $\Psi_a(x)$ (Equ. (7.3)) do not anti-commute at space-like distances (the $\Psi_a(x)$ commute!), hence these fields are <u>non-local Fermi fields.</u> However, the <u>observables</u> $F_{\mu\nu}(x)$ and $T(x_1y_2)$ <u>commute</u> for totally space-like distances, thus assuring causality.

d) This special structure of two-dimensional QED becomes only transparent by the consistent use of gauge invariant observables and state vectors. In this spirit, one should consider only <u>gauge invariant field amplitudes</u> (B.S. amplitudes). We get by direct calculation [74]:

$$\langle 0|\,T(x_1,x_2)\,|{}^{\mu}_{p}\rangle = \tilde{N}\gamma^{\mu} z_{\mu}\left(\epsilon^{\nu\rho} z_{\nu} p_{\rho} - \gamma_5 (pz)\right)\frac{\sin pz}{(pz)}\,\frac{e^{ipX}}{z^2} \tag{7.8}$$

$$X = \tfrac{1}{2}(x_1+x_2)\ , \quad z = x_1 - x_2$$

Due to the non-local structure of $T(x_1,x_2)$, the general properties of the gauge invariant field amplitudes differ considerably from those of the conventional B.S. amplitudes, on which we have based our phenomenological approach to quark confinement (cf. 2.2). At this stage, it is not yet clear how relevant this two-dimensional model is for phenomenological physics. However, it gives interesting indications for the further development of realistic quark models with strong binding.

e) Gauge invariant states describe necessarily charged particles together with their electric field. With this in mind, one may get some insight into the dynamical mechanism of charge screening by a classical consideration [76]. The electric field $E(x)$, $F_{\mu\nu}(x) = \epsilon_{\mu\nu} E(x)$ of a static point charge satisfies the equation (cf. Equ. (7.2)):

$$\frac{dE}{dx} = e\,\delta(x-y) \quad ; \quad E(x) = e\,\Theta(x-y) \tag{7.9}$$

The constant field on the right side of the charge point is responsible for the infinite energy of this state. In contrast to this, a charge dipole

$$j_0(x) = e\,(\delta(x-y_1) - \delta(x-y_2))$$ with the field

$$E(x) = e\,(\Theta(x-y_1) - \Theta(x-y_2))$$ has a finite energy,

which is proportional to the distance of the two charges. This may explain why in gauge invariant two-dimensional QED, all states with finite energy have charge zero. There acts a linear confining potential (cf. 2.1) between charged particles! It is evident that this argument relies heavily on the restriction to 1+1-dimensions.

f) A.Casher, J. Kogut, and L. Susskind [77] discussed the following 2-point function in two-dim. QED:

$$\langle 0|T S(x) S(0)|0\rangle = (2\pi^2 x^2)^{-1} \exp\{-4\pi i(\Delta_F^{(2)}(x^2,0) - \Delta_F^{(2)}(x^2,\mu^2))\}$$

$$S(x) = \overline{\Psi}(x)\,\Psi(x)$$

From this expression they gain an intuitive picture for the process $e^+e^- \rightarrow$ hadrons: In the first moment the pair, created by the annihilation of the pointlike (scalar!) photon, propagates freely. Then the field between them strongly polarizes the vacuum. Finally the propagating dipole field catches the original pair, and the polarization charge annihilates it. Thus the strong polarizability of the vacuum prevents the appearance of free charges [78].

g) We discussed the mechanism of charge screening only for a single type of charges. There are generalizations [79] to more charges, which come closer to the real problem of the screening of the colour charges, related to the non-abelian gauge group SU(3).

h) These ideas on colour charge screening derived from 1+1-dimensional QED further the preference for string-like soliton solutions (cf. 5.3c) describing the strong coupling limit of the 3-1-dimensional standard Lagrangian. These strings might have the right spacial configuration to allow the charge screening mechanism.

7.2 Does colour charge screening happen also in the long distance limit of the 3+1 dimensional standard Lagrangian? In order to investigate this question, one should keep in mind the following results of the preceeding discussion:
(i) One has to consider the standard $\mathcal{L}$ with strong coupling (cf. 4.3).
(ii) charge screening becomes transparent only if one regards strictly the requirements of gauge invariance (cf. 7.1). It is known that field theories with strong coupling might be approached with help of a lattice approximation to space [80] (and time). Hence it is natural to consider gauge invariance for a lattice field theory. It was first suggested by K. Wilson [49] to study quark confinement in such a framework. He developped a theory of quantized fields on a lattice, which possess local gauge groups [80], using the Feynman path integral method of quantum mechanics [81]. A Hamiltonian formulation of Wilson's lattice gauge theory was given by J. Kogut and L. Susskind [82]. These investigations seem to indicate the following result: String like states in which quark-antiquark pairs are linked by gauge fields form energetically preferred configurations. In the case of non-abelian gauge groups, the crudest strong coupling approximation gives that the energy of such a state is proportional to the length of the string (cf. 7.1e). Thus the lattice gauge theories support our general view on quark confinement. They may even be applied to derive the Zweig's rule [5]. On the other hand, many problems of these theories remain to be solved. In particular, there is the open question of the systematic transition to the continuum. Of course, these short remarks are meant only as a reference to those important investigations concerned with quark confinement.

8. Conclusions

These lectures had the aim to give an introduction to the problem of quark confinement and to the different models and notions invented for the solution of this fundamental puzzle of the quark theory of hadrons. Seen from a point of uncritical, synthetizing imagination, the following picture emerges:

"Similar to QED, hadron dynamics is described by a renormalizable field theory in which coloured and flavoured quark fields are coupled to gluon fields according to the principles of local, non-abelian $SU_c(3)$ gauge invariance. This theory is asymptotically free, thus reproducing essentially the results of the quark parton model. In the large distance limit the theory is instable. The strong interaction causes, via vacuum polarization, the formation of physical strings, consisting of confined $q\bar{q}$-pairs linked by gluon lines. Colour charges are completely screened in this theory. The quantum mechanics of field strings provides the starting point for the dual dynamics of mesons". [83)]

At this stage, this picture may have the quality of a fairy-tale. In these lectures we summarized the facts on which it is based, and we illustrated the theoretical concepts from which it is built with the help of very (too?) simple examples. Even so not correct in any detail, this picture may provide a general pattern of hadron dynamics, which is worthwhile to be pursued further.

There remains the question of the relation of confinement models to actual experimental results. The phenomenological approach to quark confinement with help of a model of strongly bound massive quarks allows the interpretation of data of the meson spectrum, their leptonic and hadronic decay properties. The more theoretical models like the SLAC- and MIT-bag also produce their first figures [84)] on hadron properties. But much more has to be done for a real theoretical and quantitative understanding of high energy phenomenology. As example I mention just one more problem, the problem of spallation of bags or strings, which provides the description of the hadronic decays of mesons and baryons.

I hope that all these considerations made clear how stimulating the problem of quark confinement already has been for the development of theoretical elementary particle physics.

References

1) R.K. Adair, Proceedings of the XVI. International Conference on High Energy Physics 1972, NAL, Vol. 4, p. 307

2) R. Casalbuoni, G. Domokos, S. Koevesi-Domokos, John Hopkins COO-3285-19

3) A. De Rújula, S.L. Glashow, Phys. Rev. Letters 34 (1975) 46,
T. Appelquist, H.D. Pollitzer, Phys. Rev. Letters 34 (1975) 43

4) G. Zweig, in "Symmetries in Elementary Particle Physics, ed. A. Zichichi (Academic, New York, 1965); J. Iizuka, Progr. Theor. Phys. Suppl 37-38 (1966) 21

5) H.G. Dosch, V.F. Müller, Phys. Rev. D (to appear) and HD-THEP-75-21

6) E. Eichten, K. Gottfried, T. Kinoshita, J. Kogut, K.D. Lane, T.M. Yani, Phys. Rev. Letters 34 (1975) 369

7) D. Amati, Strong interaction theory, IIe Conf. Int. sur les particules elementaires, Suppl. J. Phys. (France) 34 C 1, 12 (1973)
R. Haag, ibid.
D. Amati, M. Testa, Phys. Letters 48B (1974) 227

8) M. Gell-Mann, F.E. Low, Phys. Rev. 95 (1954) 1300
C.G. Callan, Phys. Rev. D 2 (1970) 1541
K. Symanzik, Commun. Math. Phys. 16 (1970) 48, 18 (1970) 227

9) L.P. Kadanoff, Phys. Rev. Letters 23 (1969) 1430
K.G. Wilson, Phys. Rev. B4 (1971) 3184

10) H.B. Nielsen, P. Oleson, Nucl. Phys. B61 (1973) 45
B. Zumino, CERN preprint Th. 1779 (1973)

11) A. Scott, F. Chu, and D. McLaughlin, Proc. IEEE 61 (1973) 1443

12) S. Eliezer, R. Weiner, Los Alamos Sci. Lab. - La-Ur-75-1226
R.J. Finkelstein, Hypothetical Solitons and Hadronic Structure (preprint)

13) I would like to acknowledge the help of Dr. Mellentin and the DESY Documentation Group for providing printouts of references to the relevant literature

14) G. Morpurgo, Physics 2 (1965) 95

15) J.F. Gunion, R.S. Wiley, Phys. Rev. D12 (1975) 174

16) M. Böhm, H. Joos, M. Krammer
a) Nuov. Cim. 7A (1972) 21; Nucl. Phys. B51 (1973) 397
b) in Recent developments in mathematical physics (P. Urban ed.) (Springer Verlag, Wien and New York, 1973) p.p. 3-116
c) TH-1949 CERN

17) H. Joos, Proceedings of the XVI. International Conference on High Energy Physics; NAL (1972), Vol. I, p. 199

18) K. Johnson, Phys. Rev. D7 (1972) 1101
For further development along this line:
C.M. Bender, J.E. Mandula, G.S. Guralnik, Phys. Rev. D11 (1975) 409 and literature quoted there

19) G. Preparata, Phys. Rev. D7 (1973) 2973; Nucl. Phys. B89 (1975) 445, and lectures delivered at the 1974 International School of Subnuclear Physics for further references
J.C. Polkinhorne, Nucl. Phys. B93 (1975) 515
H. Osborn, Physics Letters 58 B (1975) 111

20) H. Suura, Phys. Letters 42 B (1972) 237
DESY 1972/7, DESY 1972/47

21) H.A. Bethe, E.E. Salpeter, Phys. Rev. 84 (1951) 1232
M. Gell-Mann, F. Low, Phys. Rev. 84 (1951) 350
N. Nakanishi, Progr. Theor. Phys. Suppl. 43 (1969) 1 (for further references)

22) R. Haag, Phys. Rev. 112 (1958) 669
W. Zimmermann, Nuov. Cimento 10 (1958) 669; ibid. 13 (1959) 503; ibid. 16 (1960) 690
R. Haag, D. Kastler, J. Math. Phys. 5 (1964) 848

23) K. Symanzik, J. Math. Phys. 1 (1960) 249
Symposia on Theoretical Physics 3 (1967) 121

24) S. Mandelstam, Proc. Roy. Soc. A233 (1955) 248
R.E. Cutkosky, M. Leon, Phys. Rev. 135 B (1964) 1445

25) Models with other spin structures:
M. Krammer, Acta Phys. Austr. 40 (1974) 187

26. C.G. Wick, Phys. Rev. 96 (1954) 1124

27) M.K. Sundaresan, P.J.S. Watson, Ann. Phys. (N.Y.) 59 (1970) 375
G. Preparata, Subnuclear phenomena (Academic Press New York and London 1970) p. 240

28) R.P. Feynman, M. Kislinger, and F. Ravndal, Phys. Rev. D3 (1971) 2706
M. Markov, Nuov. Cim. Suppl. 3 (1956) 760

29) M. Böhm, M. Krammer, Phys. Letters 50B (1974) 457, Acta Phys. Austr. 41 (1975) 401
M. Böhm, H. Joos, M. Krammer, Ref. 17b, c; 32, 34
D. Flamm, P. Kielanowski, J. Sánchez Guillén, Nuov. Cim. 25A (1975) 16; Acta Phys. Austr. 41 (1975) 33; D. Flamm, A. Morales, R. Nuñez-Lagos, P. Kielanowski, J. Sánchez Guillén, Act. Phys. Austr. 42 (1975) 289

30) This is an advantage over the models by R.F. Feynman et al., Ref. 29) and the non-relativistic quarkmodel

31) M. Böhm, H. Joos, M. Krammer, Nucl. Phys. B69 (1974) 349

32) M. Gourdin, in Hadronic interactions of electrons and photons (Academic Press New York and London, 1971) p. 395

33) M. Böhm, H. Joos, M. Krammer, Acta Phys. Austr. 38 (1973) 123
Ref. 17b, c

34) H. Fritzsch, M. Gell-Mann, Proceedings of the XVI. Int. Conf. on High Energy Physics, NAL 1972, Vol. 2
R. Dashen, Proceedings of the 1975 International Symposium on Lepton and Photon Interactions at high energies

35) Y. Nambu, M.Y. Han, Phys. Rev. D10 (1974) 674
M. Krammer, D. Schildknecht, F. Steiner, Phys. Rev. D12 (1975)
B. Stech, Lectures at this Summer Institute

36) M. Gell-Mann, Acta Physica Austriaca, Suppl. IX (1972) 733
W. Bardeen, H. Fritzsch, M. Gell-Mann, in Scale and Conformal Invariance in Hadron Physics, Wiley (New York, 1973)

37) S.L. Adler, Phys. Rev. 177 (1969) 2426
J.S. Bell, R. Jackiw, Nuov. Cim. 60A (1969) 47

38) H. Fritzsch, M. Gell-Mann, and H. Leutwyler, Phys. Letters 47B (1973) 365, Ref.36

39) D.H. Perkins, Proc. of the NAL-Conference 1972, Vol. IV, p. 189

40) E.C. Poggio, H.R. Quinn, S. Weinberg, Smearing the Quark Model (Harvard preprint)

41) a) D.J. Gross, F. Wilczek, Phys. Rev. Lett. 30 (1973) 1343, Phys. Rev. D8 (1973) 3633; D9 (1974) 980
H.D. Pollitzer, Phys. Rev. Lett. 30 (1973) 1346, H. Georgi, H.D. Politzer, Phys. Rev. D9 (1974) 416; D. Bailin, A. Love, and D. Nanopoulos, Nuov. Cim. Lett. 9 (1974) 501

b) D. Gross, A Review on Asymptotic Freedom, Proceedings on the XVII Int. Conf. on High Energy Physics, London 1974, p. III-65

42) S. Weinberg, Phys. Rev. Lett. 31 (1973) 494

43) A. de Rújula, Lectures at this school

44) M. Bander, P. Thomas, Phys. Rev. D12 (1975) 1798
I.T. Drummond, R.W. Fidler, Nucl. Phys. B90 (1975) 77

45) G. Parisi, Phys. Rev. D11 (1975) 970

46) "Superconductivity", Ed. R.D. Park (Marcel Dekker Inc. N.Y. 1969) Vol. II, ch. 6 and ch. 14

47) W.A. Bardeen, M.S. Chanowitz, S.D. Drell, M. Weinstein, T.M. Yan, Phys. Rev. 11D (1975) 1094

48) A. Chodos, R.L. Jaffe, K. Johnson, C.B. Thorn, V.F. Weisskopf, Phys. Rev. 9D (1974) 3471

49) K.G. Wilson, Phys. Rev. D10 (1974) 2445

50) K.G. Wilson, Quarks and Strings on a Lattice, Report CLNS-321

51) S. Coleman, Phys. Rev. D11 (1975) 2088
S. Mandelstam, Phys. Rev. D11 (1975) 3026

52) Skyrme, Proc. Roy. Soc. A 247 (1958) 260; A262 (1961) 237

53) W. Thirring, Ann. Phys. (N.Y.) 3 (1958) 91
V. Glaser, Nuovo Cim. 9 (1958) 990
B. Klaiber, Boulder Lectures 1967; New York, Gordon Breach 1968
These references refer to the more studied massless case

54) I learned much on the difficulties of the problem from discussions with H. Lehmann

55) R. Dashen, B. Hasslacher, A. Neveu, Phys. Rev. D10 (1974) 4130, Ref. 47

56) N.H. Christ, T.D. Lee, Phys. Rev. D12 (1975) 1606

57) J. Arafune, P.G.O. Freund, C.J. Goebel, J. Math. Phys. 16 (1975) 433
S. Coleman, Lectures delivered at the 1975 International School of Subnuclear Physics "Ettore Majorana" and references there

58) M.Y. Han, Y. Nambu, Phys. Rev. 139B (1965) 1006

59) P. Vinciarelly; Nuov. Cim. Lett. 4 (1972) 905, Nucl. Phys. B89 (1975) 463, 493

60) M. Creutz, Phys. Rev. D10 (1974) 1749

61) H. Nielsen, P. Olesen, Nucl. Phys. B61 (1973) 45

62) S. Mandelstam, Physics Letters 53B (1975) 476

63) G. 't Hooft, Nucl. Phys. B79 (1974) 276

64) J.H.Swank, L. Swank, T. Derelt, Phys. Rev. D12 (1975) 1096

65) V. Alessandrini, D. Amati, M. Le Bellac, D. Olive, Phys. Reports 1c (1971) No. 6 and Literature quoted there

66) A. Casher, J. Kogut, L. Susskind, Phys. Rev. Lett. 31 (1973) 792

67) M. Creutz, Phys. Rev. 10 (1974) 2696
A. Patkos, Nucl. Phys. B97 (1975) 352
A. Jevicki, P. Senjanovic, Phys. Rev. D11 (1975) 860
S. Gasiorowicz, Phys. Rev. D12 (1975) 2526

68) R. Rajaraman, Phys. Reports 21 (1975) 227, Ref. 56

69) J. Goldstone, R. Jackiw, Phys. Rev. D11 (1975) 1486
A. Klein, F.R. Krejs to be published (UPR-0039T)

70) K. Cahill, Phys. Rev. Lett. 53B (1974) 174
P. Vinciarelli, CERN-Th 1993, 1975
W.A. Bardeen et al., Ref. 47

71) J.L. Gervais, A. Jevicki, B. Sakita, Phys. Rev. D12 (1975) 1038
Ref. 56

72) R.F. Dashen, B. Hasslacher, A. Neveu, Phys. Rev. 10D (1974) 4114, Ref. 55

73) J. Schwinger, Phys. Rev. 128 (1962) 2425; Theoretical Physics, Trieste Lectures, 1962, p. 89, I.A.E.A., Vienna (1963)

74) The precise Definitions of the normalizations, N, N, $N(x_1-x_2)$ involves problems for which we refer to the original paper Ref. 75

75) J.H. Lowenstein, J.A. Swieca, Ann. Phys. 68 (1971) 172

76) L. Susskind at Bonn

77) A. Casher, J. Kogut, and L. Susskind; Phys. Rev. D10 (1974) 732
J. Kogut, D.K. Sinclair; Phys. Rev. 10 D (1974) 4181

78) P. Vinciarelli, Physics Letters 53B (1975) 457

79) C.G. Callan, N. Coote, D.J. Gross, Two dimensional Yang-Mills Theory (preprint Princeton University)

80) K.G. Wilson, Report CLNS-321
Ref. 60, 61
G.P. Canning, D. Foerster, Nucl. Phys. B91 (1975) 151

81) R.P. Feynman, A.R. Hibbs, "Quantum mechanics and Path integrals"
New York 1965

82) J. Kogut, L. Susskind, Phys. Rev. D11 (1975) 395
Reports by A. Caroll, J. Kogut, D.K. Sinclair, L. Susskind (CLNS-325)
T. Banks, L. Susskind, J. Kogut (CLNS-318)

83) In these lectures we do not consider the analogous considerations
of the baryons

84) A. Chodos, R.L. Jaffe, K. Johnson, C.B. Thorn, Phys. Rev. D10 (1974) 2599
R.L. Jaffe, Phys. Rev. D11 (1975) 1953
E.M. Allen, Physics Letters 57B (1975) 263

CHARMED PARTICLES AND OTHER OBSERVABLE CONSEQUENCES FROM GAUGE THEORIES

by

D. V. Nanopoulos

CERN, Geneva

1. FLASH REVIEW OF THE UNDERLYING THEORY

A. Conventional Weak Interactions (< 1971)

Thanks to the efforts of generations of physicists, we had a fairly clear situation concerning the weak interactions of elementary particles.

The leptonic world consisted of e, ν_e, μ, ν_μ and the hadronic world was described successfully by three quarks p, n, λ.

All "low energy" phenomenology could be explained easily using a fairly simple Hamiltonian:

$$H_{inter.} = g\, j_\mu^+ W_\mu^- + h.c. \quad ; \quad \frac{g^2}{M_W^2} \sim G_F$$

where as usual:

$$j_\mu^+ = (j_\mu^+)_h + (j_\mu^+)_\ell$$

$$(j_\mu^+)_\ell \rightarrow \bar{e}\gamma_\mu(1-\gamma_5)\nu_e + (e \rightarrow \mu)$$

$$(j_\mu^+)_h \rightarrow \bar{p}\gamma_\mu(1-\gamma_5)n_c$$

$$(n_c \equiv n\cos\Theta_c + \lambda\sin\Theta_c)$$

So we had a nice phenomenological simple description of weak interactions, but unfortunately not a theory! Simply, our field theory was not renormalizable (e.g. higher order corrections turn out to be infinite) and also, something which is related, we had a bad high energy behavior (e.g. violation of unitarity). So, we were looking for a renormalizable, unitary field theory which would describe successfully weak interactions. The search was positive[1], such a kind of field theory seems to exist and not only satisfies our criteria of simplicity, renormalizability and unitarity, but also a great old DREAM, becomes reality: A UNIFIED THEORY OF WEAK AND ELECTROMAGNETIC INTERACTIONS[2]! The simplest way[3] to see the structure of such a theory is to play with higher order diagrams for processes like: $e^+e^- \to W^+W^-$, $\nu\bar{\nu} \to W^+W^-$. Then one sees that in order to have acceptable high energy behavior, one needs

i) Neutral currents and/or heavy leptons

ii) Relations between the forms of the couplings of the particles.

Actually ii) means that we have some group structure which is a characteristic of the so-called Yang-Mills field theory.

Then what about renormalizability?
We need a soft way to put masses in. One way to do that is to use the Higgs Mechanism[4]. Then everything is fine.
Moral: A Yang-Mills field theory with spontaneous breakdown leads to a unified theory of weak and E.M. interactions, which is renormalizable and unitary.
This brings us to:

B. Weak Interactions (> 1971)

We were lead to a Y-M. field theory with spontaneous breakdown to describe a unified theory of weak and E.M. interactions. But, unavoidably we have newcomers:

1) neutral currents and/or heavy leptons

2) Higgs scalars[5]

Is that all? No!

If we have a group structure, it means that we have not only j^+, j^- but also $[j^+, j^-] \sim j_3$, i.e. neutral currents with $\Delta S = 0$ and $\Delta S = 1$, as shown below. We have nothing against N.C. with $\Delta S = 0$ (in fact they exist!)[6] but what about $\Delta S = 1$ N.C.? We know from $K_L \to \mu^+\mu^-$, $K^+ \to \pi^+\nu\bar{\nu}$, that if they exist they are suppressed by comparison with $\Delta S = 0$ N.C. (for more on this see Section 2).

But in the conventional theory with 3 quarks, the neutral current is

$$\bar{n}_c \gamma_\mu (1-\gamma_5) n_c \equiv \bar{n}_c \gamma_L n_c \qquad (2)$$
$$= \bar{n}\gamma_L n \cos^2\theta_c + \bar{\lambda}\gamma_L \lambda \sin^2\theta_c + (\bar{n}\gamma_L\lambda + \bar{\lambda}\gamma_L n)\sin\theta_c \cos\theta_c$$

It is exactly the last term of Eq.2, which is unacceptably big. It is the evil in our theory! So we want something to avert this evil. That means we want CHARM[7)8)], which in the quark language means the introduction of at least one more quark c. The easiest way to see the need and the properties of this new quark is to look at the diagrams of Fig.1. Then we see that in order to avoid $\Delta S = 1$ N.C., and of course $\Delta S = 2$ N.C., like the $K_L - K_S$ mass difference we must have:

$$g_{np}\, g_{p\lambda} + g_{nc}\, g_{\lambda c} = 0 \tag{3}$$

Now the only solution that satisfies (3) and simultaneously creates a state orthogonal to n_c, which is coupled to the new quark c, is :

$$g_{nc} = - g_{\lambda p} \quad , \quad g_{\lambda c} = g_{pn}$$

So our current now is enlarged:

$$\begin{aligned} (j_\mu^+)_{hadr.} &= \bar{p}\,\gamma_\mu(1-\gamma_5)(n\cos\Theta_c + \lambda\sin\Theta_c) \\ &+ \bar{c}\,\gamma_\mu(1-\gamma_5)(-n\sin\Theta_c + \lambda\cos\Theta_c) \end{aligned} \tag{4}$$

This is the Cabibbo-(Glashow,-Iliopoulos-Maiani)[8)] current, for later use we put $\lambda_c \equiv \lambda\cos\Theta_c - n\sin\Theta_c$. So we succeeded in suppressing $\Delta S = 1$ and $\Delta S = 2$, but we have to introduce an extra quark c. In the formal language: SU(3) → SU(4). There is now a very nice analogy between leptons and quarks:

$$\begin{pmatrix}\nu_e \\ e\end{pmatrix}_L , \begin{pmatrix}\nu_\mu \\ \mu\end{pmatrix}_L , \quad R\text{-fields}, \; I = 0 \quad \begin{pmatrix} \psi_L \equiv \frac{1}{2}(1-\gamma_5)\psi \\ \psi_R \equiv \frac{1}{2}(1+\gamma_5)\psi \end{pmatrix}$$

$$\begin{pmatrix} p \\ n_c \end{pmatrix}_L , \begin{pmatrix} c \\ \lambda_c \end{pmatrix}_L$$

Where we have arranged the leptons and quarks as doublets under an $(SU(2))_L$ group corresponding to "weak" isospin. (The right-handed fields are singlets under "weak" isospin).

In the case of fractional quark changes, "weak" hypercharge is a good candidate to distinguish between quarks and leptons: $Y^W \equiv Q - I_3^W$ (5)
Actually in the $SU(2)_L \times U(1)$ Weinberg-Salam[2)] model (which we are going to use as a prototype throughout these lectures)

$$L_{inter.} = g\Big[\sum_i \bar{\psi}_L^i \frac{\vec{I}^W}{2} \psi_L^i\Big]\cdot\vec{W} + g'\Big[\sum_i \bar{\psi}_L^i Y^W \psi_L^i\Big] B \tag{6}$$

and as usually: $\tan\Theta_W = g'/g$.

Now the theory is renormalizable if:

1) All particles are massless.

2) ALL INTERACTIONS are invariant under $\vec{I}^W$, Y^W gauge transformations.

1) May be avoided through the Higgs mechanism[4].

2) Asks strong interactions to be invariant under "weak" $\vec{I}$, Y.

(a) (except mass terms). We know also

(b) Parity and strong $\vec{I}$ are conserved in strong interactions.

Now (a) and (b) imply that:

Strong interactions are invariant under chiral SU(4) x SU(4)

but also (c): Parity, strangeness, baryonic number are conserved in order $\alpha = \frac{1}{137}$.

It has been shown[9] that (a), (b), (c) are automatically satisfied if we consider a gauge theory of strong interactions, where the strong gauge group G_s commutes with the weak-E.M. gauge group G_W:

$$[G_s, G_W] = 0 \qquad (7)$$

Here it is worth mentioning that condition (7) is sufficient in order to satisfy (a), (b), (c), but nobody has yet proven that this condition is also necessary. A well-known example is to identify $G_s \equiv (SU(3))_{colour}$. The need to colour the quarks is well-known by now:

1) To satisfy Fermi-statistics, for quarks in the baryons.

2) To have the correct amplitude for $\pi^0 \to 2\gamma$

3) To have $R \equiv \frac{(e^+e^- \to \gamma \to \text{hadrons})}{(e^+e^- \to \mu^+\mu^-)} = 2$ and not $\frac{2}{3}$ at low energies $E_{cm} \leq 3.5$ GeV.

4) To cancel Adler anomalies.

Properties 1)2)3) have been pointed out by Gell-Mann[10] (1) had initially been discussed by Han-Nambu[11]. Then in the framework of gauge theories it was Bouchiat-Iliopoulos-Meyer[12], who found a quark structure which satisfies 1) 2) 3) and 4):

$$\underset{\text{"flavour"}}{SU(4)\downarrow}\ \overset{\text{SU(3) "colour"}\ \longrightarrow}{\begin{pmatrix} p_1 & p_2 & p_3 \\ n_1 & n_2 & n_3 \\ \lambda_1 & \lambda_2 & \lambda_3 \\ c_1 & c_2 & c_3 \end{pmatrix}} \qquad \begin{aligned} Q_{p_i} &= Q_{c_i} = 2/3 \\ Q_{n_i} &= Q_{\lambda_i} = -1/3 \end{aligned} \quad i = 1,2,3 \qquad (8)$$

From now on, I am going to assume the following:

1) As a model of weak and E.M. interactions I will take the conventional Weinberg-Salam[2] $SU(2)_L \times U(1)$ model.

2) Strong interactions also are turned on via a gauge theory where the strong gauge group is $SU(3)_{colour}$, and I will assume the standard Glashow-Iliopoulos-Maiani[8] (GIM) four quark ("flavour") model.

So strong interactions are mediated by 8 colour gluons (maybe massless). That will be the "standard model" which we are going to consider and the name of it is QUANTUM CHROMODYNAMICS(chroma is a greek word which means colour)

Before going on, two comments are in order:

i) Condition (7), makes it difficult to build models of the above form with quarks of integer charges, as for instance the Han-Nanbu[11] quarks, because the photon is not a colour-singlet and so (7) is violated.

ii) "Colour" in our language is going to be always "hidden", everything that is permitted to come out is a colour singlet. Also "colour" is respected by strong-weak and E.M. interactions. Notice the difference with the Han-Nambu "colour", where colour states may exist and colour symmetry is violated at least by electromagnetism.

I am aware of the fact that maybe the "standard model" has some experimental difficulties, but I believe it is an excellent theoretical laboratory in order to study properties of the general framework of gauge theories. If more than four quarks exist or heavy leptons which make the $SU(2)_L \times U(1) \times SU(3)_{colour}$ model wrong, at least the methodology can apply to any other model inside this general framework.

What are the merits of the "standard" model?

1) ASYMPTOTIC FREEDOM[13]: A natural explanation, within field theory, of Bjorken scaling. It also permits, at least within particular kinematical regions, the use of perturbation theory for strong interactions giving believable answers. (since $g(q^2) \xrightarrow[q^2 \to -\infty]{} 0$)

2) A "natural"[14] explanation of approximate symmetries.

In this kind of theory the form of strong interactions Lagrangian is very restricted:

$$L_{strong} \underset{e \to 0}{=} -\bar{\Psi} \gamma_\mu D_\mu^s \Psi - \frac{1}{4} F_{\alpha\mu\nu} F_\alpha^{\mu\nu} - \bar{\Psi} M \Psi \quad . \qquad (9)$$

where as usual: ($A_{\alpha\mu}$ is the gluon field)

i) $(D_\mu \psi)_\eta = \partial_\mu \psi_\eta - i(t_\alpha)_{\eta\omega} \psi_\omega A_{\alpha\mu}$

ii) $F_{\alpha\mu\nu} = \partial_\mu A_{\alpha\nu} - \partial_\nu A_{\alpha\mu} - C_{\alpha\beta\gamma} A_{\beta\mu} A_{\gamma\nu}$

iii) t_α are matrices representing the Lie algebra (with structure constants $C_{\alpha\beta\gamma}$) of the gauge group:

$$[t_\alpha, t_\beta] = i C_{\alpha\beta\gamma} t_\gamma$$

iv) M is the mass matrix which is fairly complicated.

Then from (9):

SYMMETRIES OF STRONG INTERACTIONS ≡ SYMMETRIES OF MASS MATRIX M.

Is all this of pure academic interest? NO! There is for instance a trivial application[15]: The decay $\psi' \to \psi + \eta$ has been seen[16] with a B.R. ~ 4% Now, if SU(3) symmetry breaking is small, and because η is almost an octet, it means that ψ', ψ cannot both be SU(3) singlets. Then the new heavy quarks have to be in higher representations of SU(3) ($\underline{3}$,...). But then because of the form of (9) we have to distinguish between $SU(3)_{light}$ and $SU(3)_{heavy}$, and the $\psi' \to \psi + \eta$ is still suppressed. So we have a dilemma: Either we do not understand SU(3) symmetry breaking or we have problems with gauge theories. For more details see Ref.15.

Let us summarize the merits of gauge theories:

Why do people like field theory? Because
1) Special relativity
2) Quantum theory
3) Unitarity
4) Crossing
5) Analyticity
6) Calculability

are included in a trouble free way!

Which field theory should we have? Yang-Mills field theory with a spontaneous breakdown.

Why? Because:

1) It is renormalizable[1]
2) It has the appropriate high energy behavior[3] (uniquely?)
3) It unifies weak and E.M. interactions[2]
4) If strong interactions are turned on as a gauge theory they are asymptotically free[13] which gives an explanation of Bjorken scaling (uniquely?)[17]
5) It unifies weak E.M. and strong interactions[13][23]
6) It gives a justification to current algebra[18]
7) It has "Naturality"[14] (explanation of approx. symmetries)
8) It has "Reggeization"[19] (and hence possibly a connection with dual models) (uniquely?)
9) It gives a justification of the quark model[20] with

A) quark confinement (?) and

B) 1) only triality zero states:

$qqq, q\bar{q}$ (no $qq\bar{q}\bar{q}, \cdots$)

2) exotics such as $q\bar{q}q\bar{q}, qqqq\bar{q}$ ... have high mass

3) no qq states but only $q\bar{q}$.

10) It provides a dynamical explanation[21] (?) of the $\Delta I = 1/2$ rule.

11) It gives a possible explanation[22] of the "new particles"

So this kind of theory seems to be very ambitious (lately quite an effort[24] has been made to put also gravitation in the game) and we may call it "Applied Quantum Field Theory".

MORAL: We started asking for a respectable theory of weak interactions and we end up with a unified theory of strong-electromagnetic and weak interactions.

Not BAD!

Of course we have NEWCOMERS:

1) Neutral currents and/or heavy leptons.
2) New quarks $\Longrightarrow$ new particles
3) Higgs scalars.

2. RARE DECAYS OF KAONS, K_L-K_S MASS DIFFERENCE IN GAUGE THEORIES

We saw that in order to avoid $\Delta S = 1$ N.C. and a large K_L-K_S mass difference, we were led to introduce a new quark c.

Then the current became:

$$(j_\mu^+)_{hadr.}^{C-GIM} = \bar{p}\gamma_L n_c + \bar{c}\gamma_L \lambda_c \tag{4}$$

which also means:

$$\underset{SU(3)}{\begin{pmatrix} p \\ n \\ \lambda \end{pmatrix}} \Longrightarrow \underset{SU(4)}{\begin{pmatrix} p \\ n \\ \lambda \\ c \end{pmatrix}}$$

Of course then, we increase our spectroscopy with new particles containing this new quark. Then where are these new particles? A main answer is that the c quark is very heavy and so the new particles can escape detection so far. But, fortunately or unfortunately!, things are not so arbitrary. Experimentally we know the $K_L - K_S$ mass difference and the decay $K_L \to \mu^+\mu^-$ has been seen (even if it is very small, see Table I for experimental information).

Theoretically we have a respectable (≡ renormalizable) theory and we can calculate higher order diagrams. Then, comparing theory with experiment, we are bound to learn something about the heaviness of the c quark, and then say something about the masses of the new particles.

An excellent analysis has been carried out by Madame Gaillard and Mr Lee[25)] which we follow:

i) Naive analysis

In gauge theories it is expected that $g \sim e$. Then to first order in perturbation theory the amplitude $A \sim G_F/\sqrt{2} \sim g^2/M_W^2$, and in second order

$$A \sim g^4/M_W^2 \simeq G_F \cdot \alpha \tag{10}$$

Empirically $\Delta S = 1$ $\Delta Q = 0$; $\Delta S = 2$ processes are of order $G_F^2 \Lambda^2$ (where $\Lambda \sim$ few GeV) in amplitude. Which means

$$A \sim G_F^2 \Lambda^2 \simeq G_F \cdot \alpha \frac{\Lambda^2}{M_W^2} \tag{11}$$

So comparing (10) and (11) it is obvious that we need a further suppression $\sim \frac{\Lambda^2}{M_W^2}$.

Experimentally:

$$\Gamma(K_L \to \mu^+\mu^-) \simeq 2 \cdot 10^{-5}\, \Gamma(K_L \to \gamma\gamma) \simeq 4 \cdot 10^{-9}\, \Gamma(K^+ \to \mu^+\nu)$$
$$\downarrow \qquad\qquad\qquad \downarrow \qquad\qquad\qquad \downarrow$$
$$(G_F\alpha^2)^2 \qquad\qquad G_F^2\alpha^2 \qquad\qquad G_F^2$$

Moral: We need a suppression of order α in amplitude for the decay $K_L \to \mu^+\mu^-$ but not for $K_L \to \gamma\gamma$! But, we have the GIM mechanism!

It is easy to see that if $m_c \simeq m_p$, then we have no $\Delta S = 1$ N.C. to all orders, which is obviously wrong. In this case we are in an exact SU(4) limit; let us look at some subgroups of SU(4):

$$\begin{pmatrix} p \\ n \\ \lambda \\ c \end{pmatrix} \text{ implies } \begin{pmatrix} p \\ n \end{pmatrix} \begin{pmatrix} p \\ \lambda \end{pmatrix} \begin{pmatrix} n \\ \lambda \end{pmatrix} \begin{pmatrix} c \\ p \end{pmatrix}, \ldots$$

exact SU(4) exact $\vec{I}$ $\vec{V}$ $\vec{U}$ $\vec{P}$

Let us now concentrate on the consequences of $\vec{p}$ invariance: To lowest order, the effective Lagrangian for $\Delta Q = 0, \Delta S = 1$ processes is:

$$L_{eff.} \simeq \sin\theta_c \cos\theta_c (\bar{n}\gamma_L\lambda)[\bar{p}\gamma_L p - \bar{c}\gamma_L c] + h.c. \quad (12)$$
$$\Downarrow \vec{p} = 1$$

But $K_L, \gamma, \mu^{\pm}$ are of course P-spin singlets. So $K_L \to \mu^+\mu^-, \gamma\gamma, \ldots$ in the limit of P-spin invariance are forbidden to go through (12), which is simply wrong. In other words we need $m_c \gg m_p$, which means that SU(4) is badly broken. Let us look more closely at $K_L \to \mu^+\mu^-$ and $K_L \to \gamma\gamma$ (see Fig.2a and b).

We notice the following:

Both diagrams (and independently for c and p) are convergent. The structure of the current (4) is such that the effect of the c quark is simply to replace:

$$\frac{1}{m_p^2 - K^2} \longrightarrow \frac{1}{m_p^2 - K^2} - \frac{1}{m_c^2 - K^2} \sim \frac{m_c^2 - m_p^2}{(m_q^2 - K^2)^2} \equiv \frac{\Delta m^2}{(m_q^2 - K^2)^2} \quad (13)$$

Thus, we increase the rate of convergence and the diagrams remain convergent if we replace:

$$\frac{1}{M_W^2 - q^2} \longrightarrow \frac{1}{M_W^2} \qquad \left(\text{to order } \frac{m_q^2}{M_W^2}\right) \quad (14)$$

So diagram 2a becomes 3a and diagram 2b becomes 3b.
In other words:

$$A(K_L \to \mu^+\mu^-) \sim \frac{g^4}{M_W^2} \frac{m_c^2 - m_p^2}{M_W^2} \sim G_F^2 \Delta m_q^2 \quad (a)$$

and

$$A(K_L \to \gamma\gamma) \sim \frac{g^4}{M_W^2} \frac{\Delta m_q^2}{M_q^2} \sim G_F \alpha \frac{\Delta m_q^2}{m_q^2} \quad (b)$$

(15)

Physically the difference between (15a) and (15b) is easily understandable. In the case of $K_L \to \mu^+\mu^-$ we have two W-boson propagators, i.e. true 4th-order semi-weak interaction ($\sim G_F^2$), but in $K_L \to \gamma\gamma$ we have only on W-boson propagator, i.e. 2nd order semi-weak, 2nd order E.M. interaction ($\sim G_F \cdot \alpha$)
Then in order to satisfy experiment we must have:

(a) $$\Delta m_q^2 \sim \text{few GeV}^2 \quad , \text{ i.e. } \Delta m_q^2 \ll M_W^2$$

(16)

(b) $$\frac{\Delta m_q^2}{m_q^2} \sim 1 \quad , \text{ i.e. } \frac{m_p^2}{m_c^2} \ll 1$$

which actually means: $$m^2_{p,n,\lambda} \ll m_c^2 \ll M_W^2 \quad .$$

Or in group-theoretical language:

P-spin symmetry breaking
- → large in the hadronic mass scale
- → small in the W-boson mass scale

Let us see now what we get in the "standard model".

ii) Quantitative analysis

a) $K_L \to \mu^+\mu^-$

In the free quark language, we examine the process: $\bar{n} + \lambda \to \mu^+\mu^-$ We have the contribution of two classes of diagrams (see Fig.2a and Fig.4). Then one finds that each class goes as

$$G_F \cdot \alpha \frac{m_c}{(38\,\text{GeV})^2} \ln\left(\frac{M_W}{m_c}\right) \quad . \tag{17}$$

As pointed out by Gaillard and Lee[25] a remarkable feature of the above calculation is that the leading logarithmic terms in the two classes cancel. The full expression for the amplitude is:

$$iT(K_L \to \mu^+\mu^-) = \frac{G_F^2 m_c^2}{4\pi^2} \cos\theta_c \sin\theta_c \cdot$$
$$\cdot \langle K_L | \bar{n}\gamma_L\lambda + \bar{\lambda}\gamma_L n | 0 \rangle \bar{\mu}\gamma^\mu\gamma_5\mu \quad . \tag{18}$$

Then we have :

$$\frac{\Gamma(K_L \to 2\mu)}{\Gamma(K^+ \to \mu^+ + \nu)} \simeq \frac{G_F^2 m_c^4 \cos^2\theta_c}{2\pi^4} \qquad (m_c \gg m_p) \tag{19}$$

Now if we use experimental numbers (see Table I)

$$m_c < 10\,\text{GeV} \tag{20}$$

As we will see, strong interactions effects do not change dramatically the free

quark result (19), which in turn means that the dominant mechanism for $K_L \to \mu^+\mu^-$ is the conventional one: $K_L \longrightarrow 2$ virtual γ's $\to \mu^+\mu^-$.
Actually that is the reason that (19) gives an upper bound (20) and not a definite number for the mass of the charmed quark.

b) $K_L K_S$ mass difference

In the free quark language, the $K_L \leftrightarrow K_S$ transition is described by Fig.5. Then we have:

$$\frac{m_L - m_S}{m_K} = \frac{G_F}{\sqrt{2}} \frac{\alpha}{4\pi} \left(\frac{m_c}{38\,\mathrm{GeV}} \right)^2 \cos^2\theta_c \sin^2\theta_c \langle \bar{K}^0 | [\bar{n} \tfrac{\gamma_L}{2} \lambda]^2 | K^0 \rangle \quad (21)$$

Of course, the calculation of the matrix element appearing on the right-hand side is ambiguous. One can think at least two ways:

i) Insert the vacuum between the two currents.
or ii) To relate it to $K^+ \to \pi^+\pi^0$ by PCAC and SU(3).
In any case, from the known $K_L - K_S$ mass differences one deduces:

$$m_c \simeq 1.5 - \text{few GeV} \quad (22).$$

One may go on and calculate $K_L \to \gamma\gamma$, $K_S \to \pi \ell\bar{\ell}$, $K_S \to \pi\gamma\gamma$. Everything is fine. (For more see Gaillard-Lee Ref.25). Only $K^+ \to \pi^+ e^+ e^-$ maybe a problem. But we have forgotten strong interaction effects! Before taking into account strong interactions a few comments are in order:

In the "standard model" which we have considered, one gets a very satisfactory agreement between experiment and theory concerning $\Delta S = 1$ $\Delta Q = 0$ and $\Delta S = 2$ processes. Of course, these processes are used in order to determine the mass scale of the charmed quark. Then one naturally thinks that perhaps any model in which the GIM mechanism is incorporated, is bound to give agreement between theory and experiment. Such a statement is not correct. As a counter-example let us consider the De Rujula-Georgi-Glashow model[26]. In this model the $K_L \leftrightarrow K_S$ transition is described by Fig.6.

One obtains:

$$\frac{m_L - m_S}{m_K} = \frac{G_F}{\sqrt{2}} \frac{\alpha}{4\pi} \left(\frac{m_c}{M_W} \right)^2 2 \left(\ln \frac{M_W}{m_c} - 1 \right) \frac{\cos^2\theta_c}{\sin^2\theta_W} \quad (23)$$

$$\cdot \langle \bar{K}_0 | [\bar{n}(1-\gamma_5)\lambda]^2 + [\bar{n}\, \sigma_{\mu\nu} \tfrac{1-\gamma_5}{2} \lambda]^2 | K^0 \rangle$$

We notice the following, by comparing with (21):

i) There is no $\sin^2\Theta_C$ suppression.

ii) There is a big $\ln(\frac{M_W}{m_c})$ term.

Then for reasonable values of m_c we get an $m_L - m_S$ 10-100 bigger than the experimental number. For other problems of this model see A. De Rujula's talk.[22)]

<u>Moral:</u> Model-builders should check carefully that their model satisfies all the experimental information (Table I) that we have on the $\Delta S = 1$ $\Delta Q = 0$ and $\Delta S = 2$ processes, even if they use the GIM mechanism.

Now we turn on the strong interactions effects in the aforementioned processes.

iii) <u>Strong interactions effects in Rare-K decays and $K_L - K_S$ mass difference</u>

Up to now, no strong interactions effects have been considered. In the "standard model" where strong interactions are turned on as an asymptotically free gauge theory, one may use the well-known machinery of the Wilson expansion and renormalization group equations, to calculate effects of strong interactions[27)28)29)].

a) $K_L \to \mu^+\mu^-$

The box diagram described in Fig.2a: $\lambda + \bar{n} \to W^+W^- \to \mu^+\mu^-$ is cut-off at P($\equiv$ internal loop momentum) $\simeq M_W$, for if we took out both W-propagators, we would have a logarithmically divergent integral. So the dominant contribution of the integral comes from the short distance part where all integration variables are of order $\sim \frac{1}{M_W}$. Then we may use the Wilson expansion of the relevant currents:

$$j_\mu(x)\, j_\nu^+(0) \simeq \ldots + C_{\mu\nu\rho}(x)\, m_c^2\, j^\rho(0) \qquad (24)$$

Where j_μ, j_ν are the standard Cabibbo-GIM currents (see Eq.4) and

$$j_\rho = \bar{n}\,\gamma_\rho\,\frac{1-\gamma_5}{2}\,\lambda \qquad (25)$$

Now j_μ, j_ν^+, j_λ are conserved on the light cone, which means that they have γ($\equiv$ anomalous dimensions) = 0. Then, the net effect of strong interactions is to multiply the free quark result by a term:

$$A = \left(1 + \frac{g^2}{8\pi^2}\frac{25}{3}\ln\left(\frac{M_W}{\mu}\right)\right)^{-\frac{24}{25}} \simeq \frac{1}{6.5} \tag{26}$$

where $\mu \simeq 1$ GeV is the substraction point and we have assumed

$$\frac{1}{4\pi} g^2(\mu = 1\text{GeV}) \simeq 1 \qquad M_W \simeq 100 \text{ GeV} .$$

So

$$C_{\mu\nu\rho}(x, g) = \underbrace{C_{\mu\nu\rho}(x, 0)}_{\text{free quark result}} A \tag{27}$$

Note that, as pointed out by G. G. Ross and Nanopoulos[27)] strong interactions are such that they do not destroy the cancellation of $\ln \frac{M_W^2}{m_c^2}$ which take place in the free quark model.

As we mentioned before, strong interactions do not alter dramatically the free quark model result and so the dominant mechanism for $K_L \to \mu^+\mu^-$ seems to be the conventional one: $K_L \to 2$ virtual γ's $\to \mu^+\mu^-$ Finally we turn on the K_L-K_S mass difference.

b) K_L-K_S <u>mass difference</u>

Here things are more complicated. There are two scales relevant to the problem. From <u>Fig.7</u>, we must have

$$|x-y| \simeq |z-w| \lesssim \frac{1}{M_W} \qquad \text{(a)}$$

$$\left|\frac{x+y}{2} - \frac{z+w}{2}\right| \lesssim \frac{1}{m_c} \qquad \text{(b)} \tag{28}$$

Then there are two different lines of thought how to calculate the strong interactions effect:

i) G. G. Ross and Nanopoulos[27)] assume that again here the box diagrams of Fig.5 are cut-off at $p \simeq M_W$ (corresponding to the scale set by Eq.28a). Then the net effect is to multiply the free quark result by:

$$B = \left(1 + \frac{g^2}{8\pi^2}\frac{25}{3}\ln\frac{M_W}{\mu}\right)^{-\frac{30}{25}} \simeq \frac{1}{10} \tag{29}$$

But the K_LK_S mass difference is very sensitive to the m_c, so if we use (29) we come out with $m_c \sim 5$ GeV (30), which is much bigger than the free quark result[+]: $m_c \sim 1.5$ GeV.

ii) Madame Gaillard et al.[29] and Vainstein et al.[28] have different tastes. They prefer the scale set by 28b, i. e. that the box diagrams of fig. 5 are cut-off at $p \simeq m_c$. The claim is the following:

In the case of the K_LK_S mass difference, we have two c,p propagators, which means we have convergence even with the two W-propagators out, where then the diagram of fig. 5a(b) transforms to the diagram of fig. 8a(b).

If that is the case, then the correct procedure is:

1) Find the corrections to the local J-J interaction, after removing the W-propagator.

2) Then, consider the T-ordered product of two local J-J interactions thus obtained from 1).

Then one obtains:

$$B = \frac{1}{2}\Big[\underbrace{1 + \frac{g^2}{8\pi^2}\frac{25}{3}\ln\frac{M_W}{\mu}}_{m_c^2 \leq p^2 \leq M_W^2}\Big]^{\frac{24}{25}}\Big[\underbrace{1 + \frac{g^2}{8\pi^2}\frac{27}{3}\ln\frac{m_c}{\mu}}_{\mu^2 \leq p^2 \leq m_c^2}\Big]^{-\frac{10}{9}} \simeq 0.8 \qquad (30)$$

As we notice from (30), if the relevant scale is set by (28b) then there is a negligible difference from the free quark model calculations. One of course may worry if the distance $1/m_c$ is "short" on the scale of uncharmed particles, but an a posteriori justification could be the observed scaling below charm threshold.

The main conclusions of this section are:

1) We saw that a solid analysis of $\Delta S = 1$, $\Delta Q = 0$ rare K decays and K_SK_L mass difference is indeed helpful in order to set the scale for the "charmed" quarks.

Also we may use this piece of information as a laboratory in testing different gauge models. They ought, not only to cancel $\Delta S = 1$, $\Delta Q = 0$ N.C and

[+] For comparisons of Asym. free gauge theory predictions with experiment see table 2. Notice that $B(K^+ \to \pi^+e^+e^-)$ is getting closer to the experimental number.

$K_L K_S$ mass difference to order G_F, but also to provide the right order of magnitude when confronted with experiment.

2) If strong interactions are "turned on" via an asymptotically free gauge theory, their effects on the aforementioned procedures are calculable[27,28,29] and seem to give results very near to the free quark model, as expected!!!

Actually, one may construct a renormalizable phenomenological model of SU(4) pseudoscalar mesons coupled to the Weinberg-Salam gauge bosons, where one then calculates with real particles (not quarks). The outcoming results are similar to those mentioned above.

3. QUARK SPECTROSCOPY

It is clear that the presence of a fourth quark, c , is going to enlarge our spectrum of hadrons. Here, we give a quick view of the new spectroscopy. We follow the by now classic work of Gaillard-Lee-Rosner[30] (GLR). For a very comprehensive review of the situation of charm after the discovery of new particles see John Ellis[31] talk at Schladming.

In the quark language:

SU(3) means 3 quarks $\begin{pmatrix} p \\ n \\ \lambda \end{pmatrix}$; SU(4) means 4 quarks $\begin{pmatrix} p \\ n \\ \lambda \\ c \end{pmatrix}$

Mesons are viewed as $q\bar{q}$ bound states:

$$q\bar{q} : \quad 3 \otimes \bar{3} = 8 \oplus 1 \qquad ; \qquad 4 \otimes \bar{4} = 15 \oplus 1$$

Baryons are viewed as qqq bound states:

$$qqq : \quad 3 \otimes 3 \otimes 3 = 10 \oplus \underbrace{8_A \oplus 8_S}_{\text{combination}} \oplus 1 \qquad ; \qquad 4 \otimes 4 \otimes 4 = 20 \oplus \underbrace{20'_S \oplus 20'_A}_{\text{combination}} \oplus 4$$

The electric charge operator is given by

$$Q = I_3 + \frac{Y}{2} \qquad ; \qquad Q = I_3 + \frac{Y}{2} + \frac{2}{3} C \tag{31}$$

where

$$C = \frac{1}{4}(1 - \sqrt{6}\,\lambda_{15}) = \begin{pmatrix} 0&0&0&0 \\ 0&0&0&0 \\ 0&0&0&0 \\ 0&0&0&1 \end{pmatrix} \tag{32}$$

Notice that C and of course Q, because of the presence of the unit operator in (32) are not generators of SU(4) but of U(4). So we see that mesons belong to the 15⊕1 representation of SU(4) and baryons to the 20 representation of SU(4). (See figures 9 - 11 and tables 3 - 4 - 5 from GLR paper[30]).

Actually, in the case of baryons, we have a truncated tetrahedron, which has four hexagonal faces. Each face contains an octet of baryons which transforms irreducibly under an SU(3) subgroup of SU(4), acting only on three of the four quarks:

$$\begin{pmatrix} p \\ n \\ \lambda \end{pmatrix}, \begin{pmatrix} p \\ n \\ c \end{pmatrix}, \begin{pmatrix} p \\ \lambda \\ c \end{pmatrix}, \begin{pmatrix} n \\ \lambda \\ c \end{pmatrix}$$

This observation[30)] simplifies life, when one deduces the G_A/G_V ratios for weak semileptonic transitions from an ordinary nucleon to a charmed baryon.

Notice that in table 5, there is a particle $\varphi_c \simeq c\bar{c}$ ($J^{PC} = 1^{--}$) which, if $m_c \sim 1.5$ GeV, then $m_{\varphi_c} \sim 3$ GeV.

Soon we are going to identify this φ_c with the recently discovered ψ(3.1 GeV)[16] ($\varphi_c \equiv \psi$(3.1 GeV)).

4. DECAYS OF CHARMED PARTICLES

A. Qualitative features and naive expectations (as in GLR[30)])

i) (Semi)leptonic decays

Here the precise form of the weak current is crucial. We will use the conventional Cabibbo-GIM current:

$$\begin{aligned}(j_\mu^+)_R^{C-GIM} &= \bar{p}\gamma_\mu(1-\gamma_5)(n\cos\theta_c + \lambda\sin\theta_c) \\ &\quad + \bar{c}\gamma_\mu(1-\gamma_5)(-n\sin\theta_c + \lambda\cos\theta_c)\end{aligned} \tag{4}$$

Then by glancing at it, we have the following selection rules:

			Amplitude
(1)	$\Delta C = \Delta Q = \Delta S = 1;\ \Delta I = 0$	$\propto$	$\cos\theta_c$
(2)	$\Delta C = \Delta Q = 1,\ \Delta S = 0;\ \Delta I = \frac{1}{2}$	$\propto$	$\sin\theta_c$

(33)

These selection rules imply the following dominant decays (we follow the nomenclature of charmed particles as given by GLR):

$$\left.\begin{array}{l} D^0 \to "K^-" + \text{leptons} \\ D^+ \to "\bar{K}^0" + \text{leptons} \\ F^+ \to "\eta" + \text{leptons} \end{array}\right\} \quad \Gamma \propto \cos^2\theta_c \tag{34}$$

(see figure 12)

where "A" $\equiv$ hadronic final state with SU(3) quantum numbers of A, but quite possibly multiparticle states with different spin-parity.

1) Two-body decays

Here we have an analogy with K_{ℓ_2} decays. In what follows, from now on, we make the assumptions:

a) $m_D \simeq m_F \simeq 4 m_K$

b) $f_D \simeq f_F \simeq f_K \simeq f_\pi$ (assuming that SU(4) is not spontaneously broken.)

where:
$$\langle 0 | j_\mu^{GIM} | D^+(q) \rangle = i f_D q_\mu \sin\theta_c$$
$$\langle 0 | j_\mu^{GIM} | F^+(q) \rangle = -i f_F q_\mu \cos\theta_c \tag{35}$$

Then one obtains:

$$\frac{\Gamma(D^+ \to \mu^+\nu)}{\Gamma(K^+ \to \mu^+\nu)} \simeq \frac{m_D}{m_K} \quad \text{which means: } \Gamma(D^+ \to \mu^+\nu) \simeq 2\cdot 10^8 \sec^{-1} \tag{36}$$

Also:

$$\frac{\Gamma(F^+ \to \mu^+\nu)}{\Gamma(K^+ \to \mu^+\nu)} \simeq \frac{m_F}{m_K} \operatorname{ctg}^2\theta_c \quad \text{which means: } \Gamma(F^+ \to \mu^+\nu) \simeq 4\cdot 10^9 \sec^{-1} \tag{37}$$

Notice that the decays $D^+ \to e^+\nu$ and $F^+ \to e^+\nu$ are suppressed (as in the K,π-decays) by a factor 10^{-5}, compared to the muonic decays, by the usual helicity conservation arguments.

2) Three-body decays

Here we have analogy with the K_{l_3} decays. Then one finds

$$\frac{\Gamma(D^+ \to \bar{K}^0 e^+\nu)}{\Gamma(K^0 \to \pi^- e^+\nu)} \simeq \left(\frac{m_D}{m_K}\right)^5 2 \operatorname{ctg}^2\theta_c \tag{38}$$

which implies:
$$\Gamma(D^+ \to \bar{K}^0\mu^+\nu + \bar{K}^0 e^+\nu) \simeq 3\cdot 10^{11} \sec^{-1}$$

In the same way:
$$\Gamma(F^+ \to \eta\mu^+\nu + \eta e^+\nu) \simeq 2\cdot 10^{11} \sec^{-1} \tag{39}$$

Here there is no suppression of e relative to μ. One of course may give a naive estimation of the total semileptonic width: charm $\to 1 + \nu +$ hadrons. Then we suppose that we have a picture as in fig. 13, which means that we calculate: $c \to \lambda + 1^+ + \nu$ and $c \to n + 1^+ + \nu$. But such an elementary process looks like the muon decay:

$$\mu^- \to e^- + \bar{\nu}_e + \nu_\mu$$

One then obtains:

$$\Gamma_{tot}(\text{charm} \to \ell + \nu + \text{hadrons}) = \frac{G_F^2 m_c^5}{196\,\pi^3} \tag{40}$$

and taking $m_c = 1.5$ GeV:

$$\Gamma_{tot}(\text{charm} \to \ell + \nu + \text{hadrons}) \simeq 10^{12}\ \text{sec}^{-1} \qquad (40')$$

A few comments are in order:

1) Three-body decays seem to be a large fraction, if not the dominant one of charm $\to l^+ + \nu$ + hadrons (compare (39) with (40')).

2) One may naively think that, because of the massiveness of D,F, the decay e.g.: $D^0 \to K + n\pi + l^+ + \nu$ account for a large fraction of the total rate. As Gaillard-Lee-Rosner showed, this is not the case, if at least one of the outgoing pions may considered as a soft pion. Of course, if all pions are hard, then things are different.

3) In our naiveté, we have neglected form factors. Here things are different from the K decays, because the vector-pseudoscalar mass splitting is much smaller for charmed particles than for K's. So a form factor of the form

$$f(q^2) = \frac{m_{D^*}^2}{m_{D^*}^2 - q^2} \quad \text{with } q^2 \text{ in the range } 0 \leqq q^2 \lesssim (m_D - m_K)^2$$

could lead to an appreciable enhancement with respect to the estimates given above.

ii) Nonleptonic decays[32]

Using again the conventional current (see eq. 4) we get (to lowest order):

$$\mathcal{L}_{eff} = \frac{G_F}{\sqrt{2}}\Big\{(\bar{c}\gamma_L\lambda)(\bar{n}\gamma_L p)\cos^2\theta_c + [(\bar{c}\gamma_L\lambda)(\bar{\lambda}\gamma_L p) - (\bar{c}\gamma_L n)(\bar{n}\gamma_L p)]\sin\theta_c\cos\theta_c - (\bar{c}\gamma_L n)(\bar{\lambda}\gamma_L p)\sin^2\theta_c + h.c.\Big\} \qquad (41)$$

Then we have the selection rules:

		Amplitude	
$\Delta C = \Delta S$	$\propto$	$\cos^2\theta$	(42)
$\lvert\Delta C\rvert = 1,\ \Delta S = 0$	$\propto$	$\cos\theta_c \sin\theta_c$	

So the dominant decays are:

		Γ	
$D^0 \to$ "$\bar{K}^0$" ; $D^+ \to$ "$(K^- n\pi)^+$" ; $F^+ \to$ "π^+"	$\propto$	$\cos^4\theta_c$	(43)
$D^0 \to$ "π^0", "η" ; $D^+ \to$ "π^+" ; $F^+ \to$ "K^+"	$\propto$	$\cos^2\theta_c \sin^2\theta_c$	

(see figures 14, 15)

Again here, one may calculate the total width of charm → hadrons, by viewing the decay as going through the elementary process: $c \to \lambda + p + \bar{n}$ (see fig. 16). One then obtains (using the resemblance with the muon theory):

$$\Gamma_{tot}\,(\text{charm} \to \text{hadrons}) \simeq \frac{1}{196\,\pi^3}\,(G_F\, A \cos^2\theta_c)^2\, m_c^5 \tag{44}$$

The underlying factor A represents the typical enchancement of non-leptonic decays. If for instance in our naiveté we use past experience with strange-particle decays, we may put:

$$A \simeq \frac{1}{\sin\theta_c} \tag{45}$$

Then we find (using always $m_c \simeq 1.5$ GeV):

$$\Gamma_{tot}(\text{charm} \to \text{hadrons}) \simeq 10^{13}\ \text{sec}^{-1} \tag{46}$$

A few comments are in order:

1) If we use standard current algebra techniques we may calculate particular channels, e. g. $D^0 \to K^-\pi^+$. Then one finds:

$$\Gamma(D^0 \to K^-\pi^+) \simeq 5 \cdot 10^{12}\ \text{sec}^{-1} \tag{47}$$,

which is consistent with (46). For later use, notice that

$$\text{B.R.}\,(D^0 \to K^-\pi^+) \simeq 50\,\% \tag{47'}$$

2) Comparing equations (40) and (44) we obtains:

$$\frac{\Gamma_{tot}(\text{charm} \to 1 + \nu + \text{hadrons})}{\Gamma_{tot}(\text{charm} \to \text{hadrons})} \sim 2\tan^2\theta_c \sim 8\ \% \tag{48}$$

3) An exception to the general enhancement rule may occur when the non-leptonic decay leads to an exotic final state, as for instance in the dominant decay of D^+:

$$D^+ \to (\bar{K} n\pi)^+ \quad , \qquad n = 1, 2, \cdots$$

The analogous decay, from the "old physics" is: $K^\pm \to \pi^\pm \pi^0$ where the $\pi\pi$ system must have $I_{\pi\pi} = 2$. As is well known, such decays seem to lack the enhancement factor A (see eq. 45). The two-body decay: $D^+ \to \bar{K}^0 \pi^+$ is suppressed, as noticed by Madame Gaillard and the Roma group[34], in the exact SU(3) limit by the transformation properties of the enhanced effective interaction. Also, to order $\cos^4\theta$, the non-leptonic decays of charmed mesons do not lead to states with quantum numbers of $K^\pm$ (see eq. 43). Consequently the presence of a narrow peak in a $(Kn\pi)^0$ or $(\bar{K}n\pi)^0$ distribution and its absence in $(Kn\pi)^+$ or $(Kn\pi)^-$ distribution would thus be a strong indication in favour of charmed particles.

B. Quantitative Features (More on Non-Leptonic Decays)

$\Delta S \neq 0$ decays

Experimental facts

We have at least two pieces in the weak Hamiltonian, which transforms as an $I = \frac{1}{2}$ and an $I = \frac{3}{2}$

Then we have for the amplitudes:

$$\frac{A(\Delta I = 3/2)}{A(\Delta I = 1/2)} \leqslant 5\%$$

That's what we call the "$\Delta I = 1/2$ rule".

Ask: Is the $\Delta I = 1/2$ part larger than expected or the $\Delta I = 3/2$ strongly suppressed?

Answer (?): Both! As may be seen from the ratio:

$$\frac{A_{\text{non-lep.}}(I=1/2,3/2)}{A_{\text{semi-lep.}}} \simeq (5,1/5).$$

$\Delta C \neq 0$ decays

Experimental facts

?

Analysis of the effective non-leptonic Hamiltonian[33]

$\Delta S \neq 0$:

$$H^{N.L.} = \frac{1}{2}(j_\mu j_\mu^+ + j_\mu^+ j_\mu)$$

where

$$j_\mu = \bar{p}\gamma_L n \cos\Theta_c + \bar{p}\gamma_L \lambda \sin\Theta_c$$

which transforms as an $\underline{8}$ under SU(3).

$$j_\mu \cdot j_\mu \Rightarrow 8 \otimes 8 = 1_s \oplus 8_s \oplus 27_s \oplus 8_A \oplus 10_A \oplus \overline{10}_A$$

Because H_W is symmetric:

$$(H_W)_S = 1 + 8_s + 27$$

But $\underline{1}$ cannot contribute to $\Delta S \neq 0$ decays, thus

$$(H_W)_S = 8_S + 27 .$$

Now under $SU(2)_{\text{Isospin}}$, the $\underline{8}$ has an $I = 1/2$ piece and the $\underline{27}$ an $I = 1/2$ and $I = 3/2$ piece. So the $I = 1/2$ rule implies that the $\underline{8}$ is enhanced relative to the $\underline{27}$. This is a short-handed expression of the full statement: The SU(3) $\underline{8}$ component of the effective

$\Delta C \neq 0$:

$$H^{N.L.} = \frac{1}{2}(j_\mu j_\mu^+ + j_\mu^+ j_\mu)$$

where

$$j_\mu = \bar{p}\gamma_L n_c + \bar{c}\gamma_L \lambda_c$$

which transforms as a $\underline{15}$ under SU(4).

$$j_\mu \cdot j_\mu \Rightarrow 1_s \oplus 15_s \oplus 84_s \oplus 20_s \oplus 15_A \oplus 45_A \oplus \overline{45}_A$$

Because H_W is symmetric:

$$(H_W)_S = 1 + 15_s + 20_s + 84_s$$

But $\underline{1}$ cannot contribute to $\Delta C \neq 0$ decays, thus

$$(H_W)_S = 15_S + 20_S + 84_S .$$

Also because of the form of the current 15_S is absent. Since we are not concerned with the space-time structure of the current, let us write: $j = \bar{\Psi} S \Psi$

with $\Psi = \begin{pmatrix} c \\ p \\ n \\ \lambda \end{pmatrix}$ and $S = \begin{pmatrix} 0 & u \\ 0 & 0 \end{pmatrix}$

non-leptonic Hamiltonian seems to be enhanced relative to the SU(3) 27 component.

I.e: "$\Delta I = 1/2$ rule"

where

$$u = \begin{pmatrix} -\sin\theta_c & \cos\theta_c \\ \cos\theta_c & \sin\theta_c \end{pmatrix}.$$

Then the absence of 15_S is a consequence of i) $\mathrm{Tr}S = 0$ and

ii) the anticommutator $\{S, S^+\} \propto 1$ matrix.

$$(H_w)_S = 20_S + 84_S \tag{49}$$

Now under SU(3):

$$\begin{matrix} 20_s \rightarrow & 6 + \bar{6} & + 8 \\ & |\Delta C| = 1 & \Delta C = 0 \end{matrix} \tag{50}$$

$$\begin{matrix} 84_s \rightarrow & 15 + \overline{15} & + 1 + 8 + 27 \\ & |\Delta C| = 1 & \Delta C = 0 \end{matrix} \tag{51}$$

The missing terms don't contribute because of the particular form of the currents that appear here.

Let us assume now that the symmetry of the world is SU(4). Then in order to explain the $\Delta I = 1/2$ rule for $|\Delta S| \neq 0$, $\Delta C = 0$ decays of "old physics", we have to believe that the 20 is enhanced relative to the 84. This follows immediately from eq. (50) and (51) by noticing that both 20 and 84 contain an 8 piece, but in the 20 the 8 is alone, on the contrary in the 84 the 8 is accompanied by a 27. So we have to enhance the 8 of the 20 and not of the 84, if we want to explain the usual $\Delta I = 1/2$ rule. Then the consistency of our theory with the usual $\Delta I = 1/2$ rule implies "20-plet enhancement".

What then are the consequences of the above conclusion to the $\Delta C \neq 0$ decays?

Let us consider the G.I.M. current: $j_\mu^{GIM} = \bar{c}\gamma_L\lambda\cos\theta_c - \bar{c}\gamma_L n \sin\theta_c$ which transforms as a 3 under SU(3). Then the effective Hamiltonian for $|\Delta C| = 1$ decays is:

$$H^{\Delta C = 1} = \{ j_\mu(\Delta C = 1), j_\mu^+(\Delta C = 0) \} \tag{52}$$

which transforms under SU(3) as a $3 \times 8 = 3 + 6 + 15$, where the 6 comes from the 20 and the 15 comes from the 84. So the "20-plet enhancement" implies now that the 6 is enhanced relative to the 15 i.e:

"Sextet-dominance of charm-changing decays".

It is worth mentioning that we got this result by using just SU(3) invariance.

In conclusion, we have shown the following:

1) In the SU(4) limit, the usual $\Delta I = 1/2$ rule which means octet-enhancement (in SU(3)), means 20-plet enhancement (in SU(4)).
2) SU(3) invariance implies sextet dominance of charm-changing decays.

It is clear that if we enlarge our symmetry new relations among the amplitudes of various decays are going to appear. For instance, there is another relation among the S-wave amplitudes for hyperon decays beyond the well-known SU(3) relation:

$$2S(\Xi^- \to \Lambda\pi^-) = \sqrt{3}\,S(\Sigma^+ \to p\pi^0) + S(\Lambda \to p\pi) \qquad (53)$$
$$4.08 \pm 0.04 = 4.04 \pm 0.05$$

if we assume the 20 (in SU(4)) dominance. It is:[34]

$$S(\Lambda \to p\pi^-) = \frac{1}{\sqrt{3}}\,S(\Sigma^+ \to p\pi^0) \qquad (54)$$
$$1.48 = 0.85$$

or

$$S(\Xi^- \to \Lambda\pi^-) = \frac{2}{\sqrt{3}}\,S(\Sigma^+ \to p\pi^0) \qquad (55)$$
$$2.04 = 1.71$$

It is not surprising that (54) and (55) are less well satisfied than (53), since the reason for these new relations is the antisymmetry of the $\Delta C = 0$, $|\Delta S| = 1$ weak interactions under exchange of C and P quarks, and we show in section 2 that the C-P symmetry (P-spin) is badly broken.

Up to now, using basically group theory plus the information from strange particle decays, we derived some useful properties concerning the decays of charmed particles.

We have been naive all the time, assuming that the enhancement that appears in the non-leptonic strange-particle decays, remain the same for the non-leptonic charmed-particle decays.

Non-leptonic decays of charmed particles are of great importance to the experimental searches for these particles, thus a more refined and detailed analysis of these decays seems to be justified.

The basic problem here to consider is the following: Is really the enhancement factor A (see eq. 45) the same for strange particles and for charmed particles? Then, in order to answer such a question we have to analyze carefully the theoretical arguments that exist at present in the attempts to explain the magnitude of this enhancement factor A.

May be surprising results are waiting for us, which somehow conflict with our naive expectations. Let's do It!

Here we follow Ellis-Gaillard-Nanopoulos[35].

C. Non-Enhancement[35] of Non-Leptonic Decays of High-Mass Hadrons

Phenomenological analyses of weak decays of strange particles indicate that semi-leptonic modes are weaker than non-leptonic modes and that the non-leptonic weak Hamiltonian behaves predominantly as an octet under SU(3). Theoretical explanations of these facts include:

1) short-distance behaviour and asymptotic freedom[36,37];
2) duality[38];
3) current algebra, PCAC and coloured fermion quarks[39];
4) recently, the idea[26] that the $\Delta I = 1/2$ rule is fundamental (not dynamical) and has to be built in a-priori, has been proposed (see De-Rujula's talk[22] for more on this alternative)

As usual, we start with old ideas and try to extrapolate them into the new regimes. So what I am going to do next is to investigate how far we can push the arguments concerning the weak decays of strange particles to cover the weak decays of high-mass hadrons.

1. Short Distance Behaviour and Asymptotic Freedom

The effective non-leptonic $\Delta S = \Delta C = 1$ interaction is

$$\frac{G_F}{\sqrt{2}}\left[C_- O^-(0) + C_+ O^+(0)\right]\cos^2\theta_c \tag{56}$$

where

$$O^{\mp} = \tfrac{1}{2} N\left[(\bar{c}\lambda)(\bar{n}p) \mp (\bar{c}p)(\bar{n}\lambda)\right]$$

(N indicates Zimmermann's ordering and reference to the spinor structure of the currents is suppressed). O^- (O^+) belongs to the <u>20</u> (<u>84</u>) representation of SU(4) and $C_{\mp}$ can be calculated in an asymptotically free gauge theory of coloured SU(4) quarks,

$$C_{\mp} = \left[\frac{1 + \frac{25}{6\pi}\ln M_W}{1 + \frac{25}{6\pi}\ln\mu}\right]^{+0.48 \atop -0.24} \tag{57}$$

M_W is the mass of the intermediate vector boson and μ is the subtraction point. The choice of μ is ideally: a) the effective mass of the weakly interacting quarks in the hadron, and b) above the onset of scaling. So, logically, $\mu \sim 1$ GeV for the strange particles and $\mu \sim 2$ GeV for the charmed particles. Then ($M_W \sim 100$ GeV):

$$C_{\mp} = \begin{cases} \begin{matrix} 4 \\ 0.5 \end{matrix}, & \mu = 1\,\text{GeV} \qquad \text{strange particles} \\ \begin{matrix} 1.9 \\ 0.7 \end{matrix}, & \mu = 2\,\text{GeV} \qquad \text{charmed particles} \end{cases} \tag{58}$$

So, modulo the ambiguity in the value of μ, it is clear that the (non-leptonic/leptonic) and ($\underline{20}/\underline{84}$) enhancement factors may be smaller than the case of strange-particle decays. Of course, as noted by Gaillard and Lee[21], asymptotic freedom is not enough to explain the $\Delta I = 1/2$ in K decays. Another machanism seems to be needed to enhance the matrix elements of the $\Delta I = 1/2$ operator.

This brings us to:

2. Duality Arguments

In the duality arguments of Nussinov and Rosner[38], the effective non-leptonic weak Hamiltonian is treated as a spurion S, and the decay $A \to B + \pi$ related to process

$$S + A \to B + \pi \tag{59}$$

Now, using duality, one sees that the non-exotic ($\bar{q}q$) parts of S (Fig. 17) are enhanced over exotic ($\bar{q}\bar{q}qq$) parts (Fig. 18). In the dominant $\Delta C = \Delta S = 1$ charmed particle decays, the spurion has ($\bar{c}\lambda$) ($\bar{n}p$) quantum numbers and is a $(\bar{q}q)(\bar{q}q)$ operator which does not have $\underline{16}$ plet ($\bar{q}q$) quantum numbers. In contrast for $\Delta C = 0$, $\Delta S = 1$ decays, the spurion is a $(\bar{\lambda}p)(\bar{p}n)$ operator which has non-exotic ($\bar{\lambda}n$) SU(3) quantum numbers. Thus, in the $\Delta C = 0$, $\Delta S = 1$ (or $\Delta C = 1$, $\Delta S = 0$) decays, there could be contributions from spurions looking like Figs. 19a or 19b, whereas for $\Delta C = \Delta S = 1$ only Fig. 19b can contribute. Only a non-exotic spurion of the form of Fig. 19a could be enhanced by the Nussinov-Rosner duality argument. So, the matrix elements of 0^- are not enhanced over the matrix elements of 0^+. Also, one can see[39] that current algebra, PCAC and coloured fermion quarks do not help the enhancement of $\underline{20}$ over $\underline{84}$.

Now, if we use (58), we can calculate semi-leptonic decays of charmed mesons, non-leptonic decays of charmed mesons and baryons, etc. Here, I just present the results (for details, see Ref. 35):

$$\frac{\Gamma(P_c \to \mu + \nu + \text{hadrons})}{\Gamma(P_c \to \text{all})} \simeq \frac{\Gamma(P_c \to e + \nu + \text{hadrons})}{\Gamma(P_c \to \text{all})} \simeq 15 - 20\% \tag{60}$$

(where $P_c \equiv D$ or F).

$$\begin{aligned} \Gamma(D \to K\ell\nu) &\simeq 0.3\,\Gamma(D \to \ell\nu + \text{hadrons}) \simeq 0.05\,\Gamma(D \to \text{all}) \\ \Gamma(F \to \eta\ell\nu) &\simeq 0.2\,\Gamma(F \to \ell\nu + \text{hadrons}) \simeq 0.03\,\Gamma(F \to \text{all}) \end{aligned} \tag{61}$$

(where $\ell \equiv e$ or μ), and we have taken $m_c \simeq 2$ GeV

Two-body non-leptonic decays:

$$\frac{\Gamma(D^0 \to K\pi)}{\Gamma(D^0 \to all)} \simeq 3\%\ (18\%) \qquad \frac{\Gamma(F^+ \to n\pi + K\bar{K})}{\Gamma(F^+ \to all)} \simeq 2\%\ (12\%)\ , \tag{62}$$

$$\frac{\Gamma(D^+ \to K\pi)}{\Gamma(D^+ \to all)} \simeq 2\%\ (10\%) \qquad \text{etc.}$$

Total width:

$$\Gamma_{tot}(P_c \to all) = (6 \text{ to } 8)\cdot 10^{-12}\ sec^{-1}\ \left[(1.5 \text{ to } 2)\cdot 10^{-12}\ sec^{-1}\right] \tag{63}$$

where we have taken $m_c = 2$ GeV (1.5 GeV).

A simple comparison now of (48) with (60) and (47') with (62) shows that our fears of being very naive turn out to be justified. Of course, the estimates of (60) → (63) should not be taken too seriously, however they may serve as indicators of what ball-park to expect.

Moral: Statements such as

1) semi-leptonic decays of charmed particles are only a few percent and
2) sextet-dominance (complete suppresion of the 15-plet) in charm changing decays should be taken with a pinch of salt

and therefore

Warning:

Theoretical-results coming out from the above two statements should be taken with caution, especially when we confront charm with experiment.

Actually what seems to be more realistic is[35]:

1) A total-semileptonic B.R. of the order of 20% or more is very possible and
2) maybe 1/3 of the non-leptonic decays going via the non-enhanced parts of the weak Hamiltonian.

(For the experimental consequences of such a behavior of the decays of charmed particles, see the next section).

Before closing this section, here are the main points again:

i) Non-leptonic decays should be dominant, but with perhaps 0 (30 %) semileptonic decays.

ii) Decays should predominantly contain strange particles.

iii) They are probably predominantly multibody (calculations of two-body modes give very small percentages; empirically high mass systems seem to predominantly decay into many particles).

iv) The purely leptonic decay modes are probably small.

Conclusion:

The detection of charmed particles will not be an easy task.

(For more, see next section.)

5.+ THE PRODUCTION OF CHARMED PARTICLES AND COMPARISON WITH EXPERIMENT

In this final section we will discuss how charmed particles should be produced in various different reactions. In order of decreasing expected branching ratio for producing charmed particles, we will consider e^+e^- annihilation, deep inelastic neutrino scattering, deep inelastic electroproduction and finally hadron-hadron collisions. In doing this we will be looking for answers to two questions: why have we not yet seen charmed particles? and how will we be able to see them?

5.1. e^+e^- Annihilation

The data (Fig. 20)[16] on ($e^+e^- \to$ multiparticles) indicate a clear threshold around $E_{c.m.} = 4$ GeV, where $R = \frac{(e^+e^- \to \text{multiparticles})}{(e^+e^- \to \mu^+\mu^-)}$ increases from about 2 1/2 to about 5. Above this energy a novel class of events has been identified- $e^+e^- \to e^\pm \mu^\mp$ + no other visible particles[40] - which seem to indicate that some new particles are being produced. But there are no other similar indications: no narrow bumps[41] have been seen in invariant mass plots for combinations of π's and K's, there is no dramatic change above $E_{cm} = 4$ GeV in the fraction of events with a $K^\pm$ and no dramatic change in the average charged multiplicity $\langle n_{ch} \rangle$. The only clear difference between data below and above "threshold" seems to be a steady increase[16] in the fraction of E_{cm} carried off by neutral particles, which rises from about 0.4 at $E_{cm} = 3$ GeV to about 0.5 at $E_{cm} = 5$ GeV.

Let us first consider the value of R: in models with point-like quark partons, asymptotically

$$R \to \sum_{\text{quarks}} Q_i^2 = 3\tfrac{1}{3} \tag{64}$$

in the case of charmed quarks with hidden SU(3) colour. Unfortunately experiments are carried out at finite energies: in the region $E_{cm} \leq 5$ GeV they are probably in a threshold region dominated by resonant structures, and we should not expect (5.1) to apply. In the energy region above 5 GeV it may be possible to find an applicable formula with sub-asymptotic corrections to (64). For example, in asymptotically free

+ This section was written with the collaboration of John Ellis.

gauge theories[42]

$$R \simeq 3\tfrac{1}{3}\left(1 + \frac{12}{25\ln(q^2/\mu^2)} + \cdots\right) \tag{65}$$

where μ is some normalization mass. If, as done in section 4, we take $\mu \simeq m_c \simeq m_D \simeq$ $\simeq 2$ GeV, then (65) implies $R \simeq 4$ when $q^2 \simeq 40\ \text{GeV}^2$. Formula (65) has two problems when compared with present data on R: it gives values which are too small, and it predicts that R should be decreasing, whereas if anything R is rising for 5 GeV $\lesssim E_{cm} \lesssim$ 7.5 GeV (see Fig. 20). This is perhaps the single most serious problem of the charm model at this time, and we have no convincing answer to it: perhaps it will eventually kill the model.

The $e^+e^- \to e^\pm \mu^\mp +$ no visible events also raise problems for the charm model: the cross section for these events at $E_{cm} = 4.8$ GeV is 0 (0.04 nb). It seems clear that in order to produce them there must be 2 particles in the final state which are decaying weakly: e.g. $e^+e^- \to \varrho^+\varrho^-$, $\varrho^+ \to e^+ + \ldots$, $\varrho^- \to \mu^- + \ldots$
Since the increase in $\sigma(e^+e^-$ multiparticles) above 4 GeV seems to be O(10nb), it seems that a branching ratio

$$\frac{\Gamma(\varrho^+ \to e^+ + \ldots)}{\Gamma(\varrho^+ \to \text{all})} \simeq \frac{\Gamma(\varrho^+ \to \mu^+ + \ldots)}{\Gamma(\varrho^+ \to \text{all})} \simeq \text{few \%} \tag{66}$$

is necessary to explain the data. (In addition to the $e\mu + \ldots$ events there are comparable numbers of $e^+e^+ + \ldots$ and $\mu^+\mu^- + \ldots$ events.) What could the states $\varrho^\pm$ be? They could not be pseudoscalars or scalars $\to e^+\nu$ or $\mu^+\nu$ alone, because helicity conservation requires the (e/μ) branching ratio to be $O(10^{-4})$. They could in principle be vector states ($D^{*\pm}$, $F^{*\pm}$ in the charm model) decaying into $(e^\pm\nu)$ or $(\mu^\pm\nu)$. But, momentum and angular distributions[40] prefer <u>3-body decay.</u> Then could what we see be semileptonic decays of charmed mesons, e. g. $D \to$ hadrons $+ \ell + \nu$? But, not many "hadrons + e +μ" events seem to exist, hence semileptonic decays of charmed mesons are disfavored, as an explanation of the "e +μ" events.

An alternative explanation is that we see heavy lepton(s). Certainly, heavy lepton(s) help with R (and $\langle n \rangle$, $\frac{K}{\pi}$, $\cdots$ see later).

In order to compare the experimental limits[41] on production cross section times branching ratio with the theoretical expectations for branching ratios requires an assumption about charm-production cross section[43].

We are going to assume that the difference of R below and above threshold is due to the production of charmed particles: $R_{charm} \simeq 2.5$.

Then we further assume that the charmed pseudoscalar mesons are the lightest charmed particles and that the charmed final states include one of the pairs D^+D^-, $D^o\bar{D}^o$, F^+F^-, produced with equal cross sections of 3.14 nb at $E_{cm} = 4.8$ GeV.

Then, using the experimental limits reported for the relevant mass range 1.85 to 2.40 GeV, we obtain $\frac{\Gamma(D^o \to K^-\pi^+)}{\Gamma(D^o \to all)} \leq 3\%$ which is not in contradiction with eq. 62.

Moreover, the conclusions[41] that the upper limits on $B.R(D^+ \to \bar{K}^o\pi^+$ or $K^-\pi^+\pi^+)$ violates the expectation of the "standard model" by a factor of at least 3, looks chimeric to us, because $D^+ \to \bar{K}^o\pi^+$ is forbidden in the exact SU(3) limit and $D^+ \to K^-\pi^+\pi^+$ is an exotic final state.

On the contrary, if we would like to be very optimistic we would say that it is great for the "standard model", i. e. absence of $D^+ \to \bar{K}^o\pi^+$, or $K^-\pi^+\pi^+$.

Nevertheless, charm may have problems with the K/π ratio. The temptation of using again (as in the "$e+\mu$" events) heavy lepton(s) and combine them with charm, is quite big. Then we get:

	# K's	$\langle n_{charg.} \rangle$	$E_o/E_{charg.}$	R
Charm	many	~ 5 ?	not very large	3 1/3 → 4 with asymptotic freedom
Heavy lepton	few	~ 3 ?	large	1
Combine	the same below and above threshold	~ 4	large	5

We see that a combined picture of "charm + heavy lepton" may give reasonable explanations for the K/π ratio, $\langle n_{charg.} \rangle$ "Energy crisis" ($E_o/E_{charg.} > 1/2$) and R.

An experimental way to test such a combined picture, is for instance, to look at the number of K's as a function of the multiplicity: may be low multiplicity events have almost no K's (heavy-lepton case) and high-multiplicity events have many K's (charm case).

5.2. Deep Inelastic Neutrino Production

This reaction is usually described in terms of the variables $n = |q^2|/2M\nu$ $y = \nu/E_\nu$ where $p = (M,0,0,0)$ is the momentum of the target nucleon, q is the momentum transfer from the neutrino, $\nu = q\cdot p/M$, and E_ν is the incident neutrino energy. If the quark-parton model is used to estimate the cross sections for charged current interactions, then using the conventional current (1) we find

$$\frac{d^2\sigma^\nu}{dx\,dy} = \frac{G_F^2 ME_\nu}{\pi}\left[n(x)\cos^2\Theta_c + \lambda(x)\sin^2\Theta_c + (1-y)^2\bar{p}(x)\right]$$
$$\frac{d^2\sigma^{\bar\nu}}{dx\,dy} = \frac{G_F^2 ME_\nu}{\pi}\left[(1-y)^2 p(x) + \bar{n}(x)\cos^2\Theta_c + \bar{\lambda}(x)\sin^2\Theta_c\right] \quad (67)$$

where $n(x)$ etc. are the parton distribution functions averaged over the nucleons in the target, assumed to have essentially equal numbers of neutrons and protons. At high energies in the charm model the charm-changing current (4) gives extra contributions from the elementary processes

$$\nu + n \to \mu^- + c \qquad \propto \sin^2\Theta_c$$
$$\nu + \lambda \to \mu^- + c \qquad \propto \cos^2\Theta_c \quad (68)$$

giving

$$\Delta\left(\frac{d^2\sigma^\nu}{dx\,dy}\right) = \frac{G_F^2 ME_\nu}{\pi}\left[n(x)\sin^2\Theta_c + \lambda(x)\cos^2\Theta_c + (1-y)^2\bar{c}(x)\right]$$
$$\Delta\left(\frac{d^2\sigma^{\bar\nu}}{dx\,dy}\right) = \frac{G_F^2 ME_\nu}{\pi}\left[(1-y)^2 c(x) + \bar{n}(x)\sin^2\Theta_c + \bar{\lambda}(n)\cos^2\Theta_c\right] \quad (69)$$

The parton distribution functions are not very precisely known, but some things are known and prejudices held. From the assumed symmetry of the nuclear target between neutrons and protons it follows that $n(x) = p(x)$, $\bar{n}(x) = \bar{p}(x)$. From the zero strangeness and charm of the nucleons it is clear that $\int dx\, \lambda(x) = \int dx\, \bar{\lambda}(x)$, and similarly for $c(x)$ and $\bar{c}(x)$. We assume these relations also hold locally. According to prejudice, for $x \gtrsim 0.1$ only the valence quark distributions $n(x)$ and $p(x)$ are supposed non-zero, whereas as $x \to 0$ it is believed that parton and antiparton distributions become equal (the "sea"): $p(x) \approx \bar{p}(x) = \bar{n}(x) \approx n(x)$
Further, it seems not unreasonable that $\lambda(x) \approx n(x)$ as $x \to 0$ (approximate SU(3) invariance of diffractive scattering at large $|q^2|$ while quite possibly $c(x) << \lambda(x)$.) With these assumptions about the parton distributions we can analyze the new contributions (69) in the kinematic ranges $x \gtrsim 0.1$ and $x \simeq 0$, probably $\lesssim 0.1$.

<u>$x \gtrsim 0.1$:</u> Only $n(x)$ and $p(x) \neq 0$, the only charm production in νN scattering is off the n quark, proportional to $\sin^2\Theta_c$ and independent of y: it is therefore O(4 %)of the total cross section in this kinematic range. There is no charm production in $\bar\nu N$ scattering for $x \gtrsim 0.1$.

<u>$x \to 0$:</u> The dominant charm production mechanism in this range are off λ quarks in νN scattering and off $\bar\lambda$ quarks in $\bar\nu N$ scattering. The absolute production

rates should be equal to each other and independent of y. In the limit as $x \to 0$ the production of ordinary particles in νN and $\bar{\nu} N$ scattering should also be equal,

$$\frac{d\sigma^{\nu}}{dy} \propto 1 + (1-y)^2 \quad , \quad \frac{d\sigma^{\bar{\nu}}}{dy} \propto 1 + (1-y)^2 \tag{70}$$

below charm threshold, and

$$\frac{d\sigma^{\nu}}{dy} \propto 2 + (1-y)^2 \quad , \quad \frac{d\sigma^{\bar{\nu}}}{dy} \propto 2 + (1-y)^2 \tag{71}$$

when the charm production has scaled.

How do the data compare with the expectations (67) to (71)? To within 20 % or so, the total νN and $\bar{\nu} N$ cross sections seem to keep rising linearly with E_{ν}[44] with the same coefficient throughout the CERN-Gargamelle and FNAL energy ranges. As the "sea" quarks seem to account for just a few % of the total cross section and as $\sin^2\theta_c \simeq 4$ %, the charm contribution (69) to the total cross sections is expected to be just a few %: so far so good. The differential cross section $\frac{d\sigma^{\nu,\bar{\nu}}}{dy}$ integrated over $x \gtrsim 0.1$ also qualitatively agree with the simple parton model[45]. However for $x \lesssim 0.1$ the y distributions are apparently incompatible with the parton and charm model described above: $\frac{d\sigma^{\nu}}{dy}$ and $\frac{d\sigma^{\bar{\nu}}}{dy}$ in this range are apparently independent of y, and in the ratio of about 3 : 1[45] Formulae (67) and (69) show that independently of details of the parton distributions, since $n(x) = p(x)$, $\bar{n}(x) = \bar{p}(x)$ and $\lambda(x)$ is expected $= \bar{\lambda}(x)$, the limits as $y \to 0$ of $\frac{d\sigma^{\nu}}{dy}$ and $\frac{d\sigma^{\bar{\nu}}}{dy}$ should be equal. Experimentally this does not seem to be the case[45]. If this experimental result is confirmed, it will be difficult to sustain the charm model. The HPWFNAL group also report a very large violation[45] of charge symmetry in comparing ν and $\bar{\nu}$ interactions at small q^2. Such an effect is expected in the charm model, but the experimental violation is much larger. If it is confirmed, this violation could signal the presence of other weak currents besides the GIM one.

Both the Harvard-Penn-Wisconsin and Caltech FNAL experiments have reported events[47] of the type $\nu + N \to (\mu^+\mu^-) + N$ with a branching ratio of about (1 to 2) %. These events are not likely to come from decays of neutral heavy leptons that might be produced via an ν-induced reaction:

$$L^0 \to \mu^+ + \mu^- + \nu$$

Pais and Treiman[48] considered the ratio $\langle P_- \rangle / \langle P_+ \rangle$ assuming that the opposite sign muons have the same parent and the above decay is described by a local interaction (S,P,V,A and T), they obtained the bounds:

$$0.48 \leq \frac{\langle P_- \rangle}{\langle P_+ \rangle} \leq 2.1$$

Experimentally:
$$\frac{\langle P_- \rangle}{\langle P_+ \rangle} = 3.7 \pm 0.7$$

which is well beyond the upper bound. Actually the HPWF group notes that it is statistically consistent to assume that events with $P_+ > P_-$ are caused by the $\bar{V}$ contamination. If they exclude these events, they get $\frac{\langle P_- \rangle}{\langle P_+ \rangle} = 8.5 \pm 1.7$

which means that it is very unlikely that all of the dimuons events arise from the decay of a neutral heavy lepton. Also dimuon mass distributions disprove the possibility that dimuons come from a common parent of well defined mass (heavy lepton, W, Z^0, ...). It is therefore extremely plausible that the extra muons come from decays of a new particle produced at the hadron vertex. There is a ready mechanism for these in the charm picture, whereby

$$\nu + \begin{pmatrix} n \\ \text{or} \\ \lambda \end{pmatrix} \rightarrow \mu^- + c \qquad \llcorner\!\rightarrow \mu^+ + \nu_\mu + \begin{pmatrix} n \\ \text{or} \\ \lambda \end{pmatrix} \tag{72}$$

This mechanism produces $(\mu\mu)$ pairs of opposite signs, with the μ^- usually having higher momentum. For $x \gtrsim 0.1$ the charm production would be off n quark partons, with a branching ratio for the production and decay (72).

$$\simeq \tan^2 \Theta_C \, \frac{\Gamma(\text{charm} \rightarrow \mu + \nu + \text{all})}{\Gamma(\text{charm} \rightarrow \text{all})} \simeq (\tfrac{1}{2} \text{ to } 1)\% \tag{73}$$

if we take the leptonic branching ratio to be 0(10 to 25) % as suggested in section 4. So far there is no gross inconsistency with the date: however the HPW-FNAL experiment also reports $\nu + N \rightarrow (\mu^-\mu^-) + x$ events at a few % of the $(\mu^+\mu^-)$ events. A source of $(\mu^-\mu^-)$ pairs in the charm model is from D^0-$\bar{D}^0$ mixing, which can only occur by intermediate $S = 0$ states. Decays to these states are suppressed in the conventional charm model with the GIM current (4.0) and calculations (we get something like: $\frac{\mu^-\mu^-}{\mu^+\mu^-} \sim \tan^4 \Theta_C \sim 10^{-3}$) give rates for $(\mu^-\mu^-)$ production way below the reported rate.

Another source of dimuon events of the same sign is the associated production of a charmed pair:

$$\nu + N \rightarrow \mu^- + C + \bar{C} + \cdots \qquad \llcorner\!\rightarrow \mu^- + \nu$$

While we have no way of estimating charmed pair production in neutrino reactions, strange particle pair production in neutrino reactions is known to be non-negligible (15 %). Then in order to explain the ratio $\frac{\mu^-\mu^-}{\mu^-\mu^+} \sim 0.1$, we must assume that charmed pair production is about 1 % of the total neutrino cross section above, say 40 GeV. But then, we are going to have some important consequences:

1) Trimuon events of the type

$$\nu + N \rightarrow \mu^- + c + \bar{c} + \cdots \quad (c \rightarrow \mu^- + \bar{\nu},\ \bar{c} \rightarrow \mu^+ + \nu)$$

must exist at the level of $O(10^{-1})$ of the $(\mu^-\mu^-)$ events.

2) A certain fraction of the $\mu^+\mu^-$ events come from the before mentioned procedure.

3) Existence of dimuons in deep-inelastic μ-nucleon scattering at comparable energies.

Thus, if these $(\mu^-\mu^-)$ events are confirmed and the consequences 1) - 3) are not fulfilled, it will be a tragedy for the naive charm model.

There are other data from neutrino experiments which may indicate charmed particle production. A candidate has been reported[49] for an event of the type $\nu + N \rightarrow \mu^- + \Lambda^0 + 3\pi^+ + \pi^-$, which would have $\Delta S = -\Delta Q$, contrary to the usual current (1). But consistent with charmed baryon production and decay via the current (4). The neutrino energy in this event was calculated to be 13.5 GeV if all the neutrinos above a threshold around 4 to 5 GeV are counted, and a branching ratio $\sim \tan^2\theta_c$ assumed for charmed particle production, then this experiment might have been expected to produce 1 or 2 events. The probability that a charmed particle would identify itself as this candidate is however very low, $\lesssim 10$ % and quite possibly ~ 1 %. Therefore, there are three possible interpretations of the $\Delta S = -\Delta Q$ candidate: either the charm production cross section is much larger than expected, or the experimentalists were lucky in having an exceptionally visible decay and we could wait a long time for another event, or the event has nothing to do with charm. A recent FNAL bubble chamber experiment with considerably greater statistics has apparently seen no similar events, thus apparently ruling out the first possibility. They do however see an apparent signal in $\nu + p \rightarrow \mu^- + X^{++}$, where the missing mass of the incompletely observed system X^{++} is estimated indirectly, and seems to have an apparent peak but at a mass different from that found for the $\Delta S = -\Delta Q$ candidate. Finally the Gargamelle group reports a candidate for an event of the type

$$\nu + N \rightarrow \mu^- + e^+ + (K^0 \text{ or } \Lambda^0) + \cdots$$

which could be the production and semileptonic decay of a charmed particle.

A process of particular interest is the diffractive neutrino production of F^*, which is a $1^- c\bar{\lambda}$ bound state:

$$\nu + N \rightarrow \mu^- + F^{*+} + N \ .$$

In Fig. 21 is shown a figure from the paper of Gaillard et al[51]. What is plotted is the diffractive vector and axial vector boson production cross sections as fractions of the total neutrino cross section. Roughly the ratio of the ρ and

F^* cross sections is given by:

$$\left[\frac{\sigma_{tot}(F^*N)}{\sigma_{tot}(\rho N)}\right]^2 \frac{\gamma_\rho^2}{\gamma_{F^*}^2} = \left(\frac{5\ mb}{26\ nb}\right)^2 \cdot \left(\frac{m_\rho}{m_{F^*}}\right)$$

Thus we expect that the F^* diffractive production is about 2×10^{-3} of the total neutrino cross section. However, near the effective threshold of charmed particle production, i. e. at the energy range where deep inelastic, charmed particle production cross section begins to scale, the diffractive F^* production may be important, indeed dominant, source of charmed particles in the final state.

For more details and references on this matter and the use of diffractive neutrino production of vector mesons to deduce the quantum numbers of neutral currents see ref. 51.

5.3. Deep Inelastic Electroproduction

If the nucleon contains some charmed partons then deep inelastic interactions of the type

$$\begin{pmatrix} e \\ \text{or} \\ \mu \end{pmatrix} + \begin{pmatrix} c \\ \text{or} \\ \bar{c} \end{pmatrix} \rightarrow \begin{pmatrix} e \\ \text{or} \\ \mu \end{pmatrix} + \begin{pmatrix} c \\ \text{or} \\ \bar{c} \end{pmatrix}$$

shown in Fig. 22 are possible. They presumably would not be significant at SLAC energies, but should show up at FNAL. As discussed earlier, conventional wisdom would place the c and $\bar{c}$ quarks in the "sea" at $x \lesssim 0.1$. In this region the change in the electroproduction would be

$$\frac{\Delta\sigma}{\sigma} = \frac{4/9\,(c+\bar{c})}{4/9\,(p+\bar{p}) + 1/9\,(n+\bar{n}+\lambda+\bar{\lambda})} \tag{75}$$

Assuming that $p = \bar{p} = n = \bar{n} = \lambda = \bar{\lambda}$ at $x = 0$, then

$$\frac{\Delta\sigma}{\sigma} = \frac{2}{3}\left(\frac{c}{p}\right) \quad \text{as } x \rightarrow 0 \tag{76}$$

Estimating the ratio (c/p) is not easy: presumably it is $\lesssim 1$, and probably charm production in deep inelastic is considerably easier than in photoproduction where the cross section is estimated in the next subsection to be O(1) %. An estimate of $\frac{\Delta\sigma}{\sigma}$ of the order of (10 to 20) % seems not unreasonable.

Data on deep inelastic muon scattering at FNAL seem to have qualitatively the right features: for $x \gtrsim 0.2$ the data may well decrease at high (q^2) relative to scaling, whereas for $x \lesssim 0.1$ they seem to be higher than the SLAC data by O(20) %. The data at $E_\mu = 56$ GeV and 150 GeV in the FNAL experiment seem to scale relative to one another, but both are higher than SLAC. It is very possible that this increase has nothing to do with new particle production, and may occur because the SLAC data had not scaled in this region of x.

In each event where the deep inelastic production takes place off a c (or $\bar{c}$) quark parton, there are probably at least two charmed particles in the final state. The struck c (or $\bar{c}$) quark is sent given a large momentum, while a $\bar{c}$ (or c) quark is left behind at low momentum. When the state evolves into hadrons, non-local compensation of the charm quantum number is not believed to take place. Hence the final state should contain a fast $c = +1$ (or $c = -1$) particle, and a slow $c = -1$ (or $c = +1$) particle, as indicated in Fig. 22. Either of these particles may decay semileptonically giving a dilepton event: events with equal and opposite signs for the lepton pair are equally probable. From (76) we see that at small x in deep inelastic $\mu^{\pm}$ scattering

$$\frac{\sigma(\mu^{\pm}\mu^{-})}{\sigma(\mu^{\pm})} \approx \frac{\sigma(\mu^{\pm}\mu^{+})}{\sigma(\mu^{\pm})} \approx \frac{2}{3}\left(\frac{c}{p}\right) B \tag{77}$$

where B is the semileptonic branching ratio for charmed particles, estimated in section 4 to be 0(10 to 25) %. As yet there seems to be no clear statement from the FNAL group about the rate of prompt dimuon production, except that it seems[8] to be less than in deep inelastic neutrino production.

5.4. Hadronic Colisions

The first task is to estimate the total inclusive cross section for the production of charmed particles. The inclusive charges π cross section is 0(300)mb at ISR energies (recall that each inelastic event produces of the order of 10 charged pions): the charged kaon inclusive cross section is one or two orders of magnitude less, and in qualitative agreement with the prediction of the thermodynamical model that the production rate of a particle of mass m is proportional to $e^{-m/T}$ with $T \approx m_{\pi}$. This model also predicts a p_T distribution at high energies of the form $e^{-p_T/T}$ with $T = 0(150)$ MeV, which is known to be valid for $p_T \lesssim 1$ GeV, but to break down for $p_T > 1$ GeV, and is probably replaced by some sort of power law. Let us assume without justification that the thermodynamical predictions for large p_T and heavy mass production are violated in the same way. We then get

$$\frac{\sigma_{\text{inclusive}}(m)}{\sigma_{\text{inclusive}}(\pi)} \approx \frac{\sigma^{\pi}_{\text{inclusive}}(p_T)}{\sigma^{\pi}_{\text{inclusive}}(p_T = 0)} \tag{78}$$

In the case of charm it is unclear whether we should take $p_T = m_D$ or $2\ m_D$ on the right-hand side because charmed particles have to be produced in pairs.

However, the cross section for $p + p \to 3.1 +$ anything seen at Fermilab seems to be $0(1/10)\mu b$ corresponding to choosing $p_T \approx 3$ in the formula (78) even though the majority of the final states in 3.1 production should also contain charmed-anticharmed particle pairs. Hence $0(1/10)\mu b$ should be a lower bound on the charmed particle cross section. Taking (78) with $p_T \simeq m_D \simeq 2$ GeV and using ISR inclusive date[53] on large p_T phenomena we get

$$\sigma_{incl.}(m=2) : \sigma_{incl.}(\pi) \approx 10^{-4\frac{1}{3}} \tag{79}$$

corresponding

$$\sigma_{incl.}(m=2) \approx 10\,\mu b \tag{80}$$

This may well be an overestimate, but charmed particle cross sections of $O(1)\,\mu b$ in hadronic collisions seem not impossible.

At energies $\lesssim$ 30 GeV the inclusive charmed particle production cross section is presumably considerably lower (of the rise in the 3.1 production cross section between BNL and FNAL-ISR energies). This probability combined with the expected low branching ratio for charmed particles decaying into two meson systems means that upper limits[54] in the production of narrow mass groups are not yet worrying for fans of charm.

What are the prospects of enhancing the proportion of charmed particle events by using a selective trigger on the multiparticle events? One way would be to trigger on a 3.1 particle, as mentioned above. A second possibility[30] is to look for events with a μ or e coming from the decay of one charmed particle and a K coming from the decay of the other one. However, if the preferred decay modes of charmed particles are multibody with an inclusive K/π ratio $\simeq 1/4$, these events may not be very distinctive. Another possibility[30] is to look for $(\mu - e)$ coincidences generated by leptonic decays of both the charmed particles. It may also be interesting to look at large p_T events. The cross section there is very low, but the K/π ratio is larger at large p_T than at small p_T, and the same may be true of the D/π ratio.

There are already indications that new things are happening at large p_T: the inclusive electron and muon distributions at large p_T have $e/\pi \approx \mu/\pi \approx 10^{-4}$. Is it possible they could come from the decays of charmed particles? If we use the formula

$$E\frac{d\sigma}{d^3p} \propto (m^2 + p_T^2)^{-4} \tag{81}$$

which is qualitatively valid for familiar hadrons in the central region then the decays $D \to K\ell\nu$ give

$$\left(\frac{\ell}{\pi}\right) \approx B\left(\frac{D}{\pi}\right)\left(\frac{1}{300} \leftrightarrow \frac{1}{200}\right) \qquad B \sim (5 \text{ to } 10)\% \tag{82}$$

where B is the branching ratio. With $B \sim (5 \text{ to } 10)\,\%$, the formula (82) can succeed in reproducing the observed lepton spectrum if the p_T (D/π) ratio is $O(1/2 \text{ to } 1/5)$. Other sources, such as the vector mesons ρ, ω, ϕ and 3.1, are estimated to give $\lesssim 1/2$ of the observed high p_T leptons.

5.5. Conclusions

In the first two sections we discussed the origin of charm and the constraint that the $\Delta S = 1$, $\Delta Q = 0$ processes and K_L-K_S mass difference imply.

In the next two we discussed theoretical properties of hidden charmed and visibly charmed states. Finally in this section we have gone through the predictions of the simple charm model for different experimental reactions, and tried to compare them with the data. It is appropriate to briefly list what seems to be the principal problems of the charm model in this comparison, and try to asses their gravity.

The Widths of the 3.1 and 3.7[16]

No-one has yet produced a completely acceptable explanation for the exceedingly narrow widths of the new resonances. Probably they have something to do with their high masses, but how exactly? After the initial shock people seem to have accepted and forgotten about this problem.

No Charmed Particles Have Been Seen[41,54]

Experiments such as e^+e^- annihilation now require that the branching ratios of charmed particles into two pseudoscalar mesons be $\lesssim O(2)$ % (e.g. $D^o \to K^+ \bar{\pi}$). However comparison with other systems with masses O(2) GeV, and theoretical calculations[35] suggest O(1) % as a reasonable two particle branching ratio. Hence we should not get very worried just yet.

$R \simeq 5$ is Too Big[16]

This is perhaps the biggest problem of the charm model: why is $R \sim 5$ and rising if anything, instead of $R \sim 4$ and slowly falling? This problem may well kill the simple charm model: perhaps heavy leptons are necessary.

The $e^+e^- \to e^\pm\mu^\mp + \ldots$ Events[40]

There are surprisingly many of these events, unless we accept the highest semileptonic branching ratios of charmed particles argued for in section 4. It is by no means clear that semileptonic decays of pseudoscalar mesons are consistent with the energy and angular distributions of these events. Heavy leptons?

νN and $\bar{\nu}N$ Scattering at Small x [45]

As discussed earlier in this lecture, the y-distribution of deep inelastic ν and $\bar{\nu}$ scattering data at small x are incompatible with the simple parton model with charm. Either the model or the data have to change.

The FNAL $\nu N \to (\mu\mu)$ x Events [47]

There are too many $\nu N \to (\mu^+\mu^-)$ X events unless one buys the large semi-leptonic branching ratio for charmed particles proposed in section 4. If the observed rate of $\nu N \to (\mu^-\mu^-)$ X events is confirmed but the consequences of these events in the charm model are not seen, then the simple charm model will have to be modified.

The $\Delta S = -\Delta Q$ Event in νp scattering [49]

While this event is very beautiful, it should not have seen if it is charmed particle production and decay. This experiment might have been expected to produce 1 or 2 charmed particles, but the probability they would be visible is just a few %. Either the experimentalists were very lucky or the charmed particle production cross section is much higher than expected, or the $\Delta S = -\Delta Q$ candidate has nothing to do with charm.

Directly Produced Leptons at Large p_T [59]

Even with the largest semileptonic branching ratios proposed in section 4, if the observed large p_T leptons are due to charmed particle decay, then the (D/π) ratio at large p_T must be O(1/5 to 1/2). With the more conventional lower branching ratio, the (D/π) ratio would need to be even higher: unreasonably high?

The above list is not a short one, and some of the problems may become serious enough to rule out the simple SU(4) charm model with weak interactions as in the GIM paper[8]. However, it seems that in zeroth order many of the qualitatively right things are happening - anomalous prompt lepton production in a variety of different processes, anomalous distributions in deep inelastic neutrino production, and so on. However, the phenomena sometimes look not quite right when examined more closely - too many leptons, possibly of the wrong charges, peculiar y-distributions in neutrino production, and so on. Perhaps we are asking questions of new data that are too sophisticated, or perhaps we are in for more theoretical and experimental surprises.

References

1. G.t' Hooft, Nucl. Phys. B33 (1971) 173; ibid
B35 (1971) 167.
For a recent review on gauge theories see:
J. Iliopoulos - Proceedings of the XVIIth International Conference on High Energy Physics, London 1974, edited by J.R. Smith, p. III-89 (S.R.C., Rutherford Laboratory, 1974);
B.W. Lee - Invited talk presented at the International Symposium on Lepton-Photon Interactions at High Energies, Stanford 1975, Fermilab-Conf-75/2-THY (1975).

2. S. Weinberg, Phys. Rev. Lett. 19 (1967) 1264;
A. Salam in "Elementary Particle Physics", edited by N. Svartholm, Stockholm 1968, p. 367.

3. C.H. Lewellyn-Smith, Phys. Lett. B46 (1973) 233;
J.M. Cornwall, D.N. Levin and G.Tiktopoulos, Phys. Rev. Lett. 30 (1973) 1268.

4. F. Englert and R. Brout, Phys. Rev. Lett. 13 (1964) 231;
P.W. Higgs, Phys. Lett. 12 (1964) 132, Phys. Rev. Lett. 13 (1964) 508, Phys. Rev. 145 (1966) 1156;
G.S. Guralnik, C.R. Hagen and T.W.B. Kibble, Phys. Rev. Lett. 13 (1964) 585;
T.W.B. Kibble, Phys. Rev. 155 (1967) 1554;
the phenomenon was known in non-relativistic physics, see for ex. P.W. Anderson, Phys. Rev. 112 (1958) 1900.

5. J. Ellis, M.K. Gaillard and D.V. Nanopoulos, "A Phenomenological Profile of the Higgs Boson", CERN-TH-2093 (November 1975);
for a systematic analysis of the properties and experimental consequences of the Higgs scalars.

6. For the experimental status of neutrino interactions we refer to
J. Von Krogh, "Neutrino Interactions - Experimental Results" (these proceedings);
for the theoretical status of neutral currents we refer to
J.J. Sakurai, "Neutral Currents Without Gauge Prejudices" (these proceedings).

7. Y. Hara, Phys. Rev. B134 (1964) 701;
D. Amati, H. Bacry, J. Nuyts and J. Prentki, Phys. Lett. 11 (1964) 190;
J.D. Bjorken and S.L. Glashow, Phys. Lett. 11 (1964) 255;
Z. Maki and Y. Ohnuki, Progress of theor. Physics (Kyoto) 32 (1964) 144.

8. S.L. Glashow, J. Iliopoulos and L. Maiani, Phys. Rev. D2 (1970) 1285.

9. S. Weinberg, Phys. Rev. Lett 31 (1973) 494;
D.V. Nanopoulos, Nuov. Cim. Lett. 8 (1973) 873.

10. W. Bardeen, H. Fritsch and M. Gell-Mann, CERN preprint, CERN-TH 1538 (1972).

11. M.Y. Han and Y. Nambu, Phys. Rev. B139 (1965) 1006;
Y. Nambu and M.Y. Han, Phys. Rev. D10 (1974) 674.

12. C. Bouchiat, J. Iliopoulos and Ph. Meyer, Phys. Lett. 38B (1972) 519.

13. G. 't Hooft, Nucl. Phys. B61 (1973) 455, B62 (1973) 444;
H.D. Politzer, Phys. Rev. Lett. 30 (1973) 1346;
D.J. Gross and F. Wilczek, Phys. Rev. Lett. 30 (1973) 1343;
for applications of asymptotic freedom to electroproduction and neutrino production see
D.J. Gross and F. Wilczek, Phys. Rev. D8 (1973) 3633, D9 (1974) 980;
H. Georgi and H.D. Politzer, Phys. Rev. D9 (1974) 416;
D. Bailin, A. Love and D.V. Nanopoulos, Nuov. Cim. Lett. 9 (1974) 501.

14. S. Weinberg, Phys. Rev. Lett. 29 (1972) 388, 29 (1972) 1698;
H. Georgi and S.L. Glashow, Phys. Rev. D6 (1973) 2977, D7 (1973) 2457;
H. Georgi and A. Pais, Phys. Rev. D10 (1974) 539;
for an excellent review see
S. Weinberg, Rev. of Mod. Phys. Vol. 46 (1974) 255.

15. D.V. Nanopoulos and G.G. Ross, Phys.Lett. 59B, 475 (1975)

16. For the experimental status of e^+e^- annihilation and new particles see
G. Wolf, "Recent Experimental Results on e^+e^- Annihilation with Emphasis on the New Particles" (these proceedings).

17. C.G. Callan and D.J. Gross, Phys. Rev. D8 (1973) 4383.

18. S. Weinberg, Phys. Rev. D8 (1973) 605, ibid D8 (1973) 4482.

19. M.T. Grisaru, H.J. Schnitzer and H.S. Tsao, Phys. Rev. Lett. 30 (1973) 811, Phys. Rev. D8 (1973) 4498;
also see
S.Y. Lee, J.M. Rawls and D.Y. Wong, "Reggeization as Spin 1 Particles in General Lagrangian Models" to be published;
for a possible connections of Yang-Mills' field theories with dual models see
A. Neveu and J. Scherk, Nucl. Phys. B36 (1972) 155;
J.L. Gervais and A. Neveu, Nucl. Phys. B46 (1972) 381;
see also
J. Scherk, Nucl. Phys. B31 (1971) 222.

20. H. Lipkin, Phys. Lett. B45 (1973) 267;
H. Fritzsch, M. Gell-Mann and H. Leutwyler, Phys. Lett. B47 (1975) 365.

21. M.K. Gaillard and B.W. Lee, Phys. Rev. Lett. 33 (1974) 108;
G. Altareli and L. Maiani, Phys. Lett. B52 (1974) 351.

22. For a detailed analysis of applications of the "Standard Model" to the new particles see
A. De Rujula, "Asymptotic Freedom. The Discreet Charm of the New Particles" (these proceedings).

23. J.C. Pati and A. Salam, Phys. Rev. D8 (1973) 1240;
H. Georgi and S.L. Glashow, Phys. Rev. Lett. 32 (1974) 438;
H. Georgi, H.R. Quinn and S. Weinberg, Phys. Rev. Lett. 33 (1974) 451.

24. G. t' Hooft and M.Veltman, CERN preprint, CERN-TH-1723 (1973);
S. Deser, H.S. Tsao and P. van Nieuwenhuizen, Phys. Rev. Lett. 32 (1974) 245, Phys. Rev. D10 (1974) 3337, Phys. Lett. B50 (1974) 491 etc.

25. M.K. Gaillard and B.W. Lee, Phys. Rev. D10 (1974) 897;
see also
A.I. Vainshtein and I.B. Khriplovich, JETP Letters 18 (1973) 141;
V.V. Flambaum, Novosibirsk Institute of Nucl. Phys. Prepr. (1975).

26. A. De Rujula, H. Georgi and S.L. Glashow, Phys. Rev. Lett. 35 (1975) 69, "Vector Model of the Weak Interactions", Harvard preprint.

27. D.V. Nanopoulos and G.G. Ross, Phys. Lett. B56 (1975) 279.

28. E.B. Bogomolny, V.A. Novikov and M.A. Shifman, Moscow Preprint ITEP-42 (1975);
A.I. Vainshtein, V.I. Zakharov, V.A. Novikov and M.A. Shifman, Moscow Preprint ITEP-44 (1975).

29. M.K. Gaillard, B.W. Lee and R.E. Schrock, CERN-TH 2066, Fermilab Pub 75/68-THY preprint.

30. M.K. Gaillard, B.W. Lee and J.L. Rosner, Rev. of Mod. Physics Vol. 47, (1975) 277.

31. J. Ellis, Acta Physica Austriaca Supp. XIV, p. 143 (1975).

32. G. Altareli, N. Cabibbo and L. Maiani, Nucl. Phys. B88 (1975) 285, Phys. Lett. B57 (1975) 277;
R.L. Kingsley, S.B. Treiman, F. Wilczek and A. Zee, Phys. Rev. D11 (1975) 1919, Phys. Rev. D12 (1975) 106;
A. Pais and V. Rittenberg, Phys. Rev. Lett. 34 (1975) 707;
M.B. Einhorn and C. Quigg, Fermilab-Pub-75-21/THY to be published;
Y. Iwasaki, Phys. Rev. Lett. 34 (1975) 1407.

33. For a quick comprehencive review on group theory including SU(4) see M.B. Einhorn, Fermilab-Lecture-75/1-THY/EXP. (February 1975).

34. G. Altareli et al., Y. Iwasaki, Phys. Rev. Lett. 34 (1975) 1407.

35. J. Ellis, Mary K. Gaillard, D.V. Nanopoulos, Nucl. Phys. B100, 313 (1975)

36. K. Wilson, Phys. Rev. 179 (1969) 1499.

37. M.K. Gaillard and B.W. Lee, Phys. Rev. Lett. 33 (1974) 108;
G. Altarelli and L. Maiani, Phys. Lett. 52 (1974) 351.

38. S. Nussinov and J. Rosner, Phys. Rev. Lett. 23 (1969) 1266.

39. J.C. Pati and C.H. Woo, Phys. Rev. D3 (1971)2920, and references therein.

40. M.L. Perl et al., SLAC-PUB-1626 (1975) (LBL-4220);
M.L. Perl, SLAC-PUB-1592 (1975) (T/E)

41. A.M. Boyarski et al., Phys. Rev. Lett. 35 (1975) 195.

42. T. Appelquist and H. Georgi, Phys. Rev. D8 (1973) 4000;
A. Zee, Phys. Rev. D8 (1973) 4038.

43. M.B. Einhorn and C. Quigg, Phys. Rev. Lett. 35 (1975) 1114.

44. D.C. Cundy, Proceedings XVIIth International Conference on High Energy Physics, London 1974 (Rutherford Laboratory, U.K. 1974) p IV, 131.

45. D. Perkins, Rapporteur talk at the International Symposium on Lepton and Photon Interactions, Stanford 1975, Oxford University preprint (1975).

46. V. Barger, T. Weiler and R.J.N. Phillips, Wisconsin preprint (1975).

47. See for example the talks of B.C. Barish and C. Rubbia at the International Sympsium on Lepton and Photon Interactions, Stanford (1975).

48. A. Pais and S.B. Treiman, Phys. Rev. Lett. 35, 1206 (1975)
for a clear view of the dimuon events see
B.W. Lee, FERMILAB-CON-75/78 TH.

49. E.G. Cazzoli et al., Phys. Rev. Lett. 34 (1975) 1125.

50. H. Deden et al., Phys. Lett. B58 (1975) 361.

51. M.K. Gaillard, S.A. Jackson and D.V. Nanopoulos, CERN-preprint TH-2049, Nucl. Phys. B102, 326 (1975)

52. D.J. Fox et al., Phys. Rev. Lett. 33 (1974) 1504;
K.W. Chen, talk given at the FERMILAB Muon Physics Workshop, January 1975;
K. Gottfried et al., proposal P-382 to the FERMILAB P.A.C. (1975).

53. M. Frisch, Proceedings of the XVIIth International Conference on High Energy Physics, London 1974, p. V-8.

54. For the experimental status of production of new particles in hadronic reactions see
S.C.C. Ting, "Production of J Particles in Hadron Interactions" (these proceedings).

55. As Ref. 53, p. V-41.

56. M. Bourquin and J.M. Gaillard, Phys.Lett. 58B, 191 (1975).

Acknowledgements

During the last year I have benefited a lot from discussions with John Ellis, Mary K. Gaillard, Jaques Prentki and Graham Ross, to which I would like to express my sincere thanks.

Also John Ellis and Mary K. Gaillard deserve again my thanks for their help in preparing these lectures.

Finally, I would like to thank my wife, Myrto, for her patience and encouragement while I was preparing these lectures during my summer vacation in Greece.

Tables

TABLE 1 Experimental Information on $\Delta S = 1$, Neutral Current

Decay mode: Neutral	Decay mode: Charged	$\frac{\Gamma(\text{neutral})}{\Gamma(\text{charged})}$
$K_L \to \mu^+\mu^-$	$K^+ \to \mu^+\nu_\mu$	$(4.5^{+3}_{-1.5}) \times 10^{-9}$
$K_S \to \mu^+\mu^-$		$< 6.6 \times 10^{-5}$
$K_L \to e^+e^-$	$K^+ \to e^+\nu_e$	$< 3 \times 10^{-5}$
$K^+ \to \pi^+e^+e^-$	$K^+ \to \pi^o e^+\nu_e$	$(4.8 \pm 1.7) \times 10^{-6}$
$K^+ \to \pi^+\nu\bar{\nu}$		$< 1.2 \times 10^{-5}$
$K^+ \to \pi^+\mu^+\mu^-$	$K^+ \to \pi^o\mu^+\nu_\mu$	$< 7.5 \times 10^{-5}$
$\Sigma^+ \to pe^+e^-$	$\Sigma^- \to ne^-\nu_e$	$< 10^{-2}$

TABLE 2 Comparison of Asymptotically Free Gauge Theories Predictions for 3 Quark-Quartet Model with Experiment

Process	Branching ratio Prediction (AFGT) $m^2_{\rho'} = 30\ GeV^2$	Experimental rate
$K_L \to \mu\bar{\mu}$	$\sim 10^{-8}$	10^{-8}
$K^+ \to \pi^+\nu\bar{\nu}$	3×10^{-10}	$< 5.6 \times 10^{-7}$
$K^o_L \to \pi^o\nu\bar{\nu}$	Forbidden	-
$K^o_S \to \pi^o\nu\bar{\nu}$	2×10^{-12}	-
$K_L \to \gamma\gamma$	3×10^{-5}	$(4.9 \pm 0.4)10^{-4}$
$K^+ \to \pi^+\gamma\gamma$	3×10^{-8}	$\leq (3.5)10^{-5}$
$K^o_L \to \pi^o\gamma\gamma$	$\sim 10^{-9}$	$< (2.4)10^{-4}$
$K^o_S \to \pi^o\gamma\gamma$	3×10^{-10}	-
$K^+ \to \pi^+e\bar{e}$	5×10^{-7}	3×10^{-7}
$K^o_L \to \pi^o e\bar{e}$	Strongly suppressed	-
$K^o_S \to \pi^o e\bar{e}$	2×10^{-9}	-

TABLE 3 Charmed $1/2^+$ Baryon States

Label		Quark Content	Isospin	Strangeness
C = 1	C_1^{++}	cuu	$T_z = 1$	
	C_1^+	$c(ud)_{sym}$	$T = 1,\ T_z = 0$	0
	C_1^o	cdd	$T_z = 1$	
	C_o^+	$c(ud)_{anti}$	$T = 0$	0
	S^+	$c(su)_{sym}$	$T = \frac{1}{2},\ T_z = \frac{1}{2}$	-1
	S^o	$c(sd)_{sym}$	$T_z = -\frac{1}{2}$	
	A^+	$c(su)_{anti}$	$T = \frac{1}{2},\ T_z = \frac{1}{2}$	-1
	A^o	$c(sd)_{anti}$	$T_z = -\frac{1}{2}$	
	T^o	css	$T = 0$	-2
C = 2	X_u^{++}	ccu	$T = \frac{1}{2},\ T_z = \frac{1}{2}$	0
	X_d^+	ccd	$T_z = -\frac{1}{2}$	
	X_s^+	dds	$T = 0$	-1

TABLE 4 Charmed 0^- Mesons

Label		Quark Content	Isospin	Strangeness
C = 1	D^+	$c\bar{d}$	$T = \frac{1}{2}$, $T_z = \frac{1}{2}$	0
	D^o	$c\bar{u}$	$T = \frac{1}{2}$, $T_z = -\frac{1}{2}$	
	F^+	$c\bar{s}$	T = 0	+1
C = 0	η	$\simeq \frac{1}{\sqrt{6}} (u\bar{u} + d\bar{d} - 2s\bar{s})$	T = 0	0
	η'	$\simeq \frac{1}{2} (u\bar{u} + d\bar{d} + s\bar{s} + c\bar{c})$	T = 0	0
	η'_c	$\simeq \frac{1}{\sqrt{12}} (u\bar{u} + d\bar{d} + s\bar{s} - 3c\bar{c})$	T = 0	0
C = -1	$\bar{D}^o$	$\bar{c}u$	$T = \frac{1}{2}$, $T_z = \frac{1}{2}$	0
	D^-	$\bar{c}d$	$T = \frac{1}{2}$, $T_z = -\frac{1}{2}$	
	F^-	$\bar{c}s$	T = 0	-1

TABLE 5 Charmed 1^- Mesons

C = 1	D^{*+}	
	D^{*o}	
	F^{*+}	
C = 0	ω	$\simeq \frac{1}{\sqrt{2}} (u\bar{u} - d\bar{d})$
	ϕ	$\simeq s\bar{s}$
	ϕ_c	$\simeq c\bar{c}$
C = -1	$\bar{D}^{*o}$	
	D^{*-}	
	F^{*-}	

Figure Captures

Figure 1 GIM[8] construction.

Figure 2 Diagrams for
(a) $K_L \to \mu^+\mu^-$,
(b) $K_L \to 2\gamma$.

Figure 3 Reduction of diagrams of Figure 2, after removal of the one boson propagator:
(a) $K_L \to \mu^+\mu^-$,
(b) $K_L \to 2\gamma$.

Figure 4 Another diagram contributing to $K_L \to \mu^+\mu^-$.

Figure 5 Diagrams contributing to the K_L-K_S mass difference.

Figure 6 Diagram contributing to the K_L-K_S mass difference in the De Rujula-Georgi-Glashow model[26].

Figure 7 Operator products expansion of four currents relevant to the K_L-K_S mass difference calculation. Dotted lines are colour gluons.

Figure 8 The diagrams of Figure 5 in the limit
(a) $|x - y| \to 0$, $|z| \to 0$,
(b) $|x - z| \to 0$, $|y| \to 0$.

Weight diagrams for SU(4)[30]. Shaded planes denote multiplets of $SU(3) \otimes U(1)_C$.

Figure 9 The four quarks of SU(4): the conventional SU(3) triplet consisting of u ("up"), d ("down") and s ("strange") with C = 0, and an SU(3) singlet c ("charmed") with C = 1.

Figure 10 The three-quark $1/2^+$ baryons which form a 20-plet of SU(4). The SU(3) multiplets are $\underset{\sim}{8}$(C=0), $\underset{\sim}{6} + \underset{\sim}{3}$(C=1) and $\underset{\sim}{3}$(C=2).

Figure 11 The 15-plet + singlet pseudoscalars. The SU(3) multiplets are $\underset{\sim}{3}$(C=-1), $\underset{\sim}{8} + \underset{\sim}{1}$(C=0) and $\underset{\sim}{3}$(C=+1).

Schematic representation of charmed pseudoscalar decays. The arrow points from the parent particle to a state with the quantum numbers of the final state hadronic system[30].

Figure 12 Semi-leptonic decays with $\Delta S = \Delta C$ ($\Gamma \propto \cos^2\theta_c$).

Figure 13 Inclusive semi-leptonic decay of a charmed meson.

Figure 14 Non-leptonic decays with $\Delta S = \Delta C$ ($\Gamma \propto \cos^4\theta_c$).

Figure 15 Non-leptonic decays with $\Delta S \neq \Delta C$ ($\Gamma \propto \cos^2\theta_c \ \sin^2\theta_c$).

Figure 16 Inclusive non-leptonic decay of a charmed meson.

Figure 17 $(\bar{q}q)$ spurion contribution to the decay (59).

Figure 18 $(\bar{q}q\bar{q}q)$ spurion contribution to the decay (59).

Figure 19 (a) $(\bar{q}q)$ and
(b) $(\bar{q}q\bar{q}q)$ spurions.

Figure 20 The SPEAR data[16] on $R = \dfrac{\sigma(e^+e^- \to \text{hadrons})}{\sigma(e^+e^- \to \mu^+\mu^-)}$

Figure 21 Diffractive vector and axial vector boson production cross sections as fractions of the total neutrino cross section[51].

Figure 22 Deep inelastic muon scattering of a c or $\bar{c}$ quark parton.

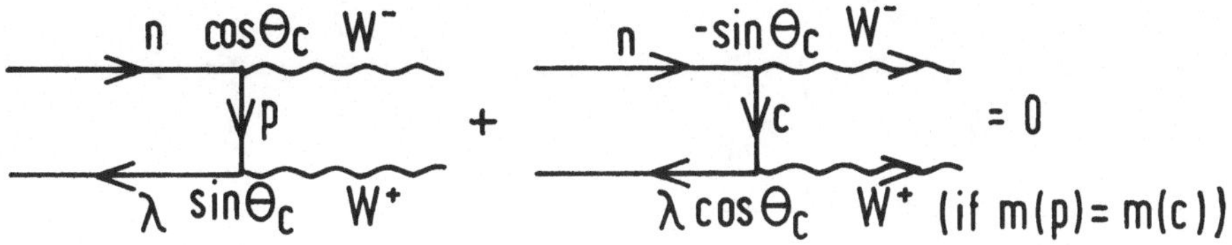

Fig.1

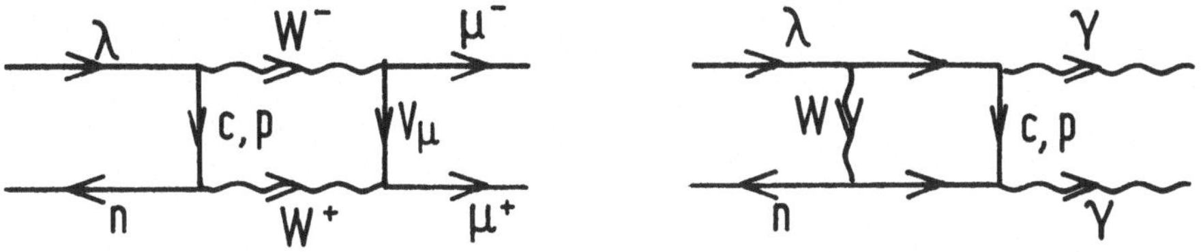

(a) (b)

Fig.2

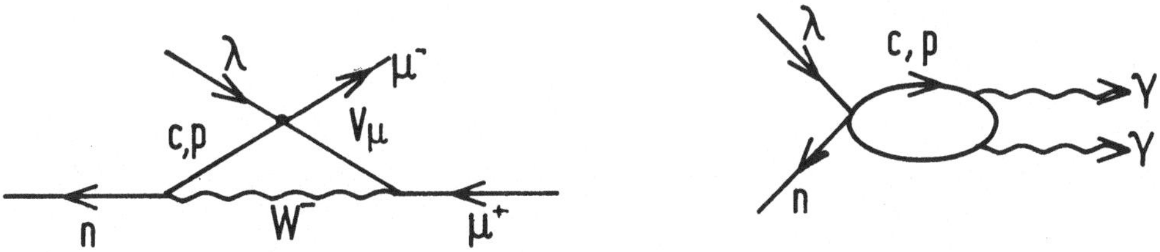

(a) (b)

Fig.3

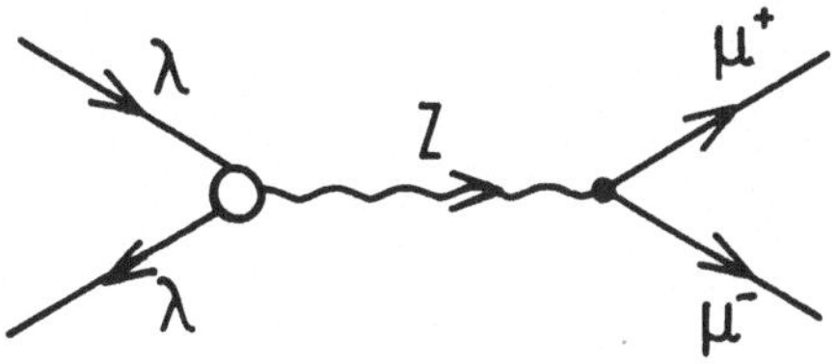

Fig.4

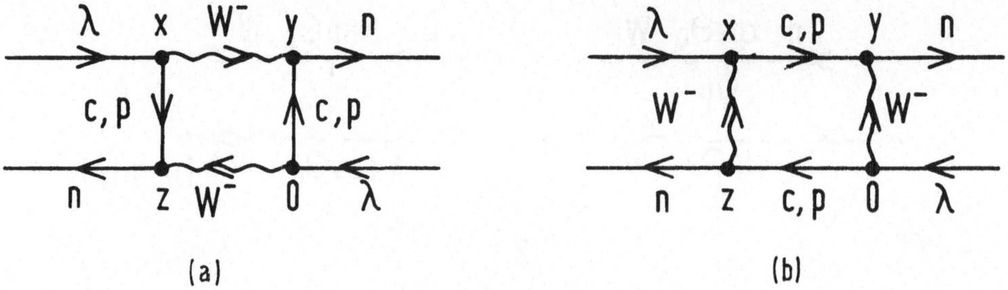

Fig.5

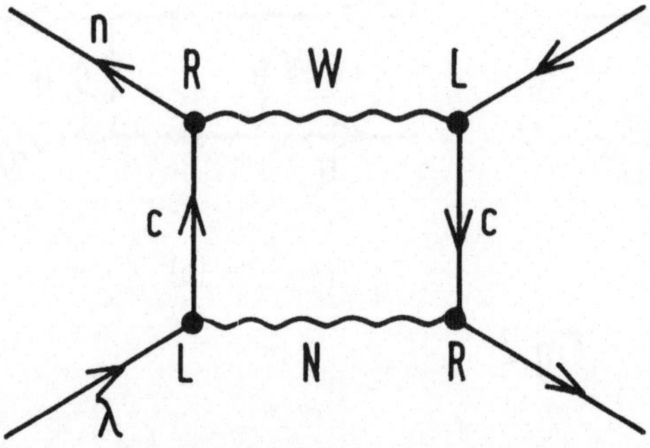

Fig.6

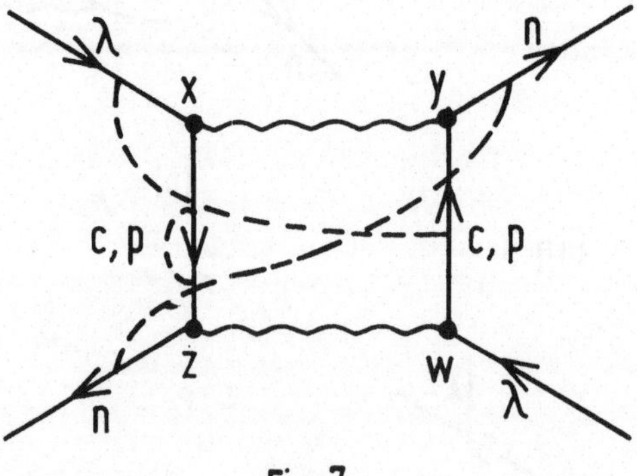

Fig.7

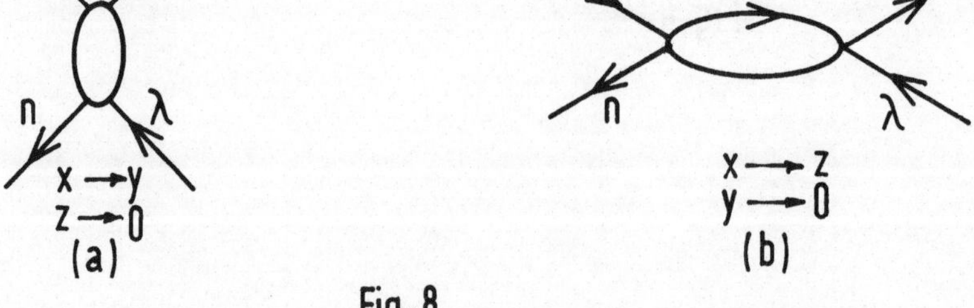

Fig. 8

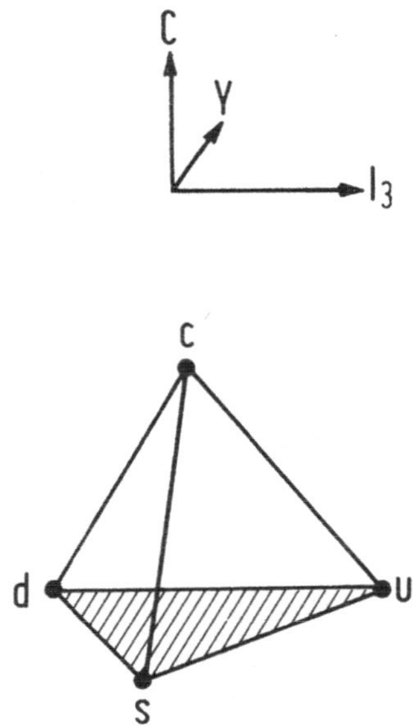

Fig.9

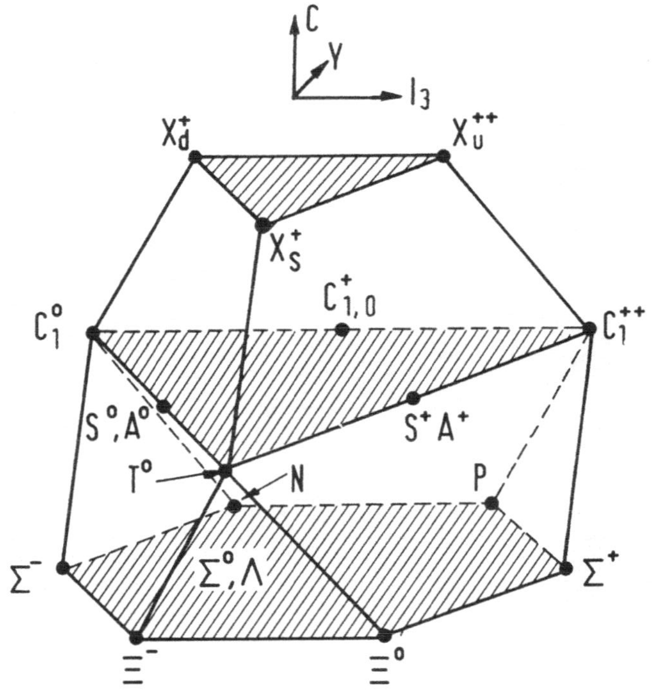

Fig.10

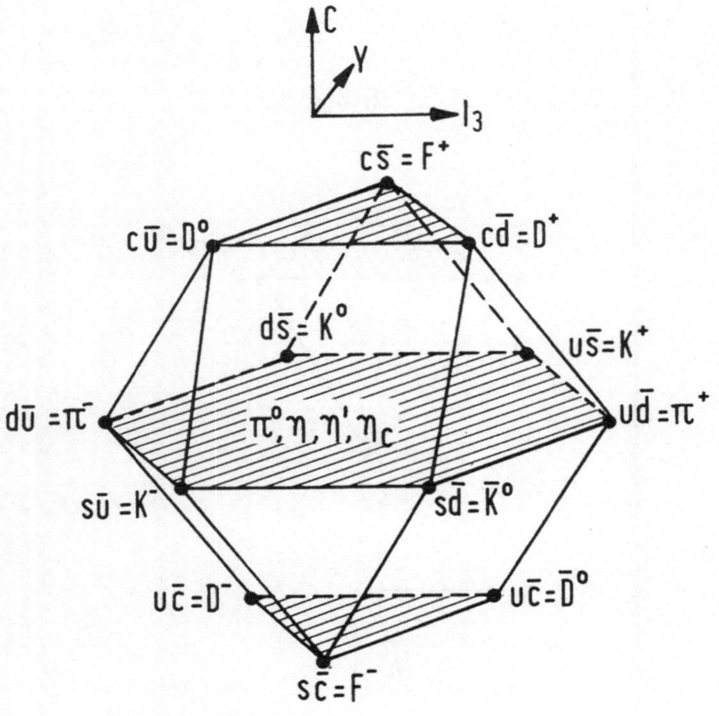
C
Y
I3
cs̄ = F+
cū = D0
cd̄ = D+
ds̄ = K0
us̄ = K+
dū = π−
π0, η, η', ηc
ud̄ = π+
sū = K−
sd̄ = K̄0
uc̄ = D−
uc̄ = D̄0
sc̄ = F−

Fig.11

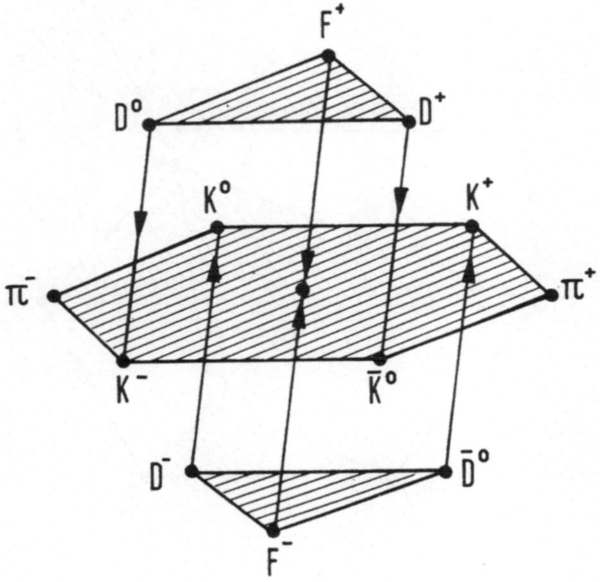
F+
D0
D+
K0
K+
π−
π+
K−
K̄0
D−
D̄0
F−

Fig.12

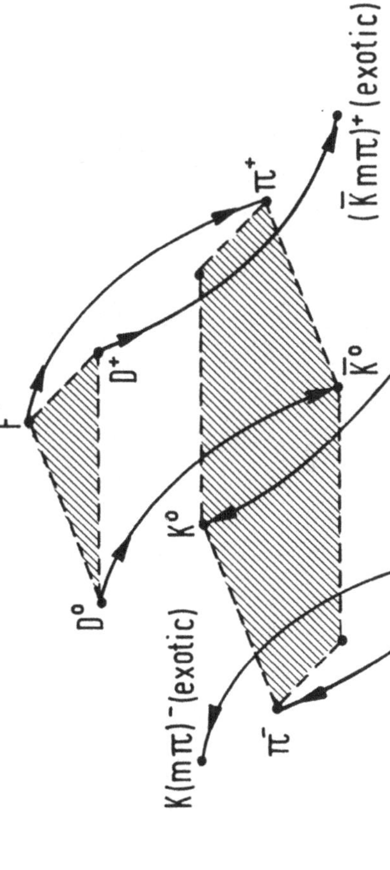
D^0
F^+
D^+
π^+
K^0
$\bar{K}^0$
$\bar{D}^0$
D^-
F^-
π^-
$K(m\pi)^-$(exotic)
$(\bar{K}m\pi)^+$(exotic)

Fig.14

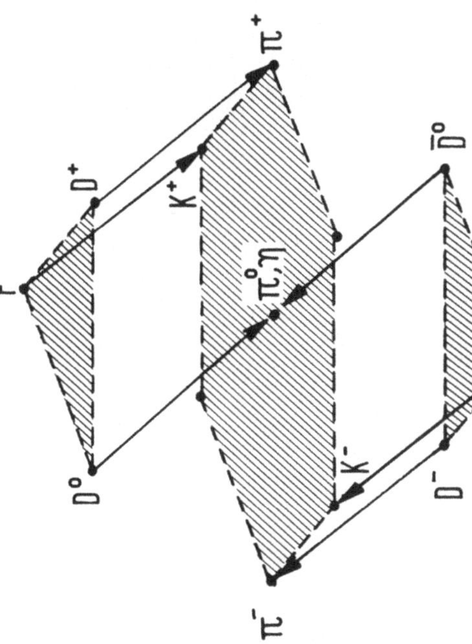
D^0
F^+
D^+
K^+
π^+
π^0,η
$\bar{D}^0$
F^-
D^-
K^-
π^-

Fig.15

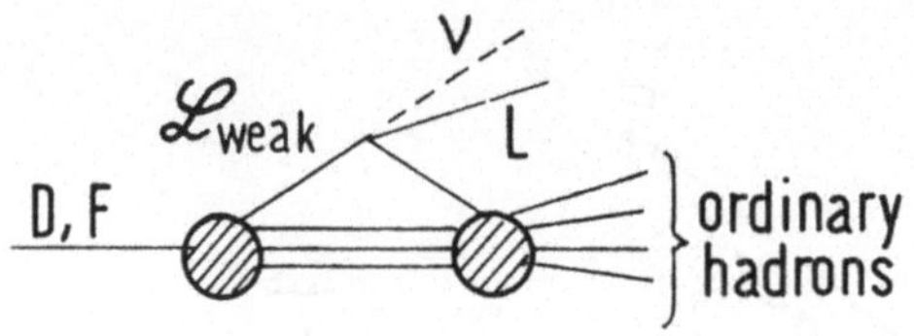

Fig. 13

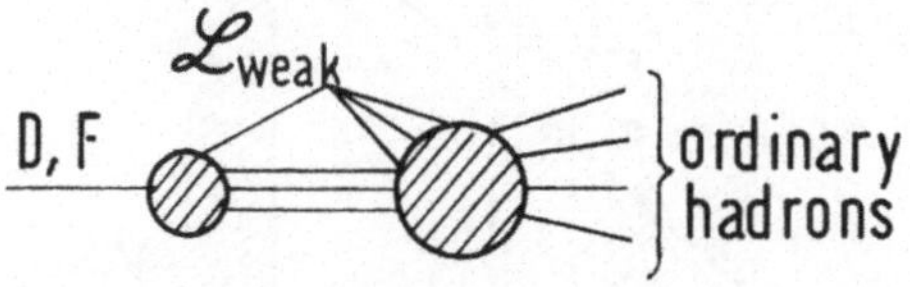

Fig. 16

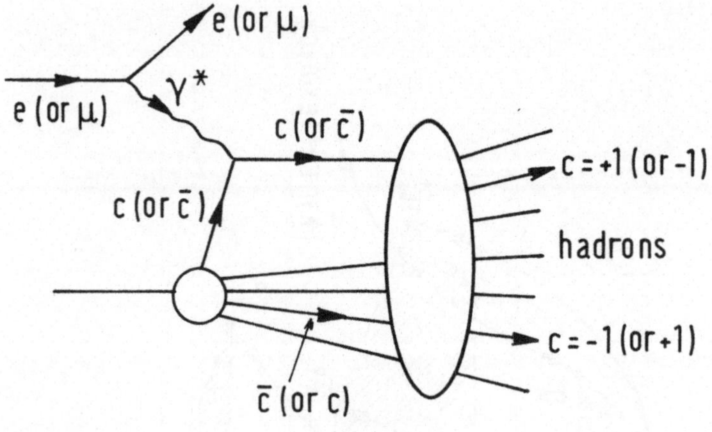

Fig. 22

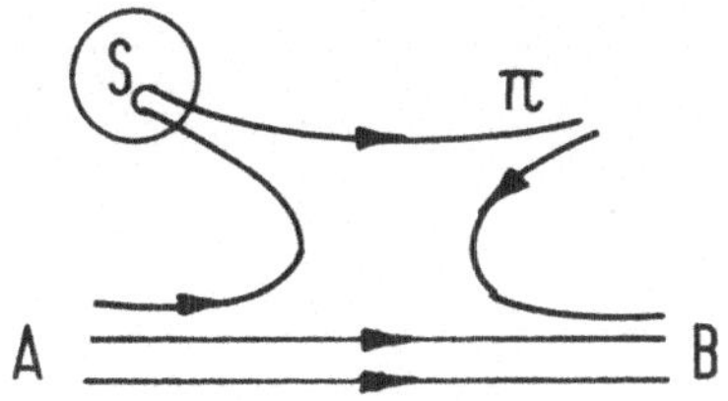

Fig.17

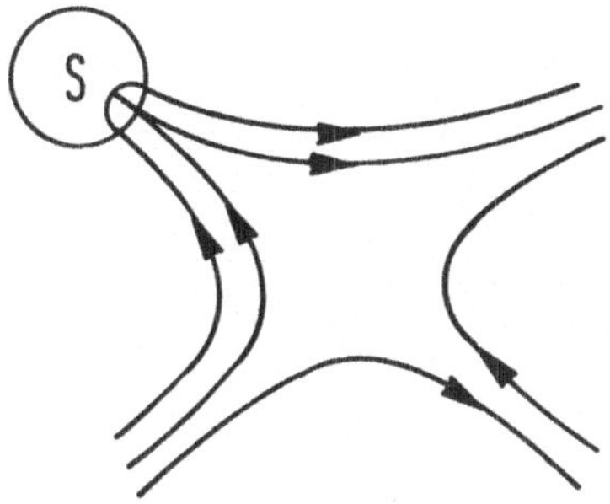

Fig.18

Fig. 19

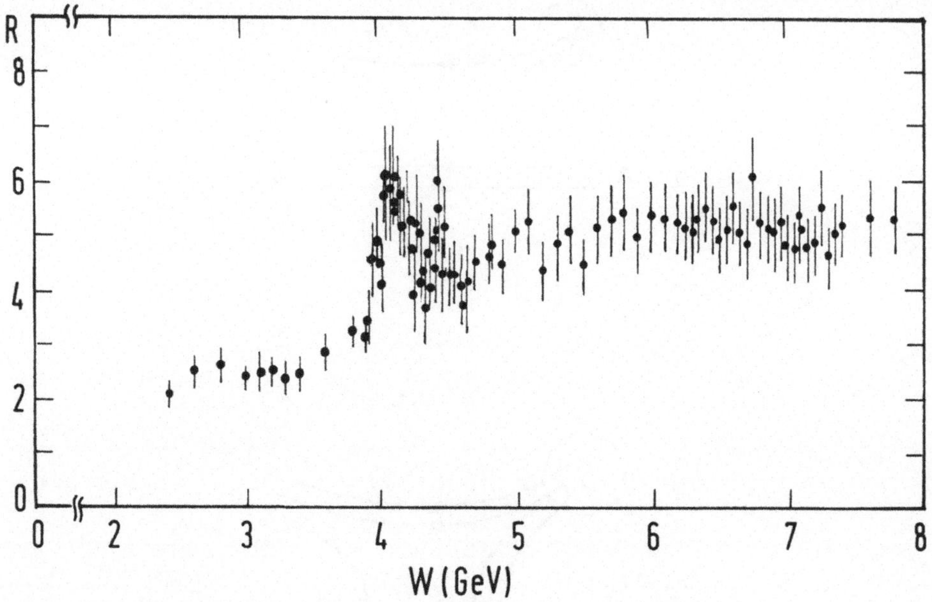

Fig.20

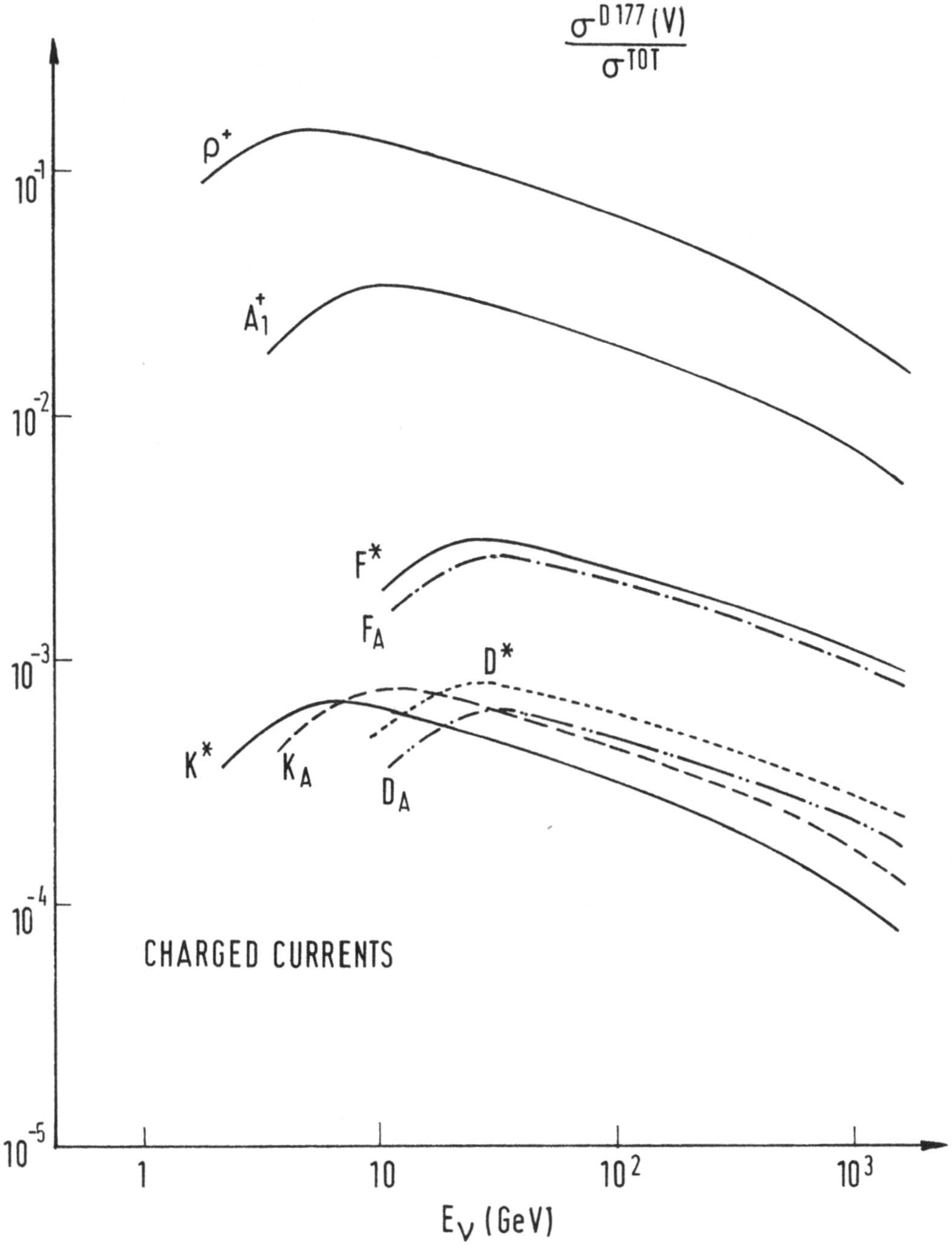

$\frac{\sigma^{D\,177}(V)}{\sigma^{TOT}}$
ρ^+
A_1^+
F^*
F_A
D^*
K^*
K_A
D_A
CHARGED CURRENTS
10^{-1}
10^{-2}
10^{-3}
10^{-4}
10^{-5}
1
10
10^2
10^3
E_ν (GeV)

Fig. 21

HADRON GEOMETRY AND QUARK COORDINATES

by

Giuliano Preparata
CERN - Geneva (Switzerland)

1. INTRODUCTION

Nobody today seems to question that the notion of matter constituents (Quarks) has developed into a flexible and powerful framework in which a multitude of experimental facts finds its natural collocation.

However precious little can be answered with any degree of confidence to the question of what really is a quark: this is precisely the central problem of high energy physics of this day.

In brief, the dilemma we are facing can be reduced to the question whether quarks are "bona fide" Quantum Fields (like, say, photons and electrons)(*), or else they are fictitious objects whose function is to carry the coordinates of a geometrical description of hadrons and their interactions(**).

It is certainly too early to say which of the two alternatives, if any, has a better chance to lead us somewhere. Thus I will not embark in the futile attempt to convince you of the "philosophical" advantages of the approach I shall follow in these lectures. As my motivation to choose an unconventional course I must mention my scepticism as to the possibility for a conventional Quantum Field Theory (QFT) to deal effectively with the formidable problems of Quark Confinement. My scepticism seems not to be shared by a great number of my colleagues [1], to their effort I can only sincerely wish good luck!

As it will become clear later, the quark is assumed to be the carrier of a very peculiar space-time representation of hadronic states. Very sketchily we are going to proceed as follows: let's take a pion; we are accustomed to think of it as an extended object roughly 1 Fermi across. Conventional QFT attributes its extension to a cloud of virtual hadrons which can couple to our pion. This picture, though suggestive, is essential hopeless: when we probe (by means, say, of electron scattering) our pion deeper and deeper we shake more and more junk off it, and it would look miraculous if some simple behaviour still manages to survive in the maze of multiparticle states we have brought to being.

We all know that this miracle actually occurs over and over again in deep

---

(*) This is the attitude of people like K. Wilson, who is trying to construct a QFT of quark confinement. See for example Ref.[1].

(**) This attitude is at the basis of the Massive Quark Model, which I have been developing since 1972 [2] .

inelastic lepton nucleon scattering from as low as 5 GeV up to 150 GeV. This suggests that the dynamics of the strong interactions even in the catastrophic highly inelastic processes remembers very clearly of some simple original structure. It is here that the quark concept, if it is worth anything, should play the crucial role of organizing hadronic matter in a well defined and simple way. In these lecture notes I shall expose the development of some simple and possibly crude ideas about quarks and hadrons into a fairly precise approach which promises to give us a lot of insight on many elusive aspects of hadrons and their interactions.(*)

Before going into a detailed exposition it will be useful to spend a few words on the basic ideas of this approach.

The extension of a hadron is attributed rather than to its virtual meson cloud to an intimate substructure somehow related with quarks. The field theoretical meson cloud ought to be a secondary aspect of this basic hadronic fact and should be put in at a later stage, when we have learnt how to deal with the interactions among hadrons.

Thus we can think of hadrons as finite space-time regions where energy, charge, spin, ... is concentrated.

In order to describe these regions we need a number of coordinates, in terms of which we can specify the relevant distributions of energy, charge, spin, etc.

Quarks are supposed to do precisely this: they supply us with the needed elements to describe the geometrical structure which we attribute to hadrons. Not only are they identified with the coordinates needed to specify extended hadrons, but they give us also the delightful opportunity to introduce the hadronic symmetries SU_2, SU_3, SU_4?... as well as a natural way of breaking them.

If we look at quarks in this way we see that we do not need confinement for the simple reason that there are no quark fields around. Outside of a hadronic space-time pocket (bag)(**) the quark notion looses entirely its meaning. Our main problem will then be to find a simple and physically correct way to characterise the structure of these bags as well as the laws governing their mutual interactions.

This is a summary of these lectures: In section 2 I shall review the Massive Quark Model as a first step towards learning how to work with quark coordinates. Simple and general ideas about the geometry of mesonic $(q\bar{q})$ states will be presented in section 3.

(*) I will mainly describe the developments of the MQM worked out in collaboration with N. Craigie - [3], [4]

(**) The reader will notice some similarities between this approach and the so called MIT-Bag.[5]

Section 4 deals with the construction of hadronic interactions and the implementation of unitarity through a perturbative expansion. In section 5 a few rather important calculations will be performed.

Some details of the structure of final states and hadronic scattering emerging from the previously developed picture are given in section 5.

2. LEARNING HOW TO WORK WITH QUARK-COORDINATES: THE MASSIVE QUARK MODEL (MQM)

The Massive Quark Model, which has been developed in the last three years, was constructed with the aim to describe deep inelastic and large transverse momentum phenomena in a realistic way. The successes of the parton model in the field of deep inelastic scattering are well known, but equally well known is the difficulty to understand why quarks do not appear in the final states, ore else why the "confinement mechanism" does not obscure the simple free field behaviour of quarks at short distances. The MQM description is built upon a number of ansatz which are assumed having in mind a realistic picture of hadrons and their interactions. As a first step toward such realistic picture one assumes that:

a) Quarks are a fundamental degree of freedom of hadronic matter; by this we mean that the description of hadrons requires the introduction of the quark notion. We can visualise the quark degree of freedom by the introduction of a spin 1/2 "field" $q_\alpha(x)$ [where α stands for a full set of space-time and internal indices]. This field, however, will have to enjoy peculiar properties if we are to exorcise quarks from final states; thus we must assume that:

b) No Quanta are associated with the quark field $q_\alpha(x)$. This ansatz introduces "by fiat" the fact that the quark is confined. Its rationalization can be founded on the discussion in the Introduction, which considers as a possible alternative to deal with the quark degree of freedom, the mere reduction of the quark notion to that of a hadronic coordinate.

In a field theoretical visualization, the assumed absence of physical states in the quark sector will traduce itself into the absence of singularities in the quark - mass variables of the quark Green's functions . The smallness of the "effective" quark masses, as indicated by the precocity of the Regge and scaling phenomena, will be implemented by assuming that the <u>Quark Green's functions are entire in the quark-mass variables and universally peaked for small values of these variables.</u>
This characterization of the behauviour in the quark-mass variables proves of extreme importance in setting up a calculational scheme for deep inelastic phenomena.

In order to be able to go further and describe deep inelastic physics we need some control on high energy behaviour. This is provided by the following ansatz:

c) Quark Green's functions are Regge behaved in the high energy regions.

This means that when energy is much bigger than the effective (see above) quark masses, quark Green's functions are dominated by Regge poles. Such a Regge behaviour is, in fact, at the origin of the analogous behaviour in usual hadronic interactions. This ansatz gives us the possibility of pouring our extensive phenomenological knowledge of high energy hadron physics into deep inelastic and large transverse momentum physics. In this respect the MQM has met with considerable success. In order to turn the MQM into a complete description of hadronic phenomena we must develop a scheme to calculate efficiently hadronic processes of all kinds. This is done in the following way:

d) One can construct the hadronic amplitudes in terms of the quark Green's functions by setting up a well defined diagrammatic expansion. The building blocks of this expansion are:

i) Meson ($q\bar{q}$) and Baryon (qqq) wave functions (Fig. 1)

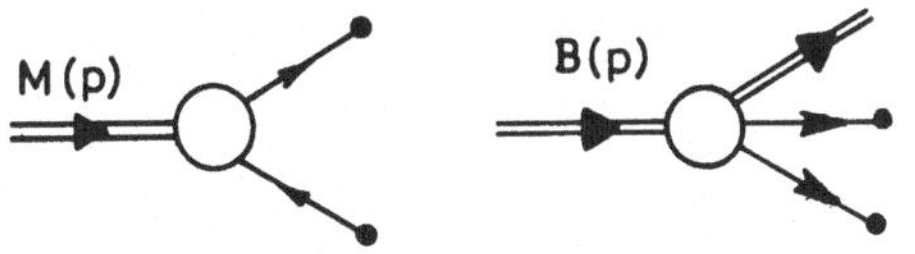

FIG. 1 Meson ($q\bar{q}$) and Baryon (qqq) wave functions

ii) elementary Green's functions (Fig. 2)

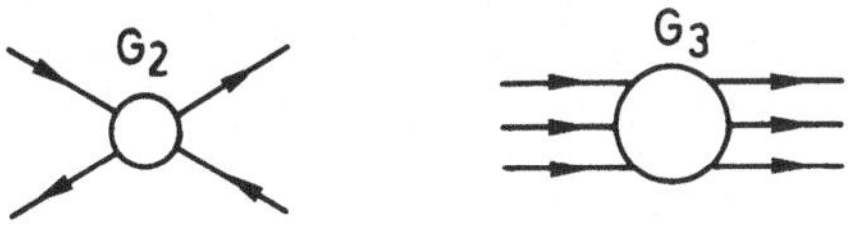

FIG. 2 The elementary Green's functions

They describe the propagation of Meson (G_2) and Baryon (G_3) systems, their discontinuities correspond to real physical states, and in particular the residues of their poles are related in the usual fashion to the wave functions previously defined;

iii) irreducible vertex functions (Fig. 3)

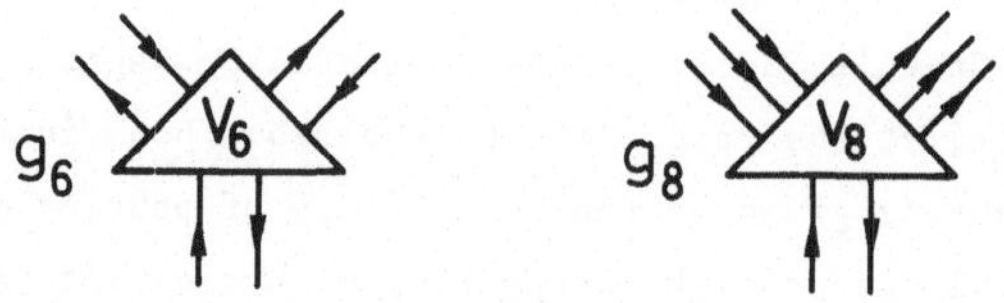

FIG.3 The irreducible vertex functions

They describe the coupling of three Meson-like systems (V_6) and two baryon-like and one meson-like system (V_8). The viability of the perturbative scheme relies on the "smallness" of the coupling constants g_6 and g_8; much in the same way as the successes of Perturbation Theory in QED are a consequence of α being so small. A short discussion of what evidence we have about g_6 and g_8 being small is in Ref.2.

The diagrammatic expansion built out of these three ingredients will thus obey crossing and unitarity order by order in g_6 and g_8, provided the elementary amplitudes do so.

Starting from these ideas one can attack a variety of problems in the realms of deep inelastic and large $p_\perp$ physics. In so doing one learns how the quark degree of freedom introduced in the way just discussed, manifests itself in a great variety of ways leaving such characteristic footprints as scaling, jets, strong interaction scaling etc.

However the MQM is not yet a completely well defined theory of the strong interactions. We lack a general and possibly simple description of the structure of hadronic states, without which it is not possible to have fully quantitative predictions. A remedy to these fundamental limitations shall be attempted in the next sections.

3. THE MESON GEOMETRY - THE $q\bar{q}$ WAVE FUNCTION

In this section the problem of introducing the mathematical object describing a meson system ($q\bar{q}$) will be attacked. Starting from heuristic arguments we shall see that it is possible to achieve a fairly precise description of mesons, their spectrum and their space-time properties.

3.1 The Bag

Innumerable facts of strong interactions physics support the notion that hadrons are extended quantum mechanical objects. The first question we want to study is thus, what kind of mathematical object is suitable to describe such extended structures?

The simplest object that we can introduce is a function (wave function) $\psi(p, x_1, x_2)$. p is the eigenvalue of the energy momentum operator, and x_1 and x_2 is the minimum set of coordinates suitable to describe an extended meson. In a conventional QFT one would have the representation

$$\psi(p; x_1, x_2) = \langle 0 | T(q(x_1) q^+(x_2)) | p \rangle \qquad (3.1)$$

which describes the "bound state" p in terms of its quark constituents. As it should be abundantly clear from the discussion in the Introduction, (3.1) must not be looked at as more than a visualisation useful to obtain some properties of $\psi(p, x_1 x_2)$.

Throughout this work quark are nothing more than a way of introducing the two basic properties of the hadrons: extension and internal quantum numbers. In this lecture notes, however, in order not to unnecessarily complicate the main developments, quarks shall be unrealistically taken as space-time scalar and SU_3 singlets.

The requirement that the wave functions (w.f.) $\psi(p,x_1x_2)$ embody a translationally invariant description of an extended hadron, leads us immediately to:

$$\psi(p; x_1 x_2) = e^{ipR} \psi(p; x) \qquad (3.2)$$

where $R = \frac{1}{2}(x_1+x_2)$ and $x = x_1 - x_2$; as it is trivial to derive i.e. from the representations (3.1).

Our next problem is to find some general and simple characterization of our w.f. (3.2).

The idea that hadrons have a finite extension can be implemented by the requirement that $\psi(p,x)$ vanishes outside a certain 4-dimensional region $R^4(p)$, characterized by the 4-momentum p of the state, and a few geometrical constants which should be adjusted to fit experimental observations. Thus we require

$$\psi(p;x) = 0 \qquad \text{for} \quad x \notin R^4(p) \; . \tag{3.3}$$

In the following the region specified by $R^4(p)$ will be often referred to as the "bag" region; for the time being it is a quite general compact space-time domain. Under the action of a Lorentz transformation Λ, which brings p into Λp, our domain will accordingly transform itself into $R^4(\Lambda p)$ comprising all points $x' = \Lambda x$ for $x \in R^4(p)$. Thus $R^4(p)$ will be uniquely defined by $R^4(M,\vec{0})$, the bag region relative to the hadron rest frame $[p \equiv (M,\vec{0})]$.

(3.3) takes care of the "confinement" of the quark coordinates; in order to go further we need to specify the way quarks "move" inside the bag $R^4(p)$. Motivated by the relevance of simple free field behaviour of quarks in LC physics, one can assume that for small $x \in R^4(p)$

$$\begin{aligned} -\Box_1 \psi(p;x_1x_2) &= m^2 \psi(p;x_1x_2) \\ -\Box_2 \psi(p;x_1x_2) &= m^2 \psi(p;x_1x_2) \end{aligned} \tag{3.4}$$

In order to give a description as simple as possible, actually this free field behaviour shall be postulated for the meson coordinates x_1 and x_2 not only for small x but over the whole bag $R^4(p)$, with the possible exception of a region close to the boundary where such a behaviour may conflict with the requirement of the continuity of the w.f.

Before going on, we stop and summarize the ideas and ansatz of this subsection:

i) We associate to a physical meson state a w.f.

$$\psi(p;x_1x_2)$$

which by translational invariance obeys (3.2);

ii) $\psi(p;x)$ is a continuous function vanishing for $x \notin R^4(p)$, where $R^4(p)$ is a compact 4-dimensional region (bag);

iii) for $x \in R^4(p)$ with the possible exception of a region close to the boundary (skin), $\psi(p;x)$ obeys the Klein-Gordon equations (3.4) where m^2 can be interpreted as the "effective" quark mass*).

---

*) I.e. the mass describing the quark propagation inside the bag $R^4(p)$.

3.2. The Eqs. of Motion and The Spectrum

The Eqs. (3.4), when use is made of the translation invariance condition (3.2), become:

$$\left(\frac{p}{2} - \frac{1}{i}\frac{\partial}{\partial x}\right)^2 \Psi(p;x) = m^2 \Psi(p;x) \tag{3.5}$$

$$\left(\frac{p}{2} + \frac{1}{i}\frac{\partial}{\partial x}\right)^2 \Psi(p;x) = m^2 \Psi(p;x) \quad ,$$

which can be cast in the following form:

$$p\cdot\frac{\partial}{\partial x}\Psi(p;x) = 0 \tag{3.6a}$$

$$\left(\frac{M^2}{4} - \Box_x\right)\Psi(p;x) = m^2\Psi(p;x) \quad . \tag{3.6b}$$

Without lack of generality, to simplify things we shall look at (3.6) in the rest frame $p \equiv (M,\vec{0})$. We get

$$\frac{\partial}{\partial t}\Psi(M;t,\vec{x}) = 0 \quad , \tag{3.7a}$$

$$\left(\frac{M^2}{4} - \frac{\partial^2}{\partial t^2} + \vec{\nabla}^2\right)\Psi(M;t,\vec{x}) = m^2\Psi(M;t,\vec{x}) \quad , \tag{3.7b}$$

for $(t,\vec{x}) \in R^4(M)$. The first important observation is that in order to be compatible with the Eqs. (3.7), the boundary R^4(M) must "factorize" into a time- and a space-bag, i.e.

$$\{ R^4(M) : \ |t| \leq R_t(M) \quad , \ |\vec{x}| \leq R_s(M) \} \tag{3.8}$$

Another very relevant observation is that (3.7a) implies that $\Psi(M;t,\vec{x})$ is time-independent, thus washing out the dynamics of the relative-time coordinate, which in realistic Bethe-Salpeter approaches has always been a source of troubles*). From the continuity requirement, however, (3.7a) cannot hold over the whole time-bag; thus we must expect the existence of a "skin" region where (3.7a) should be modified by a potential term giving a $\Psi(t)$ of the type in fig. 4.

---

*) A discussion of the problem of ghosts in B.S. equation can be found for example in Ref. 6

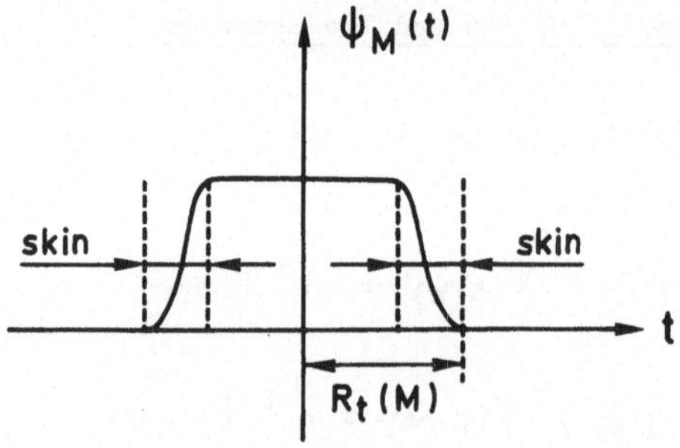

FIG.4 A possible shape of the time w.f.

An alternative treatment of the time-bag is described in the next subsection.

As for Eq. (3.7b), it will provide us with a spectrum once we impose the continuity of $\Psi(M;t,x)$ for $|\vec{x}| = R_S(M)$. Thus we get a set of basic meson states, described by the w.f.

$$\Psi_{n\ell}(M;t,\vec{x}) = \psi_M(t)\,\varphi_{M,n\ell}(\vec{x}) \tag{3.8}$$

where $\psi_M(t)$ has a shape as in FIG.4, and

$$\varphi_{M,n\ell}(\vec{x}) \equiv N_{n\ell}\, j_\ell\left(\sqrt{\tfrac{M^2}{4}-m^2}\,|\vec{x}|\right) Y_\ell^m(\theta,\varphi) \quad . \tag{3.10}$$

$N_{n\ell}$ is a normalization factor to be determined at a later stage, $j_\ell(z)$ is the spherical Bessel function of order l and Y_ℓ^m is a normalized spherical harmonic. The continuity condition yields the spectrum equation

$$\left[M_{n\ell}^2 - 4m^2\right]^{1/2} = \frac{2}{R_S(M)}\,\beta_{n,\ell} \quad , \tag{3.11}$$

where $\beta_{n,\ell}$ is the nth positive zero of the Bessel function $j_\ell(z)$; asymptotically it is well known that

$$\beta_{n,\ell} \xrightarrow[\text{large } n]{} \pi(n+\ell/2) \quad .$$

The spectrum equation (3.11) will give us definite predictions only after the function $R_S(M)$ is given. We shall discuss later a possible criterion to restrict drastically the form of $R_S(M)$.

3.3 An Alternative Way to Handle The Time Bag

We have just seen that the free field behaviour in the time coordinate (3.7a) cannot be enforced over the whole time bag $[|t| \leq R_t(M)]$ without clashing

with the requirement of continuity. FIG.4 depicts a possible shape of $\psi_M(t)$, but one can envisage in fact an alternative which appears somewhat attractive. The main feature of Eqs. (3.7) is the factorization of space and time which reflects itself in the factorization (3.9) of the w.f.
If we want to maintain this factorization over the whole bag region, without excluding the skin, we must require [see Eq.(3.7b)]

$$\frac{\partial^2}{\partial t^2} \psi_M(t) = - \omega^2 \psi_M(t) \quad , \tag{3.12}$$

which should thus replace (3.7a). The condition at the boundary $|t|=R_t(M)$, gives

$$\omega = \frac{\pi}{2R_t}(2n+1) \qquad (n=0,1,\cdots) \quad , \tag{3.13}$$

and the "quasi free" structure uniquely selects $n = 0$ in (3.13),i.e.

$$\omega = \frac{\pi}{2R_t} \quad , \tag{3.14}$$

and we obtain

$$\psi_M(t) = \frac{1}{\sqrt{R_t(M)}} \cos \frac{\pi}{2R_t(M)} t \quad . \tag{3.15}$$

The spectrum eq. (3.11) gets accordingly modified in

$$[M_{n\ell}^2 - 4m^2 + 4\omega^2]^{1/2} = \frac{2}{R_s(M)} \beta_{n,\ell} \quad , \tag{3.16}$$

where ω is given by (3.14). The spectrum now depends not only on $R_S(M)$ but also on $R_t(M)$, even though in a definitely less important way. An remarkable feature of the additional term $4\omega^2$ is its sign. It contributes to the mass $M_{n\ell}^2$ of our hadron with a sign opposite to the space term $\frac{2\beta_{n\ell}}{R_s(M)}$ thus giving us the possibility of having particles as light as π's even with realistic values of the bag constants R_s and R_t.

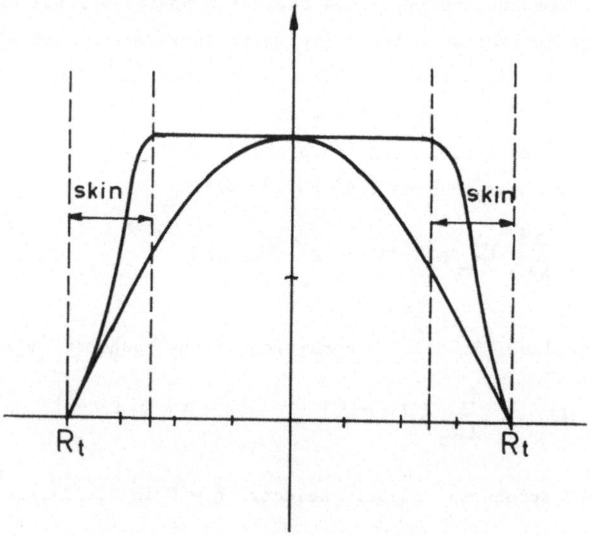

FIG.5 Comparison between $\psi_M(t)$ in FIG.4 and Eq. (3.15).

3.4. The w.f. in Momentum Space. The MQM Criterion

I shall now compute the Fourier transform of (3.9), which will provide an explicit expression of the w.f. in momentum space (FIG.6)

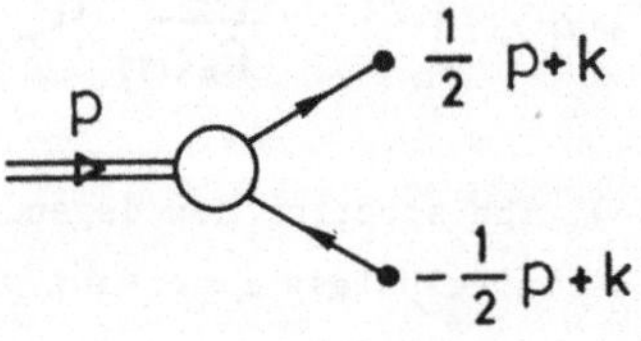

FIG.6 The Momentum space ware function

From

$$\Psi_{n\ell}(M,k) = \int d^4x\, e^{ikx}\, \Psi_{n\ell}(M;t,\vec{x}) \tag{3.17}$$

one obtains in a straightforward manner

$$\Psi_{n\ell}(M,k) = N'_{n\ell}\,(2\pi)^2\, R_t(M)^{1/2}\;\frac{\cos\left(k_0 R_t(M)\right)}{(\pi/2)^2 - \left(k_0 R_t(M)\right)^2}\;\cdot$$

$$\cdot\left(\frac{2}{\pi}\right) R_s(M)^{3/2}\;\frac{\beta_{n\ell}}{\beta_{n\ell}^2 - |\vec{k}|^2 R_s^2(M)}\; j_\ell\left(|\vec{k}| R_s(M)\right)\, Y_\ell^m(\theta_k, \varphi_k) \qquad (3.18)$$

For convenience we have introduced a new normalization factor $N'_{n\ell}$, such that

$$\int \frac{d^4k}{(2\pi)^4}\, |\Psi(M,k)|^2 = |N'_{n\ell}|^2$$

In order to appreciate the basic properties of $\Psi_{n\ell}(M,k)$ let us first express the variables K_o and $\vec{K}$ in the rest frame in terms of the quark "masses" $m_{1,2}^2 = (\frac{P}{2} \pm k)^2$. We have

$$k_0 = \frac{1}{2M}\left(m_1^2 - m_2^2\right) \qquad (3.19)$$

and

$$\vec{k}^2 - \frac{M^2}{4} = \frac{1}{4M^2}\left(m_1^2 - m_2^2\right) - \frac{1}{2}\left(m_1^2 + m_2^2\right) \quad . \qquad (3.20)$$

It is immediate to check that (3.18) is an <u>entire</u> function of the variables m_1^2 and m_2^2. As we have seen, this is one of the basic ansatz of the MQM and a general property of permanently confined quarks; this result, therefore, comes as no surprise.

But there is more to be observed about (3.18). If we look at the function

$$\frac{\cos\left(k_0 R_t(M)\right)}{(\frac{\pi}{2})^2 - \left(k_0 R_t(M)\right)^2}$$

(see FIG.7)

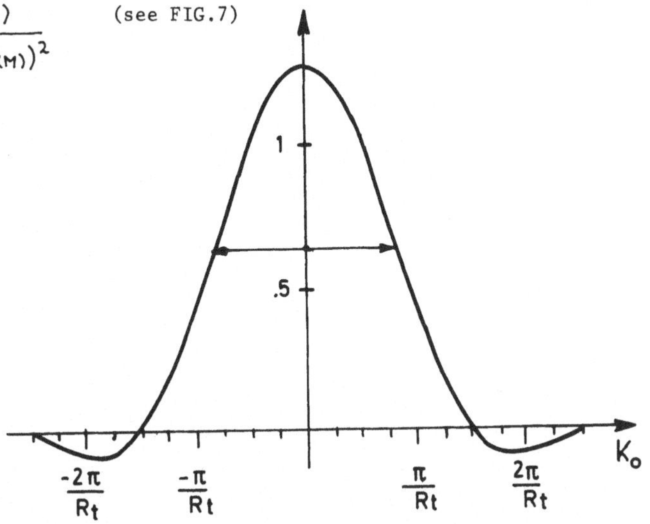

FIG.7 The K_o-distribution from Eq.(3.18)

we see that it is peaked around $K_o = 0$ with a dispersion

$$\Delta k_o \simeq \frac{\pi}{R_t(M)} \quad ; \tag{3.21}$$

and from (3.19) this means that we have a w.f. strongly peaked for $m_1^2 = m_2^2$ with a dispersion

$$\Delta\left(\frac{m_1^2 - m_2^2}{2}\right) \simeq \frac{\pi^2 M}{2 R_t(M)} \quad . \tag{3.22}$$

Again, we have recovered another fundamental assumption of the MQM [point b), sect.2] , which stipulates that the quark Green's functions are peaked at small values of the quark "masses". However the MQM can recover all the appealing aspects of deep inelastic phenomena provided such peaking is universal at high energy, i.e. the dispersions of the mass distributions do not depend on the mass of the physical state.

The universality of quark mass distributions gives us a well defined criterion to determine the functions $R_S(M)$ and $R_t(M)$ for large M. From (3.22) and (3.23) we should set

$$R_t(M) \xrightarrow[M\ \mathrm{large}]{} R^2 M \left(\frac{\pi}{2}\right) \tag{3.24}$$

$$R_s(M) \xrightarrow[M\ \mathrm{large}]{} R^2 M \quad . \tag{3.25}$$

To extrapolate (3.24) and (3.25) to any value of M, the simplest two-parameter forms will be assumed:

$$R_t^2(M_{ne}) = \left(\frac{\pi}{2}\right)^2 R_o^2 \left[\beta_{ne} + \lambda_t\right] \tag{3.26}$$

and

$$R_s^2(M_{ne}) = R_o^2 \left[\beta_{ne} + \lambda_s\right] \quad , \tag{3.27}$$

where R_o^2 is related to R^2 appearing in (3.24) and (3.25), by

$$R_o^2 = 2R^2 , \tag{3.28}$$

as can be easily checked from the spectrum equations (3.16).
We can put (3.26) and (3.27) in (3.16) now and obtain the spectrum equation

$$M_{ne}^2 = 4m^2 - \frac{4}{R_o^2(\beta_{ne} + \lambda_t)} + \frac{4\beta_{ne}^2}{R_o^2(\beta_{ne} + \lambda_s)} \quad , \tag{3.29}$$

which expresses the masses of the meson states in terms of the 4 parameters R_o, λ_t, λ_s and m^2. Asymptotically we have

$$M_{nl}^2 \longrightarrow \frac{4\beta_{nl}}{R_0^2}$$

and using the asymptotic formula for β_{nl} , we obtain the gratifying result of asymptotically parallel and linear Regge trajectories (without odd daughters). If we require a slope of the trajectories $\alpha \simeq 1\ \mathrm{GeV}^{-2}$, we obtain

$$R_0^2 \simeq 2\pi\ \mathrm{GeV}^{-2} .$$

3.5 The Structure of High Mass States. The Firesausage.

For large masses the MQM criterion of universal peaking of the quark-mass distributions, according to (3.24) and (3.25), forces the space-time extension of physical states to increase linearly with M.

An extension increasing linearly with mass along any direction would definitely be in contradiction with the very basic fact of the strong interaction, that the maximum impact parameter of a high energy collision increases at most logarithmically with energy. In fact if we do not exclude from the physical spectrum those states for which

$$l \geq l_{max} = \frac{R_\perp}{2} M , \tag{3.30}$$

where R is some typical hadronic length (~ 1 F) , we can have hadrons interacting even when their impact parameter $b \sim M$. This could happen by having the initial quarks arrange themselves in hadronic states lying on the leading Regge trajectory. (See next section).

Accordingly we shall assume that only those states are physical which obey $l \leq l_{max}$. This assumption is equivalent to the finiteness of the range of strong interactions. This allows us to introduce a suggestive picture of the space-time structure of high energy states. We can, by means of a unitary transformation, go to a basis describing approximately a high energy state as a standing $q\bar{q}$-wave in the cylindrical region of space appearing in FIG.8. We can call these $\left(\frac{l_{max}}{2}\right)^2$ degenerate

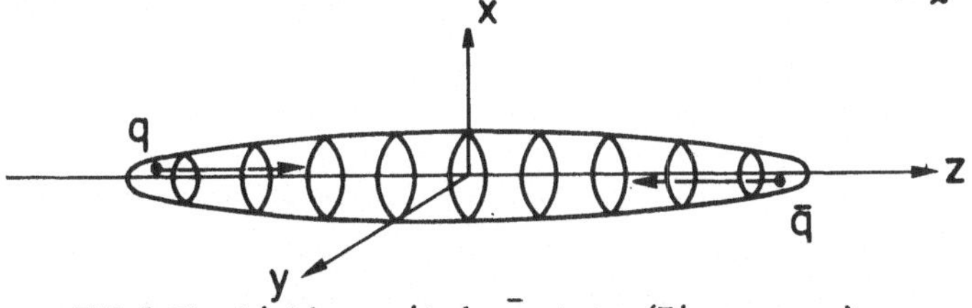

FIG.8 The highly excited $q\bar{q}$ state (Firesausage).

states of mass M, with the suggestive name of Firesausages.

3.6 The Wave function for Any Momentum: On and Off-Shell.

So far our attention has been concentrated uniquely on the hadron rest frame $p=(M,\vec{0})$. In any moving frame, the generalization of (3.8) reads

$$\left\{ R^4(p) : \frac{(xp)^2}{p^2} \leq R_t^2(M) \, , \, \left(-x^2 + \frac{(px)^2}{p^2}\right) \leq R_s^2(M) \right\} \, , \tag{3.31}$$

and we obtain the w.f. $\Psi_{n\ell}(p,K)$ by boosting on a particular direction (say z) and substituting in (3.18) for K_0 $\frac{1}{2M}(m_1^2-m_2^2)$, for $|\vec{k}|$ its expression (3.20) and for

$$\Theta = \arccos\left[\frac{Ek_3 - pk_0}{M|\vec{k}|}\right] \, , \tag{3.32}$$

where k_0 and $|\vec{k}|$ are again given by (3.19) and (3.20). Finally, by our choice of the z-axis, φ_k remains unaltered.

In performing the calculations a convenient form for wave functions of hadrons of mass $\geq$ 2 GeV is:

$$\Psi_{n\ell}(p,k) \simeq N'_{n\ell} \frac{2\pi}{R^2}(2\pi)^2 \delta_{R^2}\left[\frac{m_1^2+m_2^2}{2} - m^2\right] \delta_{R^2}\left(\frac{m_1^2-m_2^2}{2}\right) Y_\ell^n(\bar{\Theta}_k,\varphi_k) \, , \tag{3.33}$$

where $M_{n\ell}^2 \simeq \frac{2\pi}{R^2}(n+\ell/2)$ and $\delta_{R^2}(z)$ is a "smeared δ-function" whose prototype is

$$\delta_{R^2}(z) = \frac{1}{\pi}\,\frac{\sin R^2 z}{z} \, ,$$

and enjoys the following properties

$$\int_{-\infty}^{+\infty} dz\, \delta_{R^2}(z) = 1 \tag{3.34}$$

and

$$\delta_{R^2}(0) = \frac{R^2}{\pi} \, . \tag{3.35}$$

In the limit $R^2 \to \infty$, $\delta_{R^2}(z)$ obviously tends to a Dirac δ-function. For low mass hadrons (3.33) becomes a much too imprecise representation and we should use (3.18) instead.

In the next section the problem will arise of how to construct the interactions among our "bags". The quantum-mechanical nature of hadrons will force us to consider them not only as real states (on-shell) as we have been doing so far, but also as virtual objects (off-shell). It is my conviction that here lies the crux of the whole approach I am describing in these lectures. In some measure the same problem appears in conventional QFT, but there the locality of the Lagrangian and the analyticity properties of the theory suffice to pin down the off-shell behaviour without any essential ambiguity. Here, where the structure of the theory is basically nonlocal, such well defined criteria do not apply and we face the serious danger of heading towards all kinds of nonsense and inconsistencies. In order to proceed, I shall again resort to the ideas of the MQM and perform the off shell $(p^2 \neq M^2_{n\ell})$ extension of the wave function $\psi_{n\ell}(p,k)$ by substituting in (3.18)

$$k_0 \rightarrow \frac{(m_1^2-m_2^2)}{2}\frac{1}{M_{n\ell}} = \frac{(pk)}{M_{n\ell}} \tag{3.36}$$

and

$$\vec{k}^2 \rightarrow \frac{M_{n\ell}^2}{4} - m^2 + \frac{1}{4M^2_{n\ell}}(m_1^2-m_2^2)^2 - \frac{1}{2}(m_1^2+m_2^2) = M^2_{n\ell} - m^2 + (pk)^2 - \left(\frac{p^2}{4}+k^2\right) \tag{3.37}$$

(3.36) and (3.37) mean that in going off-shell we choose to have unaltered the distributions of the quark masses. Let me emphasize again that this choice is by no means an obvious one; whether it is reasonable or not we shall have to judge from the kind of hadronic world it gives rise to.

4. HADRONIC PROCESSES IN SPACE-TIME. IMPLEMENTING UNITARY IN A PERTURBATIVE WAY.

In the previous section we have explored the consequences of some simple ideas on the structure and the spectrum of meson systems. Of primary importance has been the notion of hadron wave functions, but its meaning in terms of well defined observations has so far been left quite vague. It is now time to fill this gap and to introduce a complete and precise framework for hadronic interactions. The basic elements of this framework have been presented in section 2, in the context of the MQM.

Forgetting for the time being about baryons, the MQM framework amounts to a diagrammatic expansion of the S-matrix which parallels completely that of a perturbation theory, provided we make the following identifications: (FIG.9)

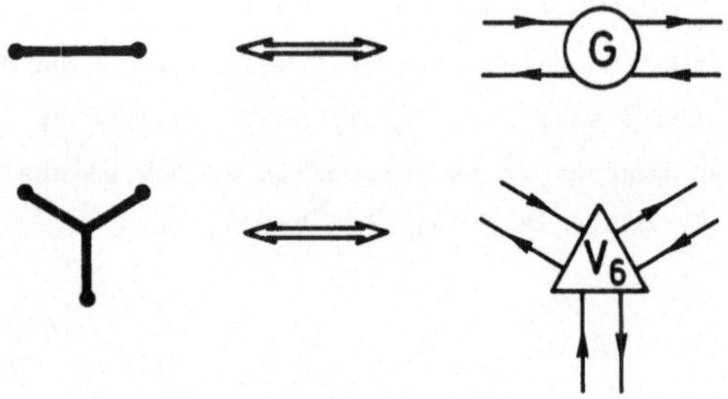

FIG.9 The Dictionary leading from $\lambda\varphi^3$ to the MQM expansion.

i) φ propagator $\Leftrightarrow$ $q\bar{q}$ Green's function G;
ii) φ^3 vertex $\Leftrightarrow$ irreducible six-point kennel V_6.

To any given order in λ, which then should be a relevant expansion parameter, we can draw the diagrams of $\lambda\varphi^3$ and then proceed to substitute the φ-propagators with the $q\bar{q}$ Green's functions and the φ^3-vertices with the irreducible kennels V_6. Crossing and unitarity will then be implemented order by order in λ, provided both G and V_6 are crossing symmetric.

The whole framework boils down to determining the structure of G, V_6 and the rules governing their juxtaposition.

4.1 The $q\bar{q}$ Green's function.

The Green's function G embodies the space-time propagation properties of a $q\bar{q}$ (qq) system. The bag states constructed in the previous section are the obvious candidates to describe such propagation. We shall therefore write the crossing symmetric Green's function as (see FIG.10)

$$G = G_s + G_t \tag{4.1}$$

where

$$G_s = \sum_{n\ell} \Psi_{n\ell}(p,k)\Psi^*_{n\ell}(p,k')\frac{1}{p^2 - m^2_{n\ell}} \tag{4.2}$$

and an analogous formula for G_t holds. G_s is built up by bag states in the s-channel, while G_t contains bag states in the t-channel.
The w.f.'s $\Psi_{n\ell}(p,k)$ in (4.2) are extended off-shell according to the prescription discussed at the end of section 3.

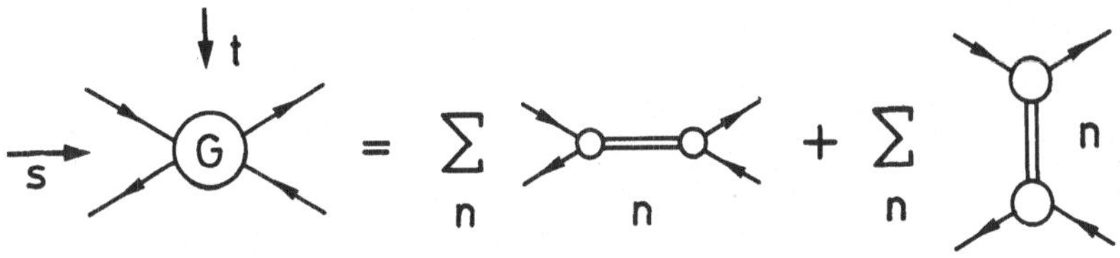

FIG.10 Construction of the Green's function G

The sum in (4.1) is actually necessary for crossing symmetry; the bag structure of the w.f's, differently from a string picture, cannot support a "dual" behaviour. We defer to section 5 a discussion of the main properties of G.

4.2 The Irreducible Kernel V_6

In determining the form of V_6, geometry will play a crucial role. Let us look at the coupling of three bag states (FIG.11)

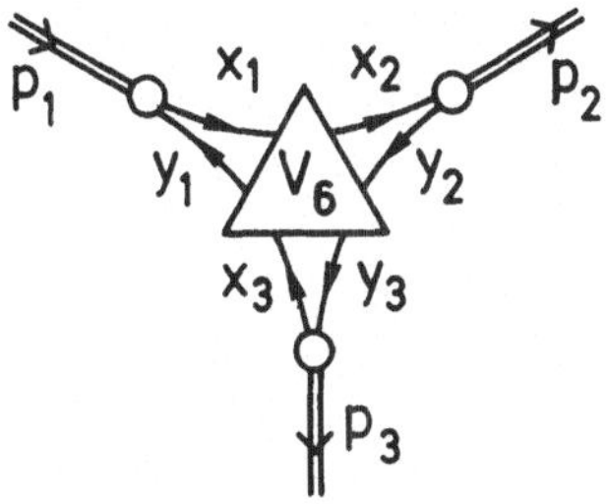

FIG.11 The coupling of three bag states

V_6 determines the probability amplitude for the quarks of one bag, say P_1, to disappear from it and reappear as the constituents of a two-bag system P_2, P_3 According to the idea pursued in these lectures that quarks are nothing but coordinates carrying the description of hadronic systems extended in small space-time domains (bags), this amplitude should be zero whenever the space-time regions spanned by the three bags do not overlap. A way of achieving this is to write:

$$V_6 [x_1 \cdot\cdot y_3] = \mu^6 \delta^4(x_1 - x_2)\delta^4(y_2 - y_3)\delta^4(x_3 - y_1) + \text{permutations} \tag{4.3}$$

where μ is a mass parameter, introduced for dimensional reasons, playing the role of a coupling constant (in the $\lambda\varphi^3$ analogy we have $\lambda \leftrightarrow \mu^6$). According to (4.3) the three bag coupling in momentum space amounts to the calculation of the diagrams depicted in FIG.12. The meaning of these diagrams is specified by a set of graphical rules which are discussed in the next subsection.

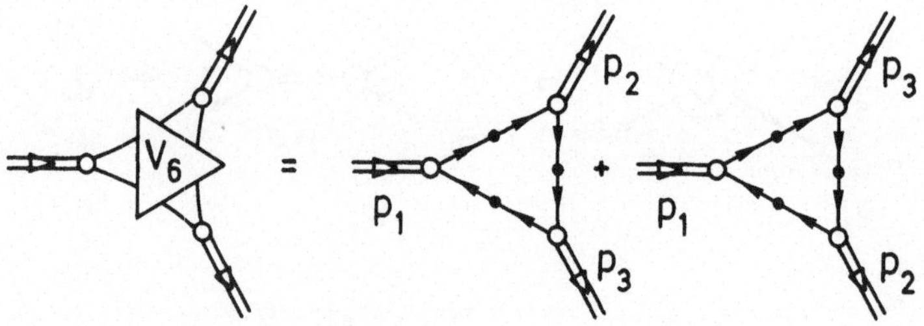

FIG.12 The diagrams corresponding to the three-bag coupling

4.3 Graphical Rules

A general quark diagram, obtained in the perturbative expansion outlined in subsection 4.1, can be reduced to a network of quark lines joining quark Green's functions and w.f's in such a way that each loop consists of no more and no less than three quark lines. The rules to evaluate any such diagrams are the following:

i) each incoming "bag" state of momentum P_i and relative momentum K_i is introduced through the w.f.'s $\psi(P_i, K_i)$, whose explicit expression is given in (3.18);

ii) each outgoing bag state of momentum P_f, and relative momentum K_f is given by $\psi^+(P_f, K_f)$;

iii) for each quark line we must multiply by μ^2;

iv) four-momentum at each two-quark hadron vertex must be conserved;

v) for each loop we have an integral $\int \frac{d^4 \ell}{(2\pi)^4}$, all integrations are over real domains according to subsection 3.6;

vi) $q\bar{q}$ Green's functions are totally connected and have discontinuities corresponding to physical hadronic states;

vii) any weak and electromagnetic current is represented by a pointlike coupling to a $q\bar{q}$ Green's function G_S (see FIG.13);

FIG.13 The current-$q\bar{q}$ coupling

the two-quark loop is calculated without multiplying by the factor μ^2 for each quark line.

Equipped with these rules we can construct amplitudes having any degree of complexity. As mentioned, the viability of this approach rests on the possibility of treating μ^2 as a perturbative parameter. It will be checked later that this is indeed the case.

4.4 The Electromagnetic Vertex

The item vii) of the graphical rules introduces in our hadronic world the very important notion of currents. The first question one can ask is whether the electromagnetic current just defined has the necessary property of being conserved. The vertex in FIG.13 has the following form:

$$\Gamma_\mu(q,k) = \sum_{n=1}^{\infty} \frac{\partial}{\partial x_\mu} \Psi_{n1}(q,x)\Big|_{x=0} \Psi^*_{n1}(q,k) \frac{1}{(m_{n1}^2 - q^2)} \qquad (4.4)$$

The current conservation is a trivial consequence of the equation

$$q^\mu \frac{\partial}{\partial x_\mu} \Psi(q,k)\Big|_{x=0} = 0$$

which was postulated in section 3. Actually the free-field equations (3.4) that have been assumed for small x will automatically guarantee all the conservation, exact or partial, which must be attributed to the weak and electromagnetic currents. Using the expression for the w.f.'s and the off-shell extrapolation procedure just discussed we can evaluate (4.4), and find quite easily that the sum diverges logarithmically. This forces us to make one subtraction, and we can write

$$\Gamma_\mu(q,k) = k_\mu F(q^2, m_1^2 m_2^2) \quad , \qquad (4.5)$$

where $F(q^2,m^2,m_2^2)$ is a function which has all the vector meson poles and has a logarithmic behaviour when $q^2 \to \infty$. The actual form of $F(q^2)$ depends on the subtraction; this is quite unpleasant and may contain the message that we are missing something important. We can only hope that when the currents will be studied in more detail this shortcoming will be remedied.

In any case, if we call $Z = F(0,m^2=m_2^2=m^2)$ we can determine the factor N'_{ne} in (3.18), by the requirement that the bag states carry the correct charge. We get in a straightforward way (*)

$$N'_{ne} = \frac{2}{Z^{1/2}}\left(\frac{1}{\mu^3}\right) = N \quad , \tag{4.6}$$

and derive the interesting result that the hadronic w.f's are proportional to the inverse of the "coupling constant" μ: an appealing property which has been recently stressed in the framework of "solitons" by a number of people.

(*) If we adopt a subtraction procedure which guarantees ρ-dominance of the e.m. current for small q^2 we would get

$$N \simeq \left(\frac{2m_\rho}{\mu^3}\right)^{1/2}$$

a result used in Ref.3.

5. A FEW IMPORTANT CALCULATIONS - THE $q\bar{q}$ GREEN'S FUNCTIONS AND THE STRUCTURE OF FINAL STATES.

This section is entirely devoted to working out some important consequences of the theory developed so far.

5.1. The $q\bar{q}$ Green's Function

We shall now derive the main properties of the $q\bar{q}$ Green's function G, whose central role has been stressed time and time again. According to (4.1) G is given by the sum of two pieces, representing the contribution of bag states in the s and in the t-channel.

i) Let's study the G_s contribution first. Let's recall (4.2):

$$G_s = \sum_{n,\ell} \Psi_{n\ell}(p,k)\Psi^*_{n\ell}(p,k)\frac{1}{s-m^2_{n\ell}} \tag{5.1}$$

where $\Psi_{n\ell}(p,k)$ is the normalized wave function corresponding to the quantum numbers n,l; and $p^2 = s$ is the centre of mass energy squared. (5.1) shall be evaluated in the high energy limit.

As it stands (5.1) contains an infinite number of poles on the real S axis; we shall discuss in a short while the mechanism by which these poles get shifted onto the second sheet, and anticipating one result of that calculation in (5.1) one should substitute

$$M_{n\ell} \rightarrow M^2_{n\ell} + i\,(M\Gamma)_{n\ell} \tag{5.2}$$

with

$$(M\Gamma)_{n\ell} \propto \text{constant}$$

Following the off-shell extrapolation discussed in the previous section and recalling (3.33), we can write

$$G_s \simeq \delta_{R^2}\left(\frac{m_1^2+m_2^2}{2}-m^2\right)\delta_{R^2}\left(\frac{m_1^2-m_2^2}{2}\right)\cdot \tag{5.3}$$

$$\cdot\,\delta_{R^2}\left(\frac{m_1'^2+m_2'^2}{2}-m^2\right)\delta_{R^2}\left(\frac{m_1'^2-m_2'^2}{2}\right)\;G_s(s,t)$$

with

$$G_s(s,t) = (2\pi)^3\left(\frac{\pi}{R^2}\right)^2\frac{4\sqrt{2}}{2\mu^6}\sum_{n\ell}\frac{(2\ell+1)\,P_\ell(\cos\theta_s)}{s-M^2_{n\ell}-i\,(M\Gamma)_{n\ell}}\;, \tag{5.4}$$

where $m^2_{1,2}$ and $m'^2_{1,2}$ are the incoming and outgoing quark masses respectively and $\cos\theta_s = (1+\frac{2t}{s})$.

As we have remarked the off-shell extrapolation employed precisely corresponds to a dispersion representation of G_s in s, keeping t and the quark masses fixed. An efficient way to compute (5.4) is to consider the imaginary part of the partial wave amplitude $a_\ell(s)$.

$$\mathrm{Im}\, a_\ell(s) \simeq (2\pi)^3 \left(\frac{\pi}{R^2}\right)^2 \frac{4\sqrt{2}}{2\mu^6} \sum_{n=n_0}^{\infty} \frac{(M\Gamma)_{n\ell}}{(s-M^2_{n\ell})^2+(M\Gamma)^2_{n\ell}} , \quad (5.5)$$

where $n_0 \simeq \frac{4\ell^2}{R_\perp^2}$ is the lowest n compatible with the ℓ cut-off (3.30). Making use of (5.2) and of the asymptotic expression

$$M^2_{n\ell} \to \frac{2\pi}{R^2}(n+\ell/2)$$

we get readily

$$\mathrm{Im}\, a_\ell(s) \xrightarrow[s\ \mathrm{large}]{} (2\pi)^4 \left(\frac{\pi}{R^2}\right)^2 \frac{\sqrt{2}}{\mu^6 z} \pi\, \Theta\left[\frac{R\sqrt{s}}{2} - \ell\right] . \quad (5.6)$$

Im $G_s(s,t)$ is then given by

$$\mathrm{Im}\, G_s(s,t) = \sum_{\ell=0}^{\infty} \mathrm{Im}\, a_\ell(s)\,(2\ell+1)\, P_\ell(\cos\Theta_s) ,$$

and making use of (5.6), we obtain (for small t)

$$\mathrm{Im}\, G_s(s,t) \xrightarrow[s\ \mathrm{large}]{} (2\pi)^4 \left(\frac{\pi}{R^2}\right)^2 \frac{\sqrt{2}}{\mu^6 z} \pi \frac{sR_\perp^2}{4} \frac{J_1(R_\perp t)}{R_\perp\sqrt{-t}} . \quad (5.7)$$

This is the "primeval Pomeron", which was at the very basis of the MQM description of deep inelastic phenomena. The full amplitude can now be calculated through a subtracted dispersion relation, and we obtain for t = 0

$$G_s(s,0) \xrightarrow[s\ \mathrm{large}]{} (2\pi)^4 \left(\frac{\pi}{R^2}\right)^2 \frac{\sqrt{2}}{\mu^6 z} \frac{sR_\perp^2}{8} \left[-\log(-s-i\epsilon)+c\right] , \quad (5.8)$$

where c is a subtraction constant.

The J-plane structure of $G_s(s,t)$ is thus a fixed pole at J=1. The t-dependence of (5.8) is typical of a grey disc of diameter R .

ii) The G_t contribution shall be expressed by a formula completely analogous to (5.1). Following the same procedure as before we get:

$$\mathcal{G}_t \simeq \delta_{R^2}\left(\frac{m_1^2+m_1'^2}{2}-m^2\right)\delta_{R^2}\left(\frac{m_1^2-m_1'^2}{2}\right)\delta_{R^2}\left(\frac{m_2^2+m_2'^2}{2}-m^2\right)\delta_{R^2}\left(\frac{m_2^2-m_2'^2}{2}\right) \quad (5.9)$$
$$\cdot\ G_t(s,t)$$

where $$G_t = (2\pi)^3\left(\frac{\pi}{R^2}\right)^2 \frac{4\sqrt{2}}{\mu^6 z} \sum_{n\ell} \frac{(2\ell+1)P_\ell(\cos\Theta_s)(-1)^\ell}{t-M^2_{n\ell}+(iM\Gamma)_{n\ell}} \quad (5.10)$$

$\cos\Theta_t = \left(1 + \frac{2s}{t-4m^2}\right)$ and the factor $(-1)^{\ell}$ comes from the crossing properties of the w.f's.

To determine the behaviour of (5.10) for large s, we use a Sommerfeld-Watson transformation:

$$\sum_{n\ell} (2\ell+1) P_\ell(-\cos\Theta_t) \frac{1}{t-m_{n\ell}^2} = \frac{1}{2i} \sum_n \int_C \frac{d\ell\,(2\ell+1)}{\sin \pi\ell} P_\ell(\cos\Theta_t) \frac{1}{t-m_n^2(\ell)} \quad (5.11)$$

where l is the usual contour encircling the positive real l axis.
Defining the trajectory function $\alpha_n(t)$ as the solution of

$$t = m_n^2\,(\alpha_n(t))$$

we have for large s and small t

$$G_t(s,t) \to \frac{4\sqrt{2}}{\mu^6 Z} (2\pi)^3 \left(\frac{\pi}{R^2}\right)^2 \frac{\alpha_1(t)(2\alpha_1(t)+1)}{\sin \pi\alpha_1(t)} \frac{\Gamma(1+2\alpha_1(t))}{\Gamma(1+\alpha_1(t))^2} \left(-\frac{s}{4m^2}\right)^{\alpha_1(t)} . \quad (5.12)$$

Thus $G_t(s,t)$ exhibits at high energy Regge behaviour, which is controlled by a set of trajectories $\alpha_n(t)$ corresponding precisely to the bag states. These trajectories are exactly exchange degenerate, a property which is crucial in order to insure the absence of physical states, discontinuity for $s < 0$, which would correspond to an unobserved non-zero triality states. This constitutes a very important self consistency check. Thus (5.12) has only a right-hand cut corresponding to zero-triality states; the question now is how to interprete physically such states. As we have seen these states arise by virtue of crossing and should be looked at as genuine physical states of a character different from simple bag states: their spectrum is continuous and the typical logarithmic Regge shrinkage shows that their "transverse dimension" increases logarithmically with energy. This strongly suggests that the discontinuity of $G_t(s,t)$ corresponds to hadrons in a configuration of a "multiperipheral" kind.

Another very important distinction between "bag" and "Regge" states is that the former build up diffractive scattering while the latter generate normal Regge behaviour.

5.2 The 3-Bag Vertex

This calculation is particularly interesting because it gives us the possibility to determine the decay properties of the $q\bar{q}$ firesausages and the consequent structure of final states. Following the graphical rules of the preceding section the diagrams in Fig. 12 are easily calculated as

$$\mathcal{M}(1\to 2+3) = (\mu^2)^3 \int \frac{d^4k}{(2\pi)^4}\, \Psi_1(p_1,k)\,\Psi_2^*(p_2,\tfrac{p_3}{2}+k)\,\Psi_3^*(p_3,-\tfrac{p_2}{2}+k) + (2\leftrightarrow 3)\ . \tag{5.13}$$

On substituting (3.13) for the wave functions we have:

$$\mathcal{M}(1\to 2+3) = (2m_\rho\mu)^{3/2}(2\pi)^2\left(\frac{2\pi}{R^2}\right)^3 \int d^4k\, \delta_{R^2}(mk_0)\,\delta_{R^2}\!\left(\vec{k}^2-\frac{m_1^2}{4}\right) \delta_R\!\left[\frac{m_2^2+m_3^2-m_1^2}{4} - m_1|\vec{p}|\cos\theta_{kp}\right] Y_{\ell_1}^{m_1}\,Y_{\ell_2}^{m_2}\,Y_{\ell_3}^{m_3} + (2\leftrightarrow 3) \tag{5.14}$$

where

$$|\vec{p}| = \frac{1}{2m_1}\left[(m_1^2-m_2^2-m_3^2)^2-4m_2^2m_3^2\right]^{1/2}$$

is the decay momentum. An analysis of the kinematic shows that the peaking of wave functions at small quark masses requires that in the rest frame of p_1 the decay momentum $\vec{p}$ is in the direction of the initial quark relative momentum $\vec{k}$. Thus calling $h_{2,3}$ the helicities of $p_{2,3}$; and integrating the two first δ_{R^2} functions in (5.14) we get

$$\mathcal{M}(1\to 2+3) \simeq 2(2m_\rho\mu)^{3/2}(2\pi)^2\left(\frac{\pi}{R^2}\right)^3\left[(2\ell_2+1)(2\ell_3+1)\right]^{1/2}\delta_{n_2 0}\,\delta_{n_3 0}\,Y_{\ell_1}^{m_1}(\theta,\varphi) \cdot \int_{-1}^{+1} dz\, \delta_{R^2}(a-bz)^4 \tag{5.15}$$

where Θ,q are the polar coordinates of the decay momentum $\vec{p}$, and

$$a = \frac{1}{4}\,(m_1^2-m_2^2-m_3^2)$$
$$b = \frac{1}{4}\,\left[(m_1^2-m_2^2-m_3^2)^2-4m_2^2m_3^2\right]^{1/2} \tag{5.16}$$

The z-integral in (5.15) is crucial in determining the decay properties of the excited bag p_1. It is not difficult to see that the matrix element strongly favors the decay of the highly excited state 1 into a light meson (say the π) and another high mass state. Setting for simplicity $m_1^2=m^2$ $m_2=0$ and $m_3^2=m'^2$, we have

$$f_{R^2}(m^2,m'^2) = \left(\frac{\pi}{R^2}\right)^3\int_{-1}^{+1} dz\, \delta_{R^2}(a-b-z)^4 = \frac{4}{\pi}\,\frac{1}{m^2-m'^2}\int_0^{R^2(m^2-m'^2)} d\xi\, \frac{\sin\xi}{\xi^4} \tag{5.17}$$

where we have explicitly substituted the prototype expression $\frac{1}{\pi}\,\frac{\sin R^2 z}{z}$ for the δ_{R^2} (z)-function.

Introducing the variable $x = \frac{m'^2}{m^2}$, a plot of (5.17) for m^2 = 10,20, and 30 GeV^2 and for $R^2 \simeq \pi\ GeV^{-2}$ is in FIG.14.

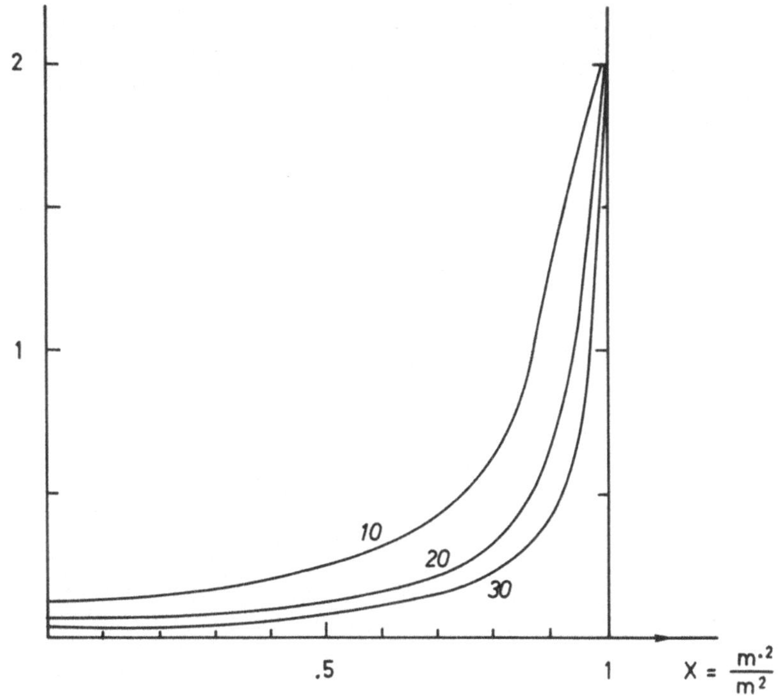

FIG.14 The function $f_{R^2}(m^2, m'^2)$ for three different values of the decaying mass.

In order to get a feeling about the magnitude of μ^2 we compute the width for the decay of the state with l=1 (which we may call ρ, with mass m_p) into two low states (π, with mass m_π). Setting $\Gamma(\text{"}\rho\text{"} \to 2\ \text{"}\pi\text{"}) \simeq 150$ MeV, we estimate $\mu \sim 300$ MeV, a quite small value when compared with a typical hadronic mass.

The results of this paragraph will constitute the starting point for the analysis of final states in hadronic collisions, which shall be performed in section 6.

6. THE STRUCTURE OF FINAL STATES, SOME IDEAS ABOUT HADRON-HADRON SCATTERING

In this last section we shall explore a little further the structure of final states emerging from the decay of a "firesausage". We shall then conclude with a brief analysis of some interesting aspects of hadron-hadron scattering.

6.1 The Linear Decay Chain

The results obtained in section 5 are very important for reaching an understanding of the structure of final states in hadronic collisions. Due to the very substantial enhancement of the decay configuration M M' + $\pi^{(*)}$ with M' close in mass to M, to a good approximation the final states evolve through a linear decay chain as depicted in Fig. 15.

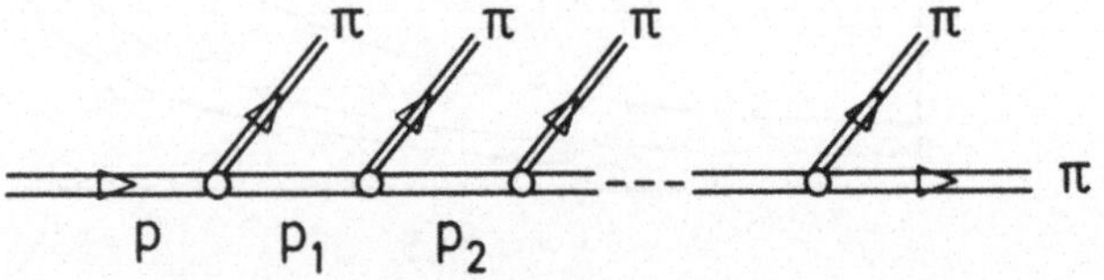

Fig. 15. The linear decay chain of an excited $q\bar{q}$ system

Such picture implies that a $q\bar{q}$ system, moving on a highly excited orbit, gets de-excited through the emission of pions while cascading down through a series of states of decreasing mass. This resembles very much the cascade decay of an excited atom through photon emission. The final states produced by this cascade will then be composed of a number of pions distributed in rapidity and transverse momentum in a characteristic way which shall be described shortly.

If we want to compare this picture with the usual quark-parton description, we notice the quite different meaning the notion of "$q\bar{q}$ sea" acquires here. Such sea is represented by a large number of low momentum pions and arises as the product of a violent collision which brings the initial hadronic system in a highly excited state.

Taking up the atomic physics analogy again, it does not appear any more

(*) Actually other low mass bags like k,ρ etc., do not give totally negligible contributions, and should be also considered.

useful to speak about the "$q\bar{q}$ sea" as a property of the initial low mass hadrons, than to try to include the "photon sea" in the wave function of the atomic ground state.

The linear decay chain problem can be solved by means of the integral equations reproduced in Fig. 16.

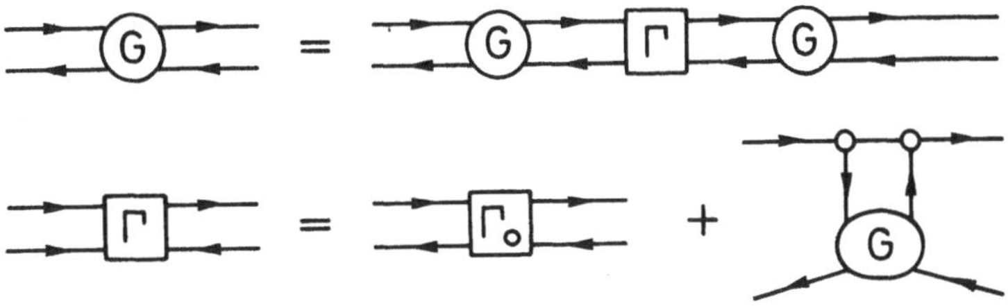

Fig. 16 The integral equations governing the decay of highly excited $q\bar{q}$ states

Here I only summarize the results: (*)

(i) if we call (M) the width of an excited state of mass M, asymptotically we get

$$M\Gamma(M) \to \lambda \frac{2\pi}{R^2} \tag{6.1}$$

where λ can be explicitly computed in terms of the parameters μ and R^2, and in this scalar model turns out to be of order unity [3], [7]. This property of λ implies that asymptotically the Green's function is smooth due to the overlap of neighbouring resonances. The spacing between neighbouring resonances, in fact, is given by

$$2M\Delta M \to \frac{2\pi}{R^2} \quad ;$$

(ii) The one-particle yield exhibits Feynman scaling;

(iii) for small x the inclusive distribution shows a plateau, thus giving rise to a multiplicity increasing logarithmically with s,

$$\langle n \rangle \simeq c \log s \tag{6.2}$$

for the charged multiplicity we calculate $c_{ch} \simeq 1$;

(*) A detailed exposition can be found in Ref.4.

(iv) there is no correlation between impact parameter and rapidity of secondary pions, and the transverse momentum is cut-off, i.e.

$$\langle k_\perp^2 \rangle \simeq \frac{\pi}{R_\pi^2} \simeq .1 \text{ GeV}^2 , \tag{6.3}$$

(v) the π's are emitted independently, thus following a Poisson distribution.

It is almost superfluous to emphasize that these results look quite close to what is going on in hadronic final states.

6.2 Hadronic Scattering at High Energies

We are not yet in a position to perform a realistic calculation of high energy scattering due to our lack of knowledge about baryonic states. However, we can try and see whether the theory developed in section 4 can reproduce the correct qualitative features of high energy hadron-hadron scattering.

The lowest order diagram in the perturbative approach is the "Born" diagram in Fig. 17.

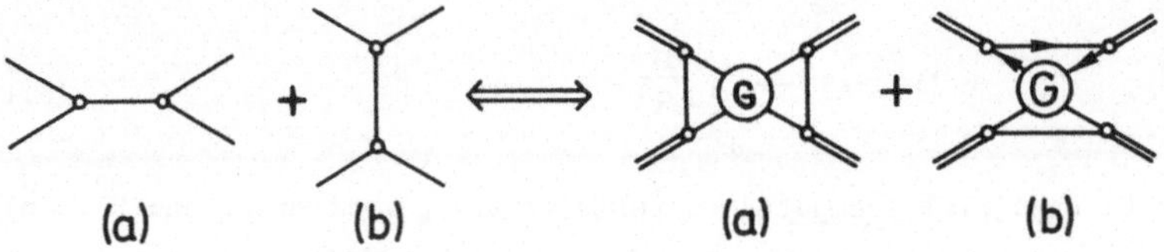

Fig. 17 Hadron-hadron scattering diagrams to lowest order

It is easy to see that both diagrams become irrelevant at high energy. Diagram (a) controls resonance formation at low energy but at high energy is suppressed by kinematics and it behaves like $\frac{1}{s}$. Diagram (b) yields a real contribution in the s-channel and at high energy, and small t is dominated by π-exchange. In order to get the diagrams at high energy we must go to 4th order in λ. The relevant diagrams appear in Fig. 18

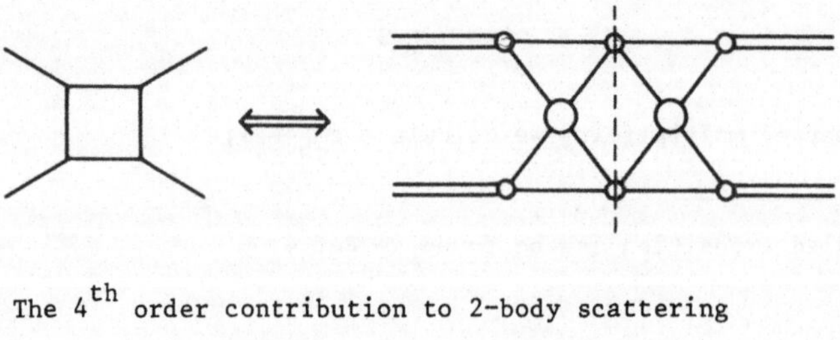

Fig.18 The 4th order contribution to 2-body scattering

A slightly involved analysis shows that at high energy the complicated diagram in Fig. 18 becomes proportional to the one appearing in Fig. 19. One should remember, however, that such diagram is not a legal quark diagram as it violates the graphical rules set up in section 4.3.

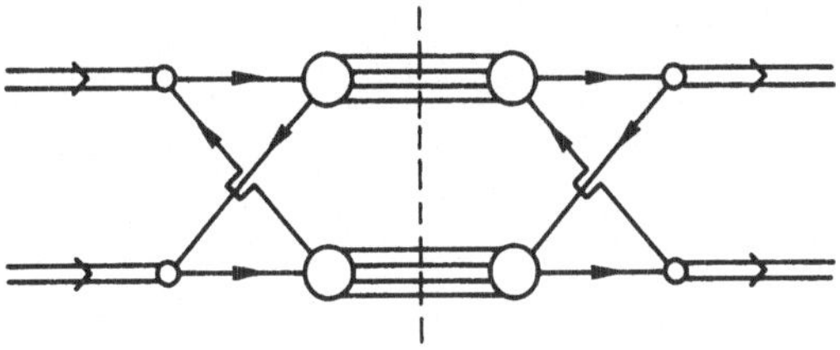

Fig. 19 The "illegal" diagram equivalent to the one in Fig.18

This diagram is calculated very easily by means of Sudakov techniques and yields at high energy a purely imaginary amplitude increasing like s.

This is the"Pomeron", which at this order appears as a fixed pole at J = 1. The secondary pions emitted through firesausage decay obey scaling, and their density in rapidity is twice (notice the two firesausages) the value in a single firesausage decay. Whereas a fixed pole at J = 1 can be perfectly tolerated in $q\bar{q}$-scattering (no two-body unitarity is present at the quark level), such as J-plane structure is incompatible with unitarity at hadronic level. A straightforward way to implement it is to sum the "multiperipheral diagrams" of the type in Fig. 20.

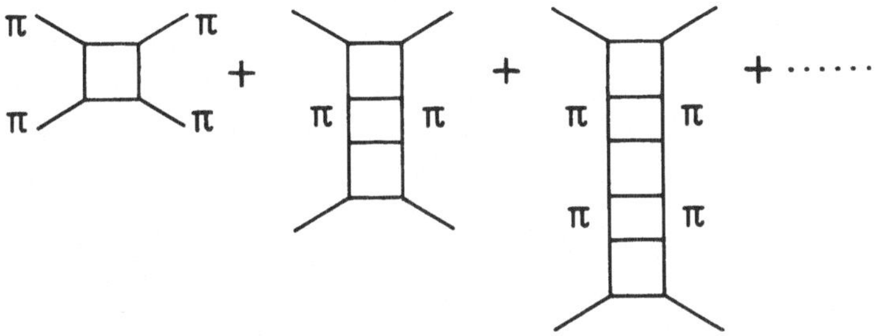

Fig. 20 Implementing t-channel unitarity

This calculation was done some time ago in the framework of the MQM by Craigie and myself [8] , and these are the results:

(i) The J = 1 fixed pole gets shifted up by approximately 10%

$$\alpha_P(0) = 1 + \epsilon \qquad (6.4)$$

$$\epsilon \simeq .06$$

(ii) It acquires a slope, which was computed as

$$\alpha'_{\mathbb{P}}(0) \simeq .3 \text{ GeV}^{-2} \qquad (6.5)$$

Again, it is almost superfluous to stress that (6.4) and (6.5) are in good shape as far as the ISR data are concerned. As a matter of principle, however, (6.4) violates the Froissart bound, and one should analyse the various absorption corrections which must bring the cross section to finally obey the Froissart bound. The smallness of ε , on the other hand, phenomenologically postpones to very high energy the need to consider such corrections.

7. CONCLUSIONS

I have tried in these lectures to present an attempt to develop the MQM into a tool as effective as possible to describe hadrodynamics. Except for the general formulation in section 4, the attention has been concentrated on the dynamics of unrealistic scalar and neutral quarks. This has been done in order to be able to determine the viability of the whole approach. I am convinced that abundant evidence has piled up that the direction followed here is worth pursuing.

The future problems of crucial importance will be the introduction of spin and internal degrees of freedom and of a consistent scheme for currents. It is on these problems that the views expressed here will find their most important test.

REFERENCES

1. K. Wilson, Lectures given at the 1975 Erice Summer School.

2. G. Preparata, A possible way to look at Hadrons: The MQM. Lectures given at the XII Course of the International School of Subnuclear Physics "E. Majorana", A. Zichichi ed. (1974)

3. G. Preparata and N.S. Craigie, A space-time Description of Quarks and Hadrons, CERN TH. 2038 (to appear in Nuclear Phys. B) (1975)

4. N.S. Craigie and G. Preparata, The structure of final states in a Space-time Description of Quarks and Extended Hadrons. CERN TH. 2056 (to appear in Nucl. Phys. B) (1975)

5. A. Chodos, R.L. Jaffe, K. Johnson, C.B. Thorn and V.F. Weisskopf, Phys. Rev. D9, 3471 (1974)

6. P. Menotti, in Subnuclear Phenomena. Proceedings of the 1969 Erice Summer School. A. Zichichi ed., Academic Press (1971)

Lecture Notes in Physics

Vol. 32: Particles, Quantum Fields and Statistical Mechanics. Proceedings 1973. Edited by M. Alexanian and A. Zepeda. V, 132 pages. 1975.

Vol. 33: Classical and Quantum Mechanical Aspects of Heavy Ion Collisions. Proceedings 1974. Edited by H. L. Harney, P. Braun-Munzinger, and C. K. Gelbke. VII, 311 pages. 1975.

Vol. 34: One-Dimensional Conductors GPS Summer School Proceedings, 1974. Edited by H. G. Schuster. VII, 371 pages. 1975.

Vol. 35: Proceedings of the Fourth International Conference on Numerical Methods in Fluid Dynamics, 1974. Edited by R. D. Richtmyer. V, 457 pages. 1975.

Vol. 36: R. Gatignol, Théorie Cinétique des Gaz à Répartition Discrète de Vitesses. II, 219 pages. 1975.

Vol. 37: Trends in Elementary Particle Theory. Proceedings 1974. Edited by H. Rollnik and K. Dietz. V, 472 pages. 1975.

Vol. 38: Dynamical Systems, Theory and Applications. Proceedings 1974. Edited by J. Moser. VI, 624 pages. 1975.

Vol. 39: International Symposium on Mathematical Problems in Theoretical Physics. Proceedings 1975. Edited by H. Araki. XII, 562 pages. 1975.

Vol. 40: Effective Interactions and Operators in Nuclei. Proceedings 1975. Edited by B. R. Barrett. XII, 339 pages. 1975.

Vol. 41: Progress in Numerical Fluid Dynamics. Proceedings 1974. Edited by H. J. Wirz. V, 471 pages. 1975.

Vol. 42: H II Regions and Related Topics. Proceedings 1975. Edited by D. Downes and T. L. Wilson. XII, 488 pages. 1975.

Vol. 43: Laser Spectroscopy. Proceedings 1975. Edited by S. Haroche, J. C. Pebay-Peyroula, T. W. Hänsch, and S. E. Harris. X, 466 pages. 1975.

Vol. 44: R. A. Breuer, Gravitational Perturbation Theory and Synchrotron Radiation. VI, 196 pages. 1975.

Vol. 45: Dynamical Concepts on Scaling Violation and the New Resonances in e^+e^- Annihilation. Edited by B. Humpert. VII, 248 pages. 1976.

Vol. 46: E. J. Flaherty, Hermitian and Kählerian Geometry in Relativity. VIII, 365 pages. 1976.

Vol. 47: Padé Approximants Method and Its Applications to Mechanics. Edited by H. Cabannes. XV, 267 pages. 1976.

Vol. 48: Interplanetary Dust and Zodiacal Light. Proceedings 1975. Edited by H. Elsässer and H. Fechtig. XII, 496 pages. 1976.

Vol. 49: W. G. Harter and C. W. Patterson, A Unitary Calculus for Electronic Orbitals. XII, 144 pages. 1976.

Vol. 50: Group Theoretical Methods in Physics. 4th International Colloquium. Nijmegen 1975. Edited by A. Janner, T. Janssen, and M. Boon. XIII. 629 pages. 1976.

Vol. 51: W. Nörenberg und H. A. Weidenmüller. Introduction to the Theory of Heavy-Ion Collisions. IX, 273 pages. 1976.

Vol. 52: M. Mladjenović, Development of Magnetic β-Ray Spectroscopy. X, 282 pages. 1976.

Vol. 53: D. J. Simms and N. M. J. Woodhouse, Lectures on Geometric Quantization. V, 166 pages. 1976.

Vol. 54: Critical Phenomena. Sitges International School on Statistical Mechanics, June 1976. Edited by J. Brey and R. B. Jones. XI, 383 pages. 1976.

Vol. 55: Nuclear Optical Model Potential. Proceedings 1976. Edited by S. Boffi and G. Passatore. VI, 221 pages. 1976.

Vol. 56: Current Induced Reactions. International Summer Institute, Hamburg 1975. Edited by J. G. Körner, G. Kramer, and D. Schildknecht. V, 553 pages. 1976.